T0202478

Lecture Notes in Computer Science 12160

Founding Editors

Gerhard Goos
  Karlsruhe Institute of Technology, Karlsruhe, Germany
Juris Hartmanis
  Cornell University, Ithaca, NY, USA

Editorial Board Members

Elisa Bertino
  Purdue University, West Lafayette, IN, USA
Wen Gao
  Peking University, Beijing, China
Bernhard Steffen ⓘ
  TU Dortmund University, Dortmund, Germany
Gerhard Woeginger ⓘ
  RWTH Aachen, Aachen, Germany
Moti Yung
  Columbia University, New York, NY, USA

More information about this series at http://www.springer.com/series/7407

Fedor V. Fomin · Stefan Kratsch ·
Erik Jan van Leeuwen (Eds.)

Treewidth, Kernels, and Algorithms

Essays Dedicated to Hans L. Bodlaender
on the Occasion of His 60th Birthday

 Springer

Editors
Fedor V. Fomin (ID)
University of Bergen
Bergen, Norway

Stefan Kratsch
Humboldt-University
Berlin, Germany

Erik Jan van Leeuwen
Utrecht University
Utrecht, The Netherlands

ISSN 0302-9743 ISSN 1611-3349 (electronic)
Lecture Notes in Computer Science
ISBN 978-3-030-42070-3 ISBN 978-3-030-42071-0 (eBook)
https://doi.org/10.1007/978-3-030-42071-0

LNCS Sublibrary: SL1 – Theoretical Computer Science and General Issues

© Springer Nature Switzerland AG 2020, corrected publication 2022
This work is subject to copyright. All rights are reserved by the Publisher, whether the whole or part of the material is concerned, specifically the rights of translation, reprinting, reuse of illustrations, recitation, broadcasting, reproduction on microfilms or in any other physical way, and transmission or information storage and retrieval, electronic adaptation, computer software, or by similar or dissimilar methodology now known or hereafter developed.
The use of general descriptive names, registered names, trademarks, service marks, etc. in this publication does not imply, even in the absence of a specific statement, that such names are exempt from the relevant protective laws and regulations and therefore free for general use.
The publisher, the authors and the editors are safe to assume that the advice and information in this book are believed to be true and accurate at the date of publication. Neither the publisher nor the authors or the editors give a warranty, expressed or implied, with respect to the material contained herein or for any errors or omissions that may have been made. The publisher remains neutral with regard to jurisdictional claims in published maps and institutional affiliations.

Cover illustration drawn by Fedor V. Fomin

This Springer imprint is published by the registered company Springer Nature Switzerland AG
The registered company address is: Gewerbestrasse 11, 6330 Cham, Switzerland

Hans L. Bodlaender (photo by Ivar Pel)

Preface

This Festschrift celebrates the contributions of Hans L. Bodlaender on the occasion of his 60th birthday. Hans has made many transformative discoveries in algorithms research, complexity theory, and graph theory, as is evidenced in the articles of this volume.

The most well-known results by Hans come from his deep and thorough investigations into the *treewidth* of graphs. Intuitively defined as a measure of how close a graph resembles a tree, treewidth can be exploited by algorithms to solve many computational problems much faster on graphs of small treewidth than is possible in general. Finding efficient algorithms to determine whether a graph actually has small treewidth has fascinated Hans for many years. One of his defining contributions in the area is a linear-time algorithm for computing treewidth on graphs of bounded treewidth. This result has influenced the field so profoundly that it has become known as *Bodlaender's theorem*. It is a truly rare feat to have a named result, which speaks to the remarkable researcher that Hans is.

Hans' interest in treewidth proved a gateway to broader investigations into graph algorithms and graph theory. His deep knowledge and brilliant problem-solving abilities led to many diverse results in these areas and paved the way for many fellow researchers to pursue similar research directions. For example, he chairs the Steering Committee of the International Workshop on Graph-Theoretic Concepts in Computer Science and organized it twice. At the same time, Hans' enthusiasm inspired students of his lectures to explore this wondrous world and this led quite a few of those students to become researchers in graph algorithms themselves.

Treewidth as a graph parameter naturally steered Hans to the blossoming field of *parameterized algorithms and complexity*. A central notion in this field is that of a kernel: effectively a polynomial-time preprocessing heuristic with provable guarantees on the reduction in instance size that it achieves. Together with Rod Downey, Mike Fellows, and Danny Hermelin, Hans developed a framework to prove limitations of kernels, in that some problems do not admit kernels that reduce a problem to size polynomial in the parameter. This framework ushered in a new era in parameterized algorithms and guided many new investigations into the kernelization complexity of computational problems. The paper won Hans and his coauthors the EATCS-IPEC Nerode Prize 2014 for outstanding papers in the area of multivariate algorithmics.

The fact that Hans enjoys mathematical and computational puzzles naturally spills into the real world, where Hans enjoys puzzles and board games. Conversely, his interest in puzzles and board games led him to define and study new problems on the computational complexity of these games. Hence, Hans sees beautiful puzzles everywhere and excudes great joy solving them every day.

The Festschrift will be presented to Hans, as a surprise, at the workshop "Graph Decompositions: Small Width, Big Challenges" at the Lorentz Center in Leiden in

April 2020. It contains short personal contributions and surveys on the topics that have defined Hans' career.

Many congratulations on your birthday, Hans! Please enjoy this Festschrift and we hope for many years of beautiful science to come!

Finally, we wish to thank the people that made this volume possible. In particular, we thank all the authors for their wonderful contributions and for reviewing and proofreading. We also thank Alfred Hofmann and Anna Kramer at Springer for making the volume possible and for their support.

January 2020

Fedor V. Fomin
Stefan Kratsch
Erik Jan van Leeuwen

Laudations

Seeing Arboretum for the (partial k-) Trees

Stefan Arnborg[1] and Andrzej Proskurowski[2(✉)]

[1] Royal Institute of Technology, Stockholm, Sweden
[2] University of Oregon, Eugene, OR, USA
andrzej@cs.uoregon.edu

Abstract. The idea of applying a dynamic programming strategy to evaluating certain objective functions on trees is fairly straightforward. The road for this idea to develop into theories of width parameters has been not so straight. Hans Bodlaender has played a major role in the process of mapping out that road. In this sentimental journey, we will recount our collective road trip over the past decades.

Collaborating with Hans: Some Remaining Wonderments

Michael R. Fellows and Frances A. Rosamond[(✉)]

University of Bergen, 5020 Bergen, Norway
{michael.fellows,frances.rosamond}@uib.no

Abstract. This paper celebrates some family adventures, three concrete open problems and several research directions that have developed over our long collaboration.

Hans Bodlaender and the Theory of Kernelization Lower Bounds

Danny Hermelin

Department of Industrial Engineering and Management,
Ben-Gurion University of the Negev, Beersheba, Israel
hermelin@bgu.ac.il

Abstract. In this short letter I give a brief subjective account of my favorite result with Hans – our kernelization lower bounds framework. The purpose of this manuscript is not to give a formal introduction to this result and the area that spawned from it, nor is it meant to be a comprehensive survey of all related and relevant results. Rather, my aim here is to informally describe the history that lead to this result from a personal perspective, and to outline Hans's role in the development of this theory into what it is today.

Algorithms, Complexity, and Hans

Jan van Leeuwen

Department of Information and Computing Sciences, Utrecht University,
Princetonplein 5, 3584 CC Utrecht, The Netherlands
J.vanLeeuwen1@uu.nl

> *"The study of algorithms is at*
> *the very heart of computer science"*
> Aho, Hopcroft, Ullman, [1], 1974, p. iii

Abstract. In this essay on the occasion of Hans Bodlaender's 60th birthday, we recount some of the early developments in the field in which Hans made his mark and of their context at Utrecht University.

Intelligent Cards for Excellent People

Gerard Tel

Utrecht University, Department of Computer Science

Abstract. We present a small, Excel-based application to print congratulation cards. The "card" automatically sorts the names of the senders, and includes the factorisation of the age of the celebrant (in days).

As a working example, we produce a card for a well-known scientist's 60th birthday. We also include some questions as food for thought.

Three levels of computing. For many people, the world of computing is a three storey building, but with an unusually long stair between the second and third floors.

At the ground floor, there is the world of *mental calculations*. Many people perform simple calculational tasks, like splitting a restaurant bill, without any supplies or devices. Few people can also perform more complicated tasks in this way, like extracting roots, and there is even a Mental Calculation World Cup[1] held every two years.

At the second floor, *electronic calculators* are a great help for many people to compute common life data, like energy consumption, holiday costs, etc. Scientific calculators can compute many functions, but are hand operated, so performing lengthy or repeated calculations can still be quite cumbersome.

The most complicated calculations, like finding treewidth or various graph optimalisations, are performed by *programmed computers* at the third floor of the calculations building. Of course, programming computers efficiently is an art mastered only by few (scientists and programmers).

In between: Excel. The gap between using a hand calculator and programming is quite large, both in the required skills and in the power of possible computations. Fortunately, there is a mechanism with intermediate difficulty and power, namely the use of *spreadsheets*, most notably, Excel. Excel offers the possibility to "program" computations in the form of formula's describing the content of each cell. The cell formula's make up a *model* of the computation, and this model allows to modify the inputs later, thus making it possible to reuse the computation.

Examples: Sorting and Factoring. Our Congratulation Card Generator[2] accepts information about the celebrant in the yellow cells on tab `YourData` and produces a personalized card on tab `Card`.

The tab `Bubble` implements Bubble Sort. It accepts (up to) ten strings or numbers in the yellow cells, and uses the names from `YourData` otherwise. The cells form a

---

[1] See https://en.wikipedia.org/wiki/Mental_calculator.

[2] See http://www.staff.science.uu.nl/~tel00101/Hans60.xlsx.

sorting network where, for example, cells C7 and D7 sort the values in C6 and D6 with a common conditional IF(C6<D6;..;..).

All cells in our example are of *constant arity*, that is, each value is calculated from a constant number (two or three mostly) of other cells. Using more advanced Excel functions, of non-constant arity, allowing to express the minimum over a range of cells, InsertionSort or SelectionSort can also be "Excelled" quite easily, but we leave this as an exercise for the reader.

The tab Factor factorises a positive integer up to 1000000. The subsequent rows implement a *recursive formulation* of factoring, where the combination *(d, n)* represents the subproblem of factoring *n* without factors smaller than *d*. If *d* divides *n*, the next line will compute the subproblem *(d, n/d)* and the factor *d* is added to its result. If *d* exceeds the root of *n*, *n* is prime so *n* is its only factor. Otherwise, the next line solves the subproblem (d^+, n), with d^+ the next possible factor.

A challenge here is that Excel only works in rows that are pre-filled with formulas, so we need to know the *maximum number of rows* needed for any allowed integer (up to 1000000). Because either *d* increases by 2 or *n* reduces by a factor of 2 or more, the difference $\sqrt{n} - d$ decreases by at least 2 in most rows, so approximately $\frac{\sqrt{1000000}}{2}$ rows should be enough.

Care should be taken when testing this factoring capability. For a 60 year old scientist, his age in days is so close to e^{10}, that we expect his age to be prime roughly once in ten days. Of course, on such days the "factorisation" will only consist of the age; the first time this happens is exactly two weeks after the 60th birthday. The number of factors reaches a temporary peak five days after the birthday.

The card prototype. The card described in this short article is available for download[3]. Apart from changing the data on which the card is based, it will change from day to day anyways because of the dependence on the addressee's age.

---

[3] See, again, http://www.staff.science.uu.nl/~tel00101/Hans60.xlsx.

Contents

About the Jubilee

Curriculum Vitae Hans L. Bodlaender

Personal Information

Name:	Hans Leo Bodlaender
Born:	April 21, 1960 in Bennekom, The Netherlands
Married to:	Brigitte J. Bodlaender-Peters
Children:	Marijke, Wim, and Annefleur

Education

1986	Ph.D. in computer science, Utrecht University
1983	Doctoral degree (equiv. M.Sc.) in mathematics, Utrecht University
1981	Candidate degree (equiv. B.Sc.) in mathematics, Utrecht University

Employment

2014 –	full professor *Algorithms and Complexity*, Department of Information and Computing Sciences, Utrecht University (part time until 2018, full time since)
2014 – 2018	full professor *Network Algorithms*, Department of Mathematics and Computer Science, Eindhoven University of Technology (part time)
2003 – 2014	associate professor, Institute of Information and Computing Sciences, Utrecht University
1987 – 2003	assistant professor, Department of Computer Science, Utrecht University
1987	postdoctoral fellow, Department of Computer Science, Massachusetts Institute of Technology
1983 – 1987	research assistant, Department of Computer Science, Utrecht University

Prizes

2014	EATCS-IPEC Nerode Prize 2014 for outstanding papers in the area of multivariate algorithmics, for the paper *On problems without polynomial kernels* with Rodney G. Downey, Michael R. Fellows and Danny Hermelin

in Journal of Computer and System Sciences, 2009
(joint with *Infeasibility of instance compression
and succinct PCPs for NP* by Lance Fortnow
and Rahul Santhanam in Journal of Computer
and System Sciences, 2011)

Thesis

1. Hans L. Bodlaender, Distributed Computing: *Structure and Complexity*, Ph.D. thesis, Utrecht University, 1986. Also appeared as CWI Tract 43, CWI, Amsterdam, 1987.

Journal Articles

1. Hans L. Bodlaender. A solution to P33. *EATCS Bulletin*, 21:200–202, 1983.
2. Jan van Leeuwen, Hans L. Bodlaender, and Harry A.G. Wijshoff. Compositions of double diagonal and cross latin squares. *Nieuw Archief voor Wiskunde*, 4:256–266, 1984.
3. Hans L. Bodlaender and Jan van Leeuwen. Simulation of large networks on smaller networks. *Information and Control*, 71(3):143–180, 1986.
4. Anneke A. Schoone, Hans L. Bodlaender, and Jan van Leeuwen. Diameter increase caused by edge deletion. *Journal of Graph Theory*, 11(3):409–427, 1987.
5. Hans L. Bodlaender. Some classes of graphs with bounded treewidth. *Bulletin of the EATCS*, 36:116–125, 1988.
6. Hans L. Bodlaender. A better lower bound for distributed leader finding in bidirectional, asynchronous rings of processors. *Inf. Process. Lett.*, 27(6):287–290, 1988.
7. Hans L. Bodlaender. The complexity of finding uniform emulations on fixed graphs. *Inf. Process. Lett.*, 29(3):137–141, 1988.
8. Hans L. Bodlaender. Achromatic Number is NP-complete for cographs and interval graphs. *Inf. Process. Lett.*, 31(3):135–138, 1989.
9. Hans L. Bodlaender. The classification of coverings of processor networks. *J. Parallel Distrib. Comput.*, 6(1):166–182, 1989.
10. Hans L. Bodlaender, Peter Gritzmann, Victor Klee, and Jan van Leeuwen. Computational complexity of norm-maximization. *Combinatorica*, 10(2):203–225, 1990.
11. Hans L. Bodlaender. The complexity of finding uniform emulations on paths and ring networks. *Inf. Comput.*, 86(1):87–106, 1990.
12. Hans L. Bodlaender and Gerard Tel. Bit-optimal election in synchronous rings. *Inf. Process. Lett.*, 36(1):53–56, 1990.
13. Hans L. Bodlaender. Polynomial algorithms for graph isomorphism and chromatic index on partial k-trees. *J. Algorithms*, 11(4):631–643, 1990.
14. Hans L. Bodlaender. On the complexity of some coloring games. *Int. J. Found. Comput. Sci.*, 2(2):133–147, 1991.
15. Hans L. Bodlaender. Some lower bound results for decentralized extrema-finding in rings of processors. *J. Comput. Syst. Sci.*, 42(1):97–118, 1991.
16. Hans L. Bodlaender. New lower bound techniques for distributed leader finding and other problems on rings of processors. *Theor. Comput. Sci.*, 81(2):237–256, 1991.
17. Hans L. Bodlaender and Dieter Kratsch. The complexity of coloring games on perfect graphs. *Theor. Comput. Sci.*, 106(2):309–326, 1992.
18. Hans L. Bodlaender. A tourist guide through treewidth. *Acta Cybern.*, 11(1–2):1–21, 1993.

19. Hans L. Bodlaender. On linear time minor tests with depth-first search. *J. Algorithms*, 14(1):1–23, 1993.

20. Hans L. Bodlaender and Ton Kloks. A simple linear time algorithm for triangulating three-colored graphs. *J. Algorithms*, 15(1):160–172, 1993.

21. Hans L. Bodlaender. Book review. *ZOR - Meth. & Mod. of OR*, 38(1):18, 1993.

22. Hans L. Bodlaender and Rolf H. Möhring. The pathwidth and treewidth of cographs. *SIAM J. Discrete Math.*, 6(2):181–188, 1993.

23. Hans L. Bodlaender. Complexity of path-forming games. *Theor. Comput. Sci.*, 110(1): 215–245, 1993.

24. Hans L. Bodlaender. Improved self-reduction algorithms for graphs with bounded treewidth. *Discrete Applied Mathematics*, 54(2–3):101–115, 1994.

25. Hans L. Bodlaender, Klaus Jansen, and Gerhard J. Woeginger. Scheduling with incompatible jobs. *Discrete Applied Mathematics*, 55(3):219–232, 1994.

26. Hans L. Bodlaender, Shlomo Moran, and Manfred K. Warmuth. The distributed bit complexity of the ring: From the anonymous to the non-anonymous case. *Inf. Comput.*, 108(1):34–50, 1994.

27. Hans L. Bodlaender. On disjoint cycles. *Int. J. Found. Comput. Sci.*, 5(1):59–68, 1994.

28. Hans L. Bodlaender, Gerard Tel, and Nicola Santoro. Trade-offs in non-reversing diameter. *Nord. J. Comput.*, 1(1):111–134, 1994.

29. Hans L. Bodlaender, Rodney G. Downey, Michael R. Fellows, Michael T. Hallett, and Harold T. Wareham. Parameterized complexity analysis in computational biology. *Computer Applications in the Biosciences*, 11(1):49–57, 1995.

30. Hans L. Bodlaender, John R. Gilbert, Hjálmtyr Hafsteinsson, and Ton Kloks. Approximating treewidth, pathwidth, frontsize, and shortest elimination tree. *J. Algorithms*, 18(2):238–255, 1995.

31. Hans L. Bodlaender, Teofilo F. Gonzalez, and Ton Kloks. Complexity aspects of two-dimensional data compression. *Nord. J. Comput.*, 2(4):462–495, 1995.

32. Hans L. Bodlaender and Michael R. Fellows. W[2]-hardness of precedence constrained K-processor scheduling. *Oper. Res. Lett.*, 18(2):93–97, 1995.

33. Hans L. Bodlaender, Ton Kloks, and Dieter Kratsch. Treewidth and pathwidth of permutation graphs. *SIAM J. Discrete Math.*, 8(4):606–616, 1995.

34. Hans L. Bodlaender, Rodney G. Downey, Michael R. Fellows, and Harold T. Wareham. The parameterized complexity of sequence alignment and consensus. *Theor. Comput. Sci.*, 147(1&2):31–54, 1995.

35. Hans L. Bodlaender and Klaus Jansen. Restrictions of graph partition problems. part I. *Theor. Comput. Sci.*, 148(1):93–109, 1995.

36. Hans L. Bodlaender and Babette L.E. de Fluiter. On intervalizing k-colored graphs for DNA physical mapping. *Discrete Applied Mathematics*, 71(1–3):55–77, 1996.

37. Hans L. Bodlaender and Ton Kloks. Efficient and constructive algorithms for the pathwidth and treewidth of graphs. *J. Algorithms*, 21(2):358–402, 1996.

38. Hans L. Bodlaender. A linear-time algorithm for finding tree-decompositions of small treewidth. *SIAM J. Comput.*, 25(6):1305–1317, 1996.

39. Hans L. Bodlaender and Babette L.E. de Fluiter. A problem on strings with an application to intervalizing colored graphs. *Bulletin of the EATCS*, 62:323–324, 1997.

40. Hans L. Bodlaender and Dimitrios M. Thilikos. Treewidth for graphs with small chordality. *Discrete Applied Mathematics*, 79(1–3):45–61, 1997.

41. Hans L. Bodlaender, Jan van Leeuwen, Richard B. Tan, and Dimitrios M. Thilikos. On interval routing schemes and treewidth. *Inf. Comput.*, 139(1):92–109, 1997.

42. Goos Kant and Hans L. Bodlaender. Triangulating planar graphs while minimizing the maximum degree. *Inf. Comput.*, 135(1):1–14, 1997.

43. Hans L. Bodlaender, Dimitrios M. Thilikos, and Koichi Yamazaki. It is hard to know when greedy is good for finding independent sets. *Inf. Process. Lett.*, 61(2):101–111, 1997.

44. Dimitrios M. Thilikos and Hans L. Bodlaender. Fast partitioning *l*-apex graphs with application to approximating maximum induced-subgraph problems. *Inf. Process. Lett.*, 61(5):227–232, 1997.

45. Hans L. Bodlaender and Joost Engelfriet. Domino treewidth. *J. Algorithms*, 24(1):94–123, 1997.

46. Hans L. Bodlaender, Ton Kloks, Dieter Kratsch, and Haiko Müller. Treewidth and minimum fill-in on *d*-trapezoid graphs. *J. Graph Algorithms Appl.*, 2(5):1–23, 1998.

47. Hans L. Bodlaender and Torben Hagerup. Parallel algorithms with optimal speedup for bounded treewidth. *SIAM J. Comput.*, 27(6):1725–1746, 1998.

48. Hans L. Bodlaender, Jitender S. Deogun, Klaus Jansen, Ton Kloks, Dieter Kratsch, Haiko Müller, and Zsolt Tuza. Rankings of graphs. *SIAM J. Discrete Math.*, 11(1):168–181, 1998.

49. Hans L. Bodlaender. A partial *k*-arboretum of graphs with bounded treewidth. *Theor. Comput. Sci.*, 209(1–2):1–45, 1998.

50. Koichi Yamazaki, Hans L. Bodlaender, Babette de Fluiter, and Dimitrios M. Thilikos. Isomorphism for graphs of bounded distance width. *Algorithmica*, 24(2):105–127, 1999.

51. Hans L. Bodlaender. A note on domino treewidth. *Discrete Mathematics & Theoretical Computer Science*, 3(4):141–150, 1999.

52. Hans L. Bodlaender, Ton Kloks, and Rolf Niedermeier. SIMPLE MAX-CUT for unit interval graphs and graphs with few *p4*s. *Electronic Notes in Discrete Mathematics*, 3:19–26, 1999.

53. Hans L. Bodlaender and Dimitrios M. Thilikos. Graphs with branchwidth at most three. *J. Algorithms*, 32(2):167–194, 1999.

54. Hans L. Bodlaender. Introduction. *Algorithmica*, 27(3):209–211, 2000.

55. Hans L. Bodlaender. The algorithmic theory of treewidth. *Electronic Notes in Discrete Mathematics*, 5:27–30, 2000.

56. Hans Zantema and Hans L. Bodlaender. Finding small equivalent decision trees is hard. *Int. J. Found. Comput. Sci.*, 11(2):343–354, 2000.

57. Hans L. Bodlaender and Klaus Jansen. On the complexity of the maximum cut problem. *Nord. J. Comput.*, 7(1):14–31, 2000.

58. Hans L. Bodlaender, Michael R. Fellows, Michael T. Hallett, Todd Wareham, and Tandy J. Warnow. The hardness of perfect phylogeny, feasible register assignment and other problems on thin colored graphs. *Theor. Comput. Sci.*, 244(1–2):167–188, 2000.

59. Hans L. Bodlaender and Babette van Antwerpen-de Fluiter. Parallel algorithms for series parallel graphs and graphs with treewidth two. *Algorithmica*, 29(4):534–559, 2001.

60. Arie M. C. A. Koster, Hans L. Bodlaender, and Stan P. M. van Hoesel. Treewidth: Computational experiments. *Electronic Notes in Discrete Mathematics*, 8:54–57, 2001.

61. Hans L. Bodlaender and Babette van Antwerpen-de Fluiter. Reduction algorithms for graphs of small treewidth. *Inf. Comput.*, 167(2):86–119, 2001.

62. Hans L. Bodlaender. A generic NP-hardness proof for a variant of graph coloring. *J. UCS*, 7(12):1114–1124, 2001.

63. Jochen Alber, Hans L. Bodlaender, Henning Fernau, Ton Kloks, and Rolf Niedermeier. Fixed parameter algorithms for DOMINATING SET and related problems on planar graphs. *Algorithmica*, 33(4):461–493, 2002.

64. Hans L. Bodlaender, Michael J. Dinneen, and Bakhadyr Khoussainov. Relaxed update and partition network games. *Fundam. Inform.*, 49(4):301–312, 2002.

65. Hans Zantema and Hans L. Bodlaender. Sizes of ordered decision trees. *Int. J. Found. Comput. Sci.*, 13(3):445–458, 2002.

66. Hans L. Bodlaender and Fedor V. Fomin. Approximation of pathwidth of outerplanar graphs. *J. Algorithms*, 43(2):190–200, 2002.

67. Hans L. Bodlaender and Dieter Kratsch. Kayles and Nimbers. *J. Algorithms*, 43(1): 106–119, 2002.

68. Hans L. Bodlaender and Udi Rotics. Computing the treewidth and the minimum fill-in with the modular decomposition. *Algorithmica*, 36(4):375–408, 2003.

69. Hans L. Bodlaender, Richard B. Tan, and Jan van Leeuwen. Finding a δ-regular supergraph of minimum order. *Discrete Applied Mathematics*, 131(1):3–9, 2003.

70. Hans L. Bodlaender. Necessary edges in k-chordalisations of graphs. *J. Comb. Optim.*, 7(3):283–290, 2003.

71. Hans L. Bodlaender, Ton Kloks, Richard B. Tan, and Jan van Leeuwen. Approximations for λ-colorings of graphs. *Comput. J.*, 47(2):193–204, 2004.

72. Hans L. Bodlaender and Gerard Tel. A note on rectilinearity and angular resolution. *J. Graph Algorithms Appl.*, 8:89–94, 2004.

73. Hans L. Bodlaender and Jan Arne Telle. Space-efficient construction variants of dynamic programming. *Nord. J. Comput.*, 11(4):374–385, 2004.

74. Hans L. Bodlaender, Hajo Broersma, Fedor V. Fomin, Artem V. Pyatkin, and Gerhard J. Woeginger. Radio labeling with preassigned frequencies. *SIAM Journal on Optimization*, 15(1):1–16, 2004.

75. Hans L. Bodlaender, Arie M. C. A. Koster, and Frank van den Eijkhof. Preprocessing rules for triangulation of probabilistic networks. *Computational Intelligence*, 21(3):286–305, 2005.

76. Hans L. Bodlaender and Fedor V. Fomin. Tree decompositions with small cost. *Discrete Applied Mathematics*, 145(2):143–154, 2005.

77. Dimitrios M. Thilikos, Maria J. Serna, and Hans L. Bodlaender. Cutwidth I: A linear time fixed parameter algorithm. *J. Algorithms*, 56(1):1–24, 2005.

78. Dimitrios M. Thilikos, Maria J. Serna, and Hans L. Bodlaender. Cutwidth II: Algorithms for partial ω-trees of bounded degree. *J. Algorithms*, 56(1):25–49, 2005.

79. Hans L. Bodlaender, Andreas Brandstädt, Dieter Kratsch, Michaël Rao, and Jeremy P. Spinrad. On algorithms for ($p5$, gem)-free graphs. *Theor. Comput. Sci.*, 349(1):2–21, 2005.

80. Hans L. Bodlaender and Fedor V. Fomin. Equitable colorings of bounded treewidth graphs. *Theor. Comput. Sci.*, 349(1):22–30, 2005.

81. Hans L. Bodlaender and Arie M. C. A. Koster. Safe separators for treewidth. *Discrete Mathematics*, 306(3):337–350, 2006.

82. Hans L. Bodlaender, Thomas Wolle, and Arie M. C. A. Koster. Contraction and treewidth lower bounds. *J. Graph Algorithms Appl.*, 10(1):5–49, 2006.

83. Irit Katriel and Hans L. Bodlaender. Online topological ordering. *ACM Trans. Algorithms*, 2(3):364–379, 2006.

84. Frank van den Eijkhof, Hans L. Bodlaender, and Arie M. C. A. Koster. Safe reduction rules for weighted treewidth. *Algorithmica*, 47(2):139–158, 2007.

85. Alexander Grigoriev and Hans L. Bodlaender. Algorithms for graphs embeddable with few crossings per edge. *Algorithmica*, 49(1):1–11, 2007.

86. Hans L. Bodlaender and Arie M. C. A. Koster. On the maximum cardinality search lower bound for treewidth. *Discrete Applied Mathematics*, 155(11):1348–1372, 2007.

87. Hans L. Bodlaender, Alexander Grigoriev, and Arie M. C. A. Koster. Treewidth lower bounds with brambles. *Algorithmica*, 51(1):81–98, 2008.

88. Hans L. Bodlaender and Arie M. C. A. Koster. Combinatorial optimization on graphs of bounded treewidth. *Comput. J.*, 51(3):255–269, 2008.

89. Hans L. Bodlaender, Corinne Feremans, Alexander Grigoriev, Eelko Penninkx, René Sitters, and Thomas Wolle. On the minimum corridor connection problem and other generalized geometric problems. *Comput. Geom.*, 42(9):939–951, 2009.

90. Hans L. Bodlaender, Rodney G. Downey, Michael R. Fellows, and Danny Hermelin. On problems without polynomial kernels. *J. Comput. Syst. Sci.*, 75(8):423–434, 2009.

91. Hans L. Bodlaender, Michael R. Fellows, and Dimitrios M. Thilikos. Derivation of algorithms for cutwidth and related graph layout parameters. *J. Comput. Syst. Sci.*, 75(4): 231–244, 2009.

92. Helmut Alt, Hans L. Bodlaender, Marc J. van Kreveld, Günter Rote, and Gerard Tel. Wooden geometric puzzles: Design and hardness proofs. *Theory Comput. Syst.*, 44(2): 160–174, 2009.

93. Frederic Dorn, Eelko Penninkx, Hans L. Bodlaender, and Fedor V. Fomin. Efficient exact algorithms on planar graphs: Exploiting sphere cut decompositions. *Algorithmica*, 58(3): 790–810, 2010.

94. Hans L. Bodlaender and Arie M. C. A. Koster. Treewidth computations I. Upper bounds. *Inf. Comput.*, 208(3):259–275, 2010.

95. Hans L. Bodlaender, Albert Hendriks, Alexander Grigoriev, and Nadejda V. Grigorieva. The valve location problem in simple network topologies. *INFORMS Journal on Computing*, 22(3):433–442, 2010.

96. Hans L. Bodlaender and Thomas C. van Dijk. A cubic kernel for feedback vertex set and loop cutset. *Theory Comput. Syst.*, 46(3):566–597, 2010.

97. Hans L. Bodlaender, Michael R. Fellows, Pinar Heggernes, Federico Mancini, Charis Papadopoulos, and Frances A. Rosamond. Clustering with partial information. *Theor. Comput. Sci.*, 411(7–9):1202–1211, 2010.

98. Hans L. Bodlaender, Michael R. Fellows, Michael A. Langston, Mark A. Ragan, Frances A. Rosamond, and Mark Weyer. Quadratic kernelization for convex recoloring of trees. Algorithmica, 61(2):362–388, 2011.

99. Hans L. Bodlaender, Pinar Heggernes, and Yngve Villanger. Faster parameterized algorithms for Minimum Fill-in. *Algorithmica*, 61(4):817–838, 2011.

100. Johan M. M. van Rooij and Hans L. Bodlaender. Exact algorithms for dominating set. *Discrete Applied Mathematics*, 159(17):2147–2164, 2011.

101. Hans L. Bodlaender, Kyohei Kozawa, Takayoshi Matsushima, and Yota Otachi. Spanning tree congestion of k-outerplanar graphs. *Discrete Mathematics*, 311(12):1040–1045, 2011.

102. Hans L. Bodlaender and Arie M. C. A. Koster. Treewidth computations II. Lower bounds. *Inf. Comput.*, 209(7):1103–1119, 2011.

103. Hans L. Bodlaender, Stéphan Thomassé, and Anders Yeo. Kernel bounds for disjoint cycles and disjoint paths. *Theor. Comput. Sci.*, 412(35):4570–4578, 2011.

104. Hans L. Bodlaender, Fedor V. Fomin, Petr A. Golovach, Yota Otachi, and Erik Jan van Leeuwen. Parameterized complexity of the spanning tree congestion problem. *Algorithmica*, 64(1):85–111, 2012.

105. Johan M. M. van Rooij and Hans L. Bodlaender. Exact algorithms for Edge Domination. *Algorithmica*, 64(4):535–563, 2012.

106. Hans L. Bodlaender, Fedor V. Fomin, Arie M. C. A. Koster, Dieter Kratsch, and Dimitrios M. Thilikos. A note on exact algorithms for vertex ordering problems on graphs. *Theory Comput. Syst.*, 50(3):420–432, 2012.

107. Hans L. Bodlaender, Petra Schuurman, and Gerhard J. Woeginger. Scheduling of pipelined operator graphs. *J. Scheduling*, 15(3):323–332, 2012.

108. Hans L. Bodlaender, Fedor V. Fomin, Arie M. C. A. Koster, Dieter Kratsch, and Dimitrios M. Thilikos. On exact algorithms for treewidth. *ACM Trans. Algorithms*, 9(1):12:1–12:23, 2012.

109. Hans L. Bodlaender and Jurriaan Hage. On switching classes, NLC-width, cliquewidth and treewidth. *Theor. Comput. Sci.*, 429:30–35, 2012.

110. Bart M. P. Jansen and Hans L. Bodlaender. Vertex cover kernelization revisited - Upper and lower bounds for a refined parameter. *Theory Comput. Syst.*, 53(2):263–299, 2013.

111. Johan M. M. van Rooij, Marcel E. van Kooten Niekerk, and Hans L. Bodlaender. Partition into triangles on bounded degree graphs. *Theory Comput. Syst.*, 52(4):687–718, 2013.

112. Hans L. Bodlaender, Bart M. P. Jansen, and Stefan Kratsch. Preprocessing for treewidth: A combinatorial analysis through kernelization. *SIAM J. Discrete Math.*, 27(4):2108–2142, 2013.

113. Hans L. Bodlaender, Bart M. P. Jansen, and Stefan Kratsch. Kernel bounds for path and cycle problems. *Theor. Comput. Sci.*, 511:117–136, 2013.

114. Nadja Betzler, Hans L. Bodlaender, Robert Bredereck, Rolf Niedermeier, and Johannes Uhlmann. On making a distinguished vertex of minimum degree by vertex deletion. *Algorithmica*, 68(3):715–738, 2014.

115. Hans L. Bodlaender, Bart M. P. Jansen, and Stefan Kratsch. Kernelization lower bounds by cross-composition. *SIAM J. Discrete Math.*, 28(1):277–305, 2014.

116. Hans L. Bodlaender, MohammadTaghi Hajiaghayi, and Giuseppe F. Italiano. Editorial. *Algorithmica*, 73(4):748–749, 2015.

117. Hans L. Bodlaender, Mohammad Taghi Hajiaghayi, and Giuseppe F. Italiano. Erratum to: Editorial. *Algorithmica*, 73(4):750, 2015.

118. Stefan Fafianie, Hans L. Bodlaender, and Jesper Nederlof. Speeding up dynamic programming with representative sets: An experimental evaluation of algorithms for steiner tree on tree decompositions. *Algorithmica*, 71(3):636–660, 2015.

119. Hans L. Bodlaender, Pinar Heggernes, and Jan Arne Telle. Recognizability equals definability for graphs of bounded treewidth and bounded chordality. *Electronic Notes in Discrete Mathematics*, 49:559–568, 2015.

120. Hans L. Bodlaender, Marek Cygan, Stefan Kratsch, and Jesper Nederlof. Deterministic single exponential time algorithms for connectivity problems parameterized by treewidth. *Inf. Comput.*, 243:86–111, 2015.

121. Hans L. Bodlaender and Marc J. van Kreveld. Google scholar makes it hard - the complexity of organizing one's publications. *Inf. Process. Lett.*, 115(12):965–968, 2015.

122. Hans L. Bodlaender, Dieter Kratsch, and Sjoerd T. Timmer. Exact algorithms for Kayles. *Theor. Comput. Sci.*, 562:165–176, 2015.

123. Hans L. Bodlaender, Fedor V. Fomin, Daniel Lokshtanov, Eelko Penninkx, Saket Saurabh, and Dimitrios M. Thilikos. (Meta) Kernelization. *J. ACM*, 63(5):44:1–44:69, 2016.

124. Hans L. Bodlaender and Johan M. M. van Rooij. Exact algorithms for intervalizing coloured graphs. *Theory Comput. Syst.*, 58(2):273–286, 2016.

125. Hans L. Bodlaender, Pål Grønås Drange, Markus S. Dregi, Fedor V. Fomin, Daniel Lokshtanov, and Michal Pilipczuk. A $c^k n$ 5-approximation algorithm for treewidth. *SIAM J. Comput.*, 45(2):317–378, 2016.

126. Hans L. Bodlaender, Stefan Kratsch, Vincent J. C. Kreuzen, O-joung Kwon, and Seongmin Ok. Characterizing width two for variants of treewidth. *Discrete Applied Mathematics*, 216:29–46, 2017.

127. Lars Jaffke, Hans L. Bodlaender, Pinar Heggernes, and Jan Arne Telle. Definability equals recognizability for k-outerplanar graphs and l-chordal partial k-trees. *Eur. J. Comb.*, 66:191–234, 2017.

128. Hans L. Bodlaender, Hirotaka Ono, and Yota Otachi. Degree-constrained orientation of maximum satisfaction: Graph classes and parameterized complexity. *Algorithmica*, 80(7): 2160–2180, 2018.

129. Hans L. Bodlaender, Hirotaka Ono, and Yota Otachi. A faster parameterized algorithm for pseudoforest deletion. *Discrete Applied Mathematics*, 236:42–56, 2018.

130. Mark de Berg, Hans L. Bodlaender, and Sándor Kisfaludi-Bak. The homogeneous broadcast problem in narrow and wide strips I: Algorithms. *Algorithmica*, 81(7): 2934–2962, 2019.

131. Mark de Berg, Hans L. Bodlaender, and Sándor Kisfaludi-Bak. The homogeneous broadcast problem in narrow and wide strips II: Lower bounds. *Algorithmica*, 81(7): 2963–2990, 2019.

132. Jianer Chen, Hans L. Bodlaender, and Virginia Vassilevska Williams. EATCS-IPEC Nerode Prize 2019 - Call for nominations. *Bulletin of the EATCS*, 127, 2019.

133. Hans L. Bodlaender and Tom C. van der Zanden. On exploring always-connected temporal graphs of small pathwidth. *Inf. Process. Lett.*, 142:68–71, 2019.

134. Tesshu Hanaka, Hans L. Bodlaender, Tom C. van der Zanden, and Hirotaka Ono. On the maximum weight minimal separator. *Theor. Comput. Sci.*, 796:294–308, 2019.

Conference Articles

1. Hans L. Bodlaender and Jan van Leeuwen. Simulations of large networks on smaller networks. *In Proceedings NGI-SION Symposium 1984*, pages 227–241, 1984.

2. Hans L. Bodlaender and Jan van Leeuwen. Uniform emulation of the shuffle-exchange network. In Uwe Pape, editor, *Proceedings of the 10th Conference on Graph Theoretic Concepts in Computer Science*, pages 19–28. Trauner Verlag, 1984.

3. Hans L. Bodlaender and Jan van Leeuwen. Simulation of large networks on smaller networks. In Kurt Mehlhorn, editor, *STACS 85, 2nd Symposium of Theoretical Aspects of Computer Science, Saarbrücken, Germany, January 3–5, 1985, Proceedings*, volume 182 of *Lecture Notes in Computer Science*, pages 47–58. Springer, 1985.

4. Anneke A. Schoone, Hans L. Bodlaender, and Jan van Leeuwen. Improved diameter bounds for altered graphs. In Gottfried Tinhofer and Gunther Schmidt, editors, *Graphtheoretic Concepts in Computer Science, International Workshop, WG '86, Bernried, Germany, June 17–19, 1986, Proceedings*, volume 246 of *Lecture Notes in Computer Science*, pages 227–236. Springer, 1986.

5. Hans L. Bodlaender and Jan van Leeuwen. New upperbounds for decentralized extrema-finding in a ring of processors. In Eli Gafni and Nicola Santoro, editors, *Proceedings of the 1st Workshop on Distributed Algorithms on Graphs*, pages 27–40. Carlton University Press, 1986.

6. Hans L. Bodlaender and Jan van Leeuwen. New upperbounds for decentralized extrema-finding in a ring of processors. In Burkhard Monien and Guy Vidal-Naquet, editors, *STACS 86, 3rd Annual Symposium on Theoretical Aspects of Computer Science, Orsay, France, January 16–18, 1986, Proceedings*, volume 210 of *Lecture Notes in Computer Science*, pages 119–129. Springer, 1986.

7. Hans L. Bodlaender. New lower bounds for distributed leader finding in asynchronous rings of processors. In Manfred Paul, editor, GI - 17. *Jahrestagung, Computerintegrierter Arbeitsplatz im Büro, München, 20–23. Oktober 1987, Proceedings*, volume 156 of *Informatik-Fachberichte*, pages 82–88. Springer, 1987.

8. Hans L. Bodlaender. Dynamic programming on graphs with bounded treewidth. In Timo Lepistö and Arto Salomaa, editors, *Automata, Languages and Programming, 15th International Colloquium, ICALP88, Tampere, Finland, July 11–15, 1988, Proceedings*, volume 317 of *Lecture Notes in Computer Science*, pages 105–118. Springer, 1988.

9. Hans L. Bodlaender. NC-algorithms for graphs with small treewidth. In Jan van Leeuwen, editor, *Graph-Theoretic Concepts in Computer Science, 14th International Workshop, WG '88, Amsterdam, The Netherlands, June 15–17, 1988, Proceedings*, volume 344 of *Lecture Notes in Computer Science*, pages 1–10. Springer, 1988.

10. Hans L. Bodlaender. New lower bound techniques for distributed leader finding and other problems on rings of processors. In *Proceedings Computing Science in the Netherlands*, pages 241–258, 1988.

11. Hans L. Bodlaender. Polynomial algorithms for graph isomorphism and chromatic index on partial k-trees. In Rolf G. Karlsson and Andrzej Lingas, editors, *SWAT 88, 1st Scandinavian Workshop on Algorithm Theory, Halmstad, Sweden, July 5–8, 1988, Proceedings*, volume 318 of *Lecture Notes in Computer Science*, pages 223–232. Springer, 1988.

12. Hans L. Bodlaender. On linear time minor tests and depth first search. In Frank K. H. A. Dehne, Jörg-Rüdiger Sack, and Nicola Santoro, editors, *Algorithms and Data Structures, Workshop WADS' 89, Ottawa, Canada, August 17–19, 1989, Proceedings*, volume 382 of *Lecture Notes in Computer Science*, pages 577–590. Springer, 1989.

13. Hans L. Bodlaender, Shlomo Moran, and Manfred K. Warmuth. The distributed bit complexity of the ring: From the anonymous to the non-anonymous case. In János Csirik, János Demetrovics, and Ferenc Gécseg, editors, *Fundamentals of Computation Theory, International Conference FCT'89, Szeged, Hungary, August 21–25, 1989, Proceedings*, volume 380 of *Lecture Notes in Computer Science*, pages 58–67. Springer, 1989.

14. Hans L. Bodlaender. Improved self-reduction algorithms for graphs with bounded treewidth. In Manfred Nagl, editor, *Graph-Theoretic Concepts in Computer Science, 15th International Workshop, WG '89, Castle Rolduc, The Netherlands, June 14–16, 1989, Proceedings*, volume 411 of *Lecture Notes in Computer Science*, pages 232–244. Springer, 1989.

15. Hans L. Bodlaender. On linear time minor tests and depth first search. In *Proceedings Computing Science in the Netherlands*, pages 15–32, 1989.

16. Hans L. Bodlaender. Classes of graphs with bounded treewidth or pathwidth. In Ulrich Faigle and Cornelis Hoede, editors, *Proceedings Twente Workshop on Graphs and Combinatorial Optimization*, pages 21–28, 1989.

17. Paul Beame and Hans L. Bodlaender. Distributed computing on transitive networks: The thorus. In Burkhard Monien and Robert Cori, editors, *STACS 89, 6th Annual Symposium on Theoretical Aspects of Computer Science, Paderborn, FRG, February 16–18, 1989, Proceedings*, volume 349 of *Lecture Notes in Computer Science*, pages 294–303. Springer, 1989.

18. Hans L. Bodlaender. On the complexity of some coloring games. In Rolf H. Möhring, editor, *Graph-Theoretic Concepts in Computer Science, 16th International Workshop, WG '90, Berlin, Germany, June 20–22, 1990, Proceedings*, volume 484 of *Lecture Notes in Computer Science*, pages 30–40. Springer, 1990.

19. Hans L. Bodlaender and Rolf H. Möhring. The pathwidth and treewidth of cographs. In John R. Gilbert and Rolf G. Karlsson, editors, *SWAT 90, 2nd Scandinavian Workshop on Algorithm Theory, Bergen, Norway, July 11–14, 1990, Proceedings*, volume 447 of *Lecture Notes in Computer Science*, pages 301–309. Springer, 1990.

20. Hans L. Bodlaender, John R. Gilbert, Ton Kloks, and Hjálmtyr Hafsteinsson. Approximating treewidth, pathwidth, and minimum elimination tree height. In Gunther Schmidt and Rudolf Berghammer, editors, *17th International Workshop, WG '91, Fischbachau, Germany, June 17–19, 1991, Proceedings*, volume 570 of *Lecture Notes in Computer Science*, pages 1–12. Springer, 1991.

21. Hans L. Bodlaender. On disjoint cycles. In Gunther Schmidt and Rudolf Berghammer, editors, *17th International Workshop, WG '91, Fischbachau, Germany, June 17–19, 1991, Proceedings*, volume 570 of *Lecture Notes in Computer Science*, pages 230–238. Springer, 1991.

22. Goos Kant and Hans L. Bodlaender. Planar graph augmentation problems (Extended abstract). In Frank K. H. A. Dehne, Jörg-Rüdiger Sack, and Nicola Santoro, editors, *Algorithms and Data Structures, 2nd Workshop WADS '91, Ottawa, Canada, August 14–16, 1991, Proceedings*, volume 519 of *Lecture Notes in Computer Science*, pages 286–298. Springer, 1991.

23. Hans L. Bodlaender and Ton Kloks. Better algorithms for the pathwidth and treewidth of graphs. In Javier Leach Albert, Burkhard Monien, and Mario Rodríguez-Artalejo, editors, *Automata, Languages and Programming, 18th International Colloquium, ICALP91, Madrid, Spain, July 8–12, 1991, Proceedings*, volume 510 of *Lecture Notes in Computer Science*, pages 544–555. Springer, 1991.

24. Jens Gustedt and Hans L. Bodlaender. The path-width of k-trees. In Neil Robertson and Paul D. Seymour, editors, *Graph Structure Theory, Proceedings of the AMS-IMS-SIAM Joint Summer Research Conference, Seattle WA, June 1991*, volume 147 of *American Mathematical Society Contemporary Mathematics*, pages 678–679, 1991.

25. Hans L. Bodlaender and Ton Kloks. A simple linear time algorithm for triangulating three-colored graphs. In Jan van Leeuwen, editor, *Proceedings Computing Science in the Netherlands*, pages 88–98, 1991.

26. Hans L. Bodlaender, Teofilo F. Gonzalez, and Ton Kloks. Complexity aspects of map compression. In James A. Storer and John H. Reif, editors, *Proceedings of the IEEE Data Compression Conference, DCC 1991, Snowbird, Utah, USA, April 8–11, 1991*, pages 287–296. IEEE Computer Society, 1991.

27. Goos Kant and Hans L. Bodlaender. Triangulating planar graphs while minimizing the maximum degree. In Otto Nurmi and Esko Ukkonen, editors, *Algorithm Theory - SWAT '92, Third Scandinavian Workshop on Algorithm Theory, Helsinki, Finland, July 8–10, 1992, Proceedings*, volume 621 of *Lecture Notes in Computer Science*, pages 258–271. Springer, 1992.

28. Ton Kloks and Hans L. Bodlaender. Testing superperfection of k-trees. In Otto Nurmi and Esko Ukkonen, editors, *Algorithm Theory - SWAT '92, Third Scandinavian Workshop on Algorithm Theory, Helsinki, Finland, July 8–10, 1992, Proceedings*, volume 621 of *Lecture Notes in Computer Science*, pages 292–303. Springer, 1992.

29. Ton Kloks and Hans L. Bodlaender. Approximating treewidth and pathwidth of some classes of perfect graphs. In Toshihide Ibaraki, Yasuyoshi Inagaki, Kazuo Iwama, Takao Nishizeki, and Masafumi Yamashita, editors, *Algorithms and Computation, Third International Symposium, ISAAC '92, Nagoya, Japan, December 16–18, 1992, Proceedings*, volume 650 of *Lecture Notes in Computer Science*, pages 116–125. Springer, 1992.

30. Hans L. Bodlaender, Michael R. Fellows, and Tandy J. Warnow. Two strikes against perfect phylogeny. In Werner Kuich, editor, *Automata, Languages and Programming, 19th International Colloquium, ICALP92, Vienna, Austria, July 13–17, 1992, Proceedings*, volume 623 of *Lecture Notes in Computer Science*, pages 273–283. Springer, 1992.

31. Hans L. Bodlaender, Klaus Jansen, and Gerhard J. Woeginger. Scheduling with incompatible jobs. In Ernst W. Mayr, editor, *Graph-Theoretic Concepts in Computer Science, 18th International Workshop, WG '92, Wiesbaden-Naurod, Germany, June 19–20, 1992, Proceedings*, volume 657 of *Lecture Notes in Computer Science*, pages 37–49. Springer, 1992.

32. Hans L. Bodlaender. Kayles on special classes of graphs - An application of Sprague-Grundy theory. In Ernst W. Mayr, editor, *Graph-Theoretic Concepts in Computer Science, 18th International Workshop, WG '92, Wiesbaden-Naurod, Germany, June 19–20, 1992, Proceedings*, volume 657 of *Lecture Notes in Computer Science*, pages 90–102. Springer, 1992.

33. Hans L. Bodlaender and Ton Kloks. A simple linear time algorithm for triangulating three-colored graphs. In Alain Finkel and Matthias Jantzen, editors, *STACS 92, 9th Annual Symposium on Theoretical Aspects of Computer Science, Cachan, France, February 13–15, 1992, Proceedings*, volume 577 of *Lecture Notes in Computer Science*, pages 415–423. Springer, 1992.

34. Ton Kloks, Hans L. Bodlaender, Haiko Müller, and Dieter Kratsch. Computing treewidth and minimum fill-in: All you need are the minimal separators. In Thomas Lengauer, editor, *Algorithms - ESA '93, First Annual European Symposium, Bad Honnef, Germany, September 30 – October 2, 1993, Proceedings*, volume 726 of *Lecture Notes in Computer Science*, pages 260–271. Springer, 1993.

35. Hans L. Bodlaender, Ton Kloks, and Dieter Kratsch. Treewidth and pathwidth of permutation graphs. In Andrzej Lingas, Rolf G. Karlsson, and Svante Carlsson, editors, *Automata, Languages and Programming, 20nd International Colloquium, ICALP93, Lund, Sweden, July 5–9, 1993, Proceedings*, volume 700 of *Lecture Notes in Computer Science*, pages 114–125. Springer, 1993.

36. Hans L. Bodlaender. On reduction algorithms for graphs with small treewidth. In Jan van Leeuwen, editor, *Graph-Theoretic Concepts in Computer Science, 19th International Workshop, WG '93, Utrecht, The Netherlands, June 16–18, 1993, Proceedings*, volume 790 of *Lecture Notes in Computer Science*, pages 45–56. Springer, 1993.

37. Hans L. Bodlaender. Dynamic algorithms for graphs with treewidth 2. In Jan van Leeuwen, editor, *Graph-Theoretic Concepts in Computer Science, 19th International Workshop, WG '93, Utrecht, The Netherlands, June 16–18, 1993, Proceedings*, volume 790 of *Lecture Notes in Computer Science*, pages 112–124. Springer, 1993.

38. Hans L. Bodlaender and Klaus Jansen. On the complexity of scheduling incompatible jobs with unit-times. In Andrzej M. Borzyszkowski and Stefan Sokolowski, editors, *Mathematical Foundations of Computer Science 1993, 18th International Symposium, MFCS '93, Gdansk, Poland, August 30 – September 3, 1993, Proceedings*, volume 711 of *Lecture Notes in Computer Science*, pages 291–300. Springer, 1993.

39. Hans L. Bodlaender. A linear time algorithm for finding tree-decompositions of small treewidth. In S. Rao Kosaraju, David S. Johnson, and Alok Aggarwal, editors, *Proceedings of the Twenty-Fifth Annual ACM Symposium on Theory of Computing, May 16–18, 1993, San Diego, CA, USA*, pages 226–234. ACM, 1993.

40. Ton Kloks, Hans L. Bodlaender, Haiko Müller, and Dieter Kratsch. Erratum: Computing treewidth and minimum fill-in: All you need are the minimal separators. In Jan van Leeuwen, editor, *Algorithms - ESA '94, Second Annual European Symposium, Utrecht, The Netherlands, September 26–28, 1994, Proceedings*, volume 855 of *Lecture Notes in Computer Science*, page 508. Springer, 1994.

41. Hans L. Bodlaender, Rodney G. Downey, Michael R. Fellows, and Harold T. Wareham. The parameterized complexity of sequence alignment and consensus. In Maxime Crochemore and Dan Gusfield, editors, *Combinatorial Pattern Matching, 5th Annual Symposium, CPM 94, Asilomar, California, USA, June 5–8, 1994, Proceedings*, volume 807 of *Lecture Notes in Computer Science*, pages 15–30. Springer, 1994.

42. Hans L. Bodlaender. A tourist guide through treewidth. In Jürgen Dassow and Alica Kelemenova, editors, *Developments in Theoretical Computer Science: Proceedings of the 7th International Meeting of Young Computer Scientists, Smolenice, 16–20 November 1992*, pages 1–20. Gordon and Breach Science Publishers, 1994.

43. Hans L. Bodlaender and Joost Engelfriet. Domino treewith (Extended abstract). In Ernst W. Mayr, Gunther Schmidt, and Gottfried Tinhofer, editors, *Graph-Theoretic Concepts in Computer Science, 20th International Workshop, WG '94, Herrsching, Germany, June 16–18, 1994, Proceedings*, volume 903 of *Lecture Notes in Computer Science*, pages 1–13. Springer, 1994.

44. Hans L. Bodlaender, Jitender S. Deogun, Klaus Jansen, Ton Kloks, Dieter Kratsch, Haiko Müller, and Zsolt Tuza. Ranking of graphs. In Ernst W. Mayr, Gunther Schmidt, and Gottfried Tinhofer, editors, *Graph-Theoretic Concepts in Computer Science, 20th International Workshop, WG '94, Herrsching, Germany, June 16–18, 1994, Proceedings*, volume 903 of *Lecture Notes in Computer Science*, pages 292–304. Springer, 1994.

45. Hans L. Bodlaender, Michael R. Fellows, and Michael T. Hallett. Beyond NP-completeness for problems of bounded width: Hardness for the W hierarchy. In Eric Backer, editor, *Proceedings CSN '94, Computing Science in the Netherlands*, pages 1–10, 1994.

46. Hans L. Bodlaender, Michael R. Fellows, and Michael T. Hallett. Beyond NP-completeness for problems of bounded width: Hardness for the W hierarchy. In Frank Thomson Leighton and Michael T. Goodrich, editors, *Proceedings of the Twenty-Sixth Annual ACM Symposium on Theory of Computing, 23–25 May 1994, Montréal, Québec, Canada*, pages 449–458. ACM, 1994.

47. Hans L. Bodlaender and Klaus Jansen. On the complexity of the maximum cut problem. In Patrice Enjalbert, Ernst W. Mayr, and Klaus W. Wagner, editors, *STACS 94, 11th Annual Symposium on Theoretical Aspects of Computer Science, Caen, France, February 24–26, 1994, Proceedings*, volume 775 of *Lecture Notes in Computer Science*, pages 769–780. Springer, 1994.

48. Hans L. Bodlaender and Babette de Fluiter. Intervalizing k-colored graphs. In Zoltán Fülöp and Ferenc Gécseg, editors, *Automata, Languages and Programming, 22nd International Colloquium, ICALP95, Szeged, Hungary, July 10–14, 1995, Proceedings*, volume 944 of *Lecture Notes in Computer Science*, pages 87–98. Springer, 1995.

49. Hans L. Bodlaender and Torben Hagerup. Parallel algorithms with optimal speedup for bounded treewidth. In Zoltán Fülöp and Ferenc Gécseg, editors, *Automata, Languages and Programming, 22nd International Colloquium, ICALP95, Szeged, Hungary, July 10–14, 1995, Proceedings*, volume 944 of *Lecture Notes in Computer Science*, pages 268–279. Springer, 1995.

50. Hans L. Bodlaender, Richard B. Tan, Dimitrios M. Thilikos, and Jan van Leeuwen. On interval routing schemes and treewidth. In Manfred Nagl, editor, *Graph-Theoretic Concepts in Computer Science, 21st International Workshop, WG '95, Aachen, Germany, June 20–22, 1995, Proceedings*, volume 1017 of *Lecture Notes in Computer Science*, pages 181–196. Springer, 1995.

51. Hans L. Bodlaender and Babette de Fluiter. Parallel algorithms for series parallel graphs. In Josep Díaz and Maria J. Serna, editors, *Algorithms - ESA '96, Fourth Annual European Symposium, Barcelona, Spain, September 25–27, 1996, Proceedings*, volume 1136 of *Lecture Notes in Computer Science*, pages 277–289. Springer, 1996.

52. Hans L. Bodlaender, Michael R. Fellows, and Patricia A. Evans. Finite-state computability of annotations of strings and trees. In Daniel S. Hirschberg and Eugene W. Myers, editors, *Combinatorial Pattern Matching, 7th Annual Symposium, CPM 96, Laguna Beach, California, USA, June 10–12, 1996, Proceedings*, volume 1075 of *Lecture Notes in Computer Science*, pages 384–391. Springer, 1996.

53. Hans L. Bodlaender and Babette de Fluiter. Reduction algorithms for constructing solutions in graphs with small treewidth. In Jin-yi Cai and C. K. Wong, editors, *Computing and Combinatorics, Second Annual International Conference, COCOON '96, Hong Kong, June 17–19, 1996, Proceedings*, volume 1090 of *Lecture Notes in Computer Science*, pages 199–208. Springer, 1996.

54. Koichi Yamazaki, Hans L. Bodlaender, Babette de Fluiter, and Dimitrios M. Thilikos. Isomorphism for graphs of bounded distance width. In Gian Carlo Bongiovanni, Daniel P. Bovet, and Giuseppe Di Battista, editors, *Algorithms and Complexity, Third Italian Conference, CIAC '97, Rome, Italy, March 12–14, 1997, Proceedings*, volume 1203 of *Lecture Notes in Computer Science*, pages 276–287. Springer, 1997.

55. Hans L. Bodlaender and Dimitrios M. Thilikos. Constructive linear time algorithms for branchwidth. In Pierpaolo Degano, Roberto Gorrieri, and Alberto Marchetti-Spaccamela, editors, *Automata, Languages and Programming, 24th International Colloquium, ICALP '97, Bologna, Italy, 7–11 July 1997, Proceedings*, volume 1256 of *Lecture Notes in Computer Science*, pages 627–637. Springer, 1997.

56. Babette de Fluiter and Hans L. Bodlaender. Parallel algorithms for treewidth two. In Rolf H. Möhring, editor, *Graph-Theoretic Concepts in Computer Science, 23rd International Workshop, WG '97, Berlin, Germany, June 18–20, 1997, Proceedings*, volume 1335 of *Lecture Notes in Computer Science*, pages 157–170. Springer, 1997.

57. Hans L. Bodlaender. Treewidth: Algorithmic techniques and results. In Igor Prívara and Peter Ruzicka, editors, *Mathematical Foundations of Computer Science 1997, 22nd International Symposium, MFCS '97, Bratislava, Slovakia, August 25–29, 1997, Proceedings*, volume 1295 of *Lecture Notes in Computer Science*, pages 19–36. Springer, 1997.

58. Linda C. van der Gaag and Hans L. Bodlaender. Comparing loop cutsets and clique trees in probabilistic inference. In Kris van Marcke and Walter Daelemans, editors, *Proceedings of the Ninth Dutch Conference on Artificial Intelligence*, pages 71–80. Nederlandse Vereniging voor Kunstmatige Intelligentie, 1997.

59. Hans L. Bodlaender and Torben Hagerup. Tree decompositions of small diameter. In Lubos Brim, Jozef Gruska, and Jirí Zlatuska, editors, *Mathematical Foundations of Computer Science 1998, 23rd International Symposium, MFCS '98, Brno, Czech Republic, August 24–28, 1998, Proceedings*, volume 1450 of *Lecture Notes in Computer Science*, pages 702–712. Springer, 1998.

60. Hans L. Bodlaender, Jens Gustedt, and Jan Arne Telle. Linear-time register allocation for a fixed number of registers. In Howard J. Karloff, editor, *Proceedings of the Ninth Annual ACM-SIAM Symposium on Discrete Algorithms, 25–27 January 1998, San Francisco, California, USA*, pages 574–583. ACM/SIAM, 1998.

61. H. N. de Ridder and Hans L. Bodlaender. Graph automorphisms with maximal projection distances. In Gabriel Ciobanu and Gheorghe Paun, editors, *Fundamentals of Computation Theory, 12th International Symposium, FCT '99, Iasi, Romania, August 30 – September 3, 1999, Proceedings*, volume 1684 of *Lecture Notes in Computer Science*, pages 204–214. Springer, 1999.

62. Ton Kloks, Hans L. Bodlaender, and Rolf Niedermeier. SIMPLE MAX-CUT for graphs with few $p4$s. In Hajo Broersma, Ulrich Faigle, Cornelis Hoede, and Johann Hurink, editors, *Proceedings 6th Twente Workshop on Graphs and Combinatorial Optimization, volume 3 of Electronic Notes in Discrete Mathematics*, pages 19–26. Elsevier Science Publishers, 1999.

63. Herman J. Haverkort and Hans L. Bodlaender. Finding a minimal tree in a polygon with its medial axis. In *Proceedings of the 11th Canadian Conference on Computational Geometry, UBC, Vancouver, British Columbia, Canada, August 15–18, 1999*, 1999.

64. Jochen Alber, Hans L. Bodlaender, Henning Fernau, and Rolf Niedermeier. Fixed parameter algorithms for PLANAR DOMINATING SET and related problems. In Magnús M. Halldórsson, editor, *Algorithm Theory - SWAT 2000, 7th Scandinavian Workshop on Algorithm Theory, Bergen, Norway, July 5–7, 2000, Proceedings*, volume 1851 of *Lecture Notes in Computer Science*, pages 97–110. Springer, 2000.

65. Dimitrios M. Thilikos, Maria J. Serna, and Hans L. Bodlaender. Constructive linear time algorithms for small cutwidth and carving-width. In D. T. Lee and Shang-Hua Teng, editors, *Algorithms and Computation, 11th International Conference, ISAAC 2000, Taipei, Taiwan, December 18–20, 2000, Proceedings*, volume 1969 of *Lecture Notes in Computer Science*, pages 192–203. Springer, 2000.

66. Hans L. Bodlaender, Ton Kloks, Richard B. Tan, and Jan van Leeuwen. λ-coloring of graphs. In Horst Reichel and Sophie Tison, editors, *STACS 2000, 17th Annual Symposium on Theoretical Aspects of Computer Science, Lille, France, February 2000, Proceedings*, volume 1770 of *Lecture Notes in Computer Science*, pages 395–406. Springer, 2000.

67. Arie M. C. A. Koster, Hans L. Bodlaender, and Stan P. M. van Hoesel. Treewidth: Computational experiments. In Hajo Broersma, Ulrich Faigle, Johann Hurink, and Stefan Pickl, editors, *1st Cologne-Twente Workshop on Graphs and Combinatorial Optimization*, volume 8 of *Electronic Notes in Discrete Mathematics*. Elsevier Science Publishers, 2001.

68. Dimitrios M. Thilikos, Maria J. Serna, and Hans L. Bodlaender. A polynomial time algorithm for the cutwidth of bounded degree graphs with small treewidth. In Friedhelm Meyer auf der Heide, editor, *Algorithms - ESA 2001, 9th Annual European Symposium, Aarhus, Denmark, August 28–31, 2001, Proceedings*, volume 2161 of *Lecture Notes in Computer Science*, pages 380–390. Springer, 2001.

69. Hans L. Bodlaender, Michael J. Dinneen, and Bakhadyr Khoussainov. On game-theoretic models of networks. In Peter Eades and Tadao Takaoka, editors, *Algorithms and Computation, 12th International Symposium, ISAAC 2001, Christchurch, New Zealand, December 19–21, 2001, Proceedings*, volume 2223 of *Lecture Notes in Computer Science*, pages 550–561. Springer, 2001.

70. Fedor V. Fomin and Hans L. Bodlaender. Approximation of pathwidth of outerplanar graphs. In Andreas Brandstädt and Van Bang Le, editors, *Graph-Theoretic Concepts in Computer Science, 27th International Workshop, WG 2001, Boltenhagen, Germany, June 14–16, 2001, Proceedings*, volume 2204 of *Lecture Notes in Computer Science*, pages 166–176. Springer, 2001.

71. Hans L. Bodlaender, Frank van den Eijkhof, and Linda C. van der Gaag. On the MPA problem in probabilistic networks. In Ben J. A. Kröse, Maarten de Rijke, Guus Schreiber, and Maarten van Someren, editors, *Proceedings of the 13th Belgium-Netherlands Conference on Artificial Intelligence BNAIC 2001*, pages 59–66, 2001.

72. Hans L. Bodlaender, Arie M. C. A. Koster, Frank van den Eijkhof, and Linda C. van der Gaag. Pre-processing for triangulation of probabilistic networks. In Ben J. A. Kröse, Maarten de Rijke, Guus Schreiber, and Maarten van Someren, editors, *Proceedings of the 13th Belgium-Netherlands Conference on Artificial Intelligence BNAIC 2001*, pages 59–66, 2001.

73. Hans L. Bodlaender, Arie M. C. A. Koster, Frank van den Eijkhof, and Linda C. van der Gaag. Pre-processing for triangulation of probabilistic networks. In Jack S. Breese and Daphne Koller, editors, *UAI '01: Proceedings of the 17th Conference in Uncertainty in Artificial Intelligence, University of Washington, Seattle, Washington, USA, August 2–5, 2001*, pages 32–39. Morgan Kaufmann, 2001.

74. Hans L. Bodlaender and Fedor V. Fomin. Tree decompositions with small cost. In Martti Penttonen and Erik Meineche Schmidt, editors, *Algorithm Theory - SWAT 2002, 8th Scandinavian Workshop on Algorithm Theory, Turku, Finland, July 3–5, 2002 Proceedings*, volume 2368 of *Lecture Notes in Computer Science*, pages 378–387. Springer, 2002.

75. Hans L. Bodlaender and Udi Rotics. Computing the treewidth and the minimum fill-in with the modular decomposition. In Martti Penttonen and Erik Meineche Schmidt, editors, *Algorithm Theory - SWAT 2002, 8th Scandinavian Workshop on Algorithm Theory, Turku, Finland, July 3–5, 2002 Proceedings*, volume 2368 of *Lecture Notes in Computer Science*, pages 388–397. Springer, 2002.

76. Hans L. Bodlaender, Hajo Broersma, Fedor V. Fomin, Artem V. Pyatkin, and Gerhard J. Woeginger. Radio labeling with pre-assigned frequencies. In Rolf H. Möhring and Rajeev Raman, editors, *Algorithms - ESA 2002, 10th Annual European Symposium, Rome, Italy, September 17–21, 2002, Proceedings*, volume 2461 of *Lecture Notes in Computer Science*, pages 211–222. Springer, 2002.

77. Frank van den Eijkhof and Hans L. Bodlaender. Safe reduction rules for weighted treewidth. In Ludek Kucera, editor, *Graph-Theoretic Concepts in Computer Science, 28th International Workshop, WG 2002, Cesky Krumlov, Czech Republic, June 13–15, 2002, Revised Papers*, volume 2573 of *Lecture Notes in Computer Science*, pages 176–185. Springer, 2002.

78. Hans L. Bodlaender, Frank van den Eijkhof, and Linda C. van der Gaag. On the complexity of the MPA problem in probabilistic networks. In Frank van Harmelen, editor, *Proceedings of the 15th Eureopean Conference on Artificial Intelligence, ECAI'2002, Lyon, France, July 2002*, pages 675–679. IOS Press, 2002.

79. Hans L. Bodlaender, Andreas Brandstädt, Dieter Kratsch, Michaël Rao, and Jeremy P. Spinrad. Linear time algorithms for some NP-complete problems on $(p_5, $ gem)-free graphs. In Andrzej Lingas and Bengt J. Nilsson, editors, *Fundamentals of Computation Theory, 14th International Symposium, FCT 2003, Malmö, Sweden, August 12–15, 2003, Proceedings*, volume 2751 of *Lecture Notes in Computer Science*, pages 61–72. Springer, 2003.

80. Hans L. Bodlaender, Michael R. Fellows, and Dimitrios M. Thilikos. Starting with nondeterminism: The systematic derivation of linear-time graph layout algorithms. In Branislav Rovan and Peter Vojtás, editors, *Mathematical Foundations of Computer Science 2003, 28th International Symposium, MFCS 2003, Bratislava, Slovakia, August 25–29, 2003, Proceedings*, volume 2747 of *Lecture Notes in Computer Science*, pages 239–248. Springer, 2003.

81. Hans L. Bodlaender, Arie M. C. A. Koster, and Thomas Wolle. Contraction and treewidth lower bounds. In Susanne Albers and Tomasz Radzik, editors, *Algorithms - ESA 2004, 12th Annual European Symposium, Bergen, Norway, September 14–17, 2004, Proceedings*, volume 3221 of *Lecture Notes in Computer Science*, pages 628–639. Springer, 2004.

82. Hans L. Bodlaender, Celina M. H. de Figueiredo, Marisa Gutierrez, Ton Kloks, and Rolf Niedermeier. Simple max-cut for split-indifference graphs and graphs with few p_4's. In Celso C. Ribeiro and Simone L. Martins, editors, *Experimental and Efficient Algorithms, Third International Workshop, WEA 2004, Angra dos Reis, Brazil, May 25–28, 2004, Proceedings*, volume 3059 of *Lecture Notes in Computer Science*, pages 87–99. Springer, 2004.

83. Hans L. Bodlaender and Arie M. C. A. Koster. On the maximum cardinality search lower bound for treewidth. In Juraj Hromkovic, Manfred Nagl, and Bernhard Westfechtel, editors, *Graph-Theoretic Concepts in Computer Science, 30th International Workshop, WG 2004, Bad Honnef, Germany, June 21–23, 2004, Revised Papers*, volume 3353 of *Lecture Notes in Computer Science*, pages 81–92. Springer, 2004.

84. Hans L. Bodlaender and Fedor V. Fomin. Equitable colorings of bounded treewidth graphs. In Jirí Fiala, Václav Koubek, and Jan Kratochvíl, editors, *Mathematical Foundations of Computer Science 2004, 29th International Symposium, MFCS 2004, Prague, Czech Republic, August 22–27, 2004, Proceedings*, volume 3153 of *Lecture Notes in Computer Science*, pages 180–190. Springer, 2004.

85. Hans L. Bodlaender and Dimitrios M. Thilikos. Computing small search numbers in linear time. In Rodney G. Downey, Michael R. Fellows, and Frank K. H. A. Dehne, editors, *Parameterized and Exact Computation, First International Workshop, IWPEC 2004, Bergen, Norway, September 14–17, 2004, Proceedings*, volume 3162 of *Lecture Notes in Computer Science*, pages 37–48. Springer, 2004.

86. Hans L. Bodlaender and Arie M. C. A. Koster. Safe seperators for treewidth. In Lars Arge, Giuseppe F. Italiano, and Robert Sedgewick, editors, *Proceedings of the Sixth Workshop on Algorithm Engineering and Experiments and the First Workshop on Analytic Algorithmics and Combinatorics, New Orleans, LA, USA, January 10, 2004*, pages 70–78. SIAM, 2004.

87. Linda C. van der Gaag, Hans L. Bodlaender, and A. J. Feelders. Monotonicity in Bayesian networks. In David Maxwell Chickering and Joseph Y. Halpern, editors, *UAI '04, Proceedings of the 20th Conference in Uncertainty in Artificial Intelligence, Banff, Canada, July 7–11, 2004*, pages 569–576. AUAI Press, 2004.

88. Frederic Dorn, Eelko Penninkx, Hans L. Bodlaender, and Fedor V. Fomin. Efficient exact algorithms on planar graphs: Exploiting sphere cut branch decompositions. In Gerth Stølting Brodal and Stefano Leonardi, editors, *Algorithms - ESA 2005, 13th Annual European Symposium, Palma de Mallorca, Spain, October 3–6, 2005, Proceedings*, volume 3669 of *Lecture Notes in Computer Science*, pages 95–106. Springer, 2005.

89. Hans L. Bodlaender, Alexander Grigoriev, and Arie M. C. A. Koster. Treewidth lower bounds with brambles. In Gerth Stølting Brodal and Stefano Leonardi, editors, *Algorithms - ESA 2005, 13th Annual European Symposium, Palma de Mallorca, Spain, October 3–6, 2005, Proceedings*, volume 3669 of *Lecture Notes in Computer Science*, pages 391–402. Springer, 2005.

90. Arie M. C. A. Koster, Thomas Wolle, and Hans L. Bodlaender. Degree-based treewidth lower bounds. In Sotiris E. Nikoletseas, editor, *Experimental and Efficient Algorithms, 4th InternationalWorkshop, WEA 2005, Santorini Island, Greece, May 10–13, 2005, Proceedings*, volume 3503 of *Lecture Notes in Computer Science*, pages 101–112. Springer, 2005.

91. Emgad H. Bachoore and Hans L. Bodlaender. New upper bound heuristics for treewidth. In Sotiris E. Nikoletseas, editor, *Experimental and Efficient Algorithms, 4th InternationalWorkshop, WEA 2005, Santorini Island, Greece, May 10–13, 2005, Proceedings*, volume 3503 of *Lecture Notes in Computer Science*, pages 216–227. Springer, 2005.

92. Alexander Grigoriev and Hans L. Bodlaender. Algorithms for graphs embeddable with few crossings per edge. In Maciej Liskiewicz and Rüdiger Reischuk, editors, *Fundamentals of Computation Theory, 15th International Symposium, FCT 2005, Lübeck, Germany, August 17–20, 2005, Proceedings*, volume 3623 of *Lecture Notes in Computer Science*, pages 378–387. Springer, 2005.

93. Irit Katriel and Hans L. Bodlaender. Online topological ordering. In *Proceedings of the Sixteenth Annual ACM-SIAM Symposium on Discrete Algorithms, SODA 2005, Vancouver, British Columbia, Canada, January 23–25, 2005*, pages 443–450. SIAM, 2005.

94. Hans L. Bodlaender. Discovering treewidth. In Peter Vojtás, Mária Bieliková, Bernadette Charron-Bost, and Ondrej Sýkora, editors, *SOFSEM 2005: Theory and Practice of Computer Science, 31st Conference on Current Trends in Theory and Practice of Computer Science, Liptovský Ján, Slovakia, January 22–28, 2005, Proceedings*, volume 3381 of *Lecture Notes in Computer Science*, pages 1–16. Springer, 2005.

95. Emgad H. Bachoore and Hans L. Bodlaender. A branch and bound algorithm for exact, upper, and lower bounds on treewidth. In Siu-Wing Cheng and Chung Keung Poon, editors, *Algorithmic Aspects in Information and Management, Second International Conference, AAIM 2006, Hong Kong, China, June 20–22, 2006, Proceedings*, volume 4041 of *Lecture Notes in Computer Science*, pages 255–266. Springer, 2006.

96. Hans L. Bodlaender, Fedor V. Fomin, Arie M. C. A. Koster, Dieter Kratsch, and Dimitrios M. Thilikos. On exact algorithms for treewidth. In Yossi Azar and Thomas Erlebach, editors, *Algorithms - ESA 2006, 14th Annual European Symposium, Zurich, Switzerland, September 11–13, 2006, Proceedings*, volume 4168 of *Lecture Notes in Computer Science*, pages 672–683. Springer, 2006.

97. Hans L. Bodlaender, Michael R. Fellows, Michael A. Langston, Mark A. Ragan, Frances A. Rosamond, and Mark Weyer. Kernelization for convex recoloring. In Hajo Broersma, Stefan S. Dantchev, Matthew Johnson, and Stefan Szeider, editors, *Algorithms and Complexity in Durham 2006 - Proceedings of the Second ACiD Workshop, 18–20 September 2006, Durham, UK*, volume 7 of *Texts in Algorithmics*, pages 23–35. King's College, London, 2006.

98. Hans L. Bodlaender, Corinne Feremans, Alexander Grigoriev, Eelko Penninkx, René Sitters, and Thomas Wolle. On the minimum corridor connection problem and other generalized geometric problems. In Thomas Erlebach and Christos Kaklamanis, editors, *Approximation and Online Algorithms, 4th International Workshop, WAOA 2006, Zurich, Switzerland, September 14–15, 2006, Revised Papers*, volume 4368 of *Lecture Notes in Computer Science*, pages 69–82. Springer, 2006.

99. Hans L. Bodlaender. Treewidth: Characterizations, applications, and computations. In Fedor V. Fomin, editor, *Graph-Theoretic Concepts in Computer Science, 32nd International Workshop, WG 2006, Bergen, Norway, June 22–24, 2006, Revised Papers*, volume 4271 of *Lecture Notes in Computer Science*, pages 1–14. Springer, 2006.

100. Emgad H. Bachoore and Hans L. Bodlaender. Weighted treewidth: Algorithmic techniques and results. In Takeshi Tokuyama, editor, *Algorithms and Computation, 18th International Symposium, ISAAC 2007, Sendai, Japan, December 17–19, 2007, Proceedings*, volume 4835 of *Lecture Notes in Computer Science*, pages 893–903. Springer, 2007.

101. Hans L. Bodlaender, Michael R. Fellows, Michael A. Langston, Mark A. Ragan, Frances A. Rosamond, and Mark Weyer. Quadratic kernelization for convex recoloring of trees. In Guohui Lin, editor, *Computing and Combinatorics, 13th Annual International Conference, COCOON 2007, Banff, Canada, July 16–19, 2007, Proceedings*, volume 4598 of *Lecture Notes in Computer Science*, pages 86–96. Springer, 2007.

102. Helmut Alt, Hans L. Bodlaender, Marc J. van Kreveld, Günter Rote, and Gerard Tel. Wooden geometric puzzles: Design and hardness proofs. In Pierluigi Crescenzi, Giuseppe Prencipe, and Geppino Pucci, editors, *Fun with Algorithms, 4th International Conference, FUN 2007, Castiglioncello, Italy, June 3–5, 2007, Proceedings*, volume 4475 of *Lecture Notes in Computer Science*, pages 16–29. Springer, 2007.

103. Johan Kwisthout, Hans L. Bodlaender, and Gerard Tel. Complexity results for local monotonicity in probabilistic networks. In Mehdi Dastani and E. de Jong, editors, *Proceedings of the 19th Belgian-Netherlands Conference on Artificial Intelligence BNAIC '07*, pages 369–370, 2007.

104. Hans L. Bodlaender, Alexander Grigoriev, Nadejda V. Grigorieva, and Albert Hendriks. The valve location problem. In Johann Hurink, Walter Kern, Gerhard F. Post, and Georg Still, editors, *Sixth Cologne Twente Workshop on Graphs and Combinatorial Optimization, University of Twente, Enschede, The Netherlands, 29–31 May, 2007*, pages 13–16. University of Twente, 2007.

105. Hans L. Bodlaender. A cubic kernel for feedback vertex set. In Wolfgang Thomas and Pascal Weil, editors, *STACS 2007, 24th Annual Symposium on Theoretical Aspects of Computer Science, Aachen, Germany, February 22–24, 2007, Proceedings*, volume 4393 of *Lecture Notes in Computer Science*, pages 320–331. Springer, 2007.

106. Hans L. Bodlaender. Treewidth: Structure and algorithms. In Giuseppe Prencipe and Shmuel Zaks, editors, *Structural Information and Communication Complexity, 14th International Colloquium, SIROCCO 2007, Castiglioncello, Italy, June 5–8, 2007, Proceedings*, volume 4474 of *Lecture Notes in Computer Science*, pages 11–25. Springer, 2007.

107. Johan Kwisthout, Hans L. Bodlaender, and Gerard Tel. Local monotonicity in probabilistic networks. In Khaled Mellouli, editor, *Symbolic and Quantitative Approaches to Reasoning with Uncertainty, 9th European Conference, ECSQARU 2007, Hammamet, Tunisia, October 31 – November 2, 2007, Proceedings*, volume 4724 of *Lecture Notes in Computer Science*, pages 548–559. Springer, 2007.

108. Hans L. Bodlaender, Richard B. Tan, Thomas C. van Dijk, and Jan van Leeuwen. Integer maximum flow in wireless sensor networks with energy constraint. In Joachim Gudmundsson, editor, *Algorithm Theory - SWAT 2008, 11th Scandinavian Workshop on Algorithm Theory, Gothenburg, Sweden, July 2–4, 2008, Proceedings*, volume 5124 of *Lecture Notes in Computer Science*, pages 102–113. Springer, 2008.

109. Hans L. Bodlaender, Pinar Heggernes, and Yngve Villanger. Faster parameterized algorithms for minimum fill-in. In Seok-Hee Hong, Hiroshi Nagamochi, and Takuro Fukunaga, editors, *Algorithms and Computation, 19th International Symposium, ISAAC 2008, Gold Coast, Australia, December 15–17, 2008. Proceedings*, volume 5369 of *Lecture Notes in Computer Science*, pages 282–293. Springer, 2008.

110. Hans L. Bodlaender, Eelko Penninkx, and Richard B. Tan. A linear kernel for the k-disjoint cycle problem on planar graphs. In Seok-Hee Hong, Hiroshi Nagamochi, and Takuro Fukunaga, editors, *Algorithms and Computation, 19th International Symposium, ISAAC 2008, Gold Coast, Australia, December 15–17, 2008. Proceedings*, volume 5369 of *Lecture Notes in Computer Science*, pages 306–317. Springer, 2008.

111. Hans L. Bodlaender, Rodney G. Downey, Michael R. Fellows, and Danny Hermelin. On problems without polynomial kernels (extended abstract). In Luca Aceto, Ivan Damgård, Leslie Ann Goldberg, Magnús M. Halldórsson, Anna Ingólfsdóttir, and Igor Walukiewicz, editors, *Automata, Languages and Programming, 35th International Colloquium, ICALP 2008, Reykjavik, Iceland, July 7–11, 2008, Proceedings, Part I: Tack A: Algorithms, Automata, Complexity, and Games*, volume 5125 of *Lecture Notes in Computer Science*, pages 563–574. Springer, 2008.

112. Hans L. Bodlaender, Alexander Grigoriev, Nadejda V. Grigorieva, and Albert Hendriks. The valve location problem in simple network topologies. In Hajo Broersma, Thomas Erlebach, Tom Friedetzky, and Daniël Paulusma, editors, *Graph-Theoretic Concepts in Computer Science, 34th International Workshop, WG 2008, Durham, UK, June 30 – July 2, 2008. Revised Papers*, volume 5344 of *Lecture Notes in Computer Science*, pages 55–65, 2008.

113. Hans L. Bodlaender, Michael R. Fellows, Pinar Heggernes, Federico Mancini, Charis Papadopoulos, and Frances A. Rosamond. Clustering with partial information. In Edward Ochmanski and Jerzy Tyszkiewicz, editors, *Mathematical Foundations of Computer Science 2008, 33rd International Symposium, MFCS 2008, Torun, Poland, August 25–29, 2008, Proceedings*, volume 5162 of *Lecture Notes in Computer Science*, pages 144–155. Springer, 2008.

114. Hans L. Bodlaender and Eelko Penninkx. A linear kernel for planar feedback vertex set. In Martin Grohe and Rolf Niedermeier, editors, *Parameterized and Exact Computation, Third International Workshop, IWPEC 2008, Victoria, Canada, May 14–16, 2008. Proceedings*, volume 5018 of *Lecture Notes in Computer Science*, pages 160–171. Springer, 2008.

115. Johan M. M. van Rooij and Hans L. Bodlaender. Exact algorithms for edge domination. In Martin Grohe and Rolf Niedermeier, editors, *Parameterized and Exact Computation, Third International Workshop, IWPEC 2008, Victoria, Canada, May 14–16, 2008. Proceedings*, volume 5018 of *Lecture Notes in Computer Science*, pages 214–225. Springer, 2008.

116. Johan M. M. van Rooij and Hans L. Bodlaender. Design by measure and conquer - A faster exact algorithm for dominating set. In Susanne Albers and Pascal Weil, editors, *STACS 2008, 25th Annual Symposium on Theoretical Aspects of Computer Science, Bordeaux, France, February 21–23, 2008, Proceedings*, volume 1 of *LIPIcs*, pages 657–668. Schloss Dagstuhl - Leibniz-Zentrum fuer Informatik, Germany, 2008.

117. Hans L. Bodlaender, Fedor V. Fomin, Daniel Lokshtanov, Eelko Penninkx, Saket Saurabh, and Dimitrios M. Thilikos. (Meta) Kernelization. In *50th Annual IEEE Symposium on Foundations of Computer Science, FOCS 2009, October 25–27, 2009, Atlanta, Georgia, USA*, pages 629–638. IEEE Computer Society, 2009.

118. Johan M. M. van Rooij, Hans L. Bodlaender, and Peter Rossmanith. Dynamic programming on tree decompositions using generalised fast subset convolution. In Amos Fiat and Peter Sanders, editors, *Algorithms - ESA 2009, 17th Annual European Symposium, Copenhagen, Denmark, September 7–9, 2009. Proceedings*, volume 5757 of *Lecture Notes in Computer Science*, pages 566–577. Springer, 2009.

119. Hans L. Bodlaender, Stéphan Thomassé, and Anders Yeo. Kernel bounds for disjoint cycles and disjoint paths. In Amos Fiat and Peter Sanders, editors, *Algorithms - ESA 2009, 17th Annual European Symposium, Copenhagen, Denmark, September 7–9, 2009. Proceedings*, volume 5757 of *Lecture Notes in Computer Science*, pages 635–646. Springer, 2009.

120. Hans L. Bodlaender. Kernelization: New upper and lower bound techniques. In Jianer Chen and Fedor V. Fomin, editors, *Parameterized and Exact Computation, 4th International Workshop, IWPEC 2009, Copenhagen, Denmark, September 10–11, 2009, Revised Selected Papers*, volume 5917 of *Lecture Notes in Computer Science*, pages 17–37. Springer, 2009.

121. Hans L. Bodlaender, Daniel Lokshtanov, and Eelko Penninkx. Planar capacitated dominating set is w[1]-hard. In Jianer Chen and Fedor V. Fomin, editors, *Parameterized and Exact Computation, 4th International Workshop, IWPEC 2009, Copenhagen, Denmark, September 10–11, 2009, Revised Selected Papers*, volume 5917 of *Lecture Notes in Computer Science*, pages 50–60. Springer, 2009.

122. Johan Kwisthout, Hans L. Bodlaender, and Linda C. van der Gaag. The necessity of bounded treewidth for efficient inference in Bayesian networks. In Helder Coelho, Rudi Studer, and Michael J. Wooldridge, editors, *ECAI 2010 - 19th European Conference on Artificial Intelligence, Lisbon, Portugal, August 16–20, 2010, Proceedings*, volume 215 of *Frontiers in Artificial Intelligence and Applications*, pages 237–242. IOS Press, 2010.

123. Yota Otachi, Hans L. Bodlaender, and Erik Jan van Leeuwen. Complexity results for the spanning tree congestion problem. In Dimitrios M. Thilikos, editor, *Graph Theoretic Concepts in Computer Science - 36th International Workshop, WG 2010, Zarós, Crete, Greece, June 28–30, 2010 Revised Papers*, volume 6410 of *Lecture Notes in Computer Science*, pages 3–14, 2010.

124. Hans L. Bodlaender, Erik Jan van Leeuwen, Johan M. M. van Rooij, and Martin Vatshelle. Faster algorithms on branch and clique decompositions. In Petr Hlinený and Antonn Kucera, editors, *Mathematical Foundations of Computer Science 2010, 35th International*

Symposium, MFCS 2010, Brno, Czech Republic, August 23–27, 2010. Proceedings, volume 6281 of *Lecture Notes in Computer Science*, pages 174–185. Springer, 2010.

125. Hans L. Bodlaender and Marc Comas. A kernel for convex recoloring of weighted forests. In Jan van Leeuwen, Anca Muscholl, David Peleg, Jaroslav Pokorný, and Bernhard Rumpe, editors, *SOFSEM 2010: Theory and Practice of Computer Science, 36th Conference on Current Trends in Theory and Practice of Computer Science, Spindleruv Mlýn, Czech Republic, January 23–29, 2010. Proceedings*, volume 5901 of *Lecture Notes in Computer Science*, pages 212–223. Springer, 2010.

126. Hans L. Bodlaender, Bart M. P. Jansen, and Stefan Kratsch. Cross-composition: A new technique for kernelization lower bounds. In Thomas Schwentick and Christoph Dürr, editors, *28th International Symposium on Theoretical Aspects of Computer Science, STACS 2011, March 10–12, 2011, Dortmund, Germany*, volume 9 of *LIPIcs*, pages 165–176. Schloss Dagstuhl - Leibniz-Zentrum fuer Informatik, 2011.

127. Bart M. P. Jansen and Hans L. Bodlaender. Vertex cover kernelization revisited: Upper and lower bounds for a refined parameter. In Thomas Schwentick and Christoph Dürr, editors, *28th International Symposium on Theoretical Aspects of Computer Science, STACS 2011, March 10–12, 2011, Dortmund, Germany*, volume 9 of *LIPIcs*, pages 177–188. Schloss Dagstuhl - Leibniz-Zentrum fuer Informatik, 2011.

128. Hans L. Bodlaender, Bart M. P. Jansen, and Stefan Kratsch. Preprocessing for treewidth: A combinatorial analysis through kernelization. In Luca Aceto, Monika Henzinger, and Jiří Sgall, editors, *Automata, Languages and Programming - 38th International Colloquium, ICALP 2011, Zurich, Switzerland, July 4–8, 2011, Proceedings, Part I*, volume 6755 of *Lecture Notes in Computer Science*, pages 437–448. Springer, 2011.

129. Hans L. Bodlaender and Dieter Kratsch. Exact algorithms for Kayles. In Petr Kolman and Jan Kratochvíl, editors, *Graph-Theoretic Concepts in Computer Science - 37th International Workshop, WG 2011, Teplá Monastery, Czech Republic, June 21–24, 2011. Revised Papers*, volume 6986 of *Lecture Notes in Computer Science*, pages 59–70. Springer, 2011.

130. Hans L. Bodlaender, Bart M. P. Jansen, and Stefan Kratsch. Kernel bounds for path and cycle problems. In Dániel Marx and Peter Rossmanith, editors, *Parameterized and Exact Computation - 6th International Symposium, IPEC 2011, Saarbrücken, Germany, September 6–8, 2011. Revised Selected Papers*, volume 7112 of *Lecture Notes in Computer Science*, pages 145–158. Springer, 2011.

131. Johan Kwisthout, Hans L. Bodlaender, and Linda C. van der Gaag. The complexity of finding kth most probable explanations in probabilistic networks. In Ivana Cerná, Tibor Gyimóthy, Juraj Hromkovic, Keith G. Jeffery, Rastislav Královic, Marko Vukolic, and Stefan Wolf, editors, *SOFSEM 2011: Theory and Practice of Computer Science - 37th Conference on Current Trends in Theory and Practice of Computer Science, Nový Smokovec, Slovakia, January 22–28, 2011. Proceedings*, volume 6543 of *Lecture Notes in Computer Science*, pages 356–367. Springer, 2011.

132. Arnold Overwijk, Eelko Penninkx, and Hans L. Bodlaender. A local search algorithm for branchwidth. In Ivana Cerná, Tibor Gyimóthy, Juraj Hromkovic, Keith G. Jeffery, Rastislav Královic, Marko Vukolic, and Stefan Wolf, editors, *SOFSEM 2011: Theory and Practice of Computer Science - 37th Conference on Current Trends in Theory and Practice of Computer Science, Nový Smokovec, Slovakia, January 22–28, 2011. Proceedings*, volume 6543 of *Lecture Notes in Computer Science*, pages 444–454. Springer, 2011.

133. Johan M. M. van Rooij, Marcel E. van Kooten Niekerk, and Hans L. Bodlaender. Partition into triangles on bounded degree graphs. In Ivana Cerná, Tibor Gyimóthy, Juraj Hromkovic, Keith G. Jeffery, Rastislav Královic, Marko Vukolic, and Stefan Wolf, editors, *SOFSEM 2011: Theory and Practice of Computer Science - 37th Conference on Current Trends in Theory and Practice of Computer Science, Nový Smokovec, Slovakia,*

January 22–28, 2011. Proceedings, volume 6543 of *Lecture Notes in Computer Science*, pages 558–569. Springer, 2011.

134. Linda C. van der Gaag and Hans L. Bodlaender. On stopping evidence gathering for diagnostic Bayesian networks. In Weiru Liu, editor, *Symbolic and Quantitative Approaches to Reasoning with Uncertainty - 11th European Conference, ECSQARU 2011, Belfast, UK, June 29 – July 1, 2011. Proceedings*, volume 6717 of *Lecture Notes in Computer Science*, pages 170–181. Springer, 2011.

135. Hans L. Bodlaender and Johan M. M. van Rooij. Exact algorithms for intervalizing colored graphs. In Alberto Marchetti-Spaccamela and Michael Segal, editors, *Theory and Practice of Algorithms in (Computer) Systems - First International ICST Conference, TAPAS 2011, Rome, Italy, April 18–20, 2011. Proceedings*, volume 6595 of *Lecture Notes in Computer Science*, pages 45–56. Springer, 2011.

136. Hans L. Bodlaender, Bart M. P. Jansen, and Stefan Kratsch. Kernel bounds for structural parameterizations of pathwidth. In Fedor V. Fomin and Petteri Kaski, editors, *Algorithm Theory - SWAT 2012 - 13th Scandinavian Symposium and Workshops, Helsinki, Finland, July 4–6, 2012. Proceedings*, volume 7357 of *Lecture Notes in Computer Science*, pages 352–363. Springer, 2012.

137. Hans L. Bodlaender. Probabilistic inference and monadic second order logic. In Jos C. M. Baeten, Thomas Ball, and Frank S. de Boer, editors, *Theoretical Computer Science - 7th IFIP TC 1/WG 2.2 International Conference, TCS 2012, Amsterdam, The Netherlands, September 26–28, 2012. Proceedings*, volume 7604 of *Lecture Notes in Computer Science*, pages 43–56. Springer, 2012.

138. Hans L. Bodlaender, Pål Grønås Drange, Markus S. Dregi, Fedor V. Fomin, Daniel Lokshtanov, and Michal Pilipczuk. An $o(c^k n)$ 5-approximation algorithm for treewidth. In *54th Annual IEEE Symposium on Foundations of Computer Science, FOCS 2013, 26–29 October, 2013, Berkeley, CA, USA*, pages 499–508. IEEE Computer Society, 2013.

139. Hans L. Bodlaender, Marek Cygan, Stefan Kratsch, and Jesper Nederlof. Deterministic single exponential time algorithms for connectivity problems parameterized by treewidth. In Fedor V. Fomin, Rusins Freivalds, Marta Z. Kwiatkowska, and David Peleg, editors, *Automata, Languages, and Programming - 40th International Colloquium, ICALP 2013, Riga, Latvia, July 8–12, 2013, Proceedings, Part I*, volume 7965 of *Lecture Notes in Computer Science*, pages 196–207. Springer, 2013.

140. Hans L. Bodlaender, Stefan Kratsch, and Vincent J. C. Kreuzen. Fixed-parameter tractability and characterizations of small special treewidth. In Andreas Brandstädt, Klaus Jansen, and Rüdiger Reischuk, editors, *Graph-Theoretic Concepts in Computer Science - 39th International Workshop, WG 2013, Lübeck, Germany, June 19–21, 2013, Revised Papers*, volume 8165 of *Lecture Notes in Computer Science*, pages 88–99. Springer, 2013.

141. Hans L. Bodlaender, Paul S. Bonsma, and Daniel Lokshtanov. The fine details of fast dynamic programming over tree decompositions. In Gregory Z. Gutin and Stefan Szeider, editors, *Parameterized and Exact Computation - 8th International Symposium, IPEC 2013, Sophia Antipolis, France, September 4–6, 2013, Revised Selected Papers*, volume 8246 of *Lecture Notes in Computer Science*, pages 41–53. Springer, 2013.

142. Stefan Fafianie, Hans L. Bodlaender, and Jesper Nederlof. Speeding up dynamic programming with representative sets - An experimental evaluation of algorithms for steiner tree on tree decompositions. In Gregory Z. Gutin and Stefan Szeider, editors, *Parameterized and Exact Computation - 8th International Symposium, IPEC 2013, Sophia Antipolis, France, September 4–6, 2013, Revised Selected Papers*, volume 8246 of *Lecture Notes in Computer Science*, pages 321–334. Springer, 2013.

143. Merel T. Rietbergen, Linda C. van der Gaag, and Hans L. Bodlaender. Provisional propagation for verifying monotonicity of Bayesian networks. In Torsten Schaub, Gerhard

Friedrich, and Barry O'Sullivan, editors, *ECAI 2014 - 21st European Conference on Artificial Intelligence, 18–22 August 2014, Prague, Czech Republic - Including Prestigious Applications of Intelligent Systems (PAIS 2014)*, volume 263 of *Frontiers in Artificial Intelligence and Applications*, pages 759–764. IOS Press, 2014.

144. Hans L. Bodlaender. Lower bounds for kernelization. In Marek Cygan and Pinar Heggernes, editors, *Parameterized and Exact Computation - 9th International Symposium, IPEC 2014, Wroclaw, Poland, September 10–12, 2014. Revised Selected Papers*, volume 8894 of *Lecture Notes in Computer Science*, pages 1–14. Springer, 2014.

145. Lars Jaffke and Hans L. Bodlaender. Definability equals recognizability for k-outerplanar graphs. In Thore Husfeldt and Iyad A. Kanj, editors, 10th International Symposium on Parameterized and Exact Computation, IPEC 2015, September 16–18, 2015, Patras, Greece, volume 43 of *LIPIcs*, pages 175–186. Schloss Dagstuhl - Leibniz-Zentrum fuer Informatik, 2015.

146. Chiel B. Ten Brinke, Frank J. P. van Houten, and Hans L. Bodlaender. Practical algorithms for linear boolean-width. In Thore Husfeldt and Iyad A. Kanj, editors, *10th International Symposium on Parameterized and Exact Computation, IPEC 2015, September 16–18, 2015, Patras, Greece,* volume 43 of *LIPIcs*, pages 187–198. Schloss Dagstuhl - Leibniz-Zentrum fuer Informatik, 2015.

147. Hans L. Bodlaender and Jesper Nederlof. Subexponential time algorithms for finding small tree and path decompositions. In Nikhil Bansal and Irene Finocchi, editors, *Algorithms - ESA 2015 - 23rd Annual European Symposium, Patras, Greece, September 14–16, 2015, Proceedings*, volume 9294 of *Lecture Notes in Computer Science*, pages 179–190. Springer, 2015.

148. Tom C. van der Zanden and Hans L. Bodlaender. PSPACE-completeness of Bloxorz and of games with 2-buttons. In Vangelis Th. Paschos and Peter Widmayer, editors, *Algorithms and Complexity - 9th International Conference, CIAC 2015, Paris, France, May 20–22, 2015. Proceedings*, volume 9079 of *Lecture Notes in Computer Science*, pages 403–415. Springer, 2015.

149. Hans L. Bodlaender, Hirotaka Ono, and Yota Otachi. A faster parameterized algorithm for pseudoforest deletion. In Jiong Guo and Danny Hermelin, editors, 11th *International Symposium on Parameterized and Exact Computation, IPEC 2016, August 24–26, 2016, Aarhus, Denmark*, volume 63 of *LIPIcs*, pages 7:1–7:12. Schloss Dagstuhl - Leibniz-Zentrum fuer Informatik, 2016.

150. Willem J. A. Pino, Hans L. Bodlaender, and Johan M. M. van Rooij. Cut and count and representative sets on branch decompositions. In Jiong Guo and Danny Hermelin, editors, *11th International Symposium on Parameterized and Exact Computation, IPEC 2016, August 24–26, 2016, Aarhus, Denmark*, volume 63 of *LIPIcs*, pages 27:1–27:12. Schloss Dagstuhl - Leibniz-Zentrum fuer Informatik, 2016.

151. Hans L. Bodlaender, Hirotaka Ono, and Yota Otachi. Degree-constrained orientation of maximum satisfaction: Graph classes and parameterized complexity. In Seok-Hee Hong, editor, *27th International Symposium on Algorithms and Computation, ISAAC 2016, December 12–14, 2016, Sydney, Australia*, volume 64 of *LIPIcs*, pages 20:1–20:12. Schloss Dagstuhl - Leibniz-Zentrum fuer Informatik, 2016.

152. Hans L. Bodlaender, Jesper Nederlof, and Tom C. van der Zanden. Subexponential time algorithms for embedding h-minor free graphs. In Ioannis Chatzigiannakis, Michael Mitzenmacher, Yuval Rabani, and Davide Sangiorgi, editors, *43rd International Colloquium on Automata, Languages, and Programming, ICALP 2016, July 11–15, 2016, Rome, Italy*, volume 55 of *LIPIcs*, pages 9:1–9:14. Schloss Dagstuhl - Leibniz-Zentrum fuer Informatik, 2016.

153. Marjan van den Akker, Hans L. Bodlaender, Thomas C. van Dijk, Han Hoogeveen, and Erik van Ommeren. Robust recoverable path using backup nodes. In Rusins Martins Freivalds, Gregor Engels, and Barbara Catania, editors, *SOFSEM 2016: Theory and Practice of Computer Science - 42nd International Conference on Current Trends in Theory and Practice of Computer Science, Harrachov, Czech Republic, January 23–28, 2016, Proceedings*, volume 9587 of *Lecture Notes in Computer Science*, pages 95–106. Springer, 2016.

154. Tom C. van der Zanden and Hans L. Bodlaender. Computing treewidth on the GPU. In Daniel Lokshtanov and Naomi Nishimura, editors, *12th International Symposium on Parameterized and Exact Computation, IPEC 2017, September 6–8, 2017, Vienna, Austria*, volume 89 of *LIPIcs*, pages 29:1–29:13. Schloss Dagstuhl - Leibniz-Zentrum fuer Informatik, 2017.

155. Hans L. Bodlaender and Tom C. van der Zanden. Improved lower bounds for graph embedding problems. In Dimitris Fotakis, Aris Pagourtzis, and Vangelis Th. Paschos, editors, *Algorithms and Complexity - 10th International Conference, CIAC 2017, Athens, Greece, May 24–26, 2017, Proceedings*, volume 10236 of *Lecture Notes in Computer Science*, pages 92–103, 2017.

156. Mark de Berg, Hans L. Bodlaender, and Sándor Kisfaludi-Bak. The homogeneous broadcast problem in narrow and wide strips. In Faith Ellen, Antonina Kolokolova, and Jörg-Rüdiger Sack, editors, *Algorithms and Data Structures - 15th International Symposium, WADS 2017, St. John's, NL, Canada, July 31 – August 2, 2017, Proceedings*, volume 10389 of *Lecture Notes in Computer Science*, pages 289–300. Springer, 2017.

157. Tesshu Hanaka, Hans L. Bodlaender, Tom C. van der Zanden, and Hirotaka Ono. On the maximum weight minimal separator. In T. V. Gopal, Gerhard Jäger, and Silvia Steila, editors, *Theory and Applications of Models of Computation - 14th Annual Conference, TAMC 2017, Bern, Switzerland, April 20–22, 2017, Proceedings*, volume 10185 of *Lecture Notes in Computer Science*, pages 304–318, 2017.

158. Hans L. Bodlaender and Tom C. van der Zanden. On the exact complexity of polyomino packing. In Hiro Ito, Stefano Leonardi, Linda Pagli, and Giuseppe Prencipe, editors, *9th International Conference on Fun with Algorithms, FUN 2018, June 13–15, 2018, La Maddalena, Italy*, volume 100 of *LIPIcs*, pages 9:1–9:10. Schloss Dagstuhl - Leibniz-Zentrum fuer Informatik, 2018.

159. Mark de Berg, Hans L. Bodlaender, Sándor Kisfaludi-Bak, and Sudeshna Kolay. An ETH-tight exact algorithm for Euclidean TSP. In Mikkel Thorup, editor, *59th IEEE Annual Symposium on Foundations of Computer Science, FOCS 2018, Paris, France, October 7–9, 2018*, pages 450–461. IEEE Computer Society, 2018.

160. Jelco M. Bodewes, Hans L. Bodlaender, Gunther Cornelissen, and Marieke van der Wegen. Recognizing hyperelliptic graphs in polynomial time. In Andreas Brandstädt, Ekkehard Köhler, and Klaus Meer, editors, *Graph-Theoretic Concepts in Computer Science - 44th International Workshop, WG 2018, Cottbus, Germany, June 27–29, 2018, Proceedings*, volume 11159 of *Lecture Notes in Computer Science*, pages 52–64. Springer, 2018.

161. Mark de Berg, Hans L. Bodlaender, Sándor Kisfaludi-Bak, Dániel Marx, and Tom C. van der Zanden. A framework for ETH-tight algorithms and lower bounds in geometric intersection graphs. In Ilias Diakonikolas, David Kempe, and Monika Henzinger, editors, *Proceedings of the 50th Annual ACM SIGACT Symposium on Theory of Computing, STOC 2018, Los Angeles, CA, USA, June 25–29, 2018*, pages 574–586. ACM, 2018.

162. Hans L. Bodlaender, Tesshu Hanaka, Yoshio Okamoto, Yota Otachi, and Tom C. van der Zanden. Subgraph isomorphism on graph classes that exclude a substructure. In Pinar Heggernes, editor, *Algorithms and Complexity - 11th International Conference, CIAC*

2019, Rome, Italy, May 27–29, 2019, Proceedings, volume 11485 of *Lecture Notes in Computer Science*, pages 87–98. Springer, 2019.

163. Hans L. Bodlaender, Sudeshna Kolay, and Astrid Pieterse. Parameterized complexity of conflict-free graph coloring. In Zachary Friggstad, Jörg-Rüdiger Sack, and Mohammad R. Salavatipour, editors, *Algorithms and Data Structures - 16th International Symposium, WADS 2019, Edmonton, AB, Canada, August 5–7, 2019, Proceedings*, volume 11646 of *Lecture Notes in Computer Science*, pages 168–180. Springer, 2019.

164. Hans L. Bodlaender, Marieke van der Wegen, and Tom C. van der Zanden. Stable divisorial gonality is in NP. In Barbara Catania, Rastislav Královic, Jerzy R. Nawrocki, and Giovanni Pighizzini, editors, *SOFSEM 2019: Theory and Practice of Computer Science - 45th International Conference on Current Trends in Theory and Practice of Computer Science, Nový Smokovec, Slovakia, January 27–30, 2019, Proceedings*, volume 11376 of *Lecture Notes in Computer Science*, pages 81–93. Springer, 2019.

Books and Book Chapters

1. Hans L. Bodlaender, editor. *Graph-Theoretic Concepts in Computer Science, 29th International Workshop, WG 2003, Elspeet, The Netherlands, June 19–21, 2003, Revised Papers*, volume 2880 of *Lecture Notes in Computer Science*. Springer, 2003.

2. Hans L. Bodlaender and Michael A. Langston, editors. *Parameterized and Exact Computation, Second International Workshop, IWPEC 2006, Zürich, Switzerland, September 13–15, 2006, Proceedings*, volume 4169 of *Lecture Notes in Computer Science*. Springer, 2006.

3. Hans L. Bodlaender. Treewidth of graphs. In Ming-Yang Kao, editor, *Encyclopedia of Algorithms - 2008 Edition*. Springer, 2008.

4. Hans L. Bodlaender, Rod Downey, Fedor V. Fomin, and Dániel Marx, editors. *The Multivariate Algorithmic Revolution and Beyond - Essays Dedicated to Michael R. Fellows on the Occasion of His 60th Birthday*, volume 7370 of *Lecture Notes in Computer Science*. Springer, 2012.

5. Hans L. Bodlaender. Fixed-parameter tractability of treewidth and pathwidth. In Hans L. Bodlaender, Rod Downey, Fedor V. Fomin, and Dániel Marx, editors, *The Multivariate Algorithmic Revolution and Beyond - Essays Dedicated to Michael R. Fellows on the Occasion of His 60th Birthday*, volume 7370 of *Lecture Notes in Computer Science*, pages 196–227. Springer, 2012.

6. Hans L. Bodlaender and Giuseppe F. Italiano, editors. *Algorithms - ESA 2013 - 21st Annual European Symposium, Sophia Antipolis, France, September 2–4, 2013. Proceedings*, volume 8125 of *Lecture Notes in Computer Science*. Springer, 2013.

7. Hans L. Bodlaender, Pinar Heggernes, and Daniel Lokshtanov. Graph modification problems (Dagstuhl seminar 14071). *Dagstuhl Reports*, 4(2):38–59, 2014.

8. Hans L. Bodlaender. Kernelization, exponential lower bounds. In *Encyclopedia of Algorithms*, pages 1013–1017. 2016.

9. Hans L. Bodlaender. Treewidth of graphs. In *Encyclopedia of Algorithms*, pages 2255–2257. 2016.

10. Kim G. Larsen, Hans L. Bodlaender, and Jean-François Raskin, editors. *42nd International Symposium on Mathematical Foundations of Computer Science, MFCS 2017, August 21–25, 2017 - Aalborg, Denmark*, volume 83 of *LIPIcs*. Schloss Dagstuhl - Leibniz-Zentrum fuer Informatik, 2017.

11. Hans L. Bodlaender and Gerhard J. Woeginger, editors. *Graph-Theoretic Concepts in Computer Science - 43rd International Workshop, WG 2017, Eindhoven, The Netherlands, June 21–23, 2017, Revised Selected Papers*, volume 10520 of *Lecture Notes in Computer Science*. Springer, 2017.

Technical Reports and arXiv

1. Hans L. Bodlaender, Harry A.G. Wijshoff, and Jan van Leeuwen. Compositions of double diagonal and cross latin squares. Technical Report RUU-CS-83-01, Department of Information and Computing Sciences, Utrecht University, 1983.
2. Hans L. Bodlaender. Uniform emulations of two different types of shuffle-exchange networks. Technical Report RUU-CS-84-09, Department of Information and Computing Sciences, Utrecht University, 1984.
3. Hans L. Bodlaender and Jan van Leeuwen. Uniform emulations of the shuffle-exchange network. Technical Report RUU-CS-84-05, Department of Information and Computing Sciences, Utrecht University, 1984.
4. Hans L. Bodlaender and Jan Leeuwen. Simulation of large networks on smaller networks. Technical Report RUU-CS-84-04, Department of Information and Computing Sciences, Utrecht University, 1984.
5. Hans L. Bodlaender and Jan Leeuwen. The minimum bisection width of (three-dimensional) blocks. Technical Report RUU-CS-84-02, Department of Information and Computing Sciences, Utrecht University, 1984.
6. A.A. Schoone, Hans L. Bodlaender, and Jan van Leeuwen. Diameter increase caused by edge deletion. Technical Report RUU-CS-85-26, Department of Information and Computing Sciences, Utrecht University, 1985.
7. Hans L. Bodlaender. Deadlock-free packet switching networks with variable packet size. Technical Report RUU-CS-85-25, Department of Information and Computing Sciences, Utrecht University, 1985.
8. Hans L. Bodlaender. Some lowerbound results for decentralized extrema-finding in rings of processors. Technical Report RUU-CS-85-22, Department of Information and Computing Sciences, Utrecht University, 1985.
9. Hans L. Bodlaender. Emulations of processor networks with buses. Technical Report RUU-CS-85-20, Department of Information and Computing Sciences, Utrecht University, 1985.
10. Hans L. Bodlaender. Finding grid embeddings with bounded maximum edge length is NP-complete. Technical Report RUU-CS-85-18, Department of Information and Computing Sciences, Utrecht University, 1985.
11. Hans L. Bodlaender and Jan van Leeuwen. New upperbounds for decentralized extrema finding in a ring of processors. Technical Report RUU-CS-85-15, Department of Information and Computing Sciences, Utrecht University, 1985.
12. Hans L. Bodlaender. The complexity of finding uniform emulations on fixed graphs. Technical Report RUU-CS-85-14, Department of Information and Computing Sciences, Utrecht University, 1985.
13. Hans L. Bodlaender. The classification of coverings of processor networks. Technical Report RUU-CS-85-11, Department of Information and Computing Sciences, Utrecht University, 1985.
14. Hans L. Bodlaender. On approximation algorithms for determining minimum cost emulations. Technical Report RUU-CS-85-10, Department of Information and Computing Sciences, Utrecht University, 1985.

15. Hans L. Bodlaender. The complexity of finding uniform emulations on paths and ring networks. Technical Report RUU-CS-85-05, Department of Information and Computing Sciences, Utrecht University, 1985.

16. Hans L. Bodlaender and Jan van Leeuwen. On the complexity of finding uniform emulations. Technical Report RUU-CS-85-04, Department of Information and Computing Sciences, Utrecht University, 1985.

17. Hans L. Bodlaender. Classes of graphs with bounded tree-width. Technical Report RUU-CS-86-22, Department of Information and Computing Sciences, Utrecht University, 1986.

18. Hans L. Bodlaender and Jan van Leeuwen. Distribution of records on a ring of processors. Technical Report RUU-CS-86-06, Department of Information and Computing Sciences, Utrecht University, 1986.

19. Hans L. Bodlaender. Dynamic programming on graphs with bounded treewidth. Technical Report RUU-CS-87-22, Department of Information and Computing Sciences, Utrecht University, 1987.

20. Hans L. Bodlaender. Polynomial algorithms for chromatic index and graph isomorphism on partial k-trees. Technical Report RUU-CS-87-17, Department of Information and Computing Sciences, Utrecht University, 1987.

21. Hans L. Bodlaender. A better lowerbound for distributed leader finding in bidirectional asynchronous rings of processors. Technical Report RUU-CS-87-13, Department of Information and Computing Sciences, Utrecht University, 1987.

22. Hans L. Bodlaender. The maximum cut and minimum cut into bounded sets problems on cographs. Technical Report RUU-CS-87-12, Department of Information and Computing Sciences, Utrecht University, 1987.

23. Hans L. Bodlaender. A new lowerbound technique for distributed extrema finding on rings of processors. Technical Report RUU-CS-87-11, Department of Information and Computing Sciences, Utrecht University, 1987.

24. Hans L. Bodlaender. Dynamic programming on graphs with bounded treewidth. Technical Report MIT/LCS/TR-394, Laboratory for Computer Science, Massachusetts Institute of Technology, 1987.

25. Hans L. Bodlaender, Shlomo Moran, and Manfred K. Warmuth. The distributed bit complexity of the ring; from the anonymous to the non-anonymous case. Technical Report RUU-CS-88-33, Department of Information and Computing Sciences, Utrecht University, 1988.

26. Paul W. Beame and Hans L. Bodlaender. Distributed computing on transitive networks; the torus. Technical Report RUU-CS-88-31, Department of Information and Computing Sciences, Utrecht University, 1988.

27. Hans L. Bodlaender. Improved self-reduction algorithms for graphs with bounded treewidth. Technical Report RUU-CS-88-29, Department of Information and Computing Sciences, Utrecht University, 1988.

28. Hans L. Bodlaender. Achromatic number is NP-complete for cographs and interval graphs. Technical Report RUU-CS-88-25, Department of Information and Computing Sciences, Utrecht University, 1988.

29. Hans L. Bodlaender. New lower bound techniques for distributed leader finding and other problems on rings of processors. Technical Report RUU-CS-88-18, Department of Information and Computing Sciences, Utrecht University, 1988.

30. Hans L. Bodlaender. Planar graphs with bounded treewidth. Technical Report RUU-CS-88-14, Department of Information and Computing Sciences, Utrecht University, 1988.

31. Hans L. Bodlaender. Nc-algorithms for graphs of small treewidth. Technical Report RUU-CS-88-04, Department of Information and Computing Sciences, Utrecht University, 1988.

32. Hans L. Bodlaender. Complexity of path-forming games. Technical Report RUU-CS-89-29, Department of Information and Computing Sciences, Utrecht University, 1989.

33. Hans L. Bodlaender. On the complexity of some coloring games. Technical Report RUU-CS-89-27, Department of Information and Computing Sciences, Utrecht University, 1989.

34. Hans L. Bodlaender, Gerard Tel, and Nicola Santoro. Trade-offs in non-reversing diameter. Technical Report RUU-CS-89-22, Department of Information and Computing Sciences, Utrecht University, 1989.

35. Hans L. Bodlaender and Gerard Tel. Bit-optimal election in synchronous rings. Technical Report RUU-CS-89-02, Department of Information and Computing Sciences, Utrecht University, 1989.

36. Hans L. Bodlaender. On linear time minor tests and depth first search. Technical Report RUU-CS-89-01, Department of Information and Computing Sciences, Utrecht University, 1989.

37. Hans L. Bodlaender. On disjoint cycles in graphs. Technical Report RUU-CS-90-29, Department of Information and Computing Sciences, Utrecht University, 1990.

38. H. Bodlaender and Ton Kloks. Fast algorithms for the Tron game on trees. Technical Report RUU-CS-90-11, Department of Information and Computing Sciences, Utrecht University, 1990.

39. Hans L. Bodlaender and Rolf H. Möhring. The pathwidth and treewidth of cographs. Technical Report RUU-CS-90-07, Department of Information and Computing Sciences, Utrecht University, 1990.

40. Hans L. Bodlaender. Kayles on special classes of graphs - An application of Sprague-Grundy theory. Technical Report RUU-CS-91-49, Department of Information and Computing Sciences, Utrecht University, 1991.

41. Hans L. Bodlaender and Klaus Jansen. Restrictions of graph partition problems. Part I. Technical Report RUU-CS-91-44, Department of Information and Computing Sciences, Utrecht University, 1991.

42. Hans L. Bodlaender and Klaus Jansen. On the complexity of the maximum cut problem. Technical Report RUU-CS-91-39, Department of Information and Computing Sciences, Utrecht University, 1991.

43. Hans L. Bodlaender, Teofilo Gonzalez, and Ton Kloks. Complexity aspects of 2-dimensional data compression. Technical Report RUU-CS-91-35, Department of Information and Computing Sciences, Utrecht University, 1991.

44. Goos Kant and Hans L. Bodlaender. Planar graph augmentation problems. Technical Report RUU-CS-91-25, Department of Information and Computing Sciences, Utrecht University, 1991.

45. Hans L. Bodlaender and Ton Kloks. A simple linear time algorithm for triangulating three-colored graphs. Technical Report RUU-CS-91-13, Department of Information and Computing Sciences, Utrecht University, 1991.

46. Hans L. Bodlaender and Dieter Kratsch. The complexity of coloring games on perfect graphs. Technical Report RUU-CS-91-03, Department of Information and Computing Sciences, Utrecht University, 1991.

47. Hans L. Bodlaender, John R. Gilbert, Hjálmtyr Hafsteinsson, and Ton Kloks. Approximating treewidth, pathwidth, and minimum elimination tree height. Technical Report

RUU-CS-91-01, Department of Information and Computing Sciences, Utrecht University, 1991.

48. Ton Kloks and Hans L. Bodlaender. Only few graphs have bounded treewidth. Technical Report RUU-CS-92-35, Department of Information and Computing Sciences, Utrecht University, 1992.

49. Hans L. Bodlaender, Ton Kloks, and Dieter Kratsch. Treewidth and pathwidth of permutation graphs. Technical Report RUU-CS-92-30, Department of Information and Computing Sciences, Utrecht University, 1992.

50. Ton Kloks and Hans L. Bodlaender. Approximating treewidth and pathwidth of some classes of perfect graphs. Technical Report RUU-CS-92-29, Department of Information and Computing Sciences, Utrecht University, 1992.

51. Hans L. Bodlaender. A linear time algorithm for finding tree-decompositions of small treewidth. Technical Report RUU-CS-92-27, Department of Information and Computing Sciences, Utrecht University, 1992.

52. Ton Kloks and Hans L. Bodlaender. On the treewidth and pathwidth of permutation graphs. Technical Report RUU-CS-92-13, Department of Information and Computing Sciences, Utrecht University, 1992.

53. Hans L. Bodlaender. A tourist guide through treewidth. Technical Report RUU-CS-92-12, Department of Information and Computing Sciences, Utrecht University, 1992.

54. Ton Kloks and Hans L. Bodlaender. Testing superperfection of k–trees. Technical Report RUU-CS-92-09, Department of Information and Computing Sciences, Utrecht University, 1992.

55. Hans L. Bodlaender, Michael R. Fellows, and Tandy J. Warnow. Two strikes against perfect phylogeny. Technical Report RUU-CS-92-08, Department of Information and Computing Sciences, Utrecht University, 1992.

56. Goos Kant and Hans L. Bodlaender. Triangslating planar graphs while minimizing the maximum degree. Technical Report RUU-CS-92-07, Department of Information and Computing Sciences, Utrecht University, 1992.

57. Hans L. Bodlaender, Klaus Jansen, and Gerhard J. Woeginger. Scheduling with incompatible jobs. *Universität Trier, Mathematik/Informatik, Forschungsbericht*, 92-09, 1992.

58. Hans L. Bodlaender and Ton Kloks. Efficient and constructive algorithms for the pathwidth and treewidth of graphs. Technical Report RUU-CS-93-27, Department of Information and Computing Sciences, Utrecht University, 1993.

59. Hans L. Bodlaender and Michael R. Fellows. W[2]-hardness of precedence constrained k-processor scheduling. Technical Report UU-CS-1994-14, Department of Information and Computing Sciences, Utrecht University, 1994.

60. Hans L. Bodlaender and Joost Engelfriet. Domino treewidth. Technical Report UU-CS-1994-11, Department of Information and Computing Sciences, Utrecht University, 1994.

61. John Shawe-Taylor, Carlos Domingo, Hans L. Bodlaender, and James Abello. Learning minor closed graph classes with membership and equivalence queries. Technical Report NC-TR-94-014, Department of Computer Science, Royal Holloway University of London, 1995.

62. Hans L. Bodlaender and Babette de Fluiter. Reduction algorithms for graphs with small treewidth. Technical Report UU-CS-1995-37, Department of Information and Computing Sciences, Utrecht University, 1995.

63. Hans L. Bodlaender, Michael R. Fellows, Michael T. Hallett, Harold T. Wareham, and Tandy J. Warnow. The hardness of problems on thin colored graphs. Technical Report

UU-CS-1995-36, Department of Information and Computing Sciences, Utrecht University, 1995.

64. Hans L. Bodlaender, Ton Kloks, Dieter Kratsch, and Heiko Müller. Treewidth and minimum fill-in on d-trapezoid graphs. Technical Report UU-CS-1995-34, Department of Information and Computing Sciences, Utrecht University, 1995.

65. Hans L. Bodlaender and Torben Hagerup. Parallel algorithms with optimal speedup for bounded treewidth. Technical Report UU-CS-1995-25, Department of Information and Computing Sciences, Utrecht University, 1995.

66. Hans L. Bodlaender and Babette de Fluiter. On intervalizing k-colored graphs for DNA physical mapping. Technical Report UU-CS-1995-20, Department of Information and Computing Sciences, Utrecht University, 1995.

67. Hans L. Bodlaender and Babette de Fluiter. Intervalizing k-colored graphs. Technical Report UU-CS-1995-15, Department of Information and Computing Sciences, Utrecht University, 1995.

68. Hans L. Bodlaender, Jitender S. Deogun, Klaus Jansen, Ton Kloks, Dieter Kratsch, Heiko Müller, and Zsolt Tuza. Rankings of graphs. Technical Report UU-CS-1995-03, Department of Information and Computing Sciences, Utrecht University, 1995.

69. Hans L. Bodlaender and Dimitrios M. Thilikos. Treewidth and small separators for graphs with small chordality. Technical Report UU-CS-1995-02, Department of Information and Computing Sciences, Utrecht University, 1995.

70. Hans L. Bodlaender, Rodney G. Downey, Michael R. Fellows, and Harold T. Wareham. The parameterized complexity of sequence alignment and consensus. Technical Report UU-CS-1995-01, Department of Information and Computing Sciences, Utrecht University, 1995.

71. Hans L. Bodlaender, Jan van Leeuwen, Richard B. Tan, and Dimitrios M. Thilikos. On interval routing schemes and treewidth. Technical Report UU-CS-1996-41, Department of Information and Computing Sciences, Utrecht University, 1996.

72. Dimitrios M. Thilikos and Hans L. Bodlaender. Fast partitioning l-apex graphs with applications to approximating maximum induced-subgraph problems. Technical Report UU-CS-1996-30, Department of Information and Computing Sciences, Utrecht University, 1996.

73. Hans L. Bodlaender, Dimitrios M. Thilikos, and Koichi Yamazaki. It is hard to know when greedy is good for finding independent sets. Technical Report UU-CS-1996-29, Department of Information and Computing Sciences, Utrecht University, 1996.

74. Hans L. Bodlaender and Babett de Fluiter. Parallel algorithms for series parallel graphs. Technical Report UU-CS-1996-13, Department of Information and Computing Sciences, Utrecht University, 1996.

75. Hans L. Bodlaender. A partial k-arboretum of graphs with bounded treewidth. Technical Report UU-CS-1996-02, Department of Information and Computing Sciences, Utrecht University, 1996.

76. Linda C. van der Gaag and Hans L. Bodlaender. Comparing loop cutsets and clique trees in probabilistic inference. Technical Report UU-CS-1997-42, Department of Information and Computing Sciences, Utrecht University, 1997.

77. Hans L. Bodlaender and Dimitrios M. Thilikos. Graphs with branchwidth at most three. Technical Report UU-CS-1997-37, Department of Information and Computing Sciences, Utrecht University, 1997.

78. Hans L. Bodlaender. Treewidth: Algorithmic results and techniques. Technical Report UU-CS-1997-31, Department of Information and Computing Sciences, Utrecht University, 1997.

79. Hans L. Bodlaender and Babette de Fluiter. Reduction algorithms for graphs of small treewidth. Technical Report UU-CS-1997-24, Department of Information and Computing Sciences, Utrecht University, 1997.

80. Babette de Fluiter and Hans L. Bodlaender. Parallel algorithms for treewidth two. Technical Report UU-CS-1997-23, Department of Information and Computing Sciences, Utrecht University, 1997.

81. Hans L. Bodlaender and Babette de Fluiter. Parallel algorithms for series parallel graphs. Technical Report UU-CS-1997-21, Department of Information and Computing Sciences, Utrecht University, 1997.

82. Koichi Yamazaki, Hans L. Bodlaender, Babette de Fluiter, and Dimitrios M. Thilikos. Isomorphism for graphs of bounded distance width. Technical Report UU-CS-1997-05, Department of Information and Computing Sciences, Utrecht University, 1997.

83. Babette de Fluiter and Hans L. Bodlaender. Intervalizing sandwich graphs. Technical Report UU-CS-1997-04, Department of Information and Computing Sciences, Utrecht University, 1997.

84. Hans L. Bodlaender. A note on domino treewidth. Technical Report UU-CS-1998-15, Department of Information and Computing Sciences, Utrecht University, 1998.

85. Hans L. Bodlaender and Dimitrios M. Thilikos. Computing small search numbers in linear time. Technical Report UU-CS-1998-05, Department of Information and Computing Sciences, Utrecht University, 1998.

86. Hans Zantema and Hans L. Bodlaender. Sizes of decision tables and decision trees. Technical Report UU-CS-1999-31, Department of Information and Computing Sciences, Utrecht University, 1999.

87. Hans Zantema and Hans L. Bodlaender. Finding small equivalent decision trees is hard. Technical Report UU-CS-1999-02, Department of Information and Computing Sciences, Utrecht University, 1999.

88. Hans L. Bodlaender and Dieter Kratsch. Kayles and Nimbers. Technical Report UU-CS-2000-42, Department of Information and Computing Sciences, Utrecht University, 2000.

89. Dimitrios M. Thilikos and Hans L. Bodlaender. Constructive linear time algorithms for branchwidth. Technical Report UU-CS-2000-38, Department of Information and Computing Sciences, Utrecht University, 2000.

90. Hans L. Bodlaender, Richard B. Tan, and Jan van Leeuwen. Finding a delta-regular supergraph of minimum order. Technical Report UU-CS-2000-29, Department of Information and Computing Sciences, Utrecht University, 2000.

91. Jochen Alber, Hans L. Bodlaender, Henning Fernau, and Rolf Niedermeier. Fixed parameter algorithms for planar dominating set. Technical Report UU-CS-2000-28, Department of Information and Computing Sciences, Utrecht University, 2000.

92. Hans L. Bodlaender. Necessary edges in k-chordalizations of graphs. Technical Report UU-CS-2000-27, Department of Information and Computing Sciences, Utrecht University, 2000.

93. Hans L. Bodlaender, Ton Kloks, Richard B. Tan, and Jan van Leeuwen. Approximations for λ-coloring of graphs. Technical Report UU-CS-2000-25, Department of Information and Computing Sciences, Utrecht University, 2000.

94. Dimitrios M. Thilikos, Maria J. Serna, and Hans L. Bodlaender. A constructive linear time algorithm for small cutwidth. Technical Report UU-CS-2000-24, Department of Information and Computing Sciences, Utrecht University, 2000.

95. Hans L. Bodlaender and Fedor V. Fomin. Approximation of pathwidth of outerplanar graphs. Technical Report UU-CS-2000-23, Department of Information and Computing Sciences, Utrecht University, 2000.

96. Arie M.C.A. Koster, Hans L. Bodlaender, and Stan P. M. van Hoesel. Treewidth: Computational experiments. Technical Report UU-CS-2001-49, Department of Information and Computing Sciences, Utrecht University, 2001.

97. Hans L. Bodlaender and Udi Rotics. Computing the treewidth and the minimum fill-in with the modular decomposition. Technical Report UU-CS-2001-22, Department of Information and Computing Sciences, Utrecht University, 2001.

98. Hans L. Bodlaender, Michael J. Dinneen, and Bakhadyr Khoussainov. Relaxed update and partition network games. Technical Report UU-CS-2001-15, Department of Information and Computing Sciences, Utrecht University, 2001.

99. Hans L. Bodlaender. A generic NP-hardness proof for a variant of graph coloring. Technical Report UU-CS-2001-08, Department of Information and Computing Sciences, Utrecht University, 2001.

100. Dimitrios M. Thilikos, Maria J. Serna, and Hans L. Bodlaender. A polynomial algorithm for the cutwidth of bounded degree graphs with small treewidth. Technical Report UU-CS-2001-04, Department of Information and Computing Sciences, Utrecht University, 2001.

101. Frank van den Eijkhof, Hans L. Bodlaender, and Arie M.C.A. Koster. Safe reduction rules for weighted treewidth. Technical Report UU-CS-2002-051, Department of Information and Computing Sciences, Utrecht University, 2002.

102. Hans L. Bodlaender, Michael R. Fellows, and Dimitrios M. Thilikos. Derivation of algorithms for cutwidth and related graph layout problems. Technical Report UU-CS-2002-032, Department of Information and Computing Sciences, Utrecht University, 2002.

103. Hans L. Bodlaender, Hajo J. Broersma, Fedor V. Fomin, Artem V. Pyatkin, and Gerhard J. Woeginer. Radio labeling with pre-assigned frequencies. Technical Report UU-CS-2002-026, Department of Information and Computing Sciences, Utrecht University, 2002.

104. Hans L. Bodlaender and Fedor V. Fomin. Tree decompositions with small cost. Technical Report UU-CS-2002-001, Department of Information and Computing Sciences, Utrecht University, 2002.

105. Hans L. Bodlaender, Andreas Brandstädt, Dieter Kratsch, Michaël Rao, and Jeremy Spinrad. On algorithms for (P_5,gem)-free graphs. Technical Report UU-CS-2003-038, Department of Information and Computing Sciences, Utrecht University, 2003.

106. Hans L. Bodlaender and Gerard Tel. Rectilinear graphs and angular resolution. Technical Report UU-CS-2003-033, Department of Information and Computing Sciences, Utrecht University, 2003.

107. Hans L. Bodlaender and Arie M. C. A. Koster. Safe separators for treewidth. Technical Report UU-CS-2003-027, Department of Information and Computing Sciences, Utrecht University, 2003.

108. Hans L. Bodlaender, Arie M. C. A. Koster, and Frank van den Eijkhof. Pre-processing rules for triangulation of probabilistic networks. Technical Report UU-CS-2003-001, Department of Information and Computing Sciences, Utrecht University, 2003.

109. Alexander Grigoriev and Hans L. Bodlaender. Algorithms for graphs embeddable with few crossings per edge. Technical Report UU-CS-2004-058, Department of Information and Computing Sciences, Utrecht University, 2004.

110. Hans L. Bodlaender and Arie M. C. A. Koster. On the maximum cardinality search lower bound for treewidth. Technical Report UU-CS-2004-053, Department of Information and Computing Sciences, Utrecht University, 2004.

111. Arie M. C. A. Koster, Thomas Wolle, and Hans L. Bodlaender. Degree-based treewidth lower bounds. Technical Report UU-CS-2004-050, Department of Information and Computing Sciences, Utrecht University, 2004.

112. Thomas Wolle, Arie M. C. A. Koster, and Hans L. Bodlaender. A note on contraction degeneracy. Technical Report UU-CS-2004-042, Department of Information and Computing Sciences, Utrecht University, 2004.

113. Emgad H. Bachoore and Hans L. Bodlaender. New upper bound heuristics for treewidth. Technical Report UU-CS-2004-036, Department of Information and Computing Sciences, Utrecht University, 2004.

114. Hans L. Bodlaender, Arie M. C. A. Koster, and Thomas Wolle. Contraction and treewidth lower bounds. Technical Report UU-CS-2004-034, Department of Information and Computing Sciences, Utrecht University, 2004.

115. Hans L. Bodlaender and Thomas Wolle. Contraction degeneracy on cographs. Technical Report UU-CS-2004-031, Department of Information and Computing Sciences, Utrecht University, 2004.

116. Hans L. Bodlaender and Jan Arne Telle. Space-efficient construction variants of dynamic programming. Technical Report UU-CS-2004-030, Department of Information and Computing Sciences, Utrecht University, 2004.

117. Thomas Wolle and Hans L. Bodlaender. A note on edge contraction. Technical Report UU-CS-2004-028, Department of Information and Computing Sciences, Utrecht University, 2004.

118. Hans L. Bodlaender and Fedor V. Fomin. Equitable colorings of bounded treewidth graphs. Technical Report UU-CS-2004-010, Department of Information and Computing Sciences, Utrecht University, 2004.

119. Hans L. Bodlaender and Thomas Wolle. A note on the complexity of network reliability problems. Technical Report UU-CS-2004-001, Department of Information and Computing Sciences, Utrecht University, 2004.

120. Hans L. Bodlaender, Alexander Grigoriev, and Arie M. C. A. Koster. Treewidth lower bounds with brambles. Technical Report UU-CS-2005-051, Department of Information and Computing Sciences, Utrecht University, 2005.

121. Hans L. Bodlaender. Discovering treewidth. Technical Report UU-CS-2005-018, Department of Information and Computing Sciences, Utrecht University, 2005.

122. Irit Katriel and Hans L. Bodlaender. Online topological ordering. Technical Report UU-CS-2005-011, Department of Information and Computing Sciences, Utrecht University, 2005.

123. Hans L. Bodlaender, Leizhen Cai, Jianer Chen, Michael R. Fellows, Jan Arne Telle, and Dániel Marx. Open problems in parameterized and exact computation - IWPEC 2006. Technical Report UU-CS-2006-052, Department of Information and Computing Sciences, Utrecht University, 2006.

124. Hans L. Bodlaender. A cubic kernel for feedback vertex set. Technical Report UU-CS-2006-042, Department of Information and Computing Sciences, Utrecht University, 2006.

125. Hans L. Bodlaender. Treewidth: Characterizations, applications, and computations. Technical Report UU-CS-2006-041, Department of Information and Computing Sciences, Utrecht University, 2006.

126. Hans L. Bodlaender, Fedor V. Fomin, Arie M. C. A. Koster, Dieter Kratsch, and Dimitrios M. Thilikos. On exact algorithms for treewidth. Technical Report UU-CS-2006-032, Department of Information and Computing Sciences, Utrecht University, 2006.

127. Hans L. Bodlaender and Dieter Kratsch. An exact algorithm for graph coloring with polynomial memory. Technical Report UU-CS-2006-015, Department of Information and Computing Sciences, Utrecht University, 2006.

128. Emgad Bachoore and Hans L. Bodlaender. Weighted treewidth: Algorithmic techniques and results. Technical Report UU-CS-2006-013, Department of Information and Computing Sciences, Utrecht University, 2006.

129. Emgad Bachoore and Hans L. Bodlaender. A branch and bound algorithm for exact, upper, and lower bounds on treewidth. Technical Report UU-CS-2006-012, Department of Information and Computing Sciences, Utrecht University, 2006.

130. Emgad H. Bachoore and Hans L. Bodlaender. Convex recoloring of leaf-colored trees. Technical Report UU-CS-2006-010, Department of Information and Computing Sciences, Utrecht University, 2006.

131. Frederic Dorn, Eelko Penninkx, Hans L. Bodlaender, and Fedor V. Fomin. Efficient exact algorithms on planar graphs: Exploiting sphere cut decompositions. Technical Report UU-CS-2006-006, Department of Information and Computing Sciences, Utrecht University, 2006.

132. Erwin M. Bakker, Hans L. Bodlaender, Richard B. Tan, and Jan van Leeuwen. Interval routing and minor-monotone graph parameters. Technical Report UU-CS-2006-001, Department of Information and Computing Sciences, Utrecht University, 2006.

133. Johan M. M. van Rooij and Hans L. Bodlaender. Exact algorithms for edge domination. Technical Report UU-CS-2007-051, Department of Information and Computing Sciences, Utrecht University, 2007.

134. Johan Kwisthout, Hans L. Bodlaender, and Gerard Tel. Complexity results for local monotonicity in probabilistic networks. Technical Report UU-CS-2007-050, Department of Information and Computing Sciences, Utrecht University, 2007.

135. Hans L. Bodlaender, Rodney G. Downey, Michael R. Fellows, and Danny Hermelin. On problems without polynomial kernels. Technical Report UU-CS-2007-046, Department of Information and Computing Sciences, Utrecht University, 2007.

136. Hans L. Bodlaender, Michael R. Fellows, Michael A. Langston, Mark A. Ragan, Frances A. Rosamond, and Mark Weyer. Quadratic kernelization of convex recoloring of trees. Technical Report UU-CS-2007-035, Department of Information and Computing Sciences, Utrecht University, 2007.

137. Hans L. Bodlaender, Corinne Feremans, Alexander Grigoriev, Eelko Penninkx, René Sitters, and Thomas Wolle. On the minimum corridor connection problem and other generalized geometric problems. Technical Report UU-CS-2007-031, Department of Information and Computing Sciences, Utrecht University, 2007.

138. Hans L. Bodlaender, Alexander Grigoriev, Nadejda V. Grigorieva, and Albert Hendriks. The valve location problem in simple network topologies. Technical Report UU-CS-2007-019, Department of Information and Computing Sciences, Utrecht University, 2007.

139. Helmut Alt, Hans L. Bodlaender, Marc van Kreveld, Günter Rote, and Gerard Tel. Wooden geometric puzzles: Design and hardness proofs. Technical Report UU-CS-2007-009, Department of Information and Computing Sciences, Utrecht University, 2007.

140. Hans L. Bodlaender, Pinar Heggernes, and Yngve Villanger. Faster parameterized algorithms for minimum fill-in. Technical Report UU-CS-2008-042, Department of Information and Computing Sciences, Utrecht University, 2008.

141. Hans L. Bodlaender, Michael R. Fellows, Pinar Heggernes, Federico Mancini, Charis Papadopoulos, and Frances Rosamond. Clustering with partial information. Technical Report UU-CS-2008-033, Department of Information and Computing Sciences, Utrecht University, 2008.

142. Hans L. Bodlaender and Arie M.C.A. Koster. Treewidth computations I: upper bounds. Technical Report UU-CS-2008-032, Department of Information and Computing Sciences, Utrecht University, 2008.

143. Hans L. Bodlaender, Stéphan Thomassé, and Anders Yeo. Analysis of data reduction: Transformations give evidence for non-existence of polynomial kernels. Technical Report UU-CS-2008-030, Department of Information and Computing Sciences, Utrecht University, 2008.

144. Hans L. Bodlaender, Erik D. Demaine, Michael R. Fellows, Jiong Guo, Danny Hermelin, Daniel Lokshtanov, Moritz Müller, Venkatesh Raman, Johan M.M. van Rooij, and Frances A. Rosamond. Open problems in parameterized and exact computation - IWPEC 2008. Technical Report UU-CS-2008-017, Department of Information and Computing Sciences, Utrecht University, 2008.

145. Hans L. Bodlaender, Richard B. Tan, Thomas C. van Dijk, and Jan van Leeuwen. Integer maximum flow in wireless sensor networks with energy constraint. Technical Report UU-CS-2008-005, Department of Information and Computing Sciences, Utrecht University, 2008.

146. Johan M. M. van Rooij and Hans L. Bodlaender. Design by measure and conquer: A faster exact algorithm for dominating set. *CoRR*, abs/0802.2827, 2008.

147. Johan M. M. van Rooij and Hans L. Bodlaender. Design by measure and conquer: Exact algorithms for dominating set. Technical Report UU-CS-2009-025, Department of Information and Computing Sciences, Utrecht University, 2009.

148. Hans L. Bodlaender, Fedor V. Fomin, Arie M. C. A. Koster, Dieter Kratsch, and Dimitrios M. Thilikos. A note on exact algorithms for vertex ordering problems on graphs. Technical Report UU-CS-2009-023, Department of Information and Computing Sciences, Utrecht University, 2009.

149. Johan Kwisthout and Hans L. Bodlaender. Conditional lower bounds on the complexity of probabilistic inference. Technical Report UU-CS-2009-018, Department of Information and Computing Sciences, Utrecht University, 2009.

150. Hans L. Bodlaender, Fedor V. Fomin, Daniel Lokshtanov, Eelko Penninkx, Saket Saurabh, and Dimitrios M. Thilikos. (Meta) Kernelization. Technical Report UU-CS-2009-012, Department of Information and Computing Sciences, Utrecht University, 2009.

151. Hans L. Bodlaender, Fedor V. Fomin, Daniel Lokshtanov, Eelko Penninkx, Saket Saurabh, and Dimitrios M. Thilikos. (Meta) Kernelization. *CoRR*, abs/0904.0727, 2009.

152. Hans L. Bodlaender, Kyohei Kozawa, Takayoshi Matsushima, and Yota Otachi. Spanning tree congestion of k-outerplanar graphs. Technical Report UU-CS-2010-027, Department of Information and Computing Sciences, Utrecht University, 2010.

153. Hans L. Bodlaender and Johan M. M. van Rooij. Exact algorithms for intervalizing colored graphs. Technical Report UU-CS-2010-024, Department of Information and Computing Sciences, Utrecht University, 2010.

154. Hans L. Bodlaender and Arie M. C. A. Koster. Treewidth computations II. lower bounds. Technical Report UU-CS-2010-022, Department of Information and Computing Sciences, Utrecht University, 2010.

155. Yota Otachi, Hans L. Bodlaender, and Erik Jan van Leeuwen. Complexity results for the spanning tree congestion problem. Technical Report UU-CS-2010-007, Department of Information and Computing Sciences, Utrecht University, 2010.

156. Johan M. M. van Rooij, Marcel E. van Kooten Niekerk, and Hans L. Bodlaender. Partitioning sparse graphs into triangles: Relations to exact satisfiability and very fast exponential time algorithms. Technical Report UU-CS-2010-005, Department of Information and Computing Sciences, Utrecht University, 2010.

157. Hans L. Bodlaender, Bart M. P. Jansen, and Stefan Kratsch. Cross-composition: A new technique for kernelization lower bounds. *CoRR*, abs/1011.4224, 2010.
158. Bart M. P. Jansen and Hans L. Bodlaender. Vertex cover kernelization revisited: Upper and lower bounds for a refined parameter. *CoRR*, abs/1012.4701, 2010.
159. Hans L. Bodlaender and Dieter Kratsch. Exact algorithms for Kayles. Technical Report UU-CS-2011-003, Department of Information and Computing Sciences, Utrecht University, 2011.
160. Hans L. Bodlaender, Bart M. P. Jansen, and Stefan Kratsch. Preprocessing for treewidth: A combinatorial analysis through kernelization. *CoRR*, abs/1104.4217, 2011.
161. Hans L. Bodlaender, Bart M. P. Jansen, and Stefan Kratsch. Kernel bounds for path and cycle problems. *CoRR*, abs/1106.4141, 2011.
162. Hans L. Bodlaender, Dieter Kratsch, and Sjoerd T. Timmer. Exact algorithms for Kayles. Technical Report UU-CS-2012-001, Department of Information and Computing Sciences, Utrecht University, 2012.
163. Hans L. Bodlaender, Bart M. P. Jansen, and Stefan Kratsch. Kernelization lower bounds by cross-composition. *CoRR*, abs/1206.5941, 2012.
164. Linda C. van der Gaag, Hans L. Bodlaender, and Ad Feelders. Monotonicity in Bayesian networks. *CoRR*, abs/1207.4160, 2012.
165. Hans L. Bodlaender, Bart M. P. Jansen, and Stefan Kratsch. Kernel bounds for structural parameterizations of pathwidth. *CoRR*, abs/1207.4900, 2012.
166. Hans L. Bodlaender, Marek Cygan, Stefan Kratsch, and Jesper Nederlof. Solving weighted and counting variants of connectivity problems parameterized by treewidth deterministically in single exponential time. *CoRR*, abs/1211.1505, 2012.
167. Hans L. Bodlaender, Arie M. C. A. Koster, Frank van den Eijkhof, and Linda C. van der Gaag. Pre-processing for triangulation of probabilistic networks. *CoRR*, abs/1301.2256, 2013.
168. Hans L. Bodlaender, Pål Grønås Drange, Markus S. Dregi, Fedor V. Fomin, Daniel Lokshtanov, and Michal Pilipczuk. A $o(c^k n)$ 5-approximation algorithm for treewidth. *CoRR*, abs/1304.6321, 2013.
169. Stefan Fafianie, Hans L. Bodlaender, and Jesper Nederlof. Speeding-up dynamic programming with representative sets - an experimental evaluation of algorithms for Steiner Tree on tree decompositions. *CoRR*, abs/1305.7448, 2013.
170. Hans L. Bodlaender and Marc J. van Kreveld. Google scholar makes it hard - The complexity of organizing one's publications. *CoRR*, abs/1410.3820, 2014.
171. Tom C. van der Zanden and Hans L. Bodlaender. PSPACE-completeness of Bloxorz and of games with 2-buttons. *CoRR*, abs/1411.5951, 2014.
172. Hans L. Bodlaender, Endre Boros, Pinar Heggernes, and Dieter Kratsch. Open problems of the lorentz workshop, "enumeration algorithms using structure". Technical Report UU-CS-2015-016, Department of Information and Computing Sciences, Utrecht University, 2015.
173. Lars Jaffke and Hans L. Bodlaender. MSOL-definability equals recognizability for Halin graphs and bounded degree k-outerplanar graphs. *CoRR*, abs/1503.01604, 2015.
174. Chiel B. Ten Brinke, Frank J. P. van Houten, and Hans L. Bodlaender. Practical algorithms for linear boolean-width. *CoRR*, abs/1509.07687, 2015.
175. Lars Jaffke and Hans L. Bodlaender. Definability equals recognizability for k-outerplanar graphs. *CoRR*, abs/1509.08315, 2015.
176. Hans L. Bodlaender and Jesper Nederlof. Subexponential time algorithms for finding small tree and path decompositions. *CoRR*, abs/1601.02415, 2016.
177. Hans L. Bodlaender and Tom C. van der Zanden. Improved lower bounds for graph embedding problems. *CoRR*, abs/1610.09130, 2016.

178. Mark de Berg, Hans L. Bodlaender, and Sándor Kisfaludi-Bak. The homogeneous broadcast problem in narrow and wide strips. *CoRR*, abs/1705.01465, 2017.

179. Jelco M. Bodewes, Hans L. Bodlaender, Gunther Cornelissen, and Marieke van der Wegen. Recognizing hyperelliptic graphs in polynomial time. *CoRR*, abs/1706.05670, 2017.

180. Tom C. van der Zanden and Hans L. Bodlaender. Computing treewidth on the GPU. *CoRR*, abs/1709.09990, 2017.

181. Federico D'Ambrosio, Gerard T. Barkema, and Hans L. Bodlaender. Optimal data structures for stochastic driven simulations. *CoRR*, abs/1802.02379, 2018.

182. Mark de Berg, Hans L. Bodlaender, Sándor Kisfaludi-Bak, Dániel Marx, and Tom C. van der Zanden. Framework for ETH-tight algorithms and lower bounds in geometric intersection graphs. *CoRR*, abs/1803.10633, 2018.

183. Johan M. M. van Rooij, Hans L. Bodlaender, Erik Jan van Leeuwen, Peter Rossmanith, and Martin Vatshelle. Fast dynamic programming on graph decompositions. *CoRR*, abs/1806.01667, 2018.

184. Mark de Berg, Hans L. Bodlaender, Sándor Kisfaludi-Bak, and Sudeshna Kolay. An ETH-tight exact algorithm for euclidean TSP. *CoRR*, abs/1807.06933, 2018.

185. Hans L. Bodlaender and Tom C. van der Zanden. On exploring temporal graphs of small pathwidth. *CoRR*, abs/1807.11869, 2018.

186. Hans L. Bodlaender, Marieke van der Wegen, and Tom C. van der Zanden. Stable divisorial gonality is in NP. *CoRR*, abs/1808.06921, 2018.

187. Hans L. Bodlaender, Benjamin A. Burton, Fedor V. Fomin, and Alexander Grigoriev. Knot diagrams of treewidth two. *CoRR*, abs/1904.03117, 2019.

188. Hans L. Bodlaender, Sudeshna Kolay, and Astrid Pieterse. Parameterized complexity of conflict-free graph coloring. *CoRR*, abs/1905.00305, 2019.

189. Hans L. Bodlaender, Lars Jaffke, and Jan Arne Telle. Typical sequences revisited - algorithms for linear orderings of series parallel digraphs. *CoRR*, abs/1905.03643, 2019.

190. Hans L. Bodlaender, Tesshu Hanaka, Yasuaki Kobayashi, Yusuke Kobayashi, Yoshio Okamoto, Yota Otachi, and Tom C. van der Zanden. Subgraph isomorphism on graph classes that exclude a substructure. *CoRR*, abs/1905.10670, 2019.

Program Committees

Chair: WG 2003, IWPEC 2006, ESA 2013 (track A), WG 2017, MFCS 2017.

Member: STACS 1993, WG 1996, CIAC 1997, WG 1997, SIROCCO 1997, WG 1998, WG 1999, ESA 1999, WG 2000, WG 2001, WG 2002, ISAAC 2002, WADS 2003, WG 2004, IWPEC 2004, WG 2005, ICALP 2005, AAIM 2006, SOFSEM 2007, WG 2007, AAIM 2007, UAI 2007, COCOA 2008, WG 2009, TAMC 2009, IWPEC 2009, SWAT 2010, ESA 2010, TAMC 2010, IPEC 2011, WALCOM 2012, STACS 2012, APEX 2012, WG 2012, IPEC 2014, FAW 2014, WG 2014, WALCOM 2015, ICALP 2015, FAW 2015, LATA 2016, WG 2016, SOFSEM 2017, LICS 2018, WALCOM 2019, WG 2019, TAMC 2019, WALCOM 2019, SODA 2019.

Editorships

2014 –   Discrete Algorithms
1997 –   Acta Cybernetica
1997 –   Discrete Mathematics and Theoretical Computer Science

2008 – 2012	Information and Computation
2000	Algorithmica, guest editor special issue

Ph.D. Students

1993	A.J.J. (Ton) Kloks (co-promotor; promotor: Jan van Leeuwen)
1993	Goos Kant (co-promotor; promotor: Jan van Leeuwen)
1997	Babette L.E. de Fluiter (co-promotor; promotor: Jan van Leeuwen)
2005	Thomas Wolle (co-promotor; promotor: Jan van Leeuwen)
2009	Johan H.P. Kwisthout (co-promotor; other co-promoter: Gerard Tel; promotor: Jan van Leeuwen and Linda C. van der Gaag)
2011	Johan M.M. van Rooij (co-promotor; promotor: Jan van Leeuwen)
2013	Bart M.P. Jansen (co-promotor; promotor Jan van Leeuwen)
2019	Tom C. van der Zanden
2019	Sándor Kisfaludi-Bak (other promotor: Mark T. de Berg)

Research Guests and Postdocs

1996 – 1998	Dimitrios M. Thilikos, postdoc
1996 – 1997	Koichi Yamazaki, postdoc
2005 – 2011	Eelko Penninkx
2009	Marc Comas, research visit
2009	Yota Otachi, research visit
2009	Manu Basavaraju, research visit
2011	Zhang Wenyan, research visit
2010 – 2012	Stefan Kratsch, postdoc
2012 – 2014	Jesper Nederlof, postdoc
2014	O-joung Kwon and Jisu Jeong, research visit
2016	Tesshu Hanaka, research visit
2019	Lars Jaffke, research visit
2019	Hisao Tamaki, sabbatical
2019	Toshiki Saitoh, sabbatical

Invited Lectures

1992	7th International Meeting of Young Computer Scientists, Smolenice
1997	International Symposium on Mathematical Foundations of Computer Science (MFCS)
2000	6th International Conference on Graph Theory, Marseille
2005	31st Annual Conference on Current Trends in Theory and Practice of Informatics (SOFSEM)

2006	32nd International Workshop on Graph-Theoretic Concepts in Computer Science (WG)
2007	14th International Colloquium on Structural Information and Communication Complexity (SIROCCO)
2008	Workshop in Graph Decomposition: Theoretical, Algorithmic and Logical Aspects, CIRM, Marseille
2009	4th International Workshop on Parameterized and Exact Computation (IWPEC), Copenhagen
2011	Theory Day of the Nederlandse Vereniging voor Theoretische Informatica (NVTI), Utrecht
2011	6th International Symposium on Parameterized and Exact Computation (IPEC), tutorial
2011	5th Workshop on Graph Classes and Width Parameters (GROW), Daejeon
2013	European Conferences on Symbolic and Quantitative Approaches to Reasoning with Uncertainty (ESQARU), Utrecht, tutorial
2014	9th International Workshop on Parameterized and Exact Computation (IWPEC), Wrocław
2016	ICT-OPEN, Apeldoorn
2017	Networks Day, Eindhoven
2019	19th Haifa Workshop on Interdisciplinary Applications of Graphs, Combinatorics and Algorithms, Haifa

Organizers

2010	WORKER 2010: Workshop on Kernelization, with Fedor V. Fomin and Saket Saurabh, Lorentz Center, Leiden
2014	Workshop "Graph modification problems", with Pinar Heggernes and Daniel Lokshtanov, Schloss Dagstuhl
2015	Workshop "Enumeration Algorithms using Structure", with Endre Boros, Pinar Heggernes, and Dieter Kratsch, Lorentz Center, Leiden
2016	Workshop "Fixed Parameter Computational Geometry", with Mark de Berg, Benjamin Burton, and Christian Knauer, Lorentz Center, Leiden
2018	Workshop "Fixed Parameter Computational Geometry", with Mark de Berg, Benjamin Burton, and Christian Knauer, Lorentz Center, Leiden
2018	NWO-JSPS joint seminar "Computations on Networks with a Tree-Structure: From Theory to Practice", Eindhoven
2020	Workshop "Fixed Parameter Computational Geometry", with Mark de Berg, Benjamin Burton, and Christian Knauer, Lorentz Center, Leiden

Grants

1989 – 1993	NWO-SION grant "Algorithms for tree-structured graphs"
1993 – 1997	NWO-SION grant "Algorithms for tree-structured graphs and their practical aspects"
1996 – 1997	Japan government grant for postdoc Koichi Yamazaki
1996 – 1998	EC Human Mobility Capital grant for postdoc Dimitrios M. Thilikos
2001 – 2006	NWO grant "Treewidth and Combinatorial Optimization" (with Stan van Hoesel)
2005 – 2009	NWO grant "Algorithmic Complexity of Probabilistic Networks"
2009 – 2013	NWO grant "KERNELS: Combinatorial Analysis of Data Reduction"
2012 – 2014	NWO grant "Space and Time Efficient Structural Improvements of Dynamic Programming Algorithms"

Research Statistics

Based on Google Scholar. Retrieved: January 2020.

Citations:	17356
h-index:	64
i10-index:	17
Top cited:	1835 (A linear time algorithm for finding tree-decompositions)
Erdös nr.:	2 (via Dieter Kratsch, Shlomo Moran, and/or Zsolt Tuza)

Short Contributions

Seeing Arboretum for the (partial k-) Trees

Stefan Arnborg[1] and Andrzej Proskurowski[2]([⊠])

[1] Royal Institute of Technology, Stockholm, Sweden
[2] University of Oregon, Eugene, OR, USA
andrzej@cs.uoregon.edu

Abstract. The idea of applying a dynamic programming strategy to evaluating certain objective functions on trees is fairly straightforward. The road for this idea to develop into theories of width parameters has been not so straight. Hans Bodlaender has played a major role in the process of mapping out that road. In this sentimental journey, we will recount our collective road trip over the past decades.

Before Hans Bodlaender stepped on the stage of partial k-trees/bounded treewidth, there was a 1982 paper by Wald and Colbourn "Computing reliability for a generalization of series-parallel networks" [16]. They introduced the idea of partial k-trees, which was then picked up in a 1984 paper by Colbourn and Proskurowski "Concurrent transmissions in broadcast networks" [12]. Unbeknown to Western researchers, Korneyenko made the connection between generalized series-parallel graphs and partial k-trees in the same year. It was reprinted ten years later in a special issue of *Discrete Applied Mathematics* (Korneyenko, "Combinatorial algorithms on a class of graphs", pp. 215–218 in [2]).

The publication of the original paper (1984) coincided with Andrzej's sabbatical leave at his Alma Mater, the Royal Institute of Technology (KTH), where he was hosted by Stefan. Stefan had developed and implemented algorithms computing reliabilities and repair time distributions for complex technical systems and felt these methods could be generalized to more complex graphs extending the notions of series-parallel and $\Delta - Y$ graphs (which already in the 19th century MacMahon, and independently Kennelly, had shown amenable to interesting combinatorial/algebraic exercises [13,14]). It appeared that also other computational problems generalizing these specific problem examples ought to be solvable by a general method.

While Arnborg and Proskurowski were exploring their common research interests, they liked the algorithmic aspects of partial k-trees [4]. They also explored recognition of partial k-trees, starting with small values of parameter k [5]. Unfortunately, "The Law of Small Numbers" applied there, as Stefan's then student Jens Lagergren showed later that the recognition idea did not extend to higher values ("The nonexistence of reduction rules giving an embedding into a k-tree", pp. 219–224 in [2].)

As sabbaticals go, Derek Corneil was at the same time visiting Grenoble, just a hop and a skip from Stockholm. Derek brought to the table Yannakakis'

© Springer Nature Switzerland AG 2020
F. V. Fomin et al. (Eds.): Bodlaender Festschrift, LNCS 12160, pp. 3–6, 2020.
https://doi.org/10.1007/978-3-030-42071-0_1

paper "Computing the Minimum Fill-In is NP-Complete" [17], and they worked together on larger values of k [1]. (It gave us a great pleasure that this work later found its way into an entry written by Hans in the Encyclopedia of Algorithms [10].)

All the while the "sabbatical group" (Stefan spent his leave in Eugene, Oregon, and they both visited Derek in Toronto, Ontario) did not forget the smaller values of the parameter k (Arnborg, Proskurowski and Corneil, "Forbidden minors characterization of partial 3-trees" [6].) This latter paper was an indirect consequence of the meeting of American Mathematical Society in Eugene in 1984, where Paul Seymour gave a talk on "Generalizing Kuratowski's theorem", in which he introduced the new to many ideas of graph minors and treewidth. After the talk, Andrzej related to Paul the closely related work on partial k-trees. A year later, Paul remarked with obvious satisfaction that "our treewidth was winning over your partial k-trees." If nothing else, our description of the concept led to the eye-catching title of a paper by Hans, cited below.

Meanwhile, Detlef Seese, then of the Karl Weierstrass Institute of DDR Science Academy, spent a sabbatical at KTH and worked with Jens Lagergren and Stefan on characterizing the computational problems for which our partial k-tree method was applicable and advantageous. At about the same time, Bruno Courcelle announced that the class of computational decision problems definable as sets of bounded tree width graphs definable in monadic second order logic can be efficiently decided in linear time for a given width if the decomposition is known. Arnborg, Lagergren, Seese noted that many related optimization and counting problems, including probability computations, could be also systematically handled [3]. Of course, the asymptotic linear time solvability of all these problems was a nice result but it meant not so much in practice since the constants were rapidly becoming astronomically large when the formula size or the width increased a little. Thus, many algorithmic problems must be studied much more in detail to see which computations are feasible, and which are not. Hans has been a discoverer or co-discoverer of several practical improvements of these methods, some as recent as 2009 [15].

Some time in the late '80s Stefan and Andrzej organized in Eugene a get-together of people interested in algorithmic aspects of bounded treewidth. They were joined by Paul Seymour, Bruno Courcelle, Steve Hedetniemi, Detlef Seese and others. Based on those discussions and the increasing general interest in this topic, they edited a special issue of *Discrete Applied Mathematics*, where the first article was "Improved self-reduction algorithm for graphs with bounded treewidth" by Hans L. Bodlaender (pages 101–115 in [2]).

By that time, Hans had already published a number of articles on the subject, including "A linear time algorithm for finding tree-decompositions of small treewidth" [7]. The "arboretum" paper appeared as journal article a few years later [9].

In the new millennium, the Catalan Mathematics Research Center (CRM at the UAB) gave Andrzej an opportunity to organize what became a biennial workshop on graph classes, optimization and width parameters (*GROW*).

It should come as no surprise that in the collection of articles based on this first meeting, there was a paper coauthored by Hans [11].

Of the more unusual encounters with Hans, Andrzej recalls one in Smolenice Castle in the woods of Slovakia in 1992, where Hans gave an invited talk that was later published as "A tourist guide through treewidth" [8].) The conference was billed as an *International Meeting of Young Computer Scientists*. It is only fitting to mention it in a Festschrift celebrating his 60th birthday – *tempus fugit!*

This essay is intended to give a snapshot of the early history of explorations into the structure of graphs and combinatorial problems and how that history intersected with the research trajectory of our friend Hans Bodlaender. Both are going strong, independently of whatever round number is associated with the years that pass.

Happy Birthday, Hans!

References

1. Arnborg, S., Corneil, D.G., Proskurowski, A.: Complexity of finding embeddings in k-trees. SIAM J. Algebraic Discrete Methods **8**(2), 277–284 (1987)
2. Arnborg, S., Hedetniemi, S.T., Proskurowski, A.: Algorithms on graphs with bounded treewidth. Discrete Appl. Math. **54**(2–3) (1994)
3. Arnborg, S., Lagergren, J., Seese, D.: Easy problems for tree-decomposable graphs. J. Algorithms **12**, 308–340 (1991)
4. Arnborg, S., Proskurowski, A.: Problems on graphs with bounded decomposability. In: Proceedings of the 17th South-Eastern Conference on Combinatorics, Graph Theory and Computing, Utilitas Mathematica, Winnipeg, Congressus Numerantium, vol. 53, pp. 167–170 (1986)
5. Arnborg, S., Proskurowski, A.: Characterization and recognition of partial 3-trees. SIAM J. Algebraic. Discrete Methods **7**(2), 305–314 (1986)
6. Arnborg, S., Proskurowski, A., Corneil, D.G.: Forbidden minors characterization of partial 3-trees. In: Proceedings of the Seventh Hungarian Colloquium on Combinatorics, Colloquia Mathematica Societatis Janos Bolyai, vol. 52, pp. 49–62 (1988)
7. Bodlaender, H.L.: A linear time algorithm for finding tree-decompositions of small treewidth. In: Proceedings of 25th Symposium on the Theory of Computing (STOC 1993), pp. 226–234 (1993)
8. Bodlaender, H.L.: A tourist guide through treewidth. Acta Cybern. **11**, 1–21 (1993)
9. Bodlaender, H.L.: A partial k-arboretum of graphs with bounded treewidth. Theor. Comput. Sci. **209**, 1–45 (1998)
10. Bodlaender, H.L.: Treewidth of graphs: 1987; Arnborg, Corneil, Proskurowski. In: Kao, M.-Y. (ed.) Encyclopedia of Algorithms, pp. 968–970. Springer, New York (2008). https://doi.org/10.1007/978-0-387-30162-4
11. Bodlaender, H.L., Fomin, F.V.: Tree decompositions with small cost. Discrete Appl. Math. **145**(2), 143–154 (2005)
12. Colbourn, C.J., Proskurowski, A.: Concurrent transmissions in broadcast networks. In: Paredaens, J. (ed.) ICALP 1984. LNCS, vol. 172, pp. 128–136. Springer, Heidelberg (1984). https://doi.org/10.1007/3-540-13345-3_11
13. Kennelly, A.E.: Equivalence of triangles and three-pointed stars in conducting networks. Electr. World Eng. **34**, 413–414 (1899)

14. MacMahon: The combination of resistances, The Electrician, 8 April (1892)
15. van Rooij, J.M.M., Bodlaender, H.L., Rossmanith, P.: Dynamic programming on tree decompositions using generalised fast subset convolution. In: Fiat, A., Sanders, P. (eds.) ESA 2009. LNCS, vol. 5757, pp. 566–577. Springer, Heidelberg (2009). https://doi.org/10.1007/978-3-642-04128-0_51
16. Wald, J.A., Colbourn, C.J.: Computing reliability for a generalization of series-parallel networks. In: Proceedings of the Twentieth Allerton Conference on Communication, Control, and Computing, pp. 25–26 (1982)
17. Yannakakis, M.: Computing the minimum fill-in is NP-complete. SIAM J. Algebraic Discrete Methods **2**(1), 77–79 (1981)

Collaborating with Hans: Some Remaining Wonderments

Michael R. Fellows and Frances A. Rosamond[✉]

University of Bergen, 5020 Bergen, Norway
{michael.fellows,frances.rosamond}@uib.no

Abstract. This paper celebrates some family adventures, three concrete open problems and several research directions that have developed over our long collaboration.

Keywords: Parameterized complexity · Functorial FPT

1 Introduction

A career's worth of collaboration with Hans has developed some fascinating themes as well as some family adventures. There are wonderful concrete unsolved problems and promising research directions that have developed in this collaboration. This note starts with an appreciation of Hans and his family, and then discusses three, of the many possible, research directions.

A Family Anecdote.

Frances and I had dinner with the family at Hans' home in Utrecht. Subsequently, we sent postcards to his two young daughters of pictures of colorful Australian parrots and kangaroos. A few years ago Hans' daughter Marijke followed him into theoretical computer science, and became a PhD student of Magnus Haldorsson in Iceland.

We suggested to Hans that we would be happy to host Marijke for a research visit at our place by the beach in Australia where we have been honoured to host many scientists at our informal Mathematical Sciences Research Institute[1]

---

[1] We have been honoured to have the following scientists, and more, visit our informal research center in Australia: Andreas Pfandler, Bart Jansen, Benny Chor, Christian Komusiewicz, Christian Sloper, D'aniel Marx, Danny Hermelin, David Juedes, Dorothea Baumeister, Faisal Abu-Khzam, Frank Dehne, Frank Neumann, Falk Hüffner, Gábor Erdélyi, Hadas Shachnai, Henning Fernau, Jens Gramm, Jochen Alber, John Plaice, Jorg Rothe, Kim Stevens, Leo Brueggeman, Ljiljana Brankovic, Marike Bodlaender, Mark Jones, Martin Lackner, Mathias Weller, Matthew Suderman, Mattias Mnich, Michael Langston, Moritz Mueller, Nathanial Watt, Noy Rotbart, Peter Rosmanith, Peter Shaw, Ralph Bottesch, René Van Bevern, Rolf Neidermeier, Ronald DeHaan, Rudiger Reischuk, Rudolph Fleischer, Sagi Snir, Saket Saurabh, Serge Gaspers, Stephan Szeider, Ulrike Stege, Venkatesh Raman, Vlad Estivill-Castro, Yiannis Koutis.

The Norwegian NFR Toppforsk PCPC Project, The Bergen (BFS) Toppforsk.

© Springer Nature Switzerland AG 2020
F. V. Fomin et al. (Eds.): Bodlaender Festschrift, LNCS 12160, pp. 7–17, 2020.
https://doi.org/10.1007/978-3-030-42071-0_2

Marijke was interested to visit for research collaboration. We responded that there was a window of opportunity of a month, and advised that she should plan on at least 10 days, and was welcome to stay for as long as she liked. She came for the month.

We recruited some other young scientists to come to our beach shack during the period. We all collaborated on the remaining concrete puzzles concerning P-time kernelization for the VERTEX COVER problem [1].

OPEN PROBLEM.
It is known, and partly due to our collective efforts by the beach, that in polynomial-time one can reduce (*kernelize*) an instance (G, k) of the parameterized VERTEX COVER problem, to a reduced instance (losing no important and constructive information) to an instance (G', k') where the minimum degree of G' is almost five. Reducing to larger minimum degree is of use in designing efficient FPT algorithms via bounded search tree FPT algorithms for this important problem.

Marijke proved to be a keen surfer, as well as theorem-prover. By surfing, we primarily mean *body-boarding*, which anyone of any age up to 105, can have fun with in five minutes. If you know what you're doing with waves and planing surfaces, you can catch and ride waves that stand-up surfers can't catch, because of superior sprinting ability, and stability due to closeness to the surface of the water. As big and radical as anything. Most people don't know this.

2   Outline of the Technical Contributions of this Note

Working with Hans over the years, we have done some publishable work together on themes that are sure to develop further. The remainder of this contribution has as a goal to explicate three of the themes with a particular focus on concrete open problems that represent them: (1) Win/wins, (2) Too much information compression, (3) The Functor Project.

3   Win/Wins

One of Hans' favorite algorithmic results was this simple little subroutine based on depth first search.

LONG CYCLE [2]:
The input is a graph G, and what we now call, a *parameter*, k. The question is whether there is in the graph G a cycle of length at least k.

One can do the following. Compute in linear time a depth first spanning tree of G. In a depth first spanning tree, any edges that are not edges of the spanning tree, are termed *back-edges*, and they must go (because of the logic of depth first search) between a vertex and an ancestor in the tree. If there is a back edge to a distant ancestor (more than k generations) then clearly there is a cycle in the graph of length at least k.

On the other hand, if all of the back edges are short, then one can simply read off from the computed depth first spanning tree of G a tree decomposition of treewidth at most k, *almost for free* in this second unresolved case. And can then proceed to do standard dynamic programming to settle the issue.

This trick (it isn't much more than that) has two reflective corollaries:

(1) Interesting results don't have to be deep. Little results that show the way forward can be the true gems of an opening, as all cavers know.
(2) Thinking abstractly, what is the essential structure of this little trick? One can rephrase the trick as a linear time FPT result. The logic of the trick can be rephrased in terms of pure (non-computational) extremal combinatorics.

For any graph G, either:

(1) G has a long cycle (greater than or equal to k), or
(2) G has bounded tree width (at most $k' = k$).

The computational trick we are celebrating here is just an FPT algorithmic connection between k and k', where k' is equal to k. But consider for a moment the horizons this opens up. Pure extremal combinatorics is full of theorems of the form:

If $\Pi(X)$ is $</ / > k$ then $\Pi'(X)$ is $< / > k'$

The connection that is known between k and k' might involve a k' that is enormously exponential in k' (think of Ramsey's Theorem connecting cliques and independent sets), and as frequently turns up in FPT algorithms, such as [3–6].

When can these *structural connections* be realized algorithmically in FPT time? We wonder (modulo ETH, SETH and other plausible hypotheses): How good can the FPT get?

This is all unexplored, because it doesn't fit very naturally into the usual paradigms of research, which partly explains why this little trick surprised everybody so much, in its time.

4   Too Much Information Compression!

It took awhile to get our heads around even the basic observations about the kernelization view of FPT. Eventually we began to ask, possibly beginning at a Dagstuhl meeting [12], when it might be that for a parameterized problem that is FPT, and therefore must have a kernel of size k' for some $k' = f(k)$ for some function f: When can we get k' bounded by a polynomial function of k? Investigating this question has been one of the most interesting collaborative projects that we have undertaken with Hans.

Kernelization has taken off almost like a well-motivated industry, because of the undisputable importance of pre-processing. There is both a history and a future. The usual form of the negative results is:

There cannot be a poly(k) kernel unless coNP ⊆ NP/poly.

This is sometimes portrayed as some sort of *exotic complexity hypothesis*. It isn't. We know now, thanks to the amazing efforts of Robertson and Seymour, that for every g the finite graphs of genus at most g are characterized by some finite number $f(g)$ of minor-minimal internal structural obstructions (like when $g = 0$, and there are two obstructions).

So it is natural to ask if $f(g)$ might be bounded by a polynomial in g. The answer is likely not, for the reason of Dinneen's Theorem. Consider the following theorem, with more or less full proof below, which should be better known, because it is another example of some of the discussion above about *Win-Wins*, coming from a perfectly natural extremal combinatorics setting.

Theorem: Dinneen. [5] *The number $f(g)$ of minor-minimal obstructions to genus g, cannot be bounded by a polynomial function of g unless $co-NP \subseteq NP/poly$.*

Proof. (Quite simple) Suppose we wish to determine whether a graph G has genus at most g. Note, importantly, that this is an NP-hard problem. Well then, if there were a polynomial in g number of obstructions that are determinative for all graphs concerning the issue of genus g, then in the case of G, then we really only need to consider the obstructions on the list that are smaller than G, which has size n. So, the advice we need (... the /*poly* part of the conclusion), to obtain a contradiction and attain the QED, is bounded by a polynomial in n. This advice will give us necessary and sufficient evidence to convict G of not being genus at most g. How do we find the evidence? That is the NP part of the NP/poly part of the conclusion of Dinneen's Theorem. We guess how to find the evidence (how the obstruction H lives in G) and check it. QED

When we started thinking about the issue of polynomial-sized kernels for FPT problems, this result of Dinneen came to mind and we were headed in this direction, but the game was leaked.

Dinneen's result shows an interesting connection between computational complexity and structural graph theory. As in the discussion above concerning win/wins, there is much more to be explored about this connection.

The main issue in the theme of this section is what to do with what happens mathematically in a situation that hypothesizes too much information compression. This is a general issue which leads back to unfinished research with Hans.

A sketch proof that GRAPH BANDWIDTH is hard for $W[t]$ for all t involving inducting up through the levels of the parameterized complexity classes $W[t]$ can be found in [7]. Hans has worked out a sketch proof that GRAPH BANDWIDTH is hard for $W[t]$ for all t, *even for trees.* The proof of this has never been written down, Hans has notes, and the task would be daunting.

OPEN PROBLEM: GRAPH BANDWIDTH is hard for $W[t]$ for all t, even for trees.

The next natural and inevitable question in the context of parameterized complexity is:

Is GRAPH BANDWIDTH in $W[P]$?

$W[P]$ is not complicated. For a parameterized problem (input size n) with declared parameter k, the issue is whether you can settle the issue with $k \log n$ bits of advice, in polynomial time.

This sort-of clean up issue was tasked to Michael Hallett, who reported after a weekend of frustration:

GRAPH BANDWIDTH is not in $W[P]$ and the reason is:
—you want $k \log n$ bits of information to polynomial-time convict. But what if G is the disjoint union of a bunch of graphs? The union can have bandwidth at most k if and only each and every one does, and you want $k \log n$ advice to convict all of them, in polynomial time?

No way! That was Mike Hallett's opinion. We agree. *Too much information compression.*

Notice that this is about *compositionality*, key to the rocket take-off of kernelization lower bounds.

OPEN PROBLEM (Still): Is GRAPH BANDWIDTH in $W[P]$?

5 The Functor Project

The Functor Project originated at the 2006 Leiden workshop on Parameterized Complexity and Computational Geometry and Topology during the Open Problems session [11]. However, its deep roots go back to the early days of the visionary graph minors project of Robertson and Seymour, which really has no equal in the history of discrete mathematics and maybe mathematics in general, as we now better understand the power of translational encoding of all kinds of mathematics into elegant combinatorial statements.

In the 1970's, the theory of NP–completeness was new and electrifying. Everyone was looking at the classic book of Garey and Johnson, their catalog of NP-hard problems, and the combinatorial reductions necessary to prove results between them [10]. Many of these problems involve, as part of the legislated input, besides the graph G, a parameter k, in modern terminology. Many of these naturally parameterized problems are polynomial time solvable for every fixed parameter value k, even though the general problem is NP hard.

The question becomes: Can you get the k out of the exponent? That is the entire quest of parameterized complexity in a nutshell.

Many of the deep issues in life are expressible in cartoons. The Downey/Fellows 1999 monograph [7] contains a cartoon about parameterized complexity that is based on the metaphor of evolution of the opposable thumb: there are some number of fingers (n), and a small number of (k) of thumbs, that can help us get a useful grip on computational complexity.

Collaborations with Hans began with understanding, digesting, and extending the work of Robertson and Seymour on graph minors and graph structure theory, related to finite graphs and computational problems. What Robertson and Seymour provided was a powerful FPT complexity classification machinery, based on two fundamental results:

Theorem A. Finite graphs are well quasi ordered by minors.

Definition. A graph H is a minor of a graph G if up to isomorphism H can be obtained from G, by deleting some vertices and edges, and contracting some other edges that remain after the deletions.

Theorem B. The parameterized computational problem defined as follows is FPT.
Input: finite graphs H and G
Parameter: H
Question: is H a minor of G?

Theorems A and B together provide powerful FPT classification machinery, for problems about finite graphs. But finite graphs are not the only kind of mathematical objects, and after a while, as the study of parameterized complexity developed, it began to be natural to ask, if there might be similar powerful FPT classification machinery for other kinds of mathematical objects.

The way that Theorem A and Theorem B work together to provide this powerful classification tool is not complicated and is quite beautiful, and illustrated concretely by Kuratowski's Theorem. Theorem A provides an internal structural finite characterization, in this case, of *planarity*, which is most naturally viewed as an external property of a finite graph—can you draw it in the plane without crossings?

Kuratowski's quite surprising theorem, tells us that if the answer is NO, that is because living inside of the graph G that we are concerned about, is, in the sense of the minor partial order, one of the two graphs H, either K(3, 3) or K(5) as part of the structure of G.

Using these two theorems in tandem to answer the question of whether a graph G is planar, we run two "order tests"—

(1) Is K(3, 3) living in G as a minor, as part of the structure of G?
(2) Is K(5) living in G as a minor, as part of the structure of G?

A key point is that the characterization of the *extrinsic* property of planarity, given by Kuratowski's Theorem, is in terms of a finite number (here, two) of intrinsic substructures.

Theorem B tells us that we may test whether a finite graph G hosts H, in FPT time: that is to say, if the graph G has n vertices, then for any graph H we can find out whether G hosts H structurally, in time $f(H).n^c$, where c is some fixed constant independent of n (the size of G) and independent of H. As per the definition of FPT. And $f(H)$ might be a really horrible function of H!

In the case of planarity, there are only tests (two H's) that we need to test G about, to resolve the issue of planarity. If either of these two obstructions are found to be living in G structurally, then we have determined, by Kuratowski's Theorem that the answer to the question of whether G is planar is NO. Because if either of them live structurally in G, then planar drawing is impossible.

For the first of the two tests, we must pay something like: $f(K(3,3)).n^3$ and for the second of the two tests we must pay something like $f(K(5)).n^3$.

We may in this way conclude that testing the planarity of a graph G, can be done in polynomial time. This is not very efficient! But it is interesting. Planarity is the same thing as being of genus 0.

For the general case, of genus k, Theorem A tells us that for every fixed value of k, there are a finite number of tests that we must make to determine the issue. And thus, the property of having genus k, for the parameter k, is FPT.

The question posed during the Leiden Workshop Open Problem session that led to the *Functor Project* research direction is the following.

Computational Geometry has many natural computational problems that are FPT, but it seems to lack powerful FPT classification technology, such as we have in the land of finite graphs by Theorem A and Theorem B described above. Can we find such tools (well quasi ordering, and FPT order tests) for finite sets of points in the plane, and perhaps other settings as well?

This seems a reasonable question to ask as parameterized complexity continues to develop and extend into different application areas of computing. The question was not resolved at the Leiden workshop, but only posed. The question has now been answered, and the structure of the answer points to exciting frontiers that are only beginning to be explored.

Theorem A': There is a natural well quasi ordering $\leq$ of finite sets of points in the plane.

Theorem B': This natural well quasi ordering has FPT order tests, meaning that the following parameterized computational problem is FPT:

PROBLEM
Instance: Two finite sets Y and X of points in the plane.
Parameter: Y
Question: Is $Y \leq X$?

But first of course, we must explain about this ordering $\leq$ of finite sets of points in the plane. Key to the Functor Project is the following parameterized complexity version of Higman's Lemma.

The FPT Higman's Lemma (Fellows, Rosamond and Sankaran).
Suppose that $(Q, \leq)$ is a well quasi order that admits FPT order tests. Then also the following are well quasi orders that admit FPT order tests:

(i) The Cartesian product $Q \times Q$, ordered by the relation: $(a, b) \leq (a', b')$ if and only if $a \leq a'$ and $b \leq b'$.

(ii) The set of all finite sequences of elements of Q, ordered by the relation: $(a_1, a_2, ...) \leq (b_1, b_2, ...)$ if and only if there is a progressive order-respecting injective map from the first sequence to the second sequence, progressive in the sense that as we go down the list of the elements of the first sequence, things are always mapped to something "further along", and by order-respecting, we mean that if a_i is mapped to b_j then in the ordering of Q, $a_i \leq b_j$.

(iii) The set of all finite subsets of elements of Q, ordered by the relation similar to (ii), but here there is no ordering, we just require an injective map that is order respecting.

The above definitions are spelled out in more formal detail in standard expositions of Higman's Lemma, such as [14].

Now to the key definition! How to put the order on finite sets of points in the plane? Let P be such a finite set. Consider growing disks around these points, of increasing radius, starting at radius zero, and making a list of the finite graphs that record how these disks intersect. Suppose P consists m points. The first graph on the list will be the empty graph (no edges) on a set of m vertices. The last graph on the list will be the complete graph on m vertices. In between, as the radius grows, occasionally we must record a new graph on the list that has more edges representing the intersections of the discs.

This is all rather functorial: the mathematical object *a finite set of points in the plane* (in the source category) is being mapped to a finite sequence of finite graphs (the target objects, in the target category). These are well quasi ordered by Higman's Lemma (ii) above (If we take Q to be, as in this example, finite sequences of finite graphs, where in the invocation of (ii) of our Lemma above, the relevant order for the injective mapping is the minor ordering of graphs (or immersions, or whatever).

Thus, the problem proposed at the workshop in Leiden is solved, in some sense. There is indeed a well quasi ordering of finite sets of points in the plane, and this ordering admits FPT order tests. That is just what we asked for! As a remark, this is all very close to Morse Theory, and recent theories concerning *prominent data*.

This is obviously only the beginning of the Functor Project. The discussion above shows a functor from some other mathematical domain, in this case the mathematical objects are finite sets of points in the plane (computational geometry) into the land of finite graphs, where thanks to Robertson and Seymour and Higman, we have this extremely powerful FPT classification machinery.

In the above account of the solution to the question asked at the Leiden workshop, many important details have been left out. With respect to the FPT Higman's Lemma, the proof of (ii) can be safely left as an exercise for the reader—you just basically go through the list of the $(a_i...)$ finite sequence of elements of Q and jump at the first chance of finding $a_i \leq b_j$, so the basic algorithmic strategy is simply dynamic programming.

The algorithmic case of the FPT Higman's Lemma (iii) is more interesting. The argument goes as follows.

FPT Higman's Lemma (iii) (Fellows, Rosamond and Sankaran). For the statement of the Lemma, see the above discussion.

Proof. Our task is to argue that if Y and X are finite subsets of a friendly well quasi order $(Q, \leq)$, then we can determine in FPT time, where the parameter is Y, if $Y \leq X$, in this case (iii) of our FPT algorithmic version of Higman's Lemma. To establish this, we use the celebrated FPT result of Alon et al. on *color-coding* [13]. This is not formally entirely fair, but suppose for purposes of exposition that Y consists of three elements of Q, which we will call a, b and c. We may consider them as colored with three colors: *red*, *blue* and *green*.
More fairly, since the parameter here is Y, we may be dealing with $k = |Y|$ colors.
The fundamental result on color coding tells us that there are then $f(k)$ colorings of X, such that, if there is any injective map from Y to X witnessing that $Y \leq X$, then for one of these 3-colorings γ, we will have $a \leq a'$, $b \leq b'$ and $c \leq c'$, where γ colors a' red, b' blue and c' green,
So we just need to go through all of the colorings, and for each color class, apply FPT order tests to see if it all works out. QED

Technically, the above is all quite sound and valid. There are two more important points that need to be discussed.

(1) Can this classification technology be used to do anything normal or new?

For example, it is known that the property of a finite set of points in the plane of being hittable by k lines, where k is the parameter, is in FPT. The issue is not FPT, *per se*. Can this new abstract technology be used to provide an easy classification of the problem being FPT? This mathematical technology is only worthy of attention if something interesting can be done with it.

The following answer to the question that was posed in Leiden is what has been achieved to date.

Yes, finite sets of points in the plane, can be well quasi ordered naturally, in a variety of ways, that admit FPT order tests. This result is thanks mostly to the FPT version of Higman's Lemma, which can be applied in many ways, and even recursively.

(2) How does this approach to developing well quasi ordering-based FPT classification machinery for computational geometry and topology, become recursive?

We do not know if the following has been published; it is known to the authors from conversations in the olden days with Robertson and Seymour, especially Neil. For purposes of further discussion we will call this:

Robertson's Conjecture. The conjecture is that not only are finite graphs well quasi ordered by the minor order, but that finite graphs which are edge labeled by elements of a WQO are well quasi ordered in the label–respecting minor order.

This raises the following FPT version of Robertson's conjecture. Let us term $(Q, \leq)$ to be *friendly* if (1) this is a well quasi order, and (2) it admits FPT order tests.

Conjecture. Finite graphs edge labeled by a friendly wqo, are friendly.

Because of the labels, this opens up enormous recursive coding capacity, if we can figure out how to use it in some concrete mathematical circumstances. We look forward to seeing how this theme develops.

6   Conclusion

This papers highlights three concrete and alluring problem areas collaborated with Hans.
These are:

(1) The horizon of exploring FPT algorithmic realizations of results in pure extremal combinatorics (however proved) that connect two structural parameters, as exemplified by the LONG CYCLE problem (Win–Wins).
(2) The horizon of exploring further the theme of *too much information compression* in extremal combinatorics modulo plausible mathematical conjectures in computational complexity theory.
(3) The horizon, beyond the heroic efforts of Robertson and Seymour, of pushing the FPT perspective which they inspired, into new realms of algorithmics, as in the *Functor Project*, which seems to us quite natural: move FPT and its tools and perspectives around, systematically.

We had a very good time working on these wonderments, Hans!

Thanks and Happy Birthday!

Mike and Fran

References

1. Downey, R.-G., Estivill-Castro, V., Fellows, M.-R., Prieto-Rodriguez, E., Rosa-mond, F.: Cutting up is hard to do: the parameterized complexity of k-cut and related problems. Electr. Notes Theor. Comput. Sci. **78**, 209–222 (2003)
2. Barefoot, C., Clark, H., Douthett, J., Entringer, R., Fellows, M.: Length 0 modulo 3 in graphs. Ann. Discret. Math. 87–101 (1991)
3. Edouard, B., Bousquet, N., Thomassé, S., Watrigant, R.: When maximum stable set can be solved in FPT time. In: 30th International Symposium on Algorithms and Computation (ISAAC), Article no. 49(22), pp. 1–49. Schloss Dagstuhl Leibniz Zentrum for Informatiks (2019)
4. Cai, L., Ye, J.: Parameterized complexity of connected induced subgraph problems. In: Gu, Q., Hell, P., Yang, B. (eds.) AAIM 2014. LNCS, vol. 8546, pp. 219–230. Springer, Cham (2014). https://doi.org/10.1007/978-3-319-07956-1_20
5. Dabrowski, K., Lozin, V., Müller, H., Rautenbach, D.: Parameterized algorithms for the independent set problem in some hereditary graph classes. In: Iliopoulos, C.S., Smyth, W.F. (eds.) IWOCA 2010. LNCS, vol. 6460, pp. 1–9. Springer, Hei-delberg (2011). https://doi.org/10.1007/978-3-642-19222-7_1
6. Downey, R.G., Fellows, M.R.: Fundamentals of Parameterized Complexity. TCS. Springer, London (2013). https://doi.org/10.1007/978-1-4471-5559-1
7. Downey, R.G., Fellows, M.R.: Parameterized Complexity. Monograph in Computer Science. Springer, New York (1999). https://doi.org/10.1007/978-1-4612-0515-9
8. Cattell, K., Dinneen, M., Downey, R., Fellows, M., Langston, M.: On computing graph minor obstruction sets. Theor. Comput. Sci. **233**(1–2), 107–127 (2000)
9. Fellows, M.R., Jaffke, L., Király, A.I., Rosamond, F.A., Weller, M.: What is known about vertex cover kernelization? In: Böckenhauer, H.-J., Komm, D., Unger, W. (eds.) Adventures Between Lower Bounds and Higher Altitudes. LNCS, vol. 11011, pp. 330–356. Springer, Cham (2018). https://doi.org/10.1007/978-3-319-98355-4_19
10. Garey, M., Johnson, D.-S.: Computers and Intractability: A Guide to the Theory of NP-Completeness. W. H. Freeman and Co., New York (1979)
11. de Berg, M., Bodlaender, H., Fellows, M., Knauer, C. (Scientific Organizers): Fixed-parameter computational geometry. In: Leiden Workshop, Lorentz Center (2016)
12. Downey, R., Fellows, M., Niedermeier, R., Rossmanith, P. (Scientific Organizers): Parameterized complexity. Dagstuhl Seminar 01311, Dagstuh (2001)
13. Alon, N., Yuster, R., Zwick, U.: Color-coding. J. Assoc. Comput. Mach. **42**(4), 844–856 (1995)
14. Higman, G.: Ordering by divisibility in abstract algebras. Proc. London Math. Soc. **3–2**(1), 326–336 (1952)

Hans Bodlaender and the Theory
of Kernelization Lower Bounds

Danny Hermelin[✉]

Department of Industrial Engineering and Management,
Ben-Gurion University of the Negev, Beersheba, Israel
hermelin@bgu.ac.il

Abstract. In this short letter I give a brief subjective account of my favorite result with Hans – our kernelization lower bounds framework. The purpose of this manuscript is not to give a formal introduction to this result and the area that spawned from it, nor is it meant to be a comprehensive survey of all related and relevant results. Rather, my aim here is to informally describe the history that lead to this result from a personal perspective, and to outline Hans's role in the development of this theory into what it is today.

1 Age of Exploration

The starting point for our journey is in the definition of a kernel, one of the central cornerstones of parameterized complexity:

Definition 1. *A kernelization algorithm (or kernel) for a parameterized problem $L \subseteq \{0,1\}^* \times \mathbb{N}$ is a polynomial time algorithm that transforms a given instance $(x,k) \in \{0,1\}^* \times \mathbb{N}$ to an instance $(x',k') \in \{0,1\}^* \times \mathbb{N}$ such that:*

- *$(x,k) \in L \iff (x',k') \in L$, and*
- *$|x'| + k' \leq f(k)$ for some computable function f.*

The function f above is referred to as the size *of the kernel.*

Thus, in essence, a kernel is a reduction from a problem onto itself and can be thought of as a formal definition of preprocessing.

It was known early on that a decidable problem was fixed-parameter tractable (FPT) iff it had a kernel, and so kernelization has also been viewed as an alternative definition of the central concept in parameterized complexity. Since every problem in FPT has a kernel, the research community focused on reducing kernel sizes as much as was known to be possible. This lead to a huge array of various methods and techniques, and to an enormous body of work geared towards such efforts. Some of these, and many other later results, can be found in the new wonderful book on kernelization [11].

During this age of exploration, it became apparent that there are quite a few problems that admit kernels of *polynomial size, i.e.*, kernels of size $f(k) = k^{O(1)}$. Many of these examples were prominent FPT problems like VERTEX COVER,

© Springer Nature Switzerland AG 2020
F. V. Fomin et al. (Eds.): Bodlaender Festschrift, LNCS 12160, pp. 18–21, 2020.
https://doi.org/10.1007/978-3-030-42071-0_3

FEEDBACK VERTEX SET, and MAX LEAF. Yet, on the other hand, there were still several FPT problems which were resisting all attempts at reducing their kernels to polynomial size. We knew of *artificial problems* which could be solved in $2^{O(k)} \cdot n^{O(1)}$ time that did not have polynomial kernels (see Section 6 in [1]), but we were not aware of similar *natural* examples. One natural candidate for such an example seemed to be the k-PATH problem, since it has quite a few elegant $2^{O(k)} \cdot n^{O(1)}$ time algorithms. Yet, despite several attempts, no one could break the trivial $2^{O(k)}$ bound on the kernel size of k-PATH; and we had no tools for showing whether this bound was indeed the best possible.

2   Technological Breakthrough

This was the state of affairs for around ten years until Hans came up with the big breakthrough in the mid 2000s: I was at the time visiting Mike Fellows in Australia when Mike received an email from Hans: In this mail, Hans described how a polynomial kernel for k-PATH would yield a certain type of compression for 3-SAT which we later called a distillation algorithm. A *distillation* algorithm for 3-SAT takes N instances of 3-SAT, each of size n, and reduces them in $(Nn)^{O(1)}$ time to an instance of size $n^{O(1)}$ which is satisfiable iff one the original N input instances is satisfiable. Since this intuitively seemed to us impossible in cases where $Nn >> n^{O(1)}$, we were all convinced that this was the first evidence for a natural problem not admitting a polynomial kernel. Later on, Fortnow and Santhanam confirmed our intuition by proving that a distillation algorithm for 3-SAT implies a collapse to the polynomial hierarchy [12].

The main component in Hans's argument about k-PATH is that: (i) k-PATH is NP-hard and thus there is a polynomial-time reduction from 3-SAT to k-PATH, and (ii) taking the disjoint union of N instances of k-PATH gives us an equivalent instance to the OR of these instances without increasing the parameter value. Thus, it became clear that his argument can be generalized to other problem which exhibit the same two properties. This lead to the concept of a *composition algorithm* which is essentially a formalization of the second property above. Thus the main component of the kernelization lower bound machinery was born:

Theorem 1 ([1,12]). *An NP-hard problem that has a composition algorithm does not admit a polynomial kernel unless the polynomial hierarchy collapses.*

It is here where I wish to point out Hans's generosity and attitude while conducting collaborative research. I was but an anonymous Masters student from Israel on a visit to Australia. Hans's mail was addressed to Mike Fellows, and he had no real reason to share his initial findings with me. Nevertheless, on Mike's account, Hans did not hesitate a second when asked whether I could join the project. Until this day I am grateful to Hans for introducing me to this wonderful topic, and this manuscript is my feeble attempt at showing gratitude.

3   Industrial Revolution

It did not take much time for the rest of parameterized complexity community
to figure out how to apply Theorem 1 to other problems and domains. The
revolution has begun! And Hans was again at the forefront.

Along with Thomassé and Yeo [5], Hans showed that a certain type of reduction (called *polynomial parameter transformation*) can be used to show that
DISJOINT PATHS and DISJOINT CYCLES do not admit polynomial kernels. This
approach was immediately picked up by Dom, Lokshtanov, and Saurabh [9] who
developed the framework even further, allowing it to apply for a multitude of
different problems. At this point it was open season, with a plethora of kernelization lower bound results starting to appear everywhere. Around the same time,
together with Jansen and Kratsch [4], Hans developed the theory even more,
developing the more modern concept of *Cross Compositions*. He was also one of
the first to apply the lower bound framework to non-standard graph parameterizations [2,3].

4   Post Revolution Era

Along the years the theory has been significantly developed. Drucker showed [10]
that the AND analog of Fortnow and Santhanam's result also holds, and thus
an AND analog of Theorem 1 can also be used for showing kernelization lower
bounds (example implications of the AND analog of Theorem 1 can already
be found in [1]). It was also shown how to use this theory to obtain *specific*
polynomial bounds, *e.g.* $k^{2-\epsilon}$ or $k^{5-\epsilon}$, for problems that do admit polynomial
kernels [6–8,14]. Kratsch [15] showed how to use the co-non-deterministic nature
of the arguments for proving Theorem 1 for obtaining polynomial-lower bounds
for a Ramsey-type of problem. Lokshtanov *et al.* [16] showed how to extend
this theory to the area of approximate kernelization. Finally, an attempt at
providing a lower bounds framework for Turing kernels, via hardness classes and
reductions, can be found in [13].

5   The World We Live in Today

Nowadays the theory of kernelization lower bounds has become a standard tool in
any researcher's toolkit working in the area of parameterized complexity. And as
I attempt to describe above, Hans has played a central role in this development,
using both his immaculate creative skills as well as his comfortable easy going
collaborative nature. It is my firm belief that we as research community are
forever in debt to him for these (and many other) contributions.

References

1. Bodlaender, H.L., Downey, R.G., Fellows, M.R., Hermelin, D.: On problems without polynomial kernels. J. Comput. Syst. Sci. **75**(8), 423–434 (2009)
2. Bodlaender, H.L., Jansen, B.M.P., Kratsch, S.: Kernel bounds for path and cycle problems. Theor. Comput. Sci. **511**, 117–136 (2013)
3. Bodlaender, H.L., Jansen, B.M.P., Kratsch, S.: Preprocessing for treewidth: a combinatorial analysis through kernelization. SIAM J. Discrete Math. **27**(4), 2108–2142 (2013)
4. Bodlaender, H.L., Jansen, B.M.P., Kratsch, S.: Kernelization lower bounds by cross-composition. SIAM J. Discrete Math. **28**(1), 277–305 (2014)
5. Bodlaender, H.L., Thomassé, S., Yeo, A.: Kernel bounds for disjoint cycles and disjoint paths. Theor. Comput. Sci. **412**(35), 4570–4578 (2011)
6. Cygan, M., Grandoni, F., Hermelin, D.: Tight kernel bounds for problems on graphs with small degeneracy. ACM Trans. Algorithms **13**(3), 43:1–43:22 (2017)
7. Dell, H., Marx, D.: Kernelization of packing problems. In: Proceedings of the 23rd Annual ACM-SIAM Symposium on Discrete Algorithms (SODA), pp. 68–81 (2012)
8. Dell, H., van Melkebeek, D.: Satisfiability allows no nontrivial sparsification unless the polynomial-time hierarchy collapses. J. ACM **61**(4), 23:1–23:27 (2014)
9. Dom, M., Lokshtanov, D., Saurabh, S.: Kernelization lower bounds through colors and ids. ACM Trans. Algorithms **11**(2), 13 (2014)
10. Drucker, A.: New limits to classical and quantum instance compression. SIAM J. Comput. **44**(5), 1443–1479 (2015)
11. Fedor, F.V., Lokshtanov, D., Saurabh, S., Zehavi, M.: Kernelization Theory of Parameterized Preprocessing. Cambridge University Press, Cambridge (2019)
12. Fortnow, L., Santhanam, R.: Infeasibility of instance compression and succinct PCPs for NP. J. Comput. Syst. Sci. **77**(1), 91–106 (2011)
13. Hermelin, D., Kratsch, S., Soltys, K., Wahlström, M., Xi, W.: A completeness theory for polynomial (turing) kernelization. Algorithmica **71**(3), 702–730 (2015)
14. Hermelin, D., Wu, X.: Weak compositions and their applications to polynomial lower bounds for kernelization. In: Proceedings of the 23rd Annual ACM-SIAM Symposium on Discrete Algorithms (SODA), pp. 104–113 (2012)
15. Kratsch, S.: Co-nondeterminism in compositions: a kernelization lower bound for a ramsey-type problem. ACM Trans. Algorithms **10**(4), 19:1–19:16 (2014)
16. Lokshtanov, D., Panolan, F., Ramanujan, M.S., Saurabh, S.: Lossy kernelization. In: Proceedings of the 49th Annual ACM SIGACT Symposium on Theory of Computing (STOC), pp. 224–237 (2017)

Algorithms, Complexity, and Hans

Jan van Leeuwen[✉]

Department of Information and Computing Sciences, Utrecht University,
Princetonplein 5, 3584 CC Utrecht, The Netherlands
J.vanLeeuwen1@uu.nl

*"The study of algorithms is at
the very heart of computer science"*
Aho, Hopcroft, Ullman, [1], 1974, p. iii

Abstract. In this essay on the occasion of Hans Bodlaender's 60th
birthday, we recount some of the early developments in the field in which
Hans made his mark and of their context at Utrecht University.

Keywords: Algorithms · Computational complexity · Graphs ·
Networks · Parameterized complexity · Structural algorithmics ·
Treewidth

Prologue

Hans Bodlaender is widely known as an expert algorithmician. From his student
years until now, not counting a postdoctoral semester at MIT in 1987, Hans
has worked at Utrecht University, for a long time in my group on Algorithmic
Systems but since 2014 as head of the new chair in Algorithms and Complexity.
From 2014 until 2018 he also held a part-time chair in Network Algorithms at
Eindhoven University of Technology. On the occasion of his 60th birthday, we
look back at some of the early trends in the areas to which Hans contributed,
and at their context at Utrecht University.

Beginnings

Hans Leo Bodlaender was born on April 21, 1960 in Bennekom, a small town
some 50 kilometers away from Utrecht in the Netherlands. There was little at
the time that would suggest a later career in computer science. The development
of programming languages was still in its infancy, home computers didn't exist
yet, and the World Wide Web was still a distant future. Knuth's first volume
on the 'The Art of Computer Programming' wasn't published until 1968, and
Cook's discovery of the theory of NP-completeness appeared only in 1971.

In 1977 I was appointed as head of, what would soon become, the first chair
in Computer Science ('Informatics') at Utrecht University. The major challenge
I had to face immediately was the development of a fully fledged teaching and
research program in computer science, in analogy to what I had seen during my

© Springer Nature Switzerland AG 2020
F. V. Fomin et al. (Eds.): Bodlaender Festschrift, LNCS 12160, pp. 22–27, 2020.
https://doi.org/10.1007/978-3-030-42071-0_4

years in the US. Lecture rooms were filled to the rim with students who wanted to know everything about computers, programming, and whatever was made possible by it.

It was in these days of pioneering activity, in 1978, that Hans Bodlaender began his studies in mathematics at Utrecht University. Like many students at the time, Hans soon discovered the attractiveness of our courses in computer science, which showed structure and method in the developing systems of, what everyone then realized, the times ahead. Mini-computers were being used everywhere, and the PC revolution was beginning. We were focusing on foundations of computer science, programming methodologies, and systems as a start, gradually broadening the spectrum of the field from there.

In the spring of 1981, Hans completed his 'candidacy exam' in mathematics (*cum laude*), with a minor in computer science. Two years later, in 1983, he completed his 'doctoral exam' in mathematics (comparable to what is now the MSc degree), with a minor in foundations of mathematics and, by special permission of the graduate student advisor in mathematics, with a specialization in computer science under my guidance, on a study of 'routing algorithms in computer networks'. I urgently needed excellent researchers like Hans in my team, which already included people like Marinus Veldhorst and Mark Overmars (who would become a leading expert in computational geometry and, later, in game design).

PhD Studies

Fortunately, I managed to keep Hans as a research assistant, from February 1984 onward in a position funded by the Dutch Research Council (ZWO, now NWO), on a project entitled 'analysis of distributed algorithms'. With the other members of the team, including e.g. Anneke Schoone, Richard Tan (visiting every summer), Gerard Tel, and Harry Wijshoff, we explored all algorithmic issues that we came across in parallel and distributed computing and tried to develop theory for them. Being a supervisor of these works was great.

In little more than three years after his doctoral exam, in November of 1986, Hans completed his PhD thesis entitled "Distributed Computing: Structure and Complexity' [2], a substantial work of 296 pages. The larger part of it was devoted to an extensive combinatorial analysis of so-called *emulations* between processor networks [3]. Soon after, his thesis would be published in the series CWI Tracts of the Centre Mathematics and Computer Science in Amsterdam. It was also an important year in Hans' personal life: in October 1986, he and Brigitte got married!

While they were anticipating a postdoctoral semester at MIT in 1987, the papers of Robertson and Seymour [12,13] on the theory of graph minors, with exciting topics like path- and tree-decomposition, landed on our desk. Stefan Arnborg, Derek Corneil, Andrzej Proskurowski, and also Detlef Seese, had already obtained a number of intricate results that were indicative of the algorithmic potential of these concepts. Before the end of 1986, Hans had written a first report, proving many new families of graphs to be of *bounded treewidth*, implying that many well-known NP-complete problems would be polynomial, or even linear-time computable for these classes [4]. The follow-up report, showing how the dynamic programming technique behind these results could be unified and extended, was completed at MIT [5].

With these results on the structural algorithmics of graphs and networks, Hans' career would take a decisive turn. Various studies on distributed computing (e.g. on the complexity of leader finding) would remain on the agenda, but gradually the emphasis in his work shifted to the analysis of hard problems on graphs of bounded treewidth. In 1988, he combined his background in parallel algorithms with the new paradigm, showing that many NP-complete problems actually are in NC when restricted to graphs of bounded treewidth [6]. The paper was presented at WG'88, the first WG workshop held in the Netherlands. (Hans organized two of the later ones, in 2003 and 2017.) Many publications would follow, also on other algorithmic questions that people would propose to Hans. However, before we continue, some more background is in order.

The Algorithmic Perspective

Until the beginning of the 20th century, the term *algorithm* was synonymous to analytic, numeric or algebraic computation. The use of electronic computers since the 1940's and the later processor networks led to new breeds of algorithmic models, structure-based algorithms and theories of computational complexity. Understanding systems became a matter of understanding their algorithms. In the opening sentence of their influential textbook from 1974, Aho, Hopcroft, and Ullman [1] aptly stated that 'the study of algorithms is at the very heart of computer science'.

Indeed, from the 1970's onward, the algorithmic perspective became a dominant paradigm for modeling processes in many fields and for uncovering the structures that must be understood to exploit them efficiently in computer applications. It is also reflected in the graduate courses many of us taught at the time, in my case alternatingly called 'Analysis of Algorithms' and 'Complexity of Algorithms' and always devoted to the newest material around. I taught on areas like *computational geometry* (1979–1980), *graph algorithms and network flow* (1980–1981), *VLSI and chip complexity* (1981–1982), *parallel algorithms* (1982–1983), *combinatorial algorithms* (1983–1984), *applied number theory and cryptographic protocols* (1984–1985), *fault-tolerant distributed computing* (1985–1986), *linear programming and combinatorial optimization* (1986–1987), and *structural complexity theory* (1987–1988). From 1983 onward, I also held yearly seminars on *distributed algorithms* and *distributed methods*. All these courses and seminars had a long-lasting impact on our research program.

Hans Bodlaender 'stepped in' right in the middle of these exciting developments. It explains his initial occupation with the structures and theories in parallel and distributed computing, about which you may have wondered. It simply was in the air! From 1986–1987 onward, the seminars on distributed methods were led jointly with Hans and Gerard Tel. After returning from his postdoctoral stay at MIT, Hans would even teach our graduate course on distributed methods. Eventually, Gerard would write his well-known textbook on distributed algorithms [15]. However, Hans turned fully to the study of problems on graphs, also of games, and later, to the challenges of parameterized complexity theory.

Research with Hans

The qualities of Hans as a discrete algorithmician were clear from the very beginning of his career. While we developed the 'algorithmic perspective' in various directions, it were usually the underlying graph- or complexity-theoretic questions that drew Hans' attention.

I remember a beautiful instance from 1985, when we reduced a question about the potential effect of faults in a network to the question what the maximum increase in diameter of a network could be if k edges would 'break' without disconnecting the network [14]. Another example was the analysis of networks that would admit 'interval routing schemes' with up to k labels per edge. Thanks to Hans we could show that the class of networks with this property was minor-closed, and *thus* linear-time recognizable, for every $k \geq 1$. We also proved tight bounds on the treewidth of these graphs, even in the planar case [8]. All imaginable cases would be covered. Many will recognize this as a typical outcome of working together with Hans!

In the summer of 1987, after his stay at MIT, Hans became an assistant professor in our department and began his extensive journey into the study of graph classes, of graph decompositions and width measures, and the utilization of these concepts in the design of highly efficient algorithms for problems that would be NP-hard in general but feasible for special classes of graphs (e.g. of bounded path- or treewidth). In 1992 he achieved a major breakthrough, by showing that for any constant k, there is linear-time algorithm that, given a graph G, determines whether the treewidth of G is at most k and, if so, finds a tree-decomposition of G with treewidth at most k [7]. As a long-sought result with far-reaching implications, it was named *Bodlaender's theorem* in Rod Downey and Mike Fellows' book on parameterized complexity [10]. In 2003, Hans was promoted to the rank of associate professor.

For many years, Hans' office would be the meeting place ('Chez Hans') of our group for the daily coffee in the morning or afternoon tea, brewed - against university rules - in his own office. We would talk about our ongoing research, new questions that had come up, the courses we taught, ideas for student projects, our travels to increasing numbers of conferences, the conferences we organized ourselves (like the second ESA, held at our National Sports Centre Papendal in 1994), about chess, and about more mundane topics as well. Hans very much preferred research, teaching, and working with students over the administrative duties that would occasionally cross his path. My son Erik Jan was one of the students who would graduate under Hans' supervision, in 2004. Later on, Hans' eldest daughter would graduate in the Algorithmic Systems group as well!

Working intensively with numerous colleagues, from Europe to Japan, Hans became a well-known theoretician. In co-supervising or supervising PhD students like Ton Kloks (PhD 1993), Goos Kant (PhD 1993), Babette van Antwerpen-de Fluiter (PhD 1997), Thomas Wolle (PhD 2005), Johan Kwisthout (PhD 2009), Johan van Rooij (PhD 2011), Bart Jansen (PhD 2013), Tom van der Zanden (PhD 2019), and Sándor Kisfaludi-Bak (PhD 2019, at Eindhoven University of Technology), always dedicated to working together with them, Hans considerably extended his scope in the field. Many deep results would be obtained in these cooperations, and were published in a large number of conference and journal publications. Hans now has by far the highest *H-index* in our department! In 2014, Hans and co-authors received the EATCS-IPEC *Nerode Prize* 'for outstanding papers in the area of multivariate algorithmics' for their paper [9], together with the authors of [11].

In 2011 I reached the mandatory retirement age, and the Algorithmic Systems group ceased to exist. In 2014, Hans was appointed to full professor on the new chair of Algorithms and Complexity, with an enthusiastic team of faculty members and PhD students who now develop the algorithmic perspective in areas like computational sustainability, public transport and network science. At the same time Hans was appointed on a temporary chair in Network Algorithms at Eindhoven University of Technology, as part of a broad inter-university research programme on the stochastic and algorithmic aspects of Networks.

Key themes in Hans' research group now include fixed parameter tractability, fine-grained complexity, kernelization, exact algorithms, computational hardness, combinatorial optimization, simulation, and various other topics in efficient algorithm design and complexity theory. Many students follow his courses, notably his graduate course on Algorithms and Networks. Many continue on, to do a MSc project under the guidance of Hans or someone from his group.

Epilogue

On this occasion of Hans' 60th birthday, we can only conclude that research in algorithms and complexity is thriving, and more extensive and important than ever before. The algorithmic perspective, now known as 'the algorithmic lens', has become a true scientific method in exploring the world [16]. Algorithms invariably are the key to automate processes and operations in any field. They must be designed and re-designed for efficiency as new computational technologies appear. Many new questions arise as algorithms need to adapt, learn, and evolve.

Congratulations, Hans, and may you continue your contributions to our field for many years to come!

Acknowledgment. I thank Gerard Tel for useful comments on a preliminary version of this essay.

References

1. Aho, A.V., Hopcroft, J.E., Ullman, J.D.: The Design and Analysis of Computer Algorithms. Addison-Wesley Publishing Company, Reading (1974)
2. Bodlaender, H.L.: Distributed computing: structure and complexity. Ph.D. thesis, Utrecht University (1986). CWI Tract 43. CWI, Amsterdam (1987)
3. Bodlaender, H.L., van Leeuwen, J.: Simulations of large networks on smaller networks. Inf. Control **71**, 143–180 (1986)
4. Bodlaender, H.L.: Classes of graphs of bounded tree-width. Technical report RUU-CS-86-22, Department of Computer Science, Utrecht University, December 1986

5. Bodlaender, H.L.: Dynamic programming on graphs with bounded treewidth. Technical report MIT/LCS/TR-394, Laboratory for Computer Science, MIT, Cambridge (1987). Automata, Languages and Programming. In: Proceedings of 15th International Colloquium (ICALP 1988). LNCS, vol. 317, pp. 105–118. Springer, Berlin (1988)
6. Bodlaender, H.L.: NC-algorithms for graphs with small treewidth. In: van Leeuwen, J. (ed.) Graph-Theoretic Concepts in Computer Science, WG 1988. LNCS, vol. 344, pp. 1–10. Springer, Berlin (1989)
7. Bodlaender, H.L.: A linear-time algorithm for finding tree-decompositions of small treewidth. In: Proceedings of the 25th Annual ACM Symposium on Theory of Computing (STOC 1993), pp. 226–234. ACM Press (1993). SIAM J. Comput. 25(6), 1305–1317 (1996)
8. Bodlaender, H.L., van Leeuwen, J., Tan, R., Thilikos, D.M.: On interval routing schemes and treewidth. Inf. Comput. 139(1), 92–109 (1997)
9. Bodlaender, H.L., Downey, R.G., Fellows, M.R., Hermelin, D.: On problems without polynomial kernels. J. Comput. Syst. Sci. 75(8), 423–434 (2009)
10. Downey, R.G., Fellows, M.R.: Parameterized Complexity. Springer, New York (1999)
11. Fortnow, L., Santhanam, R.: Infeasibility of instance compression and succinct PCPs for NP. J. Comput. Syst. Sci. 77(1), 91–106 (2011)
12. Robertson, N., Seymour, P.D.: Graph minors. I. Excluding a forest. J. Comb. Theory Ser. B 35(1), 39–61 (1983)
13. Robertson, N., Seymour, P.D.: Graph minors. II. Algorithmic aspects of tree-width. J. Algorithms 7(3), 309–322 (1986)
14. Schoone, A.S., Bodlaender, H.L., van Leeuwen, J.: Diameter increase caused by edge deletion. J. Graph Theory 11, 409–427 (1987)
15. Tel, G.: Introduction to Distributed Algorithms. Cambridge University Press, Cambridge (1994)
16. van Leeuwen, J.: A fascinating science. In: Bodlaender, H.L., Duivesteijn, W., Nijenhuis, C.J. (eds.) Fascinating for computation - 25 jaar opleiding informatica. Department of Information and Computing Sciences, Utrecht University, pp. 33–45 (2008). http://www.cs.uu.nl/groups/AD/fasc.pdf

Surveys

Lower Bounds for Dominating Set in Ball Graphs and for Weighted Dominating Set in Unit-Ball Graphs

Mark de Berg[1] and Sándor Kisfaludi-Bak[2(✉)]

[1] TU Eindhoven, Eindhoven, The Netherlands
m.t.d.berg@tue.nl
[2] Max Planck Institut für Informatik, Saarbrücken, Germany
sandor.kisfaludi-bak@mpi-inf.mpg.de

Abstract. Recently it was shown that many classic graph problems—INDEPENDENT SET, DOMINATING SET, HAMILTONIAN CYCLE, and more—can be solved in subexponential time on unit-ball graphs. More precisely, these problems can be solved in $2^{O(n^{1-1/d})}$ time on unit-ball graphs in $\mathbb{R}^d$, which is tight under ETH. The result can be generalized to intersection graphs of similarly-sized fat objects.

For INDEPENDENT SET the same running time can be achieved for non-similarly-sized fat objects, and for the weighted version of the problem. We show that such generalizations most likely are not possible for DOMINATING SET: assuming ETH, we prove that
- there is no algorithm with running time $2^{o(n)}$ for DOMINATING SET on (non-unit) ball graphs in $\mathbb{R}^3$;
- there is no algorithm with running time $2^{o(n)}$ for WEIGHTED DOMINATING SET on unit-ball graphs in $\mathbb{R}^3$;
- there is no algorithm with running time $2^{o(n)}$ for DOMINATING SET, CONNECTED DOMINATING SET, or STEINER TREE on intersections graphs of arbitrary convex (but non-constant-complexity) objects in the plane.

1 Introduction

Over the past few years the authors had the privilege to collaborate with Hans Bodlaender on various algorithmic problems on geometric intersection graphs. In this short paper, written on the occasion of Hans's 60th birthday, we further explore some of the research directions inspired by this joint research. Hans, we hope you will enjoy reading the paper and are looking forward to more collaborations. And of course: happy birthday!

Many classic optimization problems on graphs cannot be solved in subexponential time—that is, in time $2^{o(n)}$, where n is the number of vertices—on general graphs, assuming the Exponential-Time Hypothesis [10]. For planar graphs,

This research was supported by the Netherlands Organization for Scientific Research NWO under project no. 024.002.003 (NETWORKS).

© Springer Nature Switzerland AG 2020
F. V. Fomin et al. (Eds.): Bodlaender Festschrift, LNCS 12160, pp. 31–48, 2020.
https://doi.org/10.1007/978-3-030-42071-0_5

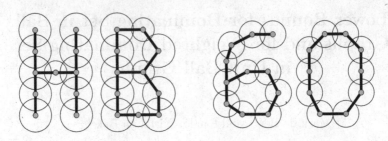

Fig. 1. An intersection graph of a set of disks in the plane.

however, subexponential algorithms are often possible. In particular, on planar graphs one can often obtain $2^{O(\sqrt{n})}$ running time. This is for example the case for INDEPENDENT SET (and, hence, VERTEX COVER), for DOMINATING SET, and for HAMILTONIAN CYCLE [9]. The fact that so many NP-hard problems can be solved in $2^{O(\sqrt{n})}$ time on planar graphs has been dubbed the *square-root phenomenon* [17]. Following recent work, we will explore to what extent this phenomenon can also be observed in certain types of geometric intersection graphs that generalize planar graphs.

The *intersection graph* $\mathcal{G}[S]$ of a set S of objects in $\mathbb{R}^d$ is the graph whose vertex set corresponds to S and that has an edge between two vertices if and only if the corresponding two objects intersect; see Fig. 1. It is well known that the class of planar graphs is equivalent to the class of contact graphs of disks in the plane [14], so contact graphs exhibit the square-root phenomenon. (A *contact graph* of disks is the intersection graph of a set S of closed disks with disjoint interiors.) Does the square-root phenomenon also arise for other types of intersection graphs? And is there a version of the square-root phenomenon for intersection graphs in higher dimensions? The answer is yes: De Berg *et al.* [2] (see also Kisfaludi-Bak's thesis [12]) recently showed that all the classic graph problems mentioned above can be solved in $2^{O(\sqrt{n})}$ time on unit-disk graphs—a *unit-disk graph* is the intersection graph of unit disks in the plane—, and in $2^{O(n^{1-1/d})}$ time on unit-ball graphs in $\mathbb{R}^d$. More generally, these problems can be solved in $2^{O(n^{1-1/d})}$ time for intersection graphs of *similarly-sized fat objects*. Here we say an object o is *fat* if there is a ball $b_{\text{in}} \subseteq o$ of radius ρ_{in} and a ball $b_{\text{out}} \supseteq o$ of radius ρ_{out} such that $\rho_{\text{in}}/\rho_{\text{out}} \geqslant \alpha$, where α is an absolute constant, and we say that a collection of objects is *similarly sized*, if the ratio of the largest and smallest object diameter is at most some absolute constant.

Algorithms with running time $2^{O(\sqrt{n})}$ on planar graphs are typically based on the Planar Separator Theorem [15,16]. This theorem states that any planar graph $\mathcal{G} = (V, E)$ has a balanced separator of size $O(\sqrt{n})$, that is, a subset $C \subset V$ of size $O(\sqrt{n})$ whose removal splits $\mathcal{G}$ into connected components with at most $\delta|V|$ vertices each, for some constant $\delta < 1$. For unit-disk graphs such a result is clearly impossible, since unit-disk graphs can have arbitrarily large cliques. However, it is always possible to find a balanced separator that consists of a small number of cliques. More precisely, De Berg *et al.* proved that any ball graph (or more generally, intersection graph of fat objects) in $\mathbb{R}^d$ admits a balanced separator $C_1 \cup \cdots \cup C_k$ such that $\sum_{i=1}^{k} \log(|C_i| + 1) = O(n^{1-1/d})$ and each C_i is a clique. This clique-based separator theorem forms the basis of an algorithmic framework [2] for obtaining algorithms with $2^{O(n^{1-1/d})}$ running time for all problems mentioned above. Interestingly, the framework only works for *unit-ball* graphs—or, more generally, intersection graphs of *similarly-sized* fat objects—even though the underlying clique-based separator exists for arbitrarily-sized balls (or more generally, for fat objects). An exception is INDEPENDENT SET, where the clique-based separator theorem immediately gives an algorithm with $2^{O(n^{1-1/d})}$ running time for arbitrarily-sized balls [2]. For INDEPENDENT SET it is also easy to extend the result to the weighted version of the problem. In addition, it is possible to relax the fatness assumption while maintaining the subexponential behavior: for INDEPENDENT SET in $\mathbb{R}^2$ there is even a subexponential algorithm for arbitrary polygons [18], while for $d \geqslant 3$ a favorable trade-off can be established between fatness and running time (in case of similarly-sized objects) that is ETH-tight [13]. In this paper we explore to what extent the framework of De Berg *et al.* [2] can be generalized. In particular, we study for which problems the restriction to similarly-sized objects is necessary. We also study whether weighted versions of the above problems can be solved in subexponential time for similarly-sized objects.

Our Results. In Sect. 2 we first argue that for CONNECTED VERTEX COVER, FEEDBACK VERTEX SET and CONNECTED FEEDBACK VERTEX SET the restriction to similarly-sized objects in not necessary: these problems admit subexponential algorithms for arbitrarily-sized fat objects in $\mathbb{R}^d$. After that we show this generalization is not possible for DOMINATING SET, CONNECTED DOMINATING SET, or STEINER TREE: already in the plane there is no algorithm with running time $2^{o(n)}$ for these problems on intersection graphs of arbitrarily-sized fat objects, assuming ETH. The fat objects we use for this lower bound are convex, but they do not have constant description complexity.

   In Sects. 3 and 4 we then turn to the main topic of this paper: the complexity of (WEIGHTED) DOMINATING SET on ball graphs in $\mathbb{R}^3$. (Recall that the DOMINATING SET problem is to decide, given a graph $\mathcal{G} = (V, E)$ and a number k, if there is a vertex subset $D \subset V$ of size at most k such that all vertices $v \in V \setminus D$ are adjacent to some vertex in D.) As our main contribution, we show that DOMINATING SET on arbitrary ball graphs in $\mathbb{R}^3$ cannot be solved in $2^{o(n)}$ time, assuming ETH. In addition, we consider WEIGHTED DOMINATING SET, where each vertex is assigned a real weight, and the goal is to find the dominating set of weight at most k. It turns out that this is considerably harder than the unweighted problem: even on unit-ball graphs in $\mathbb{R}^3$, WEIGHTED DOMINATING SET cannot be solved in $2^{o(n)}$ time, assuming ETH.

Remark. It is known that recognizing if a given graph $\mathcal{G}$ is a unit-disk graph is NP-hard [7], so conceivably certain problems are easier when the input is a set of balls inducing a ball graph, rather than just the graph itself. It is thus desirable to develop algorithms that do not need to know the set S of objects defining the intersection graph, but that work with only the graph $\mathcal{G}[S]$ as input. De Berg *et al.* [2] do this by introducing a variant of treewidth. The concept later dubbed as $\mathcal{P}$-flattened treewidth [12] is based on the idea of weighted treewidth [6]. The need for finding separators geometrically is then alleviated by using the treewidth-approximation algorithm of Bodlaender *et al.* [5]. On the other hand, lower bounds become stronger when they still apply in the geometric setting where the set S defining the intersection graph is given. All our lower bounds have this property. From now on, whenever we speak of a problem on intersection graphs, we mean this geometric version of the problem. For example, in the DOMINATING SET problem on ball graphs, we have as input a set of balls with rational coordinates and radii, and an integer k, and the question is if the intersection graph $\mathcal{G}[S]$ has a dominating set of size k.

2   Non-similarly Sized Fat Objects

In this section we consider intersection graphs of arbitrary fat objects—they are not restricted to be of similar size, and they can have complicated shapes (in particular, they need not be balls). We first argue that CONNECTED VERTEX COVER, FEEDBACK VERTEX SET, and CONNECTED FEEDBACK VERTEX SET have algorithms with running time $2^{O(n^{1-1/d})}$ for arbitrary fat objects in $\mathbb{R}^d$, even for the weighted version of these problems. Then we show that DOMINATING SET, CONNECTED DOMINATING SET, and STEINER TREE are more difficult: there are no subexponential algorithms for these problems in intersection graphs of fat objects in $\mathbb{R}^2$.

CONNECTED VERTEX COVER *and* (CONNECTED) FEEDBACK VERTEX SET. The key component in using the algorithmic machinery of De Berg *et al.* [2] for a given graph problem is to show that any clique contains at most constantly many vertices from an optimal solution. For example, a clique contains at most one vertex from a solution to INDEPENDENT SET. For FEEDBACK VERTEX SET this is not true. Here, however, we can look at the complementary problem: instead of finding a minimum-size subset of vertices whose removal destroys all cycles, we can find a maximum-size subset that forms an induced forest. Indeed, there is a feedback vertex set of size at most k if and only if there is an induced forest of size at least $n-k$, so we can concentrate on solving MAXIMUM INDUCED FOREST instead.

For MAXIMUM INDUCED FOREST we have the desired property that any solution contains at most a constant number of vertices—at most two, to be precise—from a clique. In order to solve MAXIMUM INDUCED FOREST we can now develop a standard divide-and-conquer algorithm, using the clique-based separator of [2]. Of course the two sub-problems we get, one for the subgraph "inside" the separator and one for the subgraph "outside" the separator, are not independent: a solution on the inside, together with a solution in the separator itself, puts some constraints on the solution on the outside. For this reason, one needs to keep track of the connected components created on each side. The naive algorithm would give a running time of $2^{O(n^{1-1/d}\log n)}$, which is subexponential, but not optimal. The rank-based approach of Bodlaender *et al.* [4] can be applied to improve this to $2^{O(n^{1-1/d})}$.

For CONNECTED VERTEX COVER and CONNECTED FEEDBACK VERTEX SET, we also have the property that at most constantly many vertices from any clique are present in the complement of a solution: the complement of a solution to CONNECTED VERTEX COVER contains at most one vertex per clique (since the complement is an independent set), and the complement of a solution to CONNECTED FEEDBACK VERTEX SET contains at most two vertices per clique (since, as above, the complement is an induced forest). Using the framework of De Berg *et al.* [2], and the rank-based approach [4] to handle the connectivity issues, we can again obtain algorithms with $2^{O(n^{1-1/d})}$ running time.

(CONNECTED) DOMINATING SET *and* STEINER TREE. For (CONNECTED) DOMINATING SET and STEINER TREE one cannot guarantee that the solution (or its complement) uses only a constant number of vertices from a clique. Hence, the framework from [2] cannot easily be used. As the next theorem states, a subexponential algorithm is unlikely to exist for these problems on intersection graphs of fat objects. The objects in the proof will be convex, their fatness will be arbitrarily close to 1, but they will have high (non-constant) complexity.

Theorem 1. *Let $\varepsilon > 0$ be any fixed constant. There is no $2^{o(n)}$ algorithm for* DOMINATING SET, CONNECTED DOMINATING SET, *or* STEINER TREE *in intersection graphs of convex $(1 - \varepsilon)$-fat objects in $\mathbb{R}^2$, unless the Exponential-Time Hypothesis fails.*

Proof. Our proof works in two steps. We first show that DOMINATING SET, CONNECTED DOMINATING SET, and STEINER TREE do not admit algorithms with running time $2^{o(n)}$ on split graphs (assuming ETH). (A *split graph* is a graph $\mathcal{G} = (A \cup B, E)$ on $2n$ vertices such that A induces a clique and B induces an independent set.) We then show that any split graph can be realized as the intersection graph on convex $(1 - \varepsilon)$-fat objects.

It is known that DOMINATING SET on general graphs with n vertices does not admit an algorithm with running time $2^{o(n)}$, assuming ETH [9]. This lower bound carries over to split graphs. To see this, consider an arbitrary graph $\mathcal{G} = (V, E)$ on n vertices. We create a split graph $\mathcal{G}' = (A \cup B, E')$ with $2n$ vertices, where for each $v_i \in V$ we create a vertex $a_i \in A$ and $b_i \in B$. We then add the edge (a_i, b_j) to E' if and only if $i = j$ or $(v_i, v_j) \in E$. Finally, we add all edges between vertices in A, so A induces a clique.

Note that $\mathcal{G}$ has a dominating set of size k if and only if $\mathcal{G}'$ has a dominating set of size k. To see this, note that a dominating set $D \subseteq V$ for $\mathcal{G}$ corresponds to a dominating set $D' \subseteq A$ with the same vertex indices for $\mathcal{G}'$. On the other hand, if D' is a dominating set in $\mathcal{G}'$, then any vertex $b_i \in D' \cap B$ can be exchanged with a_i, since a_i dominates a superset of the vertices dominated by b_i. If both a_i and b_i are present in D', then b_i can be removed from D'. Therefore, we can find a dominating set $D' \subset A$ of size at most k for $\mathcal{G}'$, which corresponds to a dominating set of the same size for $\mathcal{G}$.

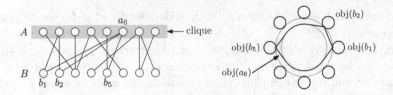

Fig. 2. Realizing a split graph as an intersection graph of fat objects. The circles γ_{in} and γ_{out}, and the points p_i, are depicted in grey. For clarity, the objects $\text{obj}(a_i)$ for $i \neq 6$ have been omitted.

From the above we can conclude that there is no $2^{o(n)}$ algorithm for DOMINATING SET on split graphs, assuming ETH. Notice that the same statement holds for CONNECTED DOMINATING SET, as the created dominating set $D' \subset A$ is connected. The statement holds for STEINER TREE as well, as we now argue. Recall that an instance of STEINER TREE is a graph $\mathcal{G} = (V, E)$, a terminal set T, and a natural number k, and the question is to decide if there is a set $W \subseteq V$ of size at most k such that $T \cup W$ induces a connected graph. Now, to solve DOMINATING SET on a given graph $\mathcal{G}$ we can create the graph $\mathcal{G}' = (A \cup B, E')$ as above, and solve STEINER TREE on $\mathcal{G}'$ with $T = B$. It is routine to check that any set W that is a solution to STEINER TREE on $\mathcal{G}'$ corresponds to a dominating set on $\mathcal{G}$, and vice versa.

It remains to show that $\mathcal{G}'$ is realizable as an intersection graph of $(1 - \varepsilon)$-fat objects in $\mathbb{R}^2$. Let γ_{out} and γ_{in} be two concentric circles, where γ_{out} has radius 1 and γ_{in} has radius $(1 - \varepsilon)$. Let $\{p_1, \ldots, p_n\}$ be n points equally spaced around γ_{out}. The object $\text{obj}(b_i)$ we create for each vertex $b_i \in B$ is a disk of radius $1/n$ that touches γ_{out} from the outside at p_i. The object $\text{obj}(a_i)$ we create for each vertex $a_i \in A$ is the convex hull of the set $\gamma_{\text{in}} \cup \{p_j \in P : (a_i, b_j) \in E'\}$; see Fig. 2. Thus $\gamma_{\text{in}} \subset \text{obj}(a_i) \subset \gamma_{\text{out}}$, and $\text{obj}(a_i)$ touches an object $\text{obj}(b_i)$ if and only if a_i is connected to b_i in $\mathcal{G}'$. Since the objects $\text{obj}(b_i)$ are pairwise disjoint, and the objects $\text{obj}(a_i)$ all intersect, the intersection graph of the created objects equals $\mathcal{G}'$. Furthermore, all objects are (at least) $(1 - \varepsilon)$-fat.   □

Remark. The objects in the above construction are disks and convex objects whose boundary consists of line segments and circular arcs. The construction can also be done with objects that are convex polygons with $O(n + 1/\varepsilon)$ vertices.

3   A Lower Bound for DOMINATING SET in Ball Graphs

In this section we prove that under the ETH, there is no subexponential algorithm for DOMINATING SET on ball graphs in $\mathbb{R}^d$, for $d \geqslant 3$. It suffices to prove this for $d = 3$, since any ball graph in $\mathbb{R}^d$ can trivially be realized as a ball graph in $\mathbb{R}^{d+1}$. We will use a reduction from a special version of 3-SAT, namely $(3, 3)$-SAT. In a $(3, 3)$-SAT problem the input formula that we want to test for satisfiability is a $(3, 3)$-CNF formula, that is, a CNF formula in which every clause has at most three literals, and every variable occurs at most three times in total.

Proposition 1 (De Berg et al. [2]). *There is no $2^{o(n)}$ algorithm for $(3, 3)$-SAT, unless the Exponential-Time Hypothesis fails.*

Our reduction from $(3, 3)$-SAT to DOMINATING SET on ball graphs works in two steps. First we convert the given $(3, 3)$-SAT instance ϕ to a graph $\mathcal{G}_\phi$ that has a dominating set of a certain size if and only if ϕ is satisfiable, and then we realize $\mathcal{G}_\phi$ as a ball graph in $\mathbb{R}^d$.

Step 1: Construction of $\mathcal{G}_\phi$. Let ϕ be a $(3, 3)$-SAT formula. We use a preprocessing step on ϕ in order to remove clauses that have only one literal the following way. If a clause has only one literal, we set its variable to satisfy the clause and delete any newly satisfied clauses or false literals that were created. We repeat this procedure as long as there are still clauses with only one literal. If at some point all literals are deleted from a clause, then that clause cannot be satisfied and we have solved the problem. In this case the reduction algorithm would return a graph $\mathcal{G}_\phi$ and number k that forms a trivial NO-instance. If all clauses are deleted because they are satisfied, then we are done as well, and we return a trivial YES-instance. If none of these two cases arises then we are left with a formula that has clauses of size two and three only, and that is satisfiable if and only if the original formula was satisfiable. With a small abuse of notation we still denote this formula by ϕ.

Let $\mathcal{X} := \{x_1, \ldots, x_n\}$ be the set of variables occurring in ϕ, let $\mathcal{C} := \{c_1, \ldots, c_s\}$ be the set of clauses in ϕ, and let $\mathcal{L} := \{l_1, \ldots, l_t\}$ be the multiset of literals occurring in ϕ. ($\mathcal{L}$ is a multiset because the same literal can occur multiple times in ϕ, as for example $\neg x_2$ does in Fig. 3.) The graph $\mathcal{G}_\phi$ is now constructed as follows; see Fig. 3 for an illustration.

- For each variable $x_j \in \mathcal{X}$ we create a *variable gadget*, which consists of three vertices forming a triangle. Two of these vertices, labeled x_j^T and x_j^F, correspond to truth assignments to x_j: selecting x_j^T into the dominating set corresponds to setting $x_j := \text{TRUE}$, and selecting x_j^F into the dominating set corresponds to setting $x_j := \text{FALSE}$. The third vertex is called the *ear* of the gadget.
- For each clause $c_k \in \mathcal{C}$ we create a *clause gadget*, consisting of a single vertex labeled c_k.
- For each literal $l_i \in \mathcal{L}$ we create a *literal gadget*, which is a path of three vertices, labeled l_i^1, l_i^2, and l_i^3, plus an extra vertex—the *ear* of the literal gadget—connected to l_i^2 and l_i^3.
- The literal gadgets are connected to the variable and clause gadgets as follows. Let c_k be the clause containing the literal l_i and let x_j be the variable corresponding to l_i. Then we add the edge (c_k, l_i^3), and we add (x_j^T, l_i^1) if $l_i = x_j$ and (x_j^F, l_i^1) if $l_i = \neg x_j$.
- Finally, we add edges between any pair of vertices in the set $\{l_i^1 : l_i \in \mathcal{L}\}$, thus creating a clique on these vertices.

Lemma 1. $\mathcal{G}_\phi$ *has a dominating set of size $n + t$ if and only if ϕ is satisfiable.*

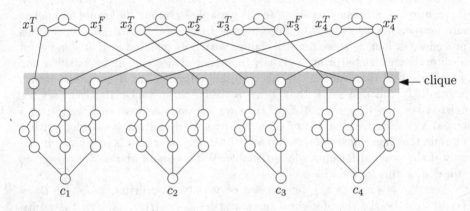

Fig. 3. Top: A variable gadget and a literal gadget. Bottom: The graph $\mathcal{G}_\phi$ for the formula $\phi := (x_1 \vee \neg x_2 \vee x_3) \wedge (x_4 \vee \neg x_1 \vee x_2) \wedge (\neg x_2 \vee \neg x_4) \wedge (\neg x_3 \vee \neg x_2 \vee x_4)$; edges in the clique on the vertices l_i^1 are omitted for clarity.

Proof. First suppose ϕ has a satisfying assignment. We can construct a dominating set D of size $n + t$ for $\mathcal{G}_\phi$ as follows. For each variable $x_j \in \mathcal{X}$ we add x_j^T to D if $x_j = \text{TRUE}$, and we add x_j^F to D if $x_j = \text{FALSE}$. In addition, for each literal $l_i \in \mathcal{L}$ we add l_i^2 to D if $l_i = \text{FALSE}$, and we add l_i^3 to D if $l_i = \text{TRUE}$. The resulting set D has size $n + t$ since we have n variables and t literal occurrences. To see that D is a dominating set, observe that any vertex in a variable gadget is dominated by the vertex selected from that gadget, and observe that the vertices l_i^2, l_i^3 and the ear in the gadget for literal l_i are all dominated by the vertex selected from that gadget. Vertex l_i^1 is also dominated, namely by a vertex in a variable gadget (if $l_i = \text{TRUE}$) or by l_i^2 (if $l_i = \text{FALSE}$). Finally, the vertex for any clause c_k must be dominated because at least one literal l_i in c_k is true, meaning that $l_i^3 \in D$.

Conversely, let D be an arbitrary dominating set of size $n + t$. The $n + t$ ears of $\mathcal{G}_\phi$ have disjoint closed neighborhoods, so to dominate all ears the set D must contain exactly one vertex of each of these closed neighborhoods. Moreover, if D contains an ear v, then we can exchange v for one of its neighbors w; the resulting set will still be dominating as w dominates a superset of the vertices dominated by v. Consequently, we may assume that D contains exactly one vertex from $\{x_j^T, x_j^F\}$ for each $1 \leqslant j \leqslant n$, and exactly one vertex from $\{l_i^2, l_i^3\}$ for each $1 \leqslant i \leqslant t$. We now set $x_j :=$ TRUE if and only if $x_j^T \in D$, for all $1 \leqslant j \leqslant n$. We claim that this assignment satisfies ϕ. To see this, consider a clause c_k. Since the vertex labeled c_k must be dominated, there is at least one literal l_i occurring in c_k such that $l_i^3 \in D$. Suppose that $l_i = x_j$; the argument for $l_i = \neg x_j$ is similar. Since $l_i^3 \in D$ we have $l_i^2 \notin D$, and since the vertex l_i^1 must be dominated this implies $x_j^T \in D$. Hence, x_j has been set to TRUE, thus satisfying c_k. □

Step 2: Realizing $\mathcal{G}_\phi$ as a Ball Graph. Next we show that the graph $\mathcal{G}_\phi$ obtained in Step 1 can be realized as a ball graph in $\mathbb{R}^3$. We need the following lemma.

Lemma 2. *Let ℓ_1 be the x-axis and let ℓ_2 be the line $(0, 0, h) + \lambda(0, 1, 0)$, for some $h \in \mathbb{R}$. Let $p_1 := (x, 0, 0)$ and $p_2 := (0, y, h)$ be arbitrary points on ℓ_1 and ℓ_2, respectively. Then there is a unique ball that touches ℓ_1 at p_1 and ℓ_2 at p_2. The center of this ball is (x, y, z_{xy}) and its radius is $r_{xy} := \sqrt{y^2 + z_{xy}^2}$, where $z_{xy} := \frac{h^2 + x^2 - y^2}{2h}$.*

Proof. Note that the center of a ball touching ℓ_1 at p_1 and ℓ_2 at p_2 must lie on each of the following planes:

  h_1: the plane through p_1 perpendicular to ℓ_1
  h_2: the plane through p_2 perpendicular to ℓ_2
  h_3: the perpendicular bisector plane of p_1 and p_2.

Since ℓ_1 and ℓ_2 are skew lines—they do not intersect and are not parallel—these three planes are pairwise non-parallel. Hence, they have a unique intersection point p, and the ball with center p and radius $|pp_1|$ has the desired properties. Verifying that the coordinates of the center and the radius of this ball are as claimed, is a routine computation which we omit. □

Next we show how to realize the graph $\mathcal{G}_\phi$ as a ball graph in $\mathbb{R}^3$. Without loss of generality, assume that the literal occurrences and clauses in the formula ϕ are indexed left to right, so that for example c_1 is the first clause and l_1 is the first literal in c_1, and c_k is the last clause and l_t is the last literal in c_k.

Let the line ℓ_1 be the x-axis, and let ℓ_2 be the line $(0, 0, h) + \lambda(0, 1, 0)$, where the height h will be defined later. The idea is that the balls for the vertices x_j^T and x_j^F will be touching ℓ_2 from above, the balls for the vertices l_i^2 will be touching ℓ_1 from below, and that Lemma 2 will then allow us to place the balls for l_i^1 such that the correct connections are realized. Next we describe in detail how the balls representing the vertices of $\mathcal{G}_\phi$ are placed. We denote the ball representing a node v by ball(v), so for instance ball(c_k) denotes the ball representing the vertex in the gadget for clause c_k.

 – The vertices labeled x_j^T and x_j^F of the variable gadget for x_j will be represented by balls ball(x_j^T) and ball(x_j^F) with centers $(0, 3j, h + \frac{1}{2})$ and $(0, 3j+1, h+\frac{1}{2})$, respectively. These balls have radius $\frac{1}{2}$, so that they touch ℓ_2 at the points $(0, 3j, h)$ and $(0, 3j+1, h)$. Note that ball(x_j^T) and ball(x_j^F) touch

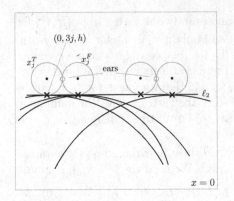 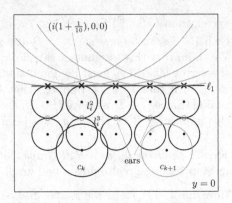

Fig. 4. Cross-section of the construction in the plane $x = 0$ (left) and $y = 0$ (right). Note that the large balls l_i^1 have their centers outside these planes. On the left, we have a literal $\neg x_j$ that occurs twice.

each other at the point $(0, 3j + \frac{1}{2}, h + \frac{1}{2})$. The ear of the gadget for x_j is a ball with that touching point as center, and with a sufficiently small radius, say $\frac{1}{20}$; see Fig. 4(left). Observe that the intersection graph of these three balls is a triangle, as required.

- For a clause c_k, the center of ball(c_k) is $(i^*(1 + \frac{1}{10}), 0, -2)$ and its radius is $9/10$, where value of i^* depends on whether c_k has two or three literals. In the former case, $i^* = i + \frac{1}{2}$, where i is the index such that c_k contains literals l_i and l_{i+1}, in the latter case $i^* = i$, where i is the index such that c_k contains literals l_{i-1}, l_i, and l_{i+1}.

- The balls representing the vertices in the gadget for literal l_i are defined as follows. The balls ball(l_i^2) and ball(l_i^3) have radius $\frac{1}{2}$ and centers $(i(1 + \frac{1}{10}), 0, -\frac{1}{2})$ and $(i(1 + \frac{1}{10}), 0, -\frac{3}{2})$, respectively. Note that if c_k is the clause containing literal l_i, then ball(l_i^3) intersects ball(c_k); see Fig. 4(right). Also note that the balls ball(l_i^2) and ball(l_i^3) touch at the point $(i(1 + \frac{1}{10}), 0, -1)$. The ball representing the ear of the gadget will be centered at that touching point and have radius $\frac{1}{20}$. Thus the intersection graph of ball(l_i^2), ball(l_i^3) and the ball representing the ear is a triangle, as required.

The crucial part of the construction is the placement of ball(l_i^1), which is done as follows. We require ball(l_i^1) to touch ℓ_1 at the point $(i(1 + \frac{1}{10}), 0, 0)$. Note that ball($l_i^2$) also touches ℓ_1 at this point, thus we have realized the edge (l_i^1, l_i^2) in $\mathcal{G}_\phi$. We also require ball(l_i^2) to touch the line ℓ_2. The touching point depends on whether l_i is a positive or a negative literal: if $l_i = x_j$ then we require ball(l_i^1) to touch ℓ_2 at $(0, 3j, h)$, and if $l_i = \neg x_j$ then we require ball(l_i^1) to touch ℓ_2 at $(0, 3j+1, h)$. In the former case it will intersect ball(x_j^T) and in the latter case it will intersect ball(x_j^F), as required. By Lemma 2 there is a unique ball touching ℓ_1 and ℓ_2 as just specified.

Let S_ϕ be the collection of balls just defined. The next lemma states that our construction realizes the graph $\mathcal{G}_\phi$ if we pick the value h specifying the x-coordinate of ℓ_2 appropriately.

Lemma 3. *Let $h := 20N^2$, where $N := \max(3n + 1, (1 + \frac{1}{10})t)$. Then the intersection graph $\mathcal{G}[S_\phi]$ induced by S_ϕ is the graph $\mathcal{G}_\phi$.*

Proof. We already argued above that most edges of $\mathcal{G}_\phi$ are also present in $\mathcal{G}[S_\phi]$, that is, if (u, v) is an edge in $\mathcal{G}_\phi$ then $\mathrm{ball}(u)$ intersects $\mathrm{ball}(v)$. The only exception are the edges forming the clique on the set $\{l_i^1 : 1 \leqslant i \leqslant t\}$. Thus it remains to show that these edges are present as well,[1] and that no spurious edges are present in $\mathcal{G}[S_\phi]$. It is straightforward to verify that any spurious edge that might arise must involve a ball $\mathrm{ball}(l_i^1)$.

Consider a ball $\mathrm{ball}(l_i^1)$. Recall that $\mathrm{ball}(l_i^1)$ touches ℓ_1 at the point $(i(1 + \frac{1}{10}), 0, 0)$, and ℓ_1 at $(0, 3j, h)$ or $(0, 3j + 1, h)$ for some $1 \leqslant j \leqslant n$ (depending on whether l_i is a positive or negative literal). Note that the first two coordinates of these touching points are in the interval $[0, N]$, and that we set $h := 20N^2$. Recall from Lemma 2 that the ball touching ℓ_1 at $(x, 0, 0)$ and ℓ_2 at $(0, y, h)$ has center (x, y, z_{xy}) and radius is $r_{xy} := \sqrt{y^2 + z_{xy}^2}$, where $z_{xy} := \frac{h^2 + x^2 - y^2}{2h}$. Thus for the z-coordinate of the center of $\mathrm{ball}(l_i^1)$ we have

$$z_{xy} = \frac{h^2 + x^2 - y^2}{2h} \in \left[\frac{h^2 - N^2}{2h}, \frac{h^2 + N^2}{2h} \right] = \left[\frac{h}{2} - \frac{1}{40}, \frac{h}{2} + \frac{1}{40} \right],$$

and for the radius of $\mathrm{ball}(l_i^1)$ we have

$$\mathrm{radius}(\mathrm{ball}(l_i^1)) = \sqrt{y^2 + z_{xy}^2} \leqslant \sqrt{N^2 + \left(\frac{h}{2} + \frac{1}{40} \right)^2}$$

$$= \sqrt{\frac{h}{20} + \left(\frac{h}{2} \right)^2 + \frac{h}{40} + \frac{1}{1600}}$$

$$< \frac{h}{2} + \frac{1}{10}.$$

Consequently, the ball is disjoint from both of the open halfspaces $z > h + 1/5$ and $z < -1/5$. These open half-spaces contain all the balls corresponding to ears and all balls $\mathrm{ball}(c_k)$ and $\mathrm{ball}(l_i^3)$. Now consider two balls $\mathrm{ball}(l_i^1)$ and $\mathrm{ball}(l_{i'}^1)$. The distance between the centers of these balls, (x, y, z_{xy}) and $(x', y', z_{x'y'})$, is at most

$$(x - x')^2 + (y - y')^2 + (z_{xy} - z_{x'y'})^2 \leqslant 2N^2 + \left(\frac{1}{20} \right)^2 < h/3 < r_{xy} + r_{x'y'}.$$

---

[1] Actually, it turns out that one can also argue that the existence of these edges is not needed for the reduction to work. We prefer to work with the specific graph $\mathcal{G}_\phi$ defined earlier, and therefore need to show that our geometric representation includes all edges in the clique.

Therefore, the set $\{\text{ball}(l_i^1) : 1 \leqslant i \leqslant t\}$ induces a clique, as desired.

It remains to be shown that $\text{ball}(l_i^1)$ is disjoint from any ball $\text{ball}(l_{i'}^2)$ for $i' \neq i$, and from any ball $\text{ball}(x_j^T)$ and $\text{ball}(x_j^F)$ except for the ball touching ℓ_2 at the same point as $\text{ball}(l_i^1)$ touches ℓ_2. Let (x, y, z_{xy}) be the center of $\text{ball}(l_i^1)$ and consider the distance of (x, y, z_{xy}) to a point $(x', 0, -\frac{1}{2})$, where $|x - x'| \geqslant 1$. (The argument is analogous for the distance of (x, y, z_{xy}) to $(0, y', h + \frac{1}{2})$, where $|y - y'| \geqslant 1$.) It is sufficient to show that this distance is larger than the sum of the ball radii, that is, we need that $\text{dist}((x, y, z_{xy}), (x', 0, -\frac{1}{2})) > r_{xy} + \frac{1}{2}$. Taking the squared distance instead, we need to show that

$$(x' - x)^2 + y^2 + \left(z_{xy} + \frac{1}{2}\right)^2 > \left(\sqrt{y^2 + z_{xy}^2} + \frac{1}{2}\right)^2,$$

which is equivalent to

$$(x' - x)^2 + z_{xy} > \sqrt{y^2 + z_{xy}^2}.$$

Since $y \leqslant N$, we have that $y^2 \leqslant N^2 = \frac{h}{20} < \frac{z_{xy}}{8}$, so we can bound the left hand side as:

$$\sqrt{y^2 + z_{xy}^2} < \sqrt{\frac{z_{xy}}{8} + z_{xy}^2} < z_{xy} + \frac{1}{4} < z_{xy} + (x' - x)^2,$$

where the last inequality follows since x and x' are distinct integers, therefore each $\text{ball}(l_i^1)$ has the correct intersections.

Finally, although the above construction uses some irrational ball radii, it is sufficient to use rational radii of precision $O(1/n)$, which results in the same intersection graph. □

Putting it all together we obtain the main result of this section.

Theorem 2. *There is no $2^{o(n)}$ algorithm for* DOMINATING SET *in ball graphs, unless the Exponential-Time Hypothesis fails.*

4   A Lower Bound for WEIGHTED DOMINATING SET in Unit-Ball Graphs

We now turn our attention to WEIGHTED DOMINATING SET for ball graphs in $\mathbb{R}^3$. Here we can even prove lower bounds for unit-ball graphs.

Our reduction is again from $(3, 3)$-SAT. Let ϕ be a $(3, 3)$-SAT formula, which we preprocess so that all clauses have size two or three as in the beginning of the proof of Theorem 2. First, we will create a vertex-weighted graph $\mathcal{G}_\phi$ that has a dominating set of a given size k if and only if ϕ is satisfiable. Then, we show that this graph can be realized as the intersection graph of unit balls in $\mathbb{R}^3$.

Step 1: Construction of $\mathcal{G}_\phi$. Let ϕ have variables $x_1, \ldots, x_n$, literals $l_1, \ldots, l_t$ (as a multiset), and clauses $c_1, \ldots, c_m$. We create a vertex-weighted graph $\mathcal{G}_\phi$ that consists of seven cliques, denoted by $C^1, \ldots, C^7$. The remaining (non-clique) edges of the graph will all be between two consecutive cliques C^p and C^{p+1}. Clique C^1 has n vertices labeled x_j^1, one corresponding to each variable. Clique C^2 has $2n$ vertices, where for each variable we have two vertices corresponding to setting the variable to true or false. These are labeled x_j^{2T} and x_j^{2F}, see Fig. 5. Cliques $C^3, \ldots, C^6$ each contain a single vertex for each literal occurrence l_i, labeled $l_i^3, l_i^4, l_i^5, l_i^6$. Finally, C^7 has a single vertex for each clause c_k ($k = 1, \ldots, m$), labeled as c_k^7. In addition to these, we add a dummy vertex to C^4 and C^6, labeled d^4 and d^6 respectively. Apart from the edges in the cliques, the other edges are defined as follows. For $j = 1 \ldots, n$, we add the edges $x_j^1 x_j^{2T}$ and $x_j^1 x_j^{2F}$. We connect the first literal vertices to the corresponding variable setting, i.e., for each positive literal $l_i = x_j$, we add $x_j^{2T} l_i^3$, and for each negative literal $l_i = \neg x_j$, we add $x_j^{2F} l_i^3$. For each literal l_i we add the edges of the path $l_i^3 l_i^4 l_i^5 l_i^6$. Finally, for $k = 1, \ldots, m$, we connect c_k^7 to l_i^6 if and only if l_i is occurs in c_k. The vertex weights in C^1, C^3, C^5, and C^7 are set to ∞, the dummy vertices get weight[2] 0, and all other vertex weights are set to 1.

Lemma 4. *The graph $\mathcal{G}_\phi$ has a dominating set of weight $n + t$ if and only if ϕ is satisfiable.*

Proof. First suppose that ϕ is satisfiable. We show how to create a dominating set D of weight $n + t$ from a satisfying assignment. For each true variable x_j, we add the vertex x_j^{2T} to D; for each false variable x_j, we add x_j^{2F} instead. For true literals, we add l_i^6 to D; for false literals we add l_i^4 to D. Finally, we add the two dummies. Note that the weight of D is $n + t$. To see that every vertex is dominated, first observe that vertices in C^1 are dominated since exactly one of their neighbors is in $D \cap C^2$, and vertices in C^7 are dominated since at least one of their literals is true, i.e., the corresponding vertex in C^6 is in D. Furthermore, there is at least one vertex selected in C^2, so C^2 is dominated as well. The dummies ensure that C^4 and C^6 are dominated. Finally, all other vertices are on some path $x_j^2 l_i^3 l_i^4 l_i^5 l_i^6$, where either x_j^2 and l_i^6 are selected (for true literals) or l_i^4 is selected (for false literals), which means that C^3 and C^5 are dominated as well.

Now suppose $\mathcal{G}_\phi$ has a dominating set D of weight $n + t$. Based on D, we show how to construct a satisfying assignment of ϕ. Note that D cannot contain vertices of infinite weight. Consequently, all vertices in C^1, C^3, C^5, and C^7 must be dominated from outside. Note that the closed neighborhood of each vertex in $C^1 \cup C^5$ consists of their own clique and two other vertices that we can call potential dominators: indeed, x_j^1 can only be dominated by x_j^{2T} or x_j^{2F}, and l_i^5 can only be dominated by l_i^4 or l_i^6. Since there are $n + t$ vertices in $C^1 \cup C^5$, and all their potential dominators are disjoint sets, it follows that D consists of exactly one vertex from each pair $\{x_j^{2T}, x_j^{2F}\} \, j = 1 \ldots n$ and $\{l_i^4, l_i^6\} \, i = 1 \ldots t$.

---

[2] Alternatively, one could give the dummies a weight of 1, and add a unique neighbor to each of them with weight ∞, which ensures that they are contained in all finite weight dominating sets.

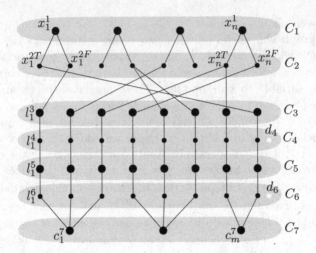

Fig. 5. The weighted graph $\mathcal{G}_\phi$. Vertices in each group C_i form a clique. Large nodes have infinite weight, the dummies d_4 and d_6 have weight 0, and all other nodes have weight 1.

Consider the assignment where x_i is set to true if and only if $x_j^{2T} \in D$. Suppose for the purpose of contradiction that a clause c_k is not satisfied. Let i be the index of a literal occurrence for which $l_i^6 \in D$; such an index exists since c_k^7 can only be dominated by a vertex from C^6. Consequently, $l_i^4 \notin D$, so in order to dominate l_i^3, it must be the case that its potential dominator in C^2 is in D. That is, if $l_i = x_j$, then $x_j^{2T} \in D$, and if $l_i = \neg x_j$, then $x_j^{2F} \in D$. But this would mean that c_k is satisfied by this literal in our assignment, which is a contradiction. $\square$

As in the proof of Theorem 2, the size of $V(\mathcal{G}_\phi)$ is $O(n)$, so it suffices to show that $\mathcal{G}_\phi$ can be realized as a unit ball graph. (For simplicity we used infinite weights in the construction, but we can use weights that are polynomial in the size of the instance as well.)

Step 2: Realizing $\mathcal{G}_\phi$ as a Unit Ball Graph.

Lemma 5. *The graph $\mathcal{G}_\phi$ can be realized as a unit ball graph.*

Proof. It will be convenient to use cylindrical coordinates: the point (x, y, z) becomes (r, ϕ, z), where $r = \sqrt{x^2 + y^2}$ and $\phi = \arctan(y/x)$. We will first define the ball centers for cliques C^2 through C^6, and then for the cliques C^1 and C^7. The balls in our construction will all have diameter 1 (rather than radius 1), so that two balls intersect if and only if their distance is at most 1.

Let $\varepsilon > 0$ be a small number that will be determined later. We place the centers of the balls for C^2 in the natural order close to each other on the z axis. More precisely, we assign the (cylindrical) coordinates $(0, 0, (2j - 1)\varepsilon)$ to the center of ball(x_j^{2T}) and $(0, 0, 2j\varepsilon)$ to the center of ball(x_j^{2F}). Without loss of generality, assume that the literal occurrences and clauses are indexed left

 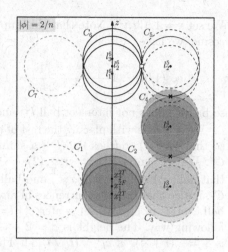

Fig. 6. Left: cliques C_2, C_3 and C_4 with blue, yellow and gray colors. The ball centers of C_2 are very close to each other so alternating shades of blue are used to distinguish the balls. Right: the intersection of the first few balls from each clique with the plane $|\phi| = 2/n$. Balls from different cliques have two touching points (indicated by squares) and two non-empty intersections (indicated by crosses) within this plane. The smaller dashed disks correspond to balls whose center is outside this plane. (Color figure online)

to right, so that l_1 is the first literal in c_1, and l_t is the last literal in the last clause c_k. We now define the centers of the balls for C^3. If $l_i = x_j$, then the center of ball(l_i^3) is center(l_i^3) := $(1, i/n, (2j - 1)\varepsilon)$; if $l_i = \neg x_j$, then we set center(l_i^3) := $(1, i/n, 2j\varepsilon)$. Observe that all balls of C^3 touch the cylinder $r = \frac{1}{2}$ at a single point, which is also a touching point with the correct ball from C^2, see Fig. 6. (Note that the picture overestimates the distances within C_2; with the correct placement the balls are indistinguishable to the naked eye).

Next, we set center(l_i^4) := $(1, i/n, 1)$ and center(l_i^5) := $(1, i/n, 2 - i\varepsilon)$. By choosing ε small enough, we can ensure that each ball ball(l_i^4) intersects only ball(l_i^3) from C^3 and ball(l_i^5) from C^5. To see this, note that the squared distance of a center $(1, \phi, z)$ to a center $(1, \phi', z')$ is

$$(1 - \cos(\phi - \phi'))^2 + \sin^2(\phi - \phi') + (z - z') = 2 - 2\cos(\phi - \phi') + (z - z')^2.$$

If $\phi = \phi'$ and $z - z' < 1$, then this distance is less than 1, so that ball(l_i^4) does intersect ball(l_i^3) and ball(l_i^5). Otherwise, the distance of center(l_i^4) to any of the centers in $C^3 \cup C^5$ is minimized if $\phi - \phi' = 1/n$ and $z - z' = 1 - t\varepsilon$. To avoid unwanted intersections, we need the following:

$$2 - 2\cos(1/n) + (1 - t\varepsilon)^2 > 1$$

which is equivalent to

$$1 - t\varepsilon + \frac{t^2\varepsilon^2}{2} > \cos(1/n).$$

We set $\varepsilon = 1/(3tn^2)$. Note that this implies $\varepsilon = \Theta(1/n^3)$. By the power series of cosine, we have

$$\cos(1/n) < 1 - \frac{1}{2n^2} + \frac{1}{24n^4} < 1 - \frac{1}{3n^2} = 1 - t\varepsilon < 1 - t\varepsilon + \frac{t^2\varepsilon^2}{2},$$

so ball(l_i^4) does not intersect ball(l_j^3) and ball(l_j^5) for any $i \neq j$.

We continue with placing the rest of the centers. Let center(l_i^6) := $(0, 0, 2 - i\varepsilon)$ so that ball(l_i^6) touches ball(l_i^5); a similar structure was used between C^3 and C^2. We set center(x_j^1) := $(r, \pi, (2j - \frac{3}{2})\varepsilon)$, where $r = \sqrt{1 - \varepsilon^2/4}$ is chosen so that ball(x_j^1) touches ball(x_j^{2T}) and ball(x_j^{2F}). Moving on to the placements for C_7, notice that literal occurrences of any given clause correspond to consecutive ball centers in C_6. We set center(c_k^7) := (r, π, z), where r and z are defined the following way. The height is $z = 2 - i\varepsilon$ if l_i is the middle literal of c_k, and it is $z = -(i + \frac{1}{2})\varepsilon$ if $c_k = (l_i \vee l_{i+1})$. The radius r is chosen to ensure that the ball of c_k touches the ball of its first and last literal: we set $r = \sqrt{1 - \varepsilon^2/4}$ and $r = \sqrt{1 - \varepsilon^2}$ for clauses of size two and three respectively. To conclude the construction, we place balls to represent the dummy vertices d_4 and d_6 at center(d_4) := $(1.5, 0.5, 1)$ and center(d_6) := $(0, 0, 2.5)$. It is routine to check that this correctly realizes $\mathcal{G}_\phi$.

Finally, although the above construction uses some irrational coordinates, one can use rational coordinates of precision $O(1/n^4)$.   □

The above construction and the lemmas imply the following theorem.

Theorem 3. *There is no $2^{o(n)}$ algorithm for* WEIGHTED DOMINATING SET *in unit ball graphs, unless the Exponential-Time Hypothesis fails.*

5   Concluding Remarks

De Berg *et al.* [2] recently presented a framework to solve many classic graph problems in subexponential time on intersections graphs of similarly-sized fat objects. We have shown that extending the framework in its full generality to arbitrary (non-similarly-sized) fat objects is impossible. More precisely, we have shown that DOMINATING SET, one of the problems that the framework for similarly-sized fat objects can handle, does not admit a subexponential algorithm on arbitrary ball graphs in $\mathbb{R}^3$, assuming ETH. Thus it seems that obtaining subexponential algorithms for arbitrary fat objects, if possible, will require rather problem-specific arguments. Similarly, obtaining subexponential algorithms for weighted problems on similarly-sized objects is not always possible: WEIGHTED DOMINATING SET does not admit a subexponential algorithm on unit-ball graphs in $\mathbb{R}^3$, assuming ETH. For arbitrary fat objects (instead of just balls), the situation is worse: even in $\mathbb{R}^2$ the problems DOMINATING SET, CONNECTED DOMINATING SET, and STEINER TREE do not admit subexponential algorithms, assuming ETH. On the positive side, we argued that if the solution to the problem at hand (or its complement) can contain at most a constant number of vertices from a clique, then the technique from De Berg *et al.* [2] may be applicable; this is for example the case for CONNECTED VERTEX COVER, FEEDBACK VERTEX SET, and CONNECTED FEEDBACK VERTEX SET.

Several questions are left open by our study. First and foremost, although we have ruled out a subexponential algorithm for DOMINATING SET in $\mathbb{R}^2$ for arbitrary fat objects, the complexity of DOMINATING SET on disk graphs remains open. Here, related work on GEOMETRIC SET COVER [8,18] leads us to believe that subexponential algorithms may be attainable for disks, possibly even for the weighted version.

Conjecture 1. There is a $2^{\tilde{O}(\sqrt{n})}$ algorithm for (WEIGHTED) DOMINATING SET in disk graphs.

A prominent problem studied in [2] that we did not tackle here is HAMILTONIAN CYCLE. Is there a subexponential algorithm for HAMILTONIAN CYCLE in disk graphs? What about ball graphs? Finally, the weighted versions of almost all of these problems remain open in $\mathbb{R}^2$, as long as we have similarly sized objects. For example, what is the complexity of vertex weighted STEINER TREE in unit-disk graphs?

References

1. de Berg, M., Bodlaender, H.L., Kisfaludi-Bak, S., Kolay, S.: An ETH-tight exact algorithm for Euclidean TSP. In: Proceedings of the 59th IEEE Symposium Foundations Computer Science (FOCS), pp. 450–461 (2018)
2. de Berg, M., Bodlaender, H.L., Kisfaludi-Bak, S., Marx, D., van der Zanden, T.C.: A framework for ETH-tight algorithms and lower bounds in geometric intersection graphs. In: Proceedings of the 50th ACM Symposium Theory Computer (STOC), pp. 574–586 (2018)
3. de Berg, M., Kisfaludi-Bak, S., Woeginger, G.: The complexity of dominating set in geometric intersection graphs. Theor. Comput. Sci. **769**, 18–31 (2019)
4. Bodlaender, H.L., Cygan, M., Kratsch, S., Nederlof, J.: Deterministic single exponential time algorithms for connectivity problems parameterized by treewidth. Inf. Comput. **243**, 86–111 (2015)
5. Bodlaender, H.L., Drange, P.G., Dregi, M.S., Fomin, F.V., Lokshtanov, D., Pilipczuk, M.: A $c^k n$ 5-approximation algorithm for treewidth. SIAM J. Comput. **45**(2), 317–378 (2016)
6. van den Eijkhof, F., Bodlaender, H.L., Koster, A.M.C.A.: Safe reduction rules for weighted treewidth. Algorithmica **47**(2), 139–158 (2007)
7. Breu, H., Kirkpatrick, D.G.: Unit disk graph recognition is NP-hard. Comput. Geom. Theory Appl. **9**, 3–24 (1998)
8. Bringmann, K., Kisfaludi-Bak, S., Pilipczuk, M., van Leeuwen, E.J.: On geometric set cover for orthants. In: Proceedings of the 27th European Symposium on Algorithms (ESA), pp. 26:1–26:18 (2019)
9. Cygan, M., et al.: Parameterized Algorithms. Springer, Heidelberg (2015). https://doi.org/10.1007/978-3-319-21275-3
10. Impagliazzo, R., Paturi, R.: On the complexity of k-SAT. J. Comput. Syst. Sci. **62**(2), 367–375 (2001)

11. Kang, R.J., Müller, T.: Sphere and dot product representations of graphs. Discret. Comput. Geom. **47**, 548–568 (2012)
12. Kisfaludi-Bak, S.: ETH-tight algorithms for geometric network problems. Ph.D. thesis, Technische Universiteit Eidnhoven (2019)
13. Kisfaludi-Bak, S., Marx, D., van der Zanden, T.C.: How does object fatness impact the complexity of packing in d dimensions? In: Proceedings of the 30th International Symposium on Algorithms and Computation (ISAAC) (2019, to appear)
14. Koebe, P.: Kontaktprobleme der konformen Abbildung (1936). Hirzel
15. Lipton, R.J., Tarjan, R.E.: A separator theorem for planar graphs. SIAM J. App. Math. **36**(2), 177–189 (1979)
16. Lipton, R.J., Tarjan, R.E.: Applications of a planar separator theorem. SIAM J. Comput. **9**(3), 615–627 (1980)
17. Marx, D.: The square root phenomenon in planar graphs. In: Proceedings of the 40th International Colloquium on Automata, Languages, and Programming (ICALP), part II, p. 28 (2013)
18. Marx, D., Pilipczuk, M.: Optimal parameterized algorithms for planar facility location problems using Voronoi diagrams. In: Bansal, N., Finocchi, I. (eds.) ESA 2015. LNCS, vol. 9294, pp. 865–877. Springer, Heidelberg (2015). https://doi.org/10.1007/978-3-662-48350-3_72

As Time Goes By: Reflections on Treewidth for Temporal Graphs

Till Fluschnik[iD], Hendrik Molter[iD], Rolf Niedermeier[iD], Malte Renken[(✉)][iD], and Philipp Zschoche[iD]

Faculty IV, Algorithmics and Computational Complexity,
Technische Universität Berlin, Berlin, Germany
{till.fluschnik,h.molter,rolf.niedermeier,
m.renken,zschoche}@tu-berlin.de

Abstract. Treewidth is arguably the most important structural graph parameter leading to algorithmically beneficial graph decompositions. Triggered by a strongly growing interest in temporal networks (graphs where edge sets change over time), we discuss fresh algorithmic views on *temporal* tree decompositions and *temporal* treewidth. We review and explain some of the recent work together with some encountered pitfalls, and we point out challenges for future research.

Keywords: Network science · Time-evolving network · Link stream · NP-hardness · Parameterized complexity · Tree decomposition · Monadic second-order logic (MSO)

1 Introduction

»*You must remember this:*« treewidth is one of the most important structural graph parameters [10], being extremely popular in parameterized algorithmics. Without the contributions of Hans Bodlaender, this would be much less so.

Intuitively, the fundamental observation behind treewidth is that many NP-hard graph problems turn easy when restricted to trees. Indeed, typically a simple bottom-up greedy algorithm from the leaves to the (arbitrarily chosen) root of the tree suffices to solve many fundamental graph problems (including VERTEX COVER and DOMINATING SET) efficiently on trees. This naturally leads to the investigation on how "tree-likeness" of graphs helps to solve problems efficiently. Fruitful results on this are provided by the concept of a tree decomposition and, correspondingly, the treewidth of a graph: if the treewidth is small, then otherwise NP-hard problems can be solved "fast". Notably, the concepts of tree decomposition and treewidth are tightly connected to the existence of small

---

Dedicated to Hans L. Bodlaender on the occasion of his 60th birthday.

The inclined reader, besides hopefully discovering interesting science, is also invited to enjoy a few quotes from a famous movie scattered around our text; the paper title is partially taken from the theme song of this movie.

© Springer Nature Switzerland AG 2020
F. V. Fomin et al. (Eds.): Bodlaender Festschrift, LNCS 12160, pp. 49–77, 2020.
https://doi.org/10.1007/978-3-030-42071-0_6

graph separators (that is, vertex sets whose deletion partitions the graph into at least two connected components) that are arranged in a tree-like structure (see Sect. 2 for formalities and an example). It is fair to say that tree decompositions currently are the most popular structural graph decompositions used in (parameterized) algorithms for (NP-hard) problems on (static) graphs. More specifically, these algorithmic results typically are "fixed-parameter tractability" results with respect to the parameter, that is, the studied problems then can be solved by an exponential-time algorithm whose exponential part exclusively depends on the treewidth of the input graph.

Computing the treewidth of a graph is NP-hard [4], even on graphs of maximum degree nine [20], but linear-time solvable if the treewidth is some fixed constant [11,15] (more specifically, the running time is $c^{k^3} n$ [11]) and 5-approximable in $(c')^k n$ time [13] on n-vertex graphs of treewidth k, for some constants $c, c' > 1$. Polynomial-time algorithms are known for several restricted graph classes [16,17,19,22]. From an algorithmic point of view, a tree decomposition of a graph typically allows for a dynamic programming approach. The twist is that these dynamic programs for many NP-hard problems run in polynomial time when the width of the tree decomposition is constant [12]. Indeed, many NP-hard problems are known to be fixed-parameter tractable when parameterized by treewidth [27,33], underpinning the reputation of treewidth as one of the most fundamental algorithmically exploitable graph parameters, as confirmed in experimental studies [14,18] (practically useful implementations for computing tree decompositions are also available [30,31]). In this work, our goal is to discuss the role treewidth currently plays in the strongly emerging field of *temporal graphs*[1]; these are graphs where the edge set may change over time. Further, we reflect on possible definitions of a temporal version of treewidth: temporal treewidth.

Temporal graphs model networks where adjacencies of vertices change over discrete time steps. In fact, many natural time-dependent networks can be modeled by temporal graphs, for instance interaction/contact networks, connection/availability networks, or bio-physical networks (see, e.g., [52, Section II]). Applications range from epidemiology over sociology to transportation. In the last decade, problems on temporal graphs gained increased attention in theoretical computer science [2,3,6,7,21,23,25,38–40,42,45–47,50,58,60,64,65].

Formally, a temporal graph $\mathcal{G} = (V, \mathcal{E}, \tau)$ consists of a vertex set V, a lifetime τ, and a set $\mathcal{E} \subseteq \binom{V}{2} \times \{1, \dots, \tau\}$ of temporal edges (that is, an edge is additionally equipped with a *time stamp*). Alternatively, a temporal graph on vertex set V and lifetime τ can also be defined as (see Fig. 1 for exemplary illustrations).

(a) a static graph G equipped with a function $\lambda \colon E(G) \to 2^{\{1,\dots,\tau\}}$ (Fig. 1(a)),
(b) a tuple $(V, E_1, \dots, E_\tau)$ (Fig. 1(b)), or
(c) a sequence of τ static graphs (called *layers*) $G_1 = (V, E_1), \dots, G_\tau = (V, E_\tau)$ (Fig. 1(c)).

---

[1] Also known as time-varying graphs, evolving graphs, link streams, or dynamic graphs where no changes on the vertex set are allowed.

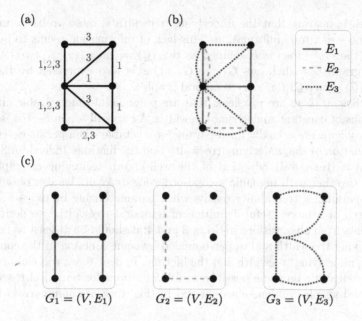

Fig. 1. Three different illustrations (a)–(c) (according to alternative definitions (a)–(c)) of a temporal graph $\mathcal{G}$ with six vertices, fourteen temporal edges, and lifetime three. If the time stamps are dropped in (a), the underlying graph is depicted.

The *underlying* (static) graph of a temporal graph $\mathcal{G} = (V, E_1, \ldots, E_\tau)$ is the graph $G_\downarrow(\mathcal{G}) = (V, E_1 \cup \cdots \cup E_\tau)$. If, for instance, contact networks are modeled, then the underlying graph gives complete information on which contacts appeared; however, it gives no information on *when* and *how often* they appeared.

The addition of temporality to the graph model significantly increases the computational complexity of many basic graph problems. For instance, consider the following: given a graph G with a designated vertex s, decide whether there is a walk in G that starts at s and reaches all vertices. (One may think of s as the starting location of a traveling salesman who wants to visit every vertex at least once.) This problem is well-known to be linear-time solvable on static graphs: a solution exists if and only if G is connected. On temporal graphs, the question is to decide whether all vertices can be reached from s by a single so-called *strict temporal walk*, that is, a walk whose sequence of temporal edges has strictly increasing time stamps. (Intuitively, strict temporal walks model the traversal of the graph at finite speed.) This problem is known as TEMPORAL EXPLORATION and proven to be NP-hard [2, 21].

When attempting to adapt the notion of treewidth to temporal graphs, one might consider the *underlying treewidth* $\mathrm{tw}_\downarrow(\mathcal{G}) = \mathrm{tw}(G_\downarrow(\mathcal{G}))$. Inherited from the loss of information in the underlying graph, this notion captures no information about time and occurrences of temporal edges. As we will see in Sect. 3, many temporal graph problems remain NP-hard even when the underlying treewidth is constant, indicating that the underlying treewidth is most probably missing useful "time-structural" information. This lack of information seems to be even larger for the *layer treewidth*, defined as $\mathrm{tw}_\infty(\mathcal{G}) := \max_{i \in \{1,\dots,\tau\}} \mathrm{tw}(G_i)$ for a temporal graph $\mathcal{G}$ with layers $G_1, \dots, G_\tau$. This is also expressed by the fact that $\mathrm{tw}_\infty(\mathcal{G}) \leq \mathrm{tw}_\downarrow(\mathcal{G})$ for every temporal graph $\mathcal{G}$.

Nevertheless, there are problems that are polynomial-time solvable on temporal graphs of constant underlying treewidth. As we will see in Sect. 4, several of these problems are actually fixed-parameter tractable when parameterized by the combination of the underlying treewidth and the lifetime. Indeed, in Sect. 5, we present a (temporal) adaption of the well-known technique of employing treewidth together with monadic second-order logic. Again, we can obtain several fixed-parameter tractability results when parameterizing by $\mathrm{tw}_\downarrow + \tau$.

In search of a more useful definition of temporal treewidth, we derive two requirements from observations in Sects. 3 and 4: it should be at least as large as the underlying treewidth and upper-bounded by some function in the combination of the underlying treewidth and the lifetime. In Sect. 6, we will elaborate on this while reflecting on some possibly useful definitions for temporal treewidth. All temporal graph problems encountered in this work are summarized in the appendix.

2   Preliminaries

»*But what about us?*« In this section, we provide some basic definitions and facts. By $\mathbb{N}$ and $\mathbb{N}_0$ we denote the natural numbers excluding and including zero, respectively. For any set A, we write $\binom{A}{k}$ for the set of all size-k subsets of A.

2.1   Static Graphs and Treewidth

Let $G = (V, E)$ be a (static) graph with vertex set V and edge set $E \subseteq \binom{V}{2}$. Alternatively, $V(G)$ and $E(G)$ also denote the vertex set and edge set of G, respectively. We write $G[W]$ for the subgraph induced by a set of vertices $W \subseteq V$ and use $G - W$ as a shorthand for $G[V \setminus W]$.

Tree Decompositions and Treewidth. In the following, we define (rooted and nice) tree decompositions and treewidth of static graphs, and we explain the connection to a cops-and-robber game.

Definition 1 (Tree Decomposition, Treewidth). *Let $G = (V, E)$ be an undirected graph. A tuple $\mathbb{T} = (T, \{B_u \mid u \in V(T)\})$ consisting of a tree T and a set of so-called bags $B_u \subseteq V$ is a tree decomposition (tdc) of G if*

(i) $\bigcup_{u \in V(T)} B_u = V$,
(ii) for every $e \in E$ there is a node $u \in V(T)$ such that $e \subseteq B_u$, and
(iii) for every $v \in V$, the graph $T[\{u \in V(T) \mid v \in B_u\}]$ is a tree.

(a)   (b)   (c)

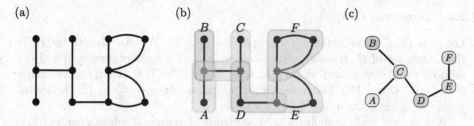

Fig. 2. (a) A graph G of treewidth two with (c) a tree decomposition of G with (b) bags A–F.

The width *of* $\mathbb{T}$ *is* width$(\mathbb{T}) := \max_{u \in V(T)} |B_u| - 1$. *The treewidth of G is the minimum width over all tree decompositions of G, that is,*

$$\text{tw}(G) = \min_{\mathbb{T} \text{ is tdc of } G} \text{width}(\mathbb{T}).$$

It follows from the definition, that for any edge $\{u, u'\}$ of T, the intersection of the corresponding bags $B_u \cap B_{u'}$ is a separator of G of size at most width$(\mathbb{T})$ (as long as $\mathbb{T}$ does not contain redundant bags). Refer to Fig. 2 for an illustrative example.

A tree decomposition $\mathbb{T} = (T, \{B_u \mid u \in V(T)\})$ is *rooted* if there is a designated node $r \in V(T)$ being the root of T (this allows to talk about children, parents, ancestors, descendants, etc. of the nodes of T). A rooted tree decomposition $\mathbb{T} = (T, \{B_u \mid u \in V(T)\})$ is *nice* if each node $u \in V(T)$ is either (i) a leaf node (u has no children), (ii) an introduce node (u has one child v with $B_v \subset B_u$ and $|B_u \setminus B_v| = 1$), (iii) a forget node (u has one child v with $B_v \supset B_u$ and $|B_v \setminus B_u| = 1$), or (iv) a join node (u has two children v, w with $B_v = B_w = B_u$). Given a tree decomposition, one can compute a corresponding nice tree decomposition in linear time [53].

Alternatively, treewidth can be defined through a cops-and-robber game [63] as follows. Let $G = (V, E)$ be an undirected graph, and $k \in \mathbb{N}$.

- At the start, the k cops choose a set $C_0 \in \binom{V}{k}$ of vertices, and then the robber chooses a vertex $r_0 \in V \setminus C_0$.
- In round $i \in \mathbb{N}$, first the cops choose $C_i \in \binom{V}{k}$, and then the robber chooses $r_i \in V \setminus C_i$ such that r_i and r_{i-1} are connected in $G - (C_i \cap C_{i-1})$.

The cops win if, after finitely many rounds, the robber is caught, that is, the robber has no vertex left to choose. The connection to treewidth is the following (which also implies an alternative definition for treewidth).

Lemma 1 ([63]). *Graph G has treewidth at most k if and only if at most $k+1$ cops win the cops-and-robber game.*

The *pathwidth* of a graph G is the minimum width over all tree decomposition $\mathbb{T} = (T, \{B_u \mid u \in V(T)\})$ with T being a path. Note that for every graph its treewidth is at most its pathwidth. For the graph in Fig. 2, the treewidth and the pathwidth are equal (take the union of the bags A and B).

2.2 Temporal Graphs

Let $\mathcal{G} = (V, \mathcal{E}, \tau)$ be a temporal graph (see Sect. 1). We also denote by $V(\mathcal{G})$ the vertex set of $\mathcal{G}$. For any vertex subset $W \subseteq V$, the temporal graph $\mathcal{G}[W]$ induced by W is defined as $(W, \{(e, t) \in \mathcal{E} \mid e \subseteq W\}, \tau)$. Further, we define $\mathcal{G} - W := \mathcal{G}[V \setminus W]$. For a subset of temporal edges $\mathcal{E}' \subseteq \mathcal{E}$, the temporal graph $\mathcal{G} - \mathcal{E}'$ is defined as $(V, \mathcal{E} \setminus \mathcal{E}', \tau)$.

A *temporal walk* is defined as a sequence of temporal edges $(\{v_1, v_2\}, t_1)$, $(\{v_2, v_3\}, t_2), \ldots, (\{v_p, v_{p+1}\}, t_p)$, each contained in $\mathcal{E}$ and $t_1 \le t_2 \le \cdots \le t_p$ (also called a v_1-v_{p+1} temporal walk when the terminals are specified). A temporal walk is called *strict* if $t_1 < t_2 < \cdots < t_p$. A *(strict) temporal path* is a (strict) temporal walk where $v_i \ne v_j$ for all $i \ne j$. A (strict) (α, β)-temporal path is a (strict) temporal path where additionally $\alpha \le t_{i+1} - t_i \le \beta$ holds.

When analyzing problems on temporal graphs, the following concept often comes in handy.

Definition 2 ((Strict) Static Expansion). *The* static expansion *of a temporal graph* $\mathcal{G} = (V, \mathcal{E}, \tau)$ *is a directed graph* $H := (V', A)$, *with vertices* $V' = \{u_{t,j} \mid v_j \in V, t \in \{1, \ldots, \tau\}\}$ *and arcs* $A = A' \cup A_{\mathrm{col}}$, *where the first set* $A' := \{(u_{t,i}, u_{t,i'}), (u_{t,i'}, u_{t,i}) \mid (\{v_i, v_{i'}\}, t) \in \mathcal{E}\}$ *contains the arcs within the layers, and the second set* $A_{\mathrm{col}} := \{(u_{t,j}, u_{t+1,j}) \mid v_j \in V, t \in \{1, \ldots, \tau - 1\}\}$ *contains the arcs connecting different layers. We refer to* A_{col} *as* column-edges *of* H.

A static expansion is called strict *if its vertex set* V' *additionally contains the vertex set* $\{u_{\tau+1,j} \mid v_j \in V\}$ *and its arc set* A' *is replaced by the set* $A'' := \{(u_{t,i}, u_{t+1,i'}), (u_{t,i'}, u_{t+1,i}) \mid (\{v_i, v_{i'}\}, t) \in \mathcal{E}\}$.

Note that (strict) temporal walks correspond exactly to walks within the (strict) static expansion. Moreover, note that *strict* static expansions are always directed *acyclic* graphs.

2.3 Parameterized Complexity

We use standard notation and terminology from parameterized complexity theory [27,33,34,43,61]. A parameterized problem with parameter k is a language $L \subseteq \{(x, k) \in \Sigma^* \times \mathbb{N}\}$ for some finite alphabet Σ. A parameterized problem L is called *fixed-parameter tractable* if every instance (x, k) can be decided for L in $f(k) \cdot |x|^{\mathcal{O}(1)}$ time, where f is some computable function only depending on k. The tool for proving that a parameterized problem is presumably not fixed-parameter tractable is to show that it is hard for the parameterized complexity class W[1]. A parameterized problem L is contained in the complexity class XP if every instance (x, k) can be decided for L in $|x|^{g(k)}$ time, where g is some computable function only depending on k. A parameterized problem L is *para-NP-hard* if the problem is NP-hard for some constant value of the parameter.

Table 1. Overview on treewidth-related results for NP-hard temporal graph problems. The problems are (RETURN-TO-BASE) TEMPORAL GRAPH EXPLORATION ((RTB-)TGE), (α, β)-TEMPORAL REACHABILITY TIME-EDGE DELETION $((\alpha, \beta)$-TRTED), REACHABILITY TEMPORAL ORDERING (RTO), MIN REACHABILITY TEMPORAL MERGING (MRTM), TEMPORAL MATCHING (TM), TEMPORAL SEPARATION (TS), and MINIMUM SINGLE-SOURCE TEMPORAL CONNECTIVITY (r-MTC). See Appendix A for respective problem definitions. "FPT", "p-NP-h", and "?" abbreviate "fixed-parameter tractable", "para-NP-hard", and "open", respectively.

Problem	Parameter					
	tw_∞ $tw_\downarrow$	τ	$\|V\|$	$tw_\downarrow + \tau$	Ref.	
TGE	NP-h for $tw_\downarrow = 2$	FPT[†]	FPT[†]	FPT[†]	[21]	
RTB-TGE	NP-h for $tw_\downarrow = 1$	FPT[†]	FPT[†]	FPT[†]	[2]	
(α, β)-TRTED	NP-h for $tw_\downarrow = 1$	p-NP-h	?	?	[36]	
$\binom{\text{Min-Max}}{\text{Max-Min}}$ RTO	NP-h for $tw_\downarrow = 1$	FPT	?	FPT	[37]	
MRTM	NP-h for $tw_\downarrow = 1$	FPT[†]	?	FPT[†]	[29]	
TM	NP-h for $tw_\downarrow = 1$	p-NP-h	?	FPT[†]	[57]	
TS	p-NP-h XP[*]	p-NP-h	FPT	FPT	[45,65]	
r-MTC	p-NP-h[†] XP[*]	?	?	FPT	[6]	

[†]Results not (explicitly) contained in the given literature reference (last column) yet being simple observations/corollaries. [*]Open whether contained in FPT.

3   Intractability for Constant Underlying Treewidth

»*I'm saying this because it's true.*« For static graphs, many NP-hard problems become polynomial-time solvable when the input graph has constant treewidth. More specifically, many such problems are fixed-parameter tractable when parameterized by treewidth; comparatively few problems are W[1]-hard yet contained in XP when parameterized by treewidth [8,32,35,44] or remain NP-hard when restricted to graphs of constant treewidth [48,55,56,62]. For temporal graphs, parametrization by the underlying treewidth leads to quite different observations: so far, few temporal problems are known to be contained in XP or even fixed-parameter tractable, while several problems remain NP-hard even if the underlying treewidth is constant (see Table 1 for an overview of the results and Appendix A for problem definitions).

Next, we try to understand a little better why constraining the parameter underlying treewidth seems to offer so little algorithmic benefit. The reductions proving NP-hardness on constant-treewidth underlying graphs have the following features in common:

- the constructed underlying graph is tree-like,
- vertices and time stamps capture structures of the input instance, and hence
- the numbers of vertices and layers are unbounded.

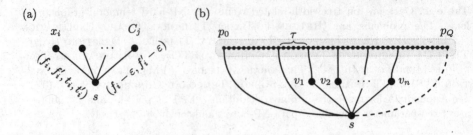

Fig. 3. Illustration of reductions behind Theorem 1 (i) and (ii). (a) A star with leaves corresponding to clauses and variables [2]. (b) A star with center s and leaves $v_1, \ldots, v_n$; a path (highlighted in gray) on $Q+1$ vertices is attached to the star [21].

In the remainder of this section we present some concrete example reductions in moderate detail. Our selected NP-hardness reductions cover problems from the fields of temporal exploration (Sect. 3.1), temporal reachability (Sect. 3.2), and temporal matching (Sect. 3.3).

3.1 Temporal Exploration

In the problem called RETURN-TO-BASE TEMPORAL GRAPH EXPLORATION (RTB-TGE), one is given a temporal graph $\mathcal{G}$ and a designated vertex s, and the task is to decide whether there is a strict temporal walk starting and ending at s that visits all vertices in $V(\mathcal{G})$. The NP-hardness of RTB-TGE follows by a simple reduction from HAMILTONIAN CYCLE (HC): given a directed graph G, decide whether there is a (simple) cycle in G that contains all vertices from G. However, from a parameterized view regarding the (underlying) treewidth, RTB-TGE is much harder than HC: while HC parameterized by treewidth is fixed-parameter tractable, for RTB-TGE we have the following.

Theorem 1 ([2, 21]). RETURN-TO-BASE TEMPORAL GRAPH EXPLORATION *is NP-hard even if*

(i) the underlying graph is a star or
(ii) each layer is a tree and the underlying graph has pathwidth at most two.

Akrida et al. [2] proved Theorem 1(i) via a reduction from 3-SAT(3), a special case of 3-SAT where each variable appears in at most three clauses. See Fig. 3(a) for an illustration. In the reduction from 3-SAT(3), a star is constructed where for each variable and clause in the input 3-SAT formula there is a leaf in the star. Moreover, each variable x_i has two unique entry time steps f_i, t_i and two unique exit time steps f_i', t_i', corresponding to setting x_i to false (entering at f_i and leaving at f_i') or true (entering at t_i and leaving at t_i') with $f_i < f_i' < t_i < t_i'$. Clearly, it is never beneficial for the explorer to linger in a leaf longer than necessary. Now assume clause C_j to contain variable x_i unnegated. Then we add to C_j an entry time step $f_i - \varepsilon$ and an exit time step $f_i' - \varepsilon$. Since

$f_i - \varepsilon < f_i < f'_i - \varepsilon < f'_i$, clause C_j can be visited at time $f_i - \varepsilon$ if and only if x_i is set to true.

By adding analogous entry and exit time steps to C_j for all its contained variables (negated or unnegated), it follows that C_j can be visited if and only if at least one of its literals is set to true.

Bodlaender and van der Zanden [21] proved Theorem 1(ii) via a reduction from RTB-TGE with the underlying graph being a star to (RTB-)TGE by adding a long path to each layer, which is connected to some of the star's leaves in such a way that each layer is a tree and the underlying graph has pathwidth at most two. See Fig. 3(b) for an illustration. In the reduction from RTB-TGE, let $\mathcal{G}$ be the temporal graph with lifetime τ and the underlying graph being a star on vertices s and $v_1, \ldots, v_n$. Then a temporal graph $\mathcal{G}'$ with lifetime $\tau' = Q + \tau + 1$ is constructed from $\mathcal{G}$ by adding a path on vertices $p_0, \ldots, p_Q$ (highlighted in gray and present in all layers), where $Q = \tau \cdot (n + 4)$, and appending $Q + 1$ layers, in which each vertex v_i is adjacent only to s. Furthermore, in each of the first τ layers, each vertex v_i is connected to some vertex on the path P if and only if v_i is the lowest numbered vertex in a connected component. This guarantees that every layer is connected. Equivalence holds since in the first τ time steps, $\mathcal{G}$ must be explored, and in the remaining $Q + 1$ time steps the path P must be explored "in one run" (starting from s, going to p_0 and ending at p_Q). In the return-to-base variant, the exploration ends with stepping from p_Q to s.

3.2 Temporal Reachability

While the problem of temporal exploration asks whether a single agent can traverse the entire graph, temporal reachability problems are concerned with the set of vertices reachable by an infinite number of agents, all starting simultaneous at a single vertex. Clearly, for any given start vertex this set can be determined by a simple search tree on the static expansion. As this setting can be understood as a model for information flow or disease spreading, a natural question is how far the set of reachable vertices can be decreased using a limited number of graph modifications; this can be understood as a measure of temporal graph connectivity. In the following, we address this question for three different types of modification operations: deletion of time-edges, reordering of layers, and merging of layers.

Deletion of Time-Edges. Enright et al. [36] investigated the (α, β)-TEMPORAL REACHABILITY TIME-EDGE DELETION $((\alpha, \beta)$-TRTED) problem: given a temporal graph $\mathcal{G} = (V, \mathcal{E}, \tau)$ and two integers $k, h \geq 0$, decide whether there is a subset $\mathcal{E}' \subseteq \mathcal{E}$ of temporal edges with $|\mathcal{E}'| \leq k$ such that in $\mathcal{G} - \mathcal{E}'$, the size of the set of vertices reachable from every vertex $s \in V$ via strict (α, β)-temporal paths is at most h. Enright et al. proved the following hardness result.

Theorem 2 ([36]). (α, β)-TEMPORAL REACHABILITY TIME-EDGE DELETION *is NP-hard even if the underlying graph consists of two stars with adjacent centers.*

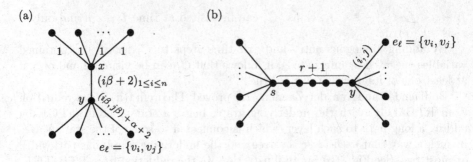

Fig. 4. Illustrations of the reductions behind Theorems 2 and 3. (a) Two stars with centers x and y and edge $\{x, y\}$; The leaves of the star centered at y one-to-one correspond to the edges of the input graph (same for (b)) [36]. (b) Two stars with centers s and y, where the centers are connected via a path of $r + 1$ vertices [37].

The proof of Theorem 2 employs a reduction from CLIQUE: given an undirected graph G and an integer r, decide whether G contains a clique (a graph where each pair of vertices is adjacent) with r vertices. In the corresponding construction, adjacencies among the vertices in the CLIQUE instance are encoded by time stamps. See Fig. 4(a) for an illustration. So suppose that a CLIQUE instance with the vertex set $\{v_1, \ldots, v_n\}$, edge set $\{e_1, \ldots, e_m\}$, and solution size r is given. The constructed underlying graph consists of two stars with m leaves each and adjacent centers x and y. The leaves of the first star are only connected to x at time step 1, thus x is the source vertex that reaches the most other vertices. For each v_i, the edge $\{x, y\}$ is present at time step $i\beta + 2$. For each edge $e_\ell = \{v_i, v_j\}$, the edge connecting y and the vertex e_ℓ (the vertex corresponding to edge e_ℓ) is present at time steps $i\beta + \alpha + 2$ and $j\beta + \alpha + 2$. Observe that reaching e_ℓ from x by a strict (α, β)-temporal path requires the edge $\{x, y\}$ to be present at time $i\beta + 2$ or at time $j\beta + 2$. Thus, if $v_{i_1}, \ldots, v_{i_r}$ form a clique on r vertices, then deleting the temporal edge set $\{(\{x, y\}, \ell\beta + 2) \mid \ell \in \{i_1, \ldots, i_r\}\}$ reduces the (α, β)-reachability of x by $\binom{r}{2}$. Conversely, reducing the (α, β)-reachability of x by $\binom{r}{2}$ with only r deletions is impossible unless a clique of size r exists in the input.

Reordering of Layers. Enright et al. [37] proved the following hardness result for the MIN-MAX REACHABILITY TEMPORAL ORDERING (MIN-MAX RTO) problem: given a temporal graph $\mathcal{G} = (V, \mathcal{E}, \tau)$ and an integer $k \in \mathbb{N}$, decide whether there is a bijection $\phi : \{1, \ldots, \tau\} \to \{1, \ldots, \tau\}$ such that the maximum reachability (that is, the maximum number of vertices any vertex can reach via a strict temporal path) in $\mathcal{G}' = (V, \{(e, \phi(t)) \mid (e, t) \in \mathcal{E}\}, \tau)$ is at most k. Correspondingly, MAX-MIN RTO is defined by exchanging "maximum" by "minimum" and "at most" by "at least".

Theorem 3 ([37]). MIN-MAX REACHABILITY TEMPORAL ORDERING *and* MAX-MIN REACHABILITY TEMPORAL ORDERING *are NP-hard even when the underlying graph is a tree obtained by connecting two stars using a path.*

Enright et al. proved Theorem 3 (similarly to the previously presented reduction by Enright et al. [36]) via a reduction from CLIQUE. See Fig. 4(b) for an illustration. In their reduction, the input consists of vertex set $\{v_1, \ldots, v_n\}$, edge set $\{e_1, \ldots, e_m\}$, and solution size r. Each layer G_i corresponds to a vertex v_i: the edge $\{y, e_\ell\}$ is present in G_i if and only if $v_i \in e_\ell$. That is, the incidence of an edge with vertex v_i is represented by the presence of that edge in layer G_i. Hence, if $v_{i_1}, \ldots, v_{i_r}$ form a clique on r vertices, then mapping $i_1, \ldots, i_r$ to the first r layers disallows s to reach $\binom{r}{2}$ leaves adjacent to y (since the s-y path contains $r + 1$ vertices).

Merging of Layers. Deligkas and Potapov [29] studied reachability minimization/maximization under certain layer-merging operations and showed hardness results on trees and paths. In this context, *merging* an interval of time stamps $M \subseteq \{1, \ldots, \tau\}$ in $\mathcal{G}$ means replacing each temporal edge (e, ℓ) with $\ell \in M$ by a new temporal edge $(e, \max(M))$. Thus, the appearance of all temporal edges within this interval M is shifted to the end of M. More precisely, Deligkas and Potapov considered the MIN REACHABILITY TEMPORAL MERGING (MRTM) problem: given a temporal graph $\mathcal{G} = (V, \mathcal{E}, \tau)$, a set of sources $S \subseteq V$, and three integers $\lambda, \mu, k \in \mathbb{N}$, decide whether there are μ disjoint intervals $M_1, \ldots, M_\mu$, each of size $|M_i \cap \{1, \ldots, \tau\}| = \lambda$, such that, after merging each of them in $\mathcal{G}$, the number of vertices reachable from S is at most k.

Theorem 4 ([29]). MIN REACHABILITY TEMPORAL MERGING *is NP-hard even when the underlying graph is a path.*

The proof of Theorem 4 employs a reduction from MAX2SAT(3), a variant of the MAX2SAT problem where each variable occurs in at most three clauses. (In the MAX2SAT problem, the goal is to find a truth assignment maximizing the number of satisfied clauses of a given 2-SAT formula).

For each clause in a given input instance, a separate subpath containing nine vertices of the underlying path is used and labeled as shown in Fig. 5(a). Here, c is the index of the clause $(x_i \vee \overline{x_j})$ and we may assume c to always be much smaller than i and j. The middle vertex s of this subpath is added to the set S of sources and we take the merge size as $\lambda = 2$. Then it is possible to either merge $\{4c, 4c + 1\}$, thus preventing s from reaching the three bottom left vertices, or to merge $\{4c + 1, 4c + 2\}$, thus blocking the three bottom right vertices, but not both (due to the disjointness condition). Hence, given a large enough number of merges, each source s can reach at most five other vertices. If we want to reduce this number to four, then one must additionally merge $\{4i, 4i + 1\}$ (thus setting x_i to true) or merge $\{4j + 1, 4j + 2\}$ (thus setting x_j to false), i.e., give an assignment satisfying clause c.

If the underlying graph is allowed to be a ternary tree, then this construction can be modified to only require a single source vertex [29].

3.3 Temporal Matching

Mertzios et al. [57] proved hardness for the TEMPORAL MATCHING (TM) problem: given a temporal graph $\mathcal{G} = (V, \mathcal{E}, \tau)$ and integers $k, \Delta \geq 0$, decide whether

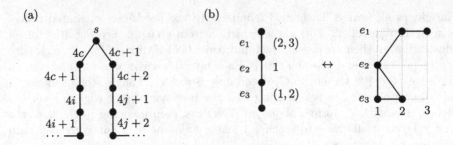

Fig. 5. Illustration of the reductions behind Theorems 4 and 5. (a) A part of the temporal graph used in the proof of Theorem 4 whose underlying graph is a path [29]. (b) A temporal graph whose underlying graph is a path (left-hand side) and its 2-temporal line graph (right-hand side; a grid is indicated by thin gray dotted lines) [57].

there is a Δ-temporal matching of cardinality at least k in $\mathcal{G}$. A Δ-*temporal matching* is a set $\mathcal{E}' \subseteq \mathcal{E}$ of temporal edges such that for every two temporal edges $(e, t), (e', t') \in \mathcal{E}'$, we have that $e \cap e' = \emptyset$ or $|t - t'| \geq \Delta$.

Theorem 5 ([57]). TEMPORAL MATCHING *is NP-hard even when the underlying graph is a path.*

The crucial observation is that solving TM on a temporal graph $\mathcal{G}$ is equivalent to solving INDEPENDENT SET on the so-called Δ-*temporal line graph* of $\mathcal{G}$, which contains a vertex for each temporal edge of $\mathcal{G}$ and has two vertices adjacent if the corresponding temporal edges cannot be both contained in a Δ-temporal matching [57, Definition 2]. For an illustration see Fig. 5(b). Moreover, if the underlying graph $\mathcal{G}_\downarrow$ is a path with m edges, then its 2-temporal line graph is an induced subgraph of a *diagonal grid graph* of size $m \times \tau$, and conversely each such grid can be obtained as a 2-temporal line graph. Here, a *diagonal grid graph* is simply a grid that additionally contains the two diagonal edges of every grid cell. Subsequently, Mertzios et al. proved that INDEPENDENT SET is NP-complete on induced subgraphs of diagonal grid graphs, thus also showing NP-hardness of TM.

4   Dynamic Programming Based on an Underlying Tree Decomposition

»*It's still the same old story...*« For many graph problems, algorithms exploiting small treewidth are dynamic programs over a corresponding tree decomposition. For temporal graph problems, few such dynamic programs are known. Yet, we present four dynamic programs known from the literature: Two XP-algorithms for two NP-hard problems, and two polynomial-time algorithms. For the former two algorithms, the running time depends exponentially on the lifetime τ (hence proving fixed-parameter tractability regarding $tw_\downarrow + \tau$ in both cases). This supports our intuition that while capturing the structure of the graph, the underlying treewidth is missing relevant time aspects.

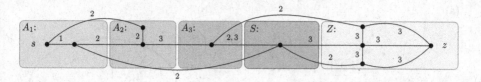

Fig. 6. The idea for the dynamic program from Theorem 6 for a temporal graph $\mathcal{G}$. Vertices in S form the temporal s-z separator, vertices in Z are not reachable from s in $\mathcal{G} - S$, and vertices in A_t are not reachable from s in $\mathcal{G} - S$ before time t.

4.1 Two XP-Algorithms

The two XP-algorithms [6,45] we sketch indeed both are FPT-algorithms regarding the combination $\mathrm{tw}_{\downarrow} + \tau$ of the underlying treewidth and the lifetime.

An XP-Algorithm for Temporal Separation. Fluschnik et al. [45] studied TEMPORAL SEPARATION, which is the problem of deciding whether all (strict) temporal paths connecting two given terminal vertices s and z in a temporal graph $\mathcal{G}$ can be destroyed by removing a set $S \subseteq V \setminus \{s, z\}$ of at most k vertices. Such a set S is called a *(strict) s-z separator*. Fluschnik et al. [45] employed dynamic programming on a given tree decomposition to prove that this problem is fixed-parameter tractable when parameterized by $\mathrm{tw}_{\downarrow} + \tau$, and is in XP when parameterized by $\mathrm{tw}_{\downarrow}$.

Theorem 6 ([45]). TEMPORAL SEPARATION *with given tree decomposition of the underlying graph is solvable in* $O((\tau + 2)^{\mathrm{tw}_{\downarrow}+2} \cdot \mathrm{tw}_{\downarrow} \cdot |V| \cdot |\mathcal{E}|)$ *time.*

The dynamic program behind Theorem 6 is based on the fact that for each vertex $v \in V \setminus \{s\}$ in a temporal graph $\mathcal{G} = (V, \mathcal{E}, \tau)$ there is a time step $t \in \{1, \dots, \tau\}$ such that v cannot be reached from $s \in V$ before t. In particular, one guesses a partition $V = A_1 \uplus A_2 \uplus \dots \uplus A_\tau \uplus S \uplus Z$ such that (i) S is a temporal s-z separator, (ii) in $\mathcal{G} - S$ no vertex contained in Z is reachable from s, and (iii) no vertex $v \in A_t$ can be reached from s before time step t, where $t \in \{1, \dots, \tau\}$. See Fig. 6 for an illustrative example.

An XP-Algorithm for Temporally Connected Subgraphs. Given a temporal graph $\mathcal{G} = (V, \mathcal{E}, \tau)$ and a designated vertex $r \in V$, a *temporally r-connected spanning subgraph* of $\mathcal{G}$ is a temporal graph $\mathcal{G}' = (V, \mathcal{E}', \tau)$ with $\mathcal{E}' \subseteq \mathcal{E}$ such that $\mathcal{G}'$ contains a temporal path from r to any other vertex $v \in V$ in $\mathcal{G}'$. The task in MINIMUM SINGLE-SOURCE TEMPORAL CONNECTIVITY (r-MTC) is to find a temporally r-connected spanning subgraph for a given vertex r such that the total weight $\sum_{(e,t) \in \mathcal{E}'} w((e, t))$ is minimized, where w is an arbitrary nonnegative weight function. Axiotis and Fotakis [6] employed dynamic programming on a nice tree decomposition [53] to prove that MINIMUM SINGLE-SOURCE TEMPORAL CONNECTIVITY (r-MTC) is fixed-parameter tractable when parameterized by $\mathrm{tw}_{\downarrow} + \tau$, and is contained in XP when parameterized by $\mathrm{tw}_{\downarrow}$ alone.

(a)   (b)

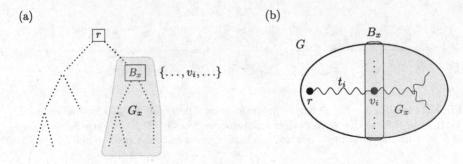

Fig. 7. Illustration to the dynamic program for r-MTC for some temporal graph with underlying graph G. (a) A (nice) tree decomposition of G is depicted with the root node's bag containing r, node x with bag $B_x = \{\ldots, v_i, \ldots\}$, and the subgraph G_x of G that is induced by all vertices contained in the bag B_x and bags of the descendants of x. (b) Graph G is depicted, with induced subgraph G_x. Moreover, a temporal r-v_i path arriving at time step t_i, and temporal paths connecting v_i to some vertices from G_x are depicted. The latter corresponds to an table entry $f(x | \ldots, a_i = 1, t_i, \ldots)$.

Theorem 7 ([6]). MINIMUM SINGLE-SOURCE TEMPORAL CONNECTIVITY *is solvable in* $\mathcal{O}(3^{\mathrm{tw}_\downarrow} \cdot (\tau + \mathrm{tw}_\downarrow)^{\mathrm{tw}_\downarrow + 1} \cdot |V|)$ *time if a nice tree decomposition of the underlying graph is given.*

The idea of the dynamic program behind Theorem 7 using a nice tree decomposition rooted at the source r is as follows (see Fig. 7 for an illustration): for each bag, the vertices contained in the bag are (bi-)partitioned into vertices connected to r (we call them *local sources*) and vertices not (yet) connected to r. Vertices in the subgraph induced by the vertices in the bag and all its descendant bags must be reachable from the local sources by a temporal path starting "late enough" (i.e., after the local source has been reached from source r). For each node x in the nice tree decomposition, a table entry $f(x \mid a_1, t_1, \ldots, a_{\mathrm{tw}_\downarrow}, t_{\mathrm{tw}_\downarrow})$ stores the minimum cost of a temporal subgraph such that in the graph induced by all the vertices in the bags of x and all its descendants, every vertex is reachable from some vertex $v_j^x \in B_x$ with $a_j = 1$ (the local sources) by some temporal path starting not before time step t_j. Hence, each such x has a table entry for each of the $2^{\mathrm{tw}_\downarrow}$ possible bipartitions, and each of the $\tau^{\mathrm{tw}_\downarrow}$ possible starting times for the temporal paths starting at local sources.

For both of the two presented problems, it appears to be crucial to guess the time steps in which a solution "touches" the corresponding bag. However, in both it is open whether the dependencies on τ (being the base of exponent $\mathrm{tw}_\downarrow$; see Theorems 6 and 7) can be avoided: Is TEMPORAL SEPARATION (TS) or MINIMUM SINGLE-SOURCE TEMPORAL CONNECTIVITY (r-MTC) fixed-parameter tractable when parameterized by $\mathrm{tw}_\downarrow$?

4.2   Two Fixed-Parameter Polynomial-Time Algorithms

The underlying tree decomposition, when part of the input, can also be used for tasks solvable in polynomial time. In this section, we present two algorithms making use of the underlying tree decomposition, one for temporal exploration, and one for computing foremost temporal walks.

Temporal Exploration. In Sect. 3, we discussed the NP-hardness of determining the exact time required to fully explore a temporal graph. However, as long as each layer of the input temporal graph is connected and the underlying treewidth is low, it can be shown that a subquadratic number of steps is always sufficient. More precisely, Erlebach et al. [38] proved the following by giving an algorithm that utilizes a given tree decomposition.

Theorem 8 ([38]). *If every layer is connected, then a temporal graph $\mathcal{G}$ can be explored in $\mathcal{O}\left(|V|^{3/2} \cdot \mathrm{tw}_\downarrow^{3/2} \cdot \log |V|\right)$ steps.*

The proof of Theorem 8 builds upon the observation that an agent needs at most $n - 1$ steps to move from any vertex to any other vertex if both of these vertices are connected in every layer. The idea is then to divide up $G_\downarrow(\mathcal{G})$ into sufficiently small subgraphs to which this observation can then be applied.

To this end, select a vertex set S as the union of $\mathcal{O}\left(\sqrt{|V| \cdot \mathrm{tw}_\downarrow}\right)$ bags of a nice tree decomposition of $G_\downarrow(\mathcal{G})$ in such a way that every connected component of $G_\downarrow(\mathcal{G}) - S$ has size at most $\mathcal{O}(\sqrt{|V|/\mathrm{tw}_\downarrow})$. If we consider a time window of $\Theta(\mathrm{tw}_\downarrow \cdot \sqrt{|V|/\mathrm{tw}_\downarrow})$ layers, then, by the pigeonhole principle, for any vertex v in any of these connected components, there is a vertex $w \in S$ that is in the same connected component in at least $\Theta(\sqrt{|V|/\mathrm{tw}_\downarrow})$ layers. Thus, by the above observation, an agent at w can reach v and return to w within $\mathcal{O}(\mathrm{tw}_\downarrow \cdot \sqrt{|V|/\mathrm{tw}_\downarrow})$ time steps.

Hence, if we use $\Theta(\mathrm{tw}_\downarrow \cdot \sqrt{|V| \cdot \mathrm{tw}_\downarrow})$ agents, then each starting at a vertex of S, we can explore $\mathcal{G}$ in at most $\mathcal{O}(\mathrm{tw}_\downarrow \cdot \sqrt{|V|/\mathrm{tw}_\downarrow} \cdot \sqrt{|V|/\mathrm{tw}_\downarrow}) = \mathcal{O}(|V|)$ steps. From this, one can derive an upper bound of $\mathcal{O}(|V|^{3/2} \cdot \mathrm{tw}_\downarrow^{3/2} \cdot \log(|V|))$ if only a single agent is used to perform these explorations sequentially.

Computing Foremost Walks. A (strict) *foremost s-z* walk is a temporal walk that arrives earliest among all s-z temporal walks. Himmel [49] proved that foremost walk queries can be answered quickly using a specific data structure that relies on a (given) underlying tree decomposition.

Theorem 9 ([49]). *There exists a data structure of size $\mathcal{O}(\mathrm{tw}_\downarrow^2 \cdot \tau \cdot |V|)$ computable in $\mathcal{O}(\mathrm{tw}_\downarrow^2 \cdot \tau^2 \cdot |V|)$ time such that one can find a foremost walk between two vertices on temporal graphs with underlying treewidth $\mathrm{tw}_\downarrow$ in $\mathcal{O}(\mathrm{tw}_\downarrow^2 \cdot \tau \cdot \log |V| \cdot \log(\mathrm{tw}_\downarrow \cdot \tau \cdot \log |V|))$ time.*

The data structure behind Theorem 9 was originally introduced by Abraham et al. [1] for computing shortest path queries in static graphs. It exploits *binary* tree decompositions of depth $\mathcal{O}(\log|V|)$. Basically, the preprocessing for the data structure computes the earliest arrival time from any vertex v to any vertex w at any possible starting time, where v and w are contained in the same bag of the tree decomposition.

5 Monadic Second-Order Logic for Temporal Graphs

»*Honest as the day is long!*« Courcelle's famous theorem states that every graph property definable in monadic second-order logic is fixed-parameter tractable when simultaneously parameterized by the treewidth of the graph and the length of the formula [26]. In this section, we review how one could lift this powerful classification tool to temporal graphs and spot some pitfalls having led to flaws in the literature.

For *monadic second-order (MSO) logic* on a graph G we need a *structure* consisting of a universe $U = V(G) \cup E(G)$ and a *vocabulary* consisting of two unary relations $V \subseteq U$ and $E \subseteq U$ containing the vertices and the edges, respectively, and two binary relations $\mathrm{adj} \subseteq U \times U$ and $\mathrm{inc} \subseteq U \times U$, where $(v, w) \in \mathrm{adj}$ if and only if $\{v, w\} \in E(G)$, and $(v, e) \in \mathrm{inc}$ if and only if $v \in e$. For a fixed finite set of *(monadic) variables*, an *atomic formula* over vocabulary ν is of the form $x_1 = x_2$ or $R(x_1, x_2)$ or $R'(x_1)$, where $R \in \{\mathrm{adj}, \mathrm{inc}\}$, $R' \in \{V, E\}$ and $x_1, x_2 \in U$. Here, $R(x_1, x_2)$ $(R'(x_1))$ evaluates to true if and only if $(x_1, x_2) \in R$ $(x_1 \in R')$. MSO *formulas* are constructed from atomic formulas using boolean operations $\neg$, $\vee$, $\wedge$ and existential and universal quantifiers $\exists$, $\forall$ over variables and set variables. For further details, refer to Courcelle and Engelfriet [26].

Example 1. The well-known CLIQUE problem can be expressed by the following MSO formula: $\exists X.(\forall x, y \in X.(V(x) \wedge V(y) \wedge \mathrm{adj}(x, y)))$.

The following, known as an optimization variant of Courcelle's theorem, connects MSO and treewidth.

Theorem 10 ([5,26]). *There exists an algorithm that, given (i) an MSO formula ρ with free monadic variables $X_1, \ldots, X_r$, (ii) an n-vertex graph G, and (iii) an affine function $\alpha(x_1, \ldots, x_r)$, finds the minimum (maximum) of $\alpha(|X_1|, \ldots, |X_r|)$ over evaluations of $X_1, \ldots, X_r$ for which formula ρ is satisfied on G. The running time of this algorithm is $f(|\rho|, \mathrm{tw}(G)) \cdot n$, where f is a computable function, $|\rho|$ is the length of ρ, and $\mathrm{tw}(G)$ is the treewidth of G.*

Having Theorem 10 at hand, we can prove that CLIQUE (see Example 1) is fixed-parameter tractable when parameterized by the treewidth of the input graph: The formula given in Example 1 has one free monadic variable and constant length c, hence, with α being the identity function, we can decide CLIQUE in $f(c, \mathrm{tw}(G)) \cdot |V|$ time.

We are aware of two successful approaches and one flawed approach to lift Theorem 10 to the temporal setting. We will survey in Sects. 5.1 and 5.2 the two successful approaches, and discuss in Sect. 5.3 the flawed approach.

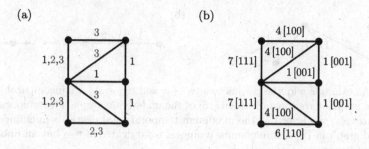

Fig. 8. (a) The temporal graph $\mathcal{G}$ from Fig. 1 and (b) its edge-labeled graph $L(\mathcal{G})$ (with labels and bit-representation in brackets) are depicted.

5.1 Using Labels

Arnborg et al. [5] showed that it is possible to apply Theorem 10 to graphs in which edges have labels from a fixed finite set, either by augmenting the graph logic to incorporate predicates describing the labels, or by representing the labels by unquantified edge set variables. Zschoche et al. [65] exploited this for temporal graphs as follows (see Fig. 8 for an example):

Define an edge-labeled graph $L(\mathcal{G})$ to be the underlying graph $G_{\downarrow}$ of a given temporal graph $\mathcal{G}$ of lifetime τ with the edge-labeling $\omega\colon E(G_{\downarrow}) \to \{1,\ldots,2^{\tau}-1\}$ such that $\omega(\{v,w\}) = \sum_{i=1}^{\tau} \mathbb{1}_{\{v,w\}\in E_i} \cdot 2^{i-1}$, where $\mathbb{1}_{\{v,w\}\in E_i} = 1$ if and only if $(\{v,w\},i) \in \mathcal{E}$, and 0 otherwise. Observe that the i-th bit of a label now expresses whether the edge is present in the i-th layer of the temporal graph. Hence, we can check whether an edge e is present in layer t using the MSO formula $\text{layer}(e,t) := \bigvee_{i=1}^{\tau} \bigvee_{j\in\sigma(i,2^{\tau}-1)} (t = i \wedge \omega(e) = j)$ of length $2^{O(\tau)}$, where $\sigma(i,z) := \{x \in \{1,\ldots,z\} \mid i\text{-th bit of } x \text{ is } 1\}$. Furthermore, we can determine whether two vertices v and w are adjacent in layer t using the MSO formula $\text{tadj}(v,w,t) := \exists e \in E.(\text{inc}(e,v) \wedge \text{inc}(e,w) \wedge \text{layer}(e,t))$ of length $2^{O(\tau)}$. Altogether, in a nutshell we get the following: If a temporal graph problem Π can be formulated by an MSO-formula which uses $\text{layer}(e,t)$ and $\text{tadj}(v,w,t)$ as black boxes, then Π is fixed-parameter tractable when parameterized by the combination of the length of the formula, the underlying treewidth, and the lifetime τ. Zschoche et al. [65] derived an MSO-formula for TEMPORAL SEPARATION (TS), where the length of the formula is upper-bounded by some function in τ. Hence, TS is fixed-parameter tractable when parameterized by the combination of the underlying treewidth and the lifetime.

5.2 Enriching the Vocabulary

Another approach, used by Enright et al. [36], can be applied to exchange the dependency on τ with a dependency on the *maximum temporal total degree* $\Delta_{\mathcal{G}}$, which is the maximum number of temporal edges incident to the same vertex in temporal graph $\mathcal{G}$. Observe that the maximum temporal total degree is at least the maximum degree of the underlying graph. Moreover, the parameters lifetime

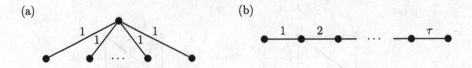

Fig. 9. An example why the parameters $tw_\downarrow + \tau$ and $tw_\downarrow + \Delta_\mathcal{G}$ are incomparable. Both temporal graphs have constant treewidth of the underlying graph. The temporal graph in (a) has only one layer but the maximum temporal total degree is unbounded. The temporal graph in (b) has maximum temporal total degree of two but an unbounded number of layers.

and maximum temporal total degree are unrelated to each other, meaning that the maximum temporal total degree can be large while the lifetime is small and vice versa (see Fig. 9 for two examples).

In a nutshell, we alter the universe and the vocabulary of the structure (we refer to this structure as *enriched*) in order to express a temporal graph problem. We add all temporal edges (e, t) of the temporal graph $\mathcal{G}$ to the universe and equip the vocabulary with two binary relation symbols $\mathcal{L}$ and $\mathcal{R}$, where

- $(e, (e, t)) \in \mathcal{L}$ if and only if e is an edge in the underlying graph and (e, t) is a temporal edge of $\mathcal{G}$, and
- $(e_1, t_1), (e_2, t_2) \in \mathcal{R}$ if and only if (e_1, t_1) and (e_2, t_2) are temporal edges where e_1 and e_2 have a vertex in common and $t_1 < t_2$.

It is easy to see that the treewidth of the Gaifman graph[2] for the enriched structure is upper-bounded by a function of the treewidth of the underlying graph of the temporal graph and the maximum temporal total degree. Hence, due to Courcelle and Engelfriet [26], if a temporal graph problem Π can be formulated by an MSO-formula in the enriched structure, then Π is fixed-parameter tractable when parameterized by the combination of the underlying treewidth, the maximum temporal total degree, and the length of that formula. Enright et al. [36] derived an MSO-formula in the enriched structure for TEMPORAL REACHABILITY EDGE DELETION, where the length of the formula depends on h (the size of the set of reachable vertices), hence proving fixed-parameter tractability for the problem when parameterized by the combination of h, the underlying treewidth, and the maximum temporal total degree.

5.3   Pitfalls in the Literature

Mans and Mathieson [54] also explored the direction of enriching the vocabulary in the context of *dynamic graphs*. In their model, vertices can (dis)appear over time as well. Furthermore, the layers are not necessarily arranged in a linear (time) ordering. Hence, their model of dynamic graphs is more general than

---

[2] In the *Gaifman graph* of a structure, there is one vertex for each element in the universe and two vertices have an edge if and only if the corresponding elements occur together in the same relation.

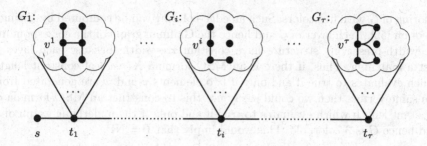

Fig. 10. Rough sketch of the Gaifman graph of a treewidth-preserving structure for a temporal graph $\mathcal{G}$ with layers $G_1, \ldots, G_\tau$. The copies v^1, v^i, v^τ are illustrated for a vertex v of $\mathcal{G}$.

temporal graphs. However, some of their results seem flawed. In the remainder of this section we discuss these flaws in the special case of temporal graphs.

Mans and Mathieson construct a so-called *treewidth-preserving structure*. Here, the universe has for each vertex v of the temporal graph τ many copies v^1, $\ldots, v^\tau$, one element t_i for each $i \in \{1, \ldots, \tau\}$, and an additional element s. Note, that there is a unary relation symbol L_v which contains an element x if and only if the element x is generated from the vertex v ($x \equiv v^t$, for some $t \in \{1, \ldots, \tau\}$). The Gaifman graph of a treewidth-preserving structure of a temporal graph is the disjoint union of the layers. Additionally, there is one long path starting at some special vertex s and then "visits" all layers in the time induced order, see Fig. 10 for an illustration.

On the good side, this keeps the treewidth of the Gaifman graph upper-bounded by a function in the maximum treewidth over all layers. Furthermore, one can still express (in MSO) time relations between elements of different layers, for example by measuring the distance to s.

On the problematic side, having two elements v and w at hand which represent vertices in some layer of the temporal graph, it seems difficult to get an MSO-formula which evaluates to true if and only if v and w are generated from the same vertex. To do so, Mans and Mathieson [54] used an expression $f_V(v) = f_V(w)$. It is unclear whether f_V is in fact part of the treewidth-preserving structure or not. Note that the length of such an expression in terms of the unary relation symbols L_v depends on the number of vertices in the temporal graph. If the expression $f_V(v) = f_V(w)$ is an short cut for an expression of size at least the number of vertices in the temporal graph, then Lemmata 13 and 17 and hence Corollaries 14–16 and 18 of Mans and Mathieson [54] break. We believe that it is rather unlikely that one can provide arguments to repair the idea of Mans and Mathieson [54] because of the following example.

Example 2. The following is a polynomial-time algorithm for the NP-complete 3-Coloring problem on graphs with maximum degree four [28]: Given a graph G with maximum degree four, construct a temporal graph $\mathcal{G}$ with five layers on the vertex set $V(G)$ such that the underlying graph of $\mathcal{G}$ is G, and each layer has treewidth one. Here, the edges in one layer of $\mathcal{G}$ correspond to one color of a edge-coloring of G with five colors. Such an edge-coloring can be computed by Vizing's theorem [59]. Each layer of $\mathcal{G}$ and hence the Gaifman graph of the corresponding treewidth-preserving structure have constant treewidth, because each layer is just a matching. Thus, if there is an MSO-formula $X(v, w)$ of constant length which evaluates to true if and only if two elements v and w are generated from the same vertex, then we could easily use this to construct an MSO-formula of constant length which evaluates to true if and only if the underlying graph of $\mathcal{G}$ and hence G is 3-colorable. This would imply that P = NP.

In private communication, we discussed our concerns with Mans and Mathieson [54]: they agreed that one cannot find a constant size MSO-formula that evaluates to true if and only if two elements v and w are generated from the same vertex, unless P = NP.

6   Possible Definitions of Temporal Treewidth

»Welcome back to the fight. This time I know our side will win.« Now we embark on the endeavor of finding useful and interesting definitions for temporal treewidth. To prepare our journey, we first briefly recapitulate how treewidth (and other structural graph parameters) have commonly been adapted for the temporal setting. In the majority of cases, structural graph parameters such as treewidth are transferred to the temporal setting in one of the following two straightforward ways:

1. Take the maximum over all layer treewidths, resulting in tw_∞.
2. Take the treewidth of the underlying graph, resulting in $\text{tw}_\downarrow$.

Since the treewidth of a graph does not increase when edges are removed, we naturally get that for any temporal graph $\mathcal{G}$ it holds that $\text{tw}_\infty(\mathcal{G}) \leq \text{tw}_\downarrow(\mathcal{G})$. We can also observe that these two variants of temporal treewidth are invariant under reordering of the layers and hence might not be considered truly temporal since they also apply to the unordered "multilayer setting".

There is a further generic way to transfer a structural graph parameter to the temporal setting. This one is particularly interesting in the context of problems that make use of Δ-time windows[3], as done in recent work on RESTLESS TEMPORAL PATHS [24], TEMPORAL CLIQUE [7,51,60,64], TEMPORAL COLORING [58], TEMPORAL MATCHING [57], and TEMPORAL VERTEX COVER [3]. In the case of treewidth we call this parameter Δ-slice treewidth[4], and as the name suggests, it depends on an additional natural number Δ that is typically part of the input or the problem specification. The Δ-slice treewidth is the maximum of the treewidths of the union graphs of all Δ-time windows, formally defined as follows:

---

[3] A Δ-time window is a set of Δ consecutive time steps.
[4] To the best of our knowledge, the concept of a "Δ-slice parameter" was introduced by Himmel et al. [51] to define a temporal version of degeneracy. It was later also used by Bentert et al. [7].

Definition 3 (Δ-Slice Treewidth). *For a temporal graph $\mathcal{G} = (V, E_1, \ldots, E_\tau)$ and a natural number $\Delta \leq \tau$, the Δ-slice treewidth $\mathrm{tw}_\Delta(\mathcal{G})$ of $\mathcal{G}$ is defined as*

$$\mathrm{tw}_\Delta(\mathcal{G}) := \max_{i \in \{1, \ldots, \tau - \Delta + 1\}} \mathrm{tw}(G_i^{(\Delta)}),$$

where $G_i^{(\Delta)} = (V, \bigcup_{j \in \{i, \ldots, i + \Delta - 1\}} E_j)$.

It is easy to see the Δ-slice treewidth interpolates between layer treewidth and underlying treewidth, hence we have that $\mathrm{tw}_\infty(\mathcal{G}) \leq \mathrm{tw}_\Delta(\mathcal{G}) \leq \mathrm{tw}_\downarrow(\mathcal{G})$, for all temporal graphs $\mathcal{G}$ and all $\Delta \leq \tau$.

In light of the known results of temporal graph problems where treewidth is used as a parameter (see Table 1 in Sect. 3), we observe that even for the largest of our established concepts of treewidth of a temporal graph, namely the underlying treewidth, we already obtain para-NP-hardness results for many temporal graph problems. Hence, temporal treewidth versions such as Δ-slice treewidth, which are upper-bounded by the underlying treewidth, are not desirable since on their own they presumably do not offer new ways to obtain tractability results.

As to islands of tractability (see Table 1 or apply Theorem 10), we find many FPT-algorithms for the combined parameter $\mathrm{tw}_\downarrow + \tau$ and for the parameter number $|V|$ of vertices for a variety of temporal graph problems. Further examples of FPT results that are not included in Table 1 are MULTISTAGE VERTEX COVER [46], RESTLESS TEMPORAL PATHS [24], and TEMPORAL COLORING [58]. This means that if we »*round up the usual suspects*« of which (combination of) parameters should lower- and upper-bound the temporal treewidth in our endeavor of finding useful definitions, then we might want to be on the look-out for something between $\mathrm{tw}_\downarrow$ and $\mathrm{tw}_\downarrow + \tau$ or something between $\mathrm{tw}_\downarrow$ and $|V|$, or something that is incomparable to the aforementioned parameters.

In the following, we are going to discuss three canonical ways to approach defining treewidth for temporal graphs:

1. Adapting tree decompositions to temporal graphs (Sect. 6.1).
2. Deriving static graphs from a temporal graph in a natural way and using the treewidth of those graphs (Sect. 6.2).
3. Looking at ways to play cops-and-robber games on temporal graphs (Sect. 6.3).

6.1   Adaptions of the Tree Decomposition

One beacon of treewidth applications has always been the tree decomposition. Hence, it is only logical to begin our quest for temporal treewidth in the decomposition territory. First, we have to ask ourselves which general properties we want a temporal tree decomposition to have. Should it be temporal as well? We could try to take inspiration from Bodlaender [9] who showed how to maintain a tree decomposition under edge additions and deletions (when the treewidth is at most two). However, it seems difficult to perform dynamic programming (which is the standard way to design FPT-algorithms for problems parameterized by treewidth) on tree decompositions that keep changing over time. Hence, we focus on *static* tree decompositions for temporal graphs, even though the idea of a temporal tree decomposition that itself is temporal as well probably deserves further consideration.

Second, we have to seek for something to put into our bags. There are two canonical choices: the vertices V of a temporal graph $\mathcal{G} = (V, \mathcal{E}, \tau)$, or its vertex appearances, that is, $V \times \{1, \ldots, \tau\}$. If we put the vertices into our bags, then it seems difficult to end up with something that is significantly different to the treewidth of the underlying graph and captures the temporal nature of the setting. So let us see what we can end up with if we put vertex appearances into the bags. We probably would want to require that for each temporal edge there is a bag that contains both endpoints of the edge, which in terms of vertex appearances would be the endpoints of the edge labeled with the time stamp of the temporal edge. However, if we stop here and add the straightforward adaptation of the third condition of tree decompositions, namely that for every vertex appearance, all bags that contain this vertex appearance should form a connected subtree, then we end up with the layer treewidth, which is something we do not want. To fix this, we may want to consider requiring every two vertex appearances with the same vertex and adjacent time stamps to be contained in at least one bag. This would surely give us something that is at least as large as the underlying treewidth and at most as large as the underlying treewidth times the lifetime. The following definition formalizes this idea.

Definition 4 (Temporal Tree Decomposition). *Let $\mathcal{G} = (V, \mathcal{E}, \tau)$ be a temporal graph. A tuple $\mathbb{T} = (T, \{B_u \mid u \in V(T)\})$ consisting of a tree T and a set of bags $B_u \subseteq V \times \{1, \ldots, \tau\}$ is a temporal tree decomposition (ttdc) of $\mathcal{G}$ if*

- *(i) $\bigcup_{u \in V(T)} B_u = V \times \{1, \ldots, \tau\}$,*
- *(ii) for every $(\{v, w\}, t) \in \mathcal{E}$ there is a node $u \in V(T)$ such that $(v, t) \in B_u$ and $(w, t) \in B_u$,*
- *(iii) for every $v \in V$ and $t \in \{1, \ldots, \tau - 1\}$ there is a node $u \in V(T)$ such that $(v, t) \in B_u$ and $(v, t+1) \in B_u$, and*
- *(iv) for every $(v, t) \in V \times \{1, \ldots, \tau\}$, the graph $T[\{u \in V(T) \mid (v, t) \in B_u\}]$ is a tree.*

The width *of $\mathbb{T}$ is* $\mathrm{width}(\mathbb{T}) := \max_{u \in V(T)} |B_u| - 1$.

As with static treewidth, this definition of a graph decomposition would give a canonical definition of a temporal treewidth: The temporal treewidth of a temporal graph $\mathcal{G}$ is the minimum width over all temporal tree decompositions of $\mathcal{G}$, that is,

$$\mathrm{ttw}(\mathcal{G}) = \min_{\mathbb{T} \text{ is ttdc of } \mathcal{G}} \mathrm{width}(\mathbb{T}).$$

As we will see in the next subsection, this definition is equivalent to using the treewidth of a certain type of static expansion of the temporal graph. Then it also will become clearer why the proposed definition gives a temporal treewidth that is at least as large as the underlying treewidth and at most as large as (roughly) the underlying treewidth times the lifetime.

6.2   Treewidth of the Static Expansion

Another direction to define a temporal version of treewidth would be to use the treewidth of static graphs as we know it, and apply it to a graph that can be naturally derived from a given temporal graph. The most canonical graph of this type is the *static expansion* (see Definition 2) of a temporal graph which, however, is typically *directed*[5]. One possibility would be to apply treewidth adaptations for directed graphs [34, Chapter 16]. Another possibility is to compute the treewidth of the undirected version of the static expansion of a temporal graph. Observe that in this case, we end up with the same temporal treewidth as in Definition 4.

Observation 1. *Let $\mathcal{G} = (V, \mathcal{E}, \tau)$ be a temporal graph and let $H = (V', A)$ be its static expansion. Let $G = (V', E)$ with $E = \{\{v, w\} \mid (v, w) \in A\}$ be the undirected static expansion of $\mathcal{G}$. Then*

$$\mathrm{ttw}(\mathcal{G}) = \mathrm{tw}(G).$$

We can check that Observation 1 is true by realizing that the bags in a temporal tree decomposition contain the vertex appearances, which are also the vertices of a static expansion. Furthermore, the edges of a static expansion connect all vertex appearances that we want to be together in at least one bag.

Using this observation, we can also check easily that the claim we made earlier holds. The precise bounds that we can show are

$$\mathrm{tw}_\downarrow(\mathcal{G}) \le \mathrm{ttw}(\mathcal{G}) \le (\mathrm{tw}_\downarrow(\mathcal{G}) + 1) \cdot \tau - 1.$$

The lower bound for the temporal treewidth follows from the fact that the underlying graph is a minor of the undirected static expansion. The upper bound follows from the observation that the following is a tree decomposition for the undirected static expansion: take a tree decomposition of the underlying graph and replace every vertex in every bag by all its appearances. This increases the size of all bags by a factor of τ.

Now we can also more easily understand how temporal graphs with very small temporal treewidth look like. A temporal graph whose temporal treewidth is one necessarily needs to have a forest as underlying graph. However, even in this case, the temporal treewidth can still be as large as $\min\{|V|, \tau\}$ if every edge appears at every time step. Take a path as underlying graph as an example where every edge appears at every time step. Then the undirected static expansion is a $|V| \times \tau$-grid. In fact, as soon as an edge appears at more than one time step, the undirected static expansion contains a cycle. Hence, a temporal graph with temporal treewidth one has a forest as underlying graph and every edge appears in exactly one time step. This seems to be a good property of temporal treewidth since many problems are indeed easy to solve on temporal graphs of this form.

---

[5] Note that there are different definitions of static expansion that are typically tailored to the applications they are used in.

6.3   Playing Cops-and-Robber Games on Temporal Graphs

Since the treewidth of a static graph can be defined via a cops-and-robber game on static graphs (see Sect. 2.1), we can also try to transfer these games to temporal graphs in a meaningful way.

Recently, Erlebach and Spooner [41] investigated a cops-and-robber game on temporal graphs with *infinite* lifetime and periodic edge appearances. Here, whenever the cops and the robber have taken their turn, time moves forward one step, and when making their moves, the cops and the robber can only use edges that are present at the current time.

The first obvious issue with this approach is that the temporal graphs we want to investigate have neither infinite lifetime nor periodic edge appearances. If the game would just stop when the lifetime finished and the robber wins if he or she does not get caught, then we would need more cops on temporal graphs with shorter lifetime. Deriving a temporal treewidth concept from this would lead to the probably undesirable property that temporal graphs with short lifetime have higher treewidth than temporal graphs with a long lifetime. To circumvent this, we could repeat the temporal graph ad infinitum. This would also make edge appearances periodic. However, then we will also get the property that the moves a robber can make in this temporal graph is a subset of the moves a robber could make in the underlying graph (or, equivalently, when all edges are always present). This means the number of cops necessary to catch a robber in this scenario is upper-bounded by the treewidth of the underlying graph—a property that we do not want to have.

Summarizing, we can say that designing cops-and-robber games on temporal graphs that lead to useful treewidth definitions seems to be a challenging task. However, since cops-and-robber games already inherently have a temporal character, maybe they are the best-suited way to define temporal treewidth.

7   Conclusion

»*Here's looking at you*«, temporal treewidth. Indeed, it is a worthwhile endeavor to explore the prospects and limitations of parameters such as treewidth transformed to the context of temporal graphs. A lot of exploration and clarification is yet to do. So let us agree, in temporal treewidth future we see. Hans, can you?

Acknowledgments. HM and MR acknowledge support by DFG, project MATE (NI 369/17). TF acknowledges support by DFG, project TORE (NI 369/18).

We thank Mark de Berg, Anne-Sophie Himmel, Frank Kammer, Sándor Kisfaludi-Bak, Erik Jan van Leeuwen, and George B. Mertzios for their constructive feedback which helped us to improve the presentation of the paper.

We further thank Bernard Mans and Luke Mathieson for helpful discussions concerning the issues presented in Sect. 5.3.

A   Temporal Graph Problem Zoo

(α, β)-TEMPORAL REACHABILITY EDGE DELETION $((\alpha, \beta)$-TRED)
Input: A temporal graph $\mathcal{G} = (V, \mathcal{E}, \tau)$ and two integers $k, h \in \mathbb{N}_0$.
Question: Is there a subset $E' \subseteq E(\mathcal{G}_\downarrow)$ of the underlying graph's edges with $|E'| \leq k$ such that in $\mathcal{G} - (E' \times \{1, \ldots, \tau\})$, the size of the set of vertices reachable from every vertex $s \in V$ via a strict (α, β)-temporal path is at most h?

(α, β)-TEMPORAL REACHABILITY TIME-EDGE DELETION $((\alpha, \beta)$-TRTED)
Input: A temporal graph $\mathcal{G} = (V, \mathcal{E}, \tau)$ and two integers $k, h \in \mathbb{N}_0$.
Question: Is there a subset $\mathcal{E}' \subseteq \mathcal{E}$ of temporal edges with $|\mathcal{E}'| \leq k$ such that in $\mathcal{G} - \mathcal{E}'$, the size of the set of vertices reachable from every vertex $s \in V$ via a strict (α, β)-temporal path is at most h?

MINIMUM SINGLE-SOURCE TEMPORAL CONNECTIVITY (r-MTC)
Input: A temporal graph $\mathcal{G} = (V, \mathcal{E}, \tau)$ with edge weights $w : \mathcal{E} \to \mathbb{Q}$, a designated vertex $r \in V$, and a number $k \in \mathbb{Q}$.
Question: Is there a temporally r-connected spanning subgraph of $\mathcal{G}$ of weight at most k?

MIN-MAX REACHABILITY TEMPORAL ORDERING (MIN-MAX RTO)
Input: A temporal graph $\mathcal{G} = (V, \mathcal{E}, \tau)$, and an integer $k \in \mathbb{N}$.
Question: Is there a bijection $\phi : \{1, \ldots, \tau\} \to \{1, \ldots, \tau\}$ such that the maximum reachability in $\mathcal{G}' = (V, \{(e, \phi(t)) \mid (e, t) \in \mathcal{E}\}, \tau)$ is at most k?

MIN REACHABILITY TEMPORAL MERGING (MRTM)
Input: A temporal graph $\mathcal{G} = (V, \mathcal{E}, \tau)$, a set of sources $S \subseteq V$, and three integers $\lambda, \mu, k \in \mathbb{N}$.
Question: Are there μ disjoint intervals $M_1, \ldots, M_\mu \subseteq \{1, \ldots, \tau\}$, each of size λ, such that, after merging each of them in $\mathcal{G}$, the number of vertices reachable from S is at most k?

RETURN-TO-BASE TEMPORAL GRAPH EXPLORATION (RTB-TGE)
Input: A temporal graph $\mathcal{G} = (V, \mathcal{E}, \tau)$ and a designated vertex $s \in V$.
Question: Is there a strict temporal walk starting and ending at s that visits all vertices in V?

TEMPORAL GRAPH EXPLORATION (TGE)
Input: A temporal graph $\mathcal{G} = (V, \mathcal{E}, \tau)$ and a designated vertex $s \in V$.
Question: Is there a strict temporal walk starting at s that visits all vertices in V?

TEMPORAL MATCHING (TM)
Input: A temporal graph $\mathcal{G} = (V, \mathcal{E}, \tau)$ and integers $k, \Delta \in \mathbb{N}_0$.
Question: Is there a set of k temporal edges $\mathcal{E}' \subseteq \mathcal{E}$ such that any pair $\{(e, t), (e', t')\} \subseteq \mathcal{E}'$ has $e \cap e' = \emptyset$ or $|t - t'| \geq \Delta$?

TEMPORAL SEPARATION (TS)
Input: A temporal graph $\mathcal{G} = (V, \mathcal{E}, \tau)$, two designated vertices $s, z \in V$, and an integer $k \in \mathbb{N}$.
Question: Is there a temporal s-z separator of size at most k in $\mathcal{G}$?

References

1. Abraham, I., Chechik, S., Delling, D., Goldberg, A.V., Werneck, R.F.: On dynamic approximate shortest paths for planar graphs with worst-case costs. In: Proceedings of the 27th Annual ACM-SIAM Symposium on Discrete Algorithms (SODA 2016), pp. 740–753. SIAM (2016)
2. Akrida, E.C., Mertzios, G.B., Spirakis, P.G.: The temporal explorer who returns to the base. In: Heggernes, P. (ed.) CIAC 2019. LNCS, vol. 11485, pp. 13–24. Springer, Cham (2019). https://doi.org/10.1007/978-3-030-17402-6_2
3. Akrida, E.C., Mertzios, G.B., Spirakis, P.G., Zamaraev, V.: Temporal vertex cover with a sliding time window. J. Comput. Syst. Sci. **107**, 108–123 (2020)
4. Arnborg, S., Corneil, D.G., Proskurowski, A.: Complexity of finding embeddings in a k-tree. SIAM J. Algebraic Discrete Methods **8**(2), 277–284 (1987)
5. Arnborg, S., Lagergren, J., Seese, D.: Easy problems for tree-decomposable graphs. J. Algorithms **12**(2), 308–340 (1991)
6. Axiotis, K., Fotakis, D.: On the size and the approximability of minimum temporally connected subgraphs. In: Proceedings of the 43rd International Colloquium on Automata, Languages, and Programming (ICALP 2016). LIPIcs, vol. 55, pp. 149:1–149:14. Schloss Dagstuhl - Leibniz-Zentrum für Informatik (2016)
7. Bentert, M., Himmel, A.S., Molter, H., Morik, M., Niedermeier, R., Saitenmacher, R.: Listing all maximal k-plexes in temporal graphs. ACM J. Exp. Algorithmics **24**(1), 1–13 (2019)
8. Betzler, N., Bredereck, R., Niedermeier, R., Uhlmann, J.: On bounded-degree vertex deletion parameterized by treewidth. Discrete Appl. Math. **160**(1–2), 53–60 (2012)
9. Bodlaender, H.L.: Dynamic algorithms for graphs with treewidth 2. In: van Leeuwen, J. (ed.) WG 1993. LNCS, vol. 790, pp. 112–124. Springer, Heidelberg (1994). https://doi.org/10.1007/3-540-57899-4_45
10. Bodlaender, H.L.: A tourist guide through treewidth. Acta Cybernetica **11**(1–2), 1–21 (1993)
11. Bodlaender, H.L.: A linear-time algorithm for finding tree-decompositions of small treewidth. SIAM J. Comput. **25**(6), 1305–1317 (1996)
12. Bodlaender, H.L.: The algorithmic theory of treewidth. Electron. Notes Discrete Math. **5**, 27–30 (2000)
13. Bodlaender, H.L., Drange, P.G., Dregi, M.S., Fomin, F.V., Lokshtanov, D., Pilipczuk, M.: A $c^k n$ 5-approximation algorithm for treewidth. SIAM J. Comput. **45**(2), 317–378 (2016)
14. Bodlaender, H.L., Hagerup, T.: Parallel algorithms with optimal speedup for bounded treewidth. SIAM J. Comput. **27**(6), 1725–1746 (1998)
15. Bodlaender, H.L., Kloks, T.: Efficient and constructive algorithms for the pathwidth and treewidth of graphs. J. Algorithms **21**(2), 358–402 (1996)
16. Bodlaender, H.L., Kloks, T., Kratsch, D.: Treewidth and pathwidth of permutation graphs. SIAM J. Discrete Math. **8**(4), 606–616 (1995)
17. Bodlaender, H.L., Kloks, T., Kratsch, D., Müller, H.: Treewidth and minimum fill-in on d-trapezoid graphs. J. Graph Algorithms Appl. **2**(5), 1–23 (1998)
18. Bodlaender, H.L., Koster, A.M.C.A.: Treewidth computations II. Lower bounds. Inf. Comput. **209**(7), 1103–1119 (2011)
19. Bodlaender, H.L., Möhring, R.H.: The pathwidth and treewidth of cographs. SIAM J. Discrete Math. **6**(2), 181–188 (1993)

20. Bodlaender, H.L., Thilikos, D.M.: Treewidth for graphs with small chordality. Discrete Appl. Math. **79**(1–3), 45–61 (1997)
21. Bodlaender, H.L., van der Zanden, T.C.: On exploring always-connected temporal graphs of small pathwidth. Inf. Process. Lett. **142**, 68–71 (2019)
22. Bouchitté, V., Todinca, I.: Treewidth and minimum fill-in: grouping the minimal separators. SIAM J. Comput. **31**(1), 212–232 (2001)
23. Casteigts, A., Flocchini, P., Quattrociocchi, W., Santoro, N.: Time-varying graphs and dynamic networks. Int. J. Parallel Emergent Distrib. Syst. **27**(5), 387–408 (2012)
24. Casteigts, A., Himmel, A.S., Molter, H., Zschoche, P.: The computational complexity of finding temporal paths under waiting time constraints. CoRR abs/1909.06437 (2019)
25. Casteigts, A., Peters, J.G., Schoeters, J.: Temporal cliques admit sparse spanners. In: Proceedings of the 46th International Colloquium on Automata, Languages, and Programming (ICALP 2019). LIPIcs, vol. 132, pp. 134:1–134:14. Schloss Dagstuhl - Leibniz-Zentrum für Informatik (2019)
26. Courcelle, B., Engelfriet, J.: Graph Structure and Monadic Second-Order Logic: A Language-Theoretic Approach. Cambridge University Press, Cambridge (2012)
27. Cygan, M., et al.: Parameterized Algorithms. Springer, Heidelberg (2015). https://doi.org/10.1007/978-3-319-21275-3
28. Dailey, D.P.: Uniqueness of colorability and colorability of planar 4-regular graphs are NP-complete. Discrete Math. **30**(3), 289–293 (1980)
29. Deligkas, A., Potapov, I.: Optimizing reachability sets in temporal graphs by delaying. In: Proceedings of the 34th AAAI Conference on Artificial Intelligence (AAAI 2020). AAAI Press (2020, to appear)
30. Dell, H., Husfeldt, T., Jansen, B.M.P., Kaski, P., Komusiewicz, C., Rosamond, F.A.: The first parameterized algorithms and computational experiments challenge. In: Proceedings of the 11th International Symposium on Parameterized and Exact Computation (IPEC 2016). LIPIcs, vol. 63, pp. 30:1–30:9. Schloss Dagstuhl - Leibniz-Zentrum für Informatik (2016)
31. Dell, H., Komusiewicz, C., Talmon, N., Weller, M.: The PACE 2017 parameterized algorithms and computational experiments challenge: the second iteration. In: Proceedings of the 12th International Symposium on Parameterized and Exact Computation (IPEC 2017). LIPIcs, vol. 89, pp. 30:1–30:12. Schloss Dagstuhl - Leibniz-Zentrum für Informatik (2017)
32. Dom, M., Lokshtanov, D., Saurabh, S., Villanger, Y.: Capacitated domination and covering: a parameterized perspective. In: Grohe, M., Niedermeier, R. (eds.) IWPEC 2008. LNCS, vol. 5018, pp. 78–90. Springer, Heidelberg (2008). https://doi.org/10.1007/978-3-540-79723-4_9
33. Downey, R.G., Fellows, M.R.: Parameterized Complexity. Monographs in Computer Science. Springer, Heidelberg (1999). https://doi.org/10.1007/978-1-4612-0515-9
34. Downey, R.G., Fellows, M.R.: Fundamentals of Parameterized Complexity. Springer, London (2013). https://doi.org/10.1007/978-1-4471-5559-1
35. Dvořák, P., Knop, D.: Parameterized complexity of length-bounded cuts and multicuts. Algorithmica **80**(12), 3597–3617 (2018)
36. Enright, J., Meeks, K., Mertzios, G., Zamaraev, V.: Deleting edges to restrict the size of an epidemic in temporal networks. In: Proceedings of the 44nd International Symposium on Mathematical Foundations of Computer Science (MFCS 2019), pp. 57:1–57:15. LIPIcs, Schloss Dagstuhl - Leibniz-Zentrum für Informatik (2019)

37. Enright, J., Meeks, K., Skerman, F.: Changing times to optimise reachability in temporal graphs. CoRR abs/1802.05905 (2018)
38. Erlebach, T., Hoffmann, M., Kammer, F.: On temporal graph exploration. In: Halldórsson, M.M., Iwama, K., Kobayashi, N., Speckmann, B. (eds.) ICALP 2015. LNCS, vol. 9134, pp. 444–455. Springer, Heidelberg (2015). https://doi.org/10.1007/978-3-662-47672-7_36. Updated version available at https://arxiv.org/abs/1504.07976v2
39. Erlebach, T., Kammer, F., Luo, K., Sajenko, A., Spooner, J.T.: Two moves per time step make a difference. In: Proceedings of the 46th International Colloquium on Automata, Languages, and Programming (ICALP 2019). LIPIcs, vol. 132, pp. 141:1–141:14. Schloss Dagstuhl - Leibniz-Zentrum für Informatik (2019)
40. Erlebach, T., Spooner, J.T.: Faster exploration of degree-bounded temporal graphs. In: Proceedings of the 43rd International Symposium on Mathematical Foundations of Computer Science (MFCS 2018). LIPIcs, vol. 117, pp. 36:1–36:13. Schloss Dagstuhl - Leibniz-Zentrum für Informatik (2018)
41. Erlebach, T., Spooner, J.T.: A game of cops and robbers on graphs with periodic edge-connectivity. CoRR abs/1908.06828 (2019)
42. Flocchini, P., Mans, B., Santoro, N.: On the exploration of time-varying networks. Theor. Comput. Sci. **469**, 53–68 (2013)
43. Flum, J., Grohe, M.: Parameterized Complexity Theory. TTCSAES, vol. XIV. Springer, Heidelberg (2006). https://doi.org/10.1007/3-540-29953-X
44. Fluschnik, T., Kratsch, S., Niedermeier, R., Sorge, M.: The parameterized complexity of the minimum shared edges problem. J. Comput. Syst. Sci. **106**, 23–48 (2019)
45. Fluschnik, T., Molter, H., Niedermeier, R., Renken, M., Zschoche, P.: Temporal graph classes: a view through temporal separators. Theor. Comput. Sci. **806**, 197–218 (2020)
46. Fluschnik, T., Niedermeier, R., Rohm, V., Zschoche, P.: Multistage vertex cover. In: Proceedings of the 14th International Symposium on Parameterized and Exact Computation (IPEC 2019). LIPIcs, vol. 148, pp. 14:1–14:14. Schloss Dagstuhl - Leibniz-Zentrum für Informatik (2019)
47. Froese, V., Jain, B., Niedermeier, R., Renken, M.: Comparing temporal graphs using dynamic time warping. In: Cherifi, H., Gaito, S., Mendes, J.F., Moro, E., Rocha, L.M. (eds.) COMPLEX NETWORKS 2019. SCI, vol. 882, pp. 469–480. Springer, Cham (2020). https://doi.org/10.1007/978-3-030-36683-4_38
48. Gassner, E.: The Steiner forest problem revisited. J. Discrete Algorithms **8**(2), 154–163 (2010)
49. Himmel, A.S.: Algorithmic investigations into temporal paths. Master thesis, TU Berlin, April 2018
50. Himmel, A.-S., Bentert, M., Nichterlein, A., Niedermeier, R.: Efficient computation of optimal temporal walks under waiting-time constraints. In: Cherifi, H., Gaito, S., Mendes, J.F., Moro, E., Rocha, L.M. (eds.) COMPLEX NETWORKS 2019. SCI, vol. 882, pp. 494–506. Springer, Cham (2020). https://doi.org/10.1007/978-3-030-36683-4_40
51. Himmel, A.S., Molter, H., Niedermeier, R., Sorge, M.: Adapting the Bron-Kerbosch algorithm for enumerating maximal cliques in temporal graphs. Soc. Netw. Anal. Min. **7**(1), 35:1–35:16 (2017)
52. Holme, P., Saramäki, J.: Temporal networks. CoRR abs/1108.1780 (2011)
53. Kloks, T.: Treewidth, Computations and Approximations. LNCS, vol. 842. Springer, Heidelberg (1994). https://doi.org/10.1007/BFb0045375

54. Mans, B., Mathieson, L.: On the treewidth of dynamic graphs. Theor. Comput. Sci. **554**, 217–228 (2014)

55. Marx, D.: NP-completeness of list coloring and precoloring extension on the edges of planar graphs. J. Graph Theory **49**(4), 313–324 (2005)

56. Marx, D.: Complexity results for minimum sum edge coloring. Discrete Appl. Math. **157**(5), 1034–1045 (2009)

57. Mertzios, G.B., Molter, H., Niedermeier, R., Zamaraev, V., Zschoche, P.: Computing maximum matchings in temporal graphs. CoRR abs/1905.05304 (2019). To appear in Proceedings of the 37th International Symposium on Theoretical Aspects of Computer Science (STACS 2020), Schloss Dagstuhl - Leibniz-Zentrum fuer Informatik. LIPIcs, vol. 154, pp. 27:1–27:14 (2020)

58. Mertzios, G.B., Molter, H., Zamaraev, V.: Sliding window temporal graph coloring. In: Proceedings of the 33rd AAAI Conference on Artificial Intelligence (AAAI 2019), pp. 7667–7674. AAAI Press (2019)

59. Misra, J., Gries, D.: A constructive proof of Vizing's theorem. Inf. Process. Lett. **41**(3), 131–133 (1992)

60. Molter, H., Niedermeier, R., Renken, M.: Enumerating isolated cliques in temporal networks. In: Cherifi, H., Gaito, S., Mendes, J.F., Moro, E., Rocha, L.M. (eds.) COMPLEX NETWORKS 2019. SCI, vol. 882, pp. 519–531. Springer, Cham (2020). https://doi.org/10.1007/978-3-030-36683-4_42

61. Niedermeier, R.: Invitation to Fixed-Parameter Algorithms. Oxford University Press, Oxford (2006)

62. Nishizeki, T., Vygen, J., Zhou, X.: The edge-disjoint paths problem is NP-complete for series-parallel graphs. Discrete Appl. Math. **115**(1–3), 177–186 (2001)

63. Seymour, P.D., Thomas, R.: Graph searching and a min-max theorem for treewidth. J. Comb. Theory Series B **58**(1), 22–33 (1993)

64. Viard, T., Latapy, M., Magnien, C.: Computing maximal cliques in link streams. Theor. Comput. Sci. **609**, 245–252 (2016)

65. Zschoche, P., Fluschnik, T., Molter, H., Niedermeier, R.: The complexity of finding small separators in temporal graphs. J. Comput. Syst. Sci. **107**, 72–92 (2020)

Possible and Impossible Attempts to Solve the Treewidth Problem via ILPs

Alexander Grigoriev[✉][iD]

Maastricht University School of Business and Economics,
Maastricht, The Netherlands
a.grigoriev@maastrichtuniversity.nl

Abstract. We survey a number of integer programming formulations for the pathwidth and treewidth problems. The attempts to find good formulations for the problems span the period of 15 years, yet without any true success. Nevertheless, some formulations provide potentially useful frameworks for attacking these notorious problems. Some others are just curious and interesting fruits of mathematical imagination.

Keywords: Treewidth · Pathwidth · Integer programming

1 Introduction

According to unverified and, most likely, unverifiable rumors, the first integer programming formulation for the treewidth problem has been developed by Hans Bodlaender and Arie Koster on the train after a WG conference about fifteen years ago. Arie brought the problem to the Maastricht mathematical programming community with a clear call to design such an integer programming formulation that, on a modern computer with contemporary ILP packages, it would outperform the state-of-the-art dynamic programming algorithms. The battle of models and algorithms had began.

To the large disappointment of the community, all formulations designed so far are far from being a reasonable tool to solve the treewidth or the pathwidth problem. Some of the formulations exhibit a high level of symmetry and corresponding integrality gaps are huge. Some others successfully fight the symmetry, but at the cost of large formulation size. None of the integer linear programs presented in this survey is getting close to a solution for either problem on graphs of thirty vertices. Nevertheless, a variety of formulations, some intriguing properties of the integer linear programs and the very generic mathematical approach to pathwidth/treewidth make this research interesting and potentially valuable.

In this paper we present four different integer programs for pathwidth and treewidth problems. We are aware that there are other very nice and closely related integer programming formulations, e.g., by Hicks [7,10]. For this work, we selected the simplest and most intuitive formulations directly based on the definitions of treewidth and pathwidth. We present linear programs in a didactic

© Springer Nature Switzerland AG 2020
F. V. Fomin et al. (Eds.): Bodlaender Festschrift, LNCS 12160, pp. 78–88, 2020.
https://doi.org/10.1007/978-3-030-42071-0_7

and scholarly way to popularize and boost this line of research. The paper is meant as a sequel to the surveys on treewidth by Bodlaender and Koster [2,3]. We hope to co-author an extended and re-worked version of this survey with Hans and Arie.

For now, we cut short on the introduction by postponing all formal definitions to the subsequent sections. This is done because different integer programming formulations are based on different, though equivalent, definitions of pathwidth/treewidth parameters. We do assume that the reader is familiar with basic graph theory and terms: linear programming, integrality gap, primal problem, dual problem, separation, optimization, approximation, branch-and-bound algorithm, branch-and-price algorithm. Through the rest of the paper, we consider an undirected simple graph $G = (V, E)$ on n vertices and m edges. The subjects of our investigation, graph parameters pathwidth and treewidth are referred as $\mathbf{pw}(G)$ and $\mathbf{tw}(G)$, respectively.

2 An Interval Representation of Pathwidth

As a warm-up, we present a simple and extremely compact mathematical programming formulation of the pathwidth problem based on an interval representation of pathwidth.

Definition 1 (Non-conventional definition of pathwidth). *Given a graph* $G = (V, E)$, *for each vertex* $v \in V$ *specify a real interval* $[s_v, t_v]$, $s_v < t_v$, *such that for any edge* $(v, u) \in E$ *intervals* $[s_v, t_v]$ *and* $[s_u, t_u]$ *have a non-empty intersection. Let* I *be a set of such intervals. The pathwidth* $\mathbf{pw}(G)$ *is the minimum, over the choice of* I, *of the maximum number of mutually intersecting intervals in* I *minus 1.*

Without loss of generality, we may assume that $s_v, t_v \in \{0, 1, \dots, n\}$. For all $v \in V$ and for all $i \in \{0, 1, \dots, n\}$, let us introduce two binary variables x_{vi} and y_{vi}. Let variable x_{vi} take value 1 if $s_v = i$ and 0 otherwise. Let variable y_{vi} take value 1 if $t_v = i$ and 0 otherwise. Then, the mixed-integer programming formulation for the pathwidth problem based on Definition 1 reads:

$$\mathbf{pw}(G) = \min_{w \geq 0, x, y} w - 1 \tag{1}$$

$$w \geq \sum_{v \in V} \sum_{i=0}^{j} (x_{vi} - y_{vi}) \quad \forall j \in \{0, 1, \dots, n\}; \tag{2}$$

$$\sum_{i=0}^{n} x_{vi} = 1 \quad \forall v \in V; \tag{3}$$

$$\sum_{i=0}^{n} y_{vi} = 1 \quad \forall v \in V; \tag{4}$$

$$\sum_{i=0}^{j} x_{vi} \geq \sum_{i=0}^{j} y_{vi} \quad \forall v \in V, j \in \{0, 1, \ldots, n\}; \tag{5}$$

$$\sum_{i=0}^{j} x_{vi} \geq \sum_{i=0}^{j} y_{ui} \quad \forall (v, u) \in E, j \in \{0, 1, \ldots, n\}; \tag{6}$$

$$x_{vi}, y_{vi} \in \{0, 1\} \quad \forall v \in V, j \in \{0, 1, \ldots, n\}, \tag{7}$$

where Eq. (2) counts the number of intersecting intervals at every point $j \in \{0, 1, \ldots, n\}$ and reports the maximum number of intersections over all points; Eqs. (3) and (4) assign the interval boundaries for every $v \in V$; Eq. (5) stands for $s_v < t_v$ for all $v \in V$; and Eq. (6) guarantees that for any $(v, u) \in E$ the intervals $[s_v, t_v]$ and $[s_u, t_u]$ have a nonempty intersection. The number of constraints with Eqs. (5) and (6) can be replaced (respectively) by

$$\sum_{i=0}^{n}(i+1)\, x_{vi} \leq \sum_{i=0}^{n}(i+1)\, y_{vi}, \quad \forall v \in V; \tag{8}$$

$$\sum_{i=0}^{n}(i+1)\, x_{vi} \leq \sum_{i=0}^{n}(i+1)\, y_{ui}, \quad \forall (v, u) \in E. \tag{9}$$

This model was tested on standard ILP packages and it can tackle instances with up to 16–18 vertices. The drawback of the formulation is obvious: in the optimal solution to the linear relaxation all values of x and y for every vertex are uniformly spread on the interval $\{0, 1, \ldots, n\}$. To close such a huge integrality gap, the underlying branch-and-bound algorithm creates an enormous number of nodes in the branching tree.

Interestingly, there is a well-known way to fight the symmetry in this case. For all $v \in V$, instead of using many x and many y variables when determining an interval $[s_v, t_v]$, one may straightforwardly use only two continuous variables s_v and t_v. Then, any set of intervals satisfying inequalities $s_v \leq t_u$ for all $(v, u) \in E$, is a feasible interval representation I in the Definition 1. Unfortunately, to model the objective (pathwidth) we need step functions $f_v(x) : \mathbb{R} \to \{0, 1\}$ taking value 1 if $x \in [s_v, t_v]$ and 0 otherwise. We are not aware of any techniques to model such step functions without blowing up the number of variables. Fortunately, in data analytics and statistics, such step functions are often and successfully approximated with sigmoid functions:

$$\hat{f}_v(x) = \frac{1}{1 + e^{-\alpha(x - s_v)}} + \frac{1}{1 + e^{\alpha(x - t_v)}} - 1, \quad \forall i \in V, \tag{10}$$

where $\alpha > 0$ is an adjustable parameter to tune up the approximation. Then, for large values of α, a very good proxy to the pathwidth of G is the solution to the following mathematical program.

$$\min_{w \geq 0, \mu, \sigma} w \tag{11}$$

$$w \geq \sum_{v \in V} \hat{f}_v(k) \ \ \forall k \in \{0, 1, \ldots, n\} \tag{12}$$

$$s_v \leq t_u \ \ \forall (v, u) \in E; \tag{13}$$

$$1 \leq s_v + 1 \leq t_v \leq n \ \forall i \in V. \tag{14}$$

Notice, this non-linear mathematical program is very compact (small size) and based on continuous variables. Locally optimal solutions to this problem can be found by any generic non-linear solver in almost no-time. However, the sum of sigmoids is a highly multi-modal function and solutions returned by the solvers are often far from the global optimum. On the other hand, it would be very interesting to see how well contemporary multi-modal optimization techniques, see e.g. [8], can tackle this problem.

3   Perfect Elimination Ordering

In this section we present the original integer programming formulation of the treewidth problem designed by Bodlaender and Koster [1] in a train. This formulation is based on the notion of a perfect elimination ordering.

Definition 2 (Non-conventional definition of treewidth). *A perfect elimination ordering (PEO) in a graph is an ordering of the vertices of the graph such that, for each vertex $v \in V$, v and the neighbors of v that occur after v in this order form a clique. The treewidth of G is one less than the number of vertices in a maximum clique after adding edges in a PEO chosen to minimize this clique size. Further, the edges from E are referred to regular edges, while the edges added in the construction of a PEO are referred to fill-in edges.*

To design the corresponding ILP, we introduce the following set of variables. For every $v, u \in V$, let

$$x_{vu} = \begin{cases} 1, & \text{if } v \text{ is ordered before } u \text{ in PEO} \\ 0, & \text{otherwise} \end{cases}$$

$$y_{vu} = \begin{cases} 1, & \text{if } v \text{ is ordered before } u \text{ in PEO} \\ & \text{and } (v, u) \text{ is a regular or a fill-in edge} \\ 0, & \text{otherwise} \end{cases}$$

Then, the following mixed integer linear program addresses the treewidth problem:

$$\mathbf{tw}(G) = \min_{w \geq 0, x, y} w - 1 \tag{15}$$

$$w \geq \sum_{u \in V: \ u \neq v} y_{vu} \ \ \forall v \in V; \tag{16}$$

$$x_{vu} + x_{uv} = 1 \ \forall v, u \in V; \tag{17}$$

$$x_{vu} + x_{uw} - x_{vw} \leq 1 \ \forall v, u, w \in V; \tag{18}$$

$$y_{vu} \leq x_{vu} \ \forall v, u \in V; \tag{19}$$

$$y_{vu} = x_{vu} \ \forall (v, u) \in E; \tag{20}$$

$$x_{uw} + y_{vu} + y_{vw} - y_{uw} \leq 2 \ \forall v, u, w \in V; \tag{21}$$

$$x_{vu} \in \{0, 1\}, y_{vu} \in \{0, 1\} \ \forall v, u \in V. \tag{22}$$

Here, Eq. (15) counts the size of the largest clique after adding edges in PEO. Constraints (16) and (17) form the classic ordering polytope, where Equations (16) assign the precedence for every pair of vertices and Eqs. (17) are the transitivity constraints. Restrictions (18) are needed to guarantee that only vertices occurring ahead in the order are counted in the clique. Equations (19) make sure that the regular edges are counted in the clique. Equations (20) guarantee that any two neighbors of a vertex, which are ahead in the order, are adjacent by a regular or a fill-in edge.

It is a folklore in the mathematical programming community that the linear relaxation of the integer program (16)–(17) is quite poor: assigning to all x-variables values $1/2$ is feasible. The consequence for the linear relaxation of MIP (14)–(21) is that letting $y_{vu} = 1/2$ for all $(v, u) \in E$, we receive the lower bound for $\mathbf{tw}(G)$ of at most $\delta_{\max}/2$, where $\delta_{\max}$ is the maximum degree of a vertex in G. Then, the integrality gap can be arbitrarily high, e.g., in a $(g \times g)$-grid graph the maximum degree of a vertex is 4, while the treewidth is $g = \sqrt{n}$.

The advantage of the formulations based on PEO is in the general improvements made for the ordering polytopes. For instance, Rao and Richa [9] and Bornstein and Vempala [4] suggested using spreading metrics to revise and improve the ordering polytopes. Particularly, Bornstein and Vempala [4] see the vertex ordering as a flow. Particularly, let $f_v^{u,w}$ be the flow from v to w that goes through the vertex u, and for $v, u \in V$ let $d(v, u)$ be the distance in the order between v and u. Then, their "new" ordering polytope is defined by the following constraints:

$$f_u^{v,u} + f_v^{u,v} = 1 \ \forall v, u \in V; \tag{23}$$

$$f_w^{v,u} + f_w^{u,v} + f_v^{u,w} + f_v^{w,u} + f_u^{v,w} + f_u^{w,v} = 1 \ \forall v, u, w \in V; \tag{24}$$

$$d(v, u) = \sum_{w \in V} f_w^{v,u} \ \forall v, u \in V; \tag{25}$$

$$d(v, u) + d(u, w) \geq d(v, w) \ \forall v, u, w \in V; \tag{26}$$

$$\sum_{v, u \in S} d(v, u) \geq \binom{|S|}{3} \ \forall S \subset V. \tag{27}$$

Constraints (22) establish the order for a pair of vertices. Constraints (23) do the same for any triplet of vertices. Constraints (24) define the distance between any two vertices v and u based on the number of vertices the flow from v to u has to pass. Constraints (25) are standard triangle inequalities. Inequalities (26) are spreading inequalities introduced by Rao and Richa [9]. The latter inequalities are significantly strengthening the formulation. Though the number of these constraints is exponential, a violated inequality can be found in polynomial time implying that separation and optimization problems on this polytope are polynomially solvable.

It is remarkable that based on the new ordering polytope Bornstein and Vempala [4] were able to construct a $O(\log^2 n)$-approximation algorithm for the pathwidth problem, which is the best known approximation for pathwidth. Moreover, their approximation technique is perfectly applicable to many other vertex ordering problems [4]. Usotskaya, in her PhD thesis [11], introduced a formulation that combines the programs (14)–(21) and (22)–(26) when modelling the treewidth. The disadvantage of the "new" ordering polytope and the combined program modelling the treewidth by Usotskaya [11] is the size of the model and the usage of separation machinery. Directly using these formulations in standard packages is practically impossible.

4 Tree Drawing Formulation

The following definition of the treewidth is even less conventional, but very intuitive and insightful. We learned this definition from the talk of Demaine at ISAAC 2002.

Definition 3 (Even less conventional definition of treewidth). *Draw a backbone tree T in the plane. For each $v \in V$ identify a subtree T_v, which is a connected subgraph of T. Consider a set I of these subtrees such that for every $(v, u) \in E$ the subtrees T_v and T_u have non-empty intersection. The treewidth of G is the minimum over the choice of the backbone tree T and the set I of the maximum number of mutually intersecting subtrees minus one.*

In the graph drawing community, there are many nice results addressing tree drawings in the plane. For instance, the following theorem allows to draw a binary tree in a small size grid in a very structured way.

Theorem 1 (Crescenzi et al. [5]). *Any binary tree on n vertices can be properly embedded in the $n \times (1 + \log n)$ grid in the (rectilinear) right-down fashion, i.e., if a grid point (i, j) is a part of the tree embedding, it is forbidden to simultaneously use grid edges $(i, j + 1)$ and $(i - 1, j)$ in the tree embedding.*

This theorem allows us to model backbone and subtree drawing in an integer linear program. Let Y be the set of points in the $n \times (1 + \log n)$ grid and let Z be the set of edges in the grid. By definition of a right-down drawing, the backbone tree and its subtrees have unique left upper corners in their respective embeddings. For every $v \in V$ and for every $y \in Y$ we introduce the variables:

$$s_{vy} = \begin{cases} 1, & \text{if } y \text{ is the unique left upper corner of } T_v; \\ 0, & \text{otherwise,} \end{cases}$$

$$u_{vy} = \begin{cases} 1, & \text{if } y \text{ belongs to the embedding of } T_v; \\ 0, & \text{otherwise.} \end{cases}$$

We also introduce the variables for every vertex $v \in V$ and every grid edge $e \in Z$:

$$x_{ve} = \begin{cases} 1, & \text{if } e \text{ is a part of the } T_v \text{ embedding;} \\ 0, & \text{otherwise.} \end{cases}$$

Then, according to Definition 3, the following mixed-integer program computes the treewidth:

$$\mathbf{tw}(G) = \min_{w \geq 0, s, u, x} w - 1 \tag{28}$$

$$w \geq \sum_{v \in V} u_{vy}, \ \forall y \in Y; \tag{29}$$

$$\sum_{y \in Y} s_{vy} = 1, \ \forall v \in V; \tag{30}$$

$$x_{v(y_\ell, y)} + x_{u(y, y_u)} \leq 1, \ \forall v, u \in V, y \in Y; \tag{31}$$

$$u_{vy} = s_{vy} + x_{v(y_\ell, y)} + x_{v(y, y_u)}, \ \forall v \in V, y \in Y; \tag{32}$$

$$x_{v(y, y_r)} \leq u_{vy}, \ \forall v \in V, y \in Y; \tag{33}$$

$$x_{v(y_b, y)} \leq u_{vy}, \ \forall v \in V, y \in Y; \tag{34}$$

$$s_{uy} + \sum_{y' \in Y: y' \npreceq y, y \npreceq y'} s_{vy'} \leq 1, \ \forall (v, u) \in E, y \in Y; \tag{35}$$

$$s_{uy} + \sum_{y' \preceq y} s_{vy'} \leq 1 + u_{vy}, \ \forall (v, u) \in E, y \in Y; \tag{36}$$

$$s_{vy}, u_{vy} \in \{0, 1\}, \ \forall v \in V, y \in Y; \tag{37}$$

$$x_{v,e} \in \{0, 1\}, \ \forall v \in V, e \in Z, \tag{38}$$

where y_l, y_u, y_r, y_b are the left, upper, right and bottom neighbors of $y \in Y$ in the grid, and the precedence $y' \preceq y$ means $y' \in Y$ is situated on the grid not to the right and not lower than $y \in Y$. Constraint (28) counts how many subtrees are met in a point $y \in Y$ and returns the maximum number of intersecting subtrees over all points in Y. Constraint (29) assigns a unique left upper corner for every subtree. Condition (30) preserves the right-down drawing. Constraints (31)–(33) guarantee connectedness of subtree drawing. Conditions (34) make sure that for every edge $(v, u) \in E$ either the unique left upper corner of T_v is to the left and higher than the left upper corner of T_u or vice versa. Notice, with the right-down drawing, two trees can meet only if the unique left upper corner of one is to the left and higher than the left upper corner of another, or vice versa. Moreover, if two trees of such drawing are met, the unique left upper corner of one of them is an element of the other, and Condition (35) requires that for every $(v, u) \in E$ the corresponding subtrees T_v and T_u meet each other in one of their unique left upper corners.

This formulation is quite large in size and it was not properly tested. The advantage of this formulation is that it has clear geometric flavour which can be exploit in future, e.g., setting up valid inequalities to strengthen the formulation.

5   Set Partitioning Formulation

In this section we present a set partitioning formulation for pathwidth. In the mathematical programming community, similar formulations are often called *configuration* ILPs. This formulation is based on a classic definition of pathwidth.

Definition 4. *Given a graph $G = (V, E)$, a path decomposition of G is a pair (X, P) such that*

- $X \subseteq 2^{|V|}$ *and P is a path on X;*
- *For any edge $(v, u) \in E$ there is $x \in X$ such that $\{v, u\} \subseteq x$;*
- *For any vertex $v \in V$, if $v \in x$ and $v \in x'$, then $v \in x''$ for all nodes x'' on the path from x to x' in P.*

The pathwidth $\mathbf{pw}(G)$ is the minimum over (X, P) maximum size $x \in X$ minus 1. For convenience, we refer to $x \in X$ as a bag of the path decomposition.

Let P be a path on n nodes. Let S be the set of all non-empty subsets of V, which are potential bags of the path decomposition. As usual for configuration ILPs, for every $s \in S$ and every $t \in P$, we introduce the following variable:

$$x_{st} = \begin{cases} 1, \text{ if } s \text{ it is the } t\text{-th bag in } P; \\ 0, \text{ otherwise;} \end{cases}$$

Then, the following mixed-integer program models the patwidth:

$$\mathbf{pw}(G) = \min_{w \geq 0, x} w - 1 \tag{39}$$

$$w \geq \sum_{s \in S} |s|\, x_{st}, \ \forall t \in P; \tag{40}$$

$$\sum_{s \in S} x_{st} = 1, \ \forall t \in P; \tag{41}$$

$$\sum_{t \in P} \sum_{s \in S : v \in s} x_{st} \geq 1, \ \forall v \in V; \tag{42}$$

$$\sum_{t \in P} \sum_{s \in S : e \subseteq s} x_{st} \geq 1, \ \forall e \in E; \tag{43}$$

$$1 + \sum_{s \in S : v \in s} x_{st} \geq \sum_{s \in S : v \in s} x_{sq} + \sum_{s \in S : v \in s} x_{sr}, \ \forall v \in V, \ \forall q \prec t \prec r, \tag{44}$$

$$x_{st} \in \{0, 1\}, \ \forall s \in S, t \in P, \tag{45}$$

where $q \prec t \prec r$ simply means that in path P node t lies between q and r. Here, Constraint (39) captures the maximum size of a bag. Constraint (40) assigns a specific subset of vertices to a bag of the path decomposition. Conditions (41) and (42) make sure that every vertex and every edge of G are present in some bag of the decomposition. Inequalities (43) guarantee the third property of the path decomposition, see Definition 4.

After rearrangements and redundancy removal, the program reads:

$$\mathbf{pw}(G) = \min_{w \geq 0, x} w - 1 \tag{46}$$

$$w \geq \sum_{s \in S} |s| \, x_{st}, \ \forall t \in P; \tag{47}$$

$$\sum_{s \in S} x_{st} = 1, \ \forall t \in P; \tag{48}$$

$$\sum_{t \in P} \sum_{s \in S: e \subseteq s} x_{st} \geq 1, \ \forall e \in E; \tag{49}$$

$$\sum_{s \in S: v \in s} (x_{sq} + x_{sr} - x_{st}) \leq 1, \ \forall v \in V, \ \forall q \prec t \prec r; \tag{50}$$

$$x_{st} \in \{0, 1\}, \ \forall s \in S, t \in P. \tag{51}$$

The program (45)–(50) has exponentially many variables. The immediate question is whether the linear relaxation of such huge program can be solved in polynomial time. To address this question, consider the dual program:

$$\max_{\varphi, \lambda, \mu, \psi \geq 0} \sum_{e \in E} \mu_e - \sum_{t \in P} \lambda_t - \sum_{v \in V} \sum_{t \in P} \psi_{vt} \tag{52}$$

$$\sum_{t \in P} \varphi_t \leq 1; \tag{53}$$

$$\sum_{e \in E: e \subseteq s} \mu_e \leq |s| \varphi_t + \lambda_t + \sum_{v \in V: v \in s} (\psi_{vt} - \psi_{vt^-} - \psi_{vt^+}), \ \forall s \in S, t \in P, \tag{54}$$

where t^-, t^+ are predecessor and successor of t in P; and $\varphi, \lambda, \mu, \psi$ are the dual variables for constraints (46), (47), (48), (49), respectively. Clearly, if one can find a violated inequality in (53) in polynomial time, the linear relaxation of the program (45)–(50) can be efficiently solved. Using $\xi_{vt} = \psi_{vt^-} + \psi_{vt^+} - \psi_{vt} - \varphi_t$, we can reformulate (53) as follows

$$\sum_{e \in E: e \subseteq s} \mu_e + \sum_{v \in V: v \in s} \xi_{vt} \leq \lambda_t, \ \ \ \ \forall s \in S, t \in P. \tag{55}$$

Essentially, given a dual solution $(\varphi, \lambda, \mu, \psi)$ and $t \in P$, we have to find a set of vertices s maximizing $\sum_{e \in E: e \subseteq s} \mu_e + \sum_{v \in V: v \in s} \xi_{vt}$. If there is $t \in P$ such that $\sum_{e \in E: e \subseteq s} \mu_e + \sum_{v \in V: v \in s} \xi_{vt} > \lambda_t$ then we have found a violated inequality in (54).

In graph theoretic terms, the pricing problem (finding violated inequalities) can be formulated as follows. Given a graph G with vertex weights $\xi_v, v \in V$, and non-negative edge weights $\mu_e, e \in E$, we have to find an induced subgraph G' of G such that the total weight of all edges and vertices in that induced subgraph is maximized. This problem resembles the well-studied MAXIMUM DENSITY SUB-GRAPH problem (with no size restriction for the subgraph).

Theorem 2 (Goldberg [6]). *Given a graph G, let $\mu_e, e \in E$, be weights of the edges and $\xi_v, v \in V$, be weights of the nodes. Assume $\mu_e \geq 0$ for all $e \in E$. The density of a graph H is defined as*

$$\frac{\sum_{e \in E(H)} \mu_e + \sum_{v \in V(H)} \xi_v}{|V(H)|}.$$

One can find a densest induced subgraph H of G in time $O(n^6)$.

The only difference between the pricing problem and MAXIMUM DENSITY SUBGRAPH is the division by $|V(H)|$. A straightforward open question arises: can the Goldberg's algorithm be adjusted in such a way that it solves the pricing problem? Then, we would know whether the linear relaxation of the primal problem is solvable in polynomial time. Moreover, we would be able to design branch-and-price algorithms solving the pathwidth problem.

The advantages of the presented formulation are many-fold. First, the formulation should be treated as a framework rather than the final ILP. It allows to accommodate many interesting valid inequalities. As a straightforward example, consider the following *clique* inequality. For any clique C in G it holds that

$$\sum_{t \in P} \sum_{s \in S : C \subseteq s} x_{st} \geq 1.$$

Second, if a backbone tree (see the previous section) is given, the presented program can find an optimal set of bags for such a backbone tree, not only for the path. Third, the program clearly allows implementation of randomized algorithms, e.g., $P[v \in t] = \sum_{s \in S : v \in s} x'_{st}$ could easily serve as the probability of a random assignment of a vertex to a bag, where x'_{st} is a solution to the linear relaxation. The negative aspects of the formulations are also prominent: high integrality gap, large size and unclear complexity status of the pricing problem.

6   Conclusions

It becomes clear that development of practical LP-based algorithms for path-width and treewidth problems is still in a premature stage. The seemingly negative message can easily be turned to the positive one: Many interesting open questions and numerous ways to improve the mixed integer programs make this research challenging and very promising. So, "... always look at the bright side of life ..." — this is what we've learned from Hans Bodlaender and a bit from Monty Python.

References

1. Bodlaender, H.L.:Personal communications (2004)
2. Bodlaender, H.L., Koster, A.M.C.A.: Treewidth computations I. Upper bounds. Inf. Comput. **208**(3), 259–275 (2010). https://doi.org/10.1016/j.ic.2009.03.008

3. Bodlaender, H.L., Koster, A.M.C.A.: Treewidth computations II. Lower bounds. Inf. Comput. **209**(7), 1103–1119 (2011). https://doi.org/10.1016/j.ic.2011.04.003
4. Bornstein, C.F., Vempala, S.: Flow metrics. Theor. Comput. Sci. **321**(1), 13–24 (2004). https://doi.org/10.1016/j.tcs.2003.05.003
5. Crescenzi, P., Di Battista, G., Piperno, A.: A note on optimal area algorithms for upward drawings of binary trees. Comput. Geom. **2**, 187–200 (1992). https://doi.org/10.1016/0925-7721(92)90021-J
6. Goldberg, A.V.: Finding a maximum density subgraph. Technical report, UCB/CSD-84-171. EECS Department, University of California, Berkeley (1984). http://www2.eecs.berkeley.edu/Pubs/TechRpts/1984/5956.html
7. Margulies, S., Ma, J., Hicks, I.V.: The Cunningham-Geelen method in practice: branch-decompositions and integer programming. INFORMS J. Comput. **25**(4), 599–610 (2013). https://doi.org/10.1287/ijoc.1120.0524
8. Preuss, M.: Multimodal Optimization by Means of Evolutionary Algorithms. Natural Computing Series. Springer, Cham (2015). https://doi.org/10.1007/978-3-319-07407-8
9. Rao, S., Richa, A.W.: New approximation techniques for some linear ordering problems. SIAM J. Comput. **34**(2), 388–404 (2004). https://doi.org/10.1137/S0097539702413197
10. Sonuc, S.B., Smith, J.C., Hicks, I.V.: A branch-and-price-and-cut method for computing an optimal bramble. Discrete Optim. **18**, 168–188 (2015). https://doi.org/10.1016/j.disopt.2015.09.005
11. Usotskaya, N.: Exploiting geometric properties in combinatorial optimization. Ph.D. thesis, Maastricht University, Maastricht, The Netherlands (2011)

Crossing Paths with Hans Bodlaender: A Personal View on Cross-Composition for Sparsification Lower Bounds

Bart M. P. Jansen(✉)

Eindhoven University of Technology, Eindhoven, The Netherlands
b.m.p.jansen@tue.nl

Abstract. On the occasion of Hans Bodlaender's 60th birthday, I give
a personal account of our history and work together on the technique
of cross-composition for kernelization lower bounds. I present several
simple new proofs for polynomial kernelization lower bounds using cross-
composition, interlaced with personal anecdotes about my time as Hans'
PhD student at Utrecht University. Concretely, I will prove that VERTEX
COVER, FEEDBACK VERTEX SET, and the H-Factor problem for every
graph H that has a connected component of at least three vertices, do
not admit kernels of $\mathcal{O}(n^{2-\varepsilon})$ bits when parameterized by the number
of vertices n for any $\varepsilon > 0$, unless NP $\subseteq$ coNP/poly. These lower bounds
are obtained by elementary gadget constructions, in particular avoiding
the use of the Packing Lemma by Dell and van Melkebeek.

Keywords: Cross-composition · Kernelization lower bounds · Graph
problems

1 Getting Acquainted

1.1 Our Meeting

Hans Bodlaender saved me from becoming a computer-game programmer.[1] After
having spent my high-school years programming a Star Wars-themed shooter, I
decided to enroll in the Computer Science program at Utrecht University. My
main motivation at the time: it was the only university in the Netherlands to
offer a master's degree in Game Design.

---

[1] For this special occasion, I will allow myself to write in first person.

This project has received funding from the European Research Council (ERC) under
the European Union's Horizon 2020 research and innovation programme (grant agreement No. 803421, ReduceSearch).

The original version of this chapter was revised: this chapter was previously published
non-open access. The correction to this chapter is available at
https://doi.org/10.1007/978-3-030-42071-0_19

© The Author(s) 2020, corrected publication 2022
F. V. Fomin et al. (Eds.): Bodlaender Festschrift, LNCS 12160, pp. 89–111, 2020.
https://doi.org/10.1007/978-3-030-42071-0_8

My undergraduate years passed by, during which I had my first encounters with Hans during his Algorithms course in 2005. I recall a lecture on dynamic programming, which dealt with the PRETTY PRINTING problem of dividing words over lines to minimize the sum of the squares of the unused spaces on each line. Hans mentioned that a greedy strategy does not work for this problem; you really need to use dynamic programming. A fellow student asked if one can at least use a greedy strategy to determine what the optimal number of lines is: can it be that an optimal layout uses more lines than the greedy minimum? Rather than immediately giving the answer, Hans gave the question back to us students. It fascinated me, and the remainder of the first half of the lecture passed me by as I worked out an example showing the answer to be *yes*. During the break, I showed my construction to Hans. I watched proudly as he shared my example with the rest of the class after the break. This was my first personal interaction with Hans, and a sample of how he piqued my interest in algorithmic questions. But at that time, in my second year of study, I had no inkling as to how much he would later affect my career.

The real epiphany came two years later, when I was already enrolled in the master's program on Game Design. I was following Hans' course *Algorithms and Networks*, which covered some basic concepts of parameterized complexity, when it hit me: I like programming computer games because you have to be very *efficient with your computations*: otherwise your game is going to be either slow or ugly, and in either case people will not enjoy it. The *game* aspect was ultimately not so appealing to me. So I decided to change course dramatically.

All the gaming courses I had followed up to that point were moved into the elective space of my new master's program *Applied Computing Science*. I asked Hans to supervise my master's thesis project on kernelization, to which he quickly agreed. He lent me a copy of Downey and Fellows' first textbook [18] on parameterized complexity to read up on the relevant background. I remember a distinct feeling of shock when seeing the title of Section 6.3: "Bodlaender's Theorem" in the table of contents. Up to that point, I had no idea that the friendly and humble algorithms professor at my home university was the internationally recognized authority on treewidth!

Several months into my final project, Hans told me that he was writing a grant proposal to acquire funding to study kernelization, and asked me if I would mind being mentioned in the proposal as a qualified candidate for the PhD position he was requesting. A couple of months later, nearing the end of my master thesis project, I had proven my first kernelization results when Hans told me the good news: his NWO TOP grant "KERNELS: Combinatorial Analysis of Data Reduction" was funded. That led to my easiest job interview ever, which consisted of two lines. Hans: "So, do you still want to become a PhD student?" to which my answer was a resounding "Yes"! I never looked back.

1.2 Our Work on Kernelization Lower Bounds

I started working on my PhD under Hans' supervision in 2009, investigating the power and limitations of efficient and provably effective preprocessing. When

Stefan Kratsch joined the research group in 2010, we were completed into a trio that studied kernelization during the day and played boardgames at night.[2] Kernelization theorems were discussed at *Chez Hans*, the nickname that Hans' office earned for the clandestine coffee machine that served many of our colleagues.[3]

Building on a series of papers that had just come out [3,4,22,23], we spent a lot of time working on hardness proofs showing that small kernels do not exist under certain complexity-theoretic assumptions. For this contribution to the festschrift on account of Hans' 60th birthday, I therefore decided to write about kernelization lower bounds based on the framework of cross-composition [5,7] developed by Hans, Stefan, and myself. The purpose of the technical content of this article is to show a number of new and elegant kernelization lower-bound proofs based on cross-composition, showing that VERTEX COVER, H-FACTOR, and FEEDBACK VERTEX SET parameterized by the number of vertices n cannot be efficiently reduced to equivalent instances on $\mathcal{O}(n^{2-\varepsilon})$ bits for any $\varepsilon > 0$, unless NP $\subseteq$ coNP/poly. The first two yield new proofs for existing theorems, but the third result is new. The proofs are elementary, based on the cross-composition framework with simple gadgeteering. All lower bounds will follow from combining a suitable choice of starting problem for the reduction, together with gadgets to deactivate most parts of the composed instance. In particular, all these lower bounds can be proven without having to resort to the Packing Lemma of Dell and van Melkebeek [16].

1.3 Organization

In Sect. 2 I introduce the prerequisite definitions of parameterized complexity and kernelization, together with the framework of cross-composition. The remainder of the article is organized into three case studies. Each case study deals with one problem, discussing the relevant related work before presenting a kernelization lower bound using cross-composition. Section 3 deals with parameterized complexity's fruit fly, VERTEX COVER. More general vertex-deletion problems such as FEEDBACK VERTEX SET are considered in Sect. 4. Section 5 focuses on the H-FACTOR problem. Finally, I give some concluding reflections in Sect. 6.

2 Kernelization and Lower Bounds

2.1 Kernelization

Kernelization investigates how a complexity *parameter* contributes to the difficulty of an input to an algorithmic decision problem, and therefore uses notions

---

[2] I have particularly fond memories of Robo Rally, or *applied algorithmic game theory* as Hans liked to put it, in which you plan a sequence of actions for your robot in the hope of it being the first to visit all checkpoints. More often than not, it ends up pushed into a pit or shot at by competing robots which simultaneously carry out the instructions that their governing players programmed.

[3] Hans did not drink coffee until his fifties, strategically saving the caffeine-filled theorem-producing beverage until it was needed. I am following in his footsteps and have never had a single cup of coffee.

from parameterized complexity [13,19–21]. Fix a finite alphabet Σ used to encode problem inputs, such as $\Sigma = \{0,1\}$. A *parameterized problem* is a subset $Q \subseteq \Sigma^* \times \mathbb{N}$, which contains the pairs (x,k) for which x is the encoding of an input to Q whose parameter value is k, and for which the answer to the encoded question is YES. Intuitively, a *kernelization* for a parameterized problem Q is a polynomial-time preprocessing algorithm that reduces any input (x,k) to an equivalent one, whose size and parameter are bounded solely in terms of the complexity parameter k, independent of the size $|x|$ of the original input. The lower bounds I present turn out to work against the more general notion of *generalized kernelization*, which essentially investigates whether inputs of one parameterized problem Q can be efficiently reduced to small instances of another problem Q'. I therefore need the following formal definition.

Definition 1. *Let $Q, Q' \subseteq \Sigma^* \times \mathbb{N}$ be parameterized problems and let $h \colon \mathbb{N} \to \mathbb{N}$ be a computable function. A generalized kernel for Q into Q' of size $h(k)$ is an algorithm that, on input $(x,k) \in \Sigma^* \times \mathbb{N}$, takes time polynomial in $|x| + k$ and outputs an instance (x',k') such that:*

1. *$|x'|$ and k' are bounded by $h(k)$, and*
2. *$(x',k') \in Q'$ if and only if $(x,k) \in Q$.*

The algorithm is a kernel *for Q if $Q' = Q$. It is a* polynomial (generalized) kernel *if $h(k)$ is a polynomial.*

Much of the initial research on kernelization lower bounds [4,6,8,17] dealt with the distinction between polynomial and super-polynomial kernel sizes. Later, a refinement of the tools made it possible to also give lower bounds on the degree of the polynomial that bounds the kernel size, for problems that *do* admit a polynomial-size kernel. In this article, I will focus on the latter kind of lower bounds. For a historical overview of the development of kernelization lower bounds, I refer to the introductory sections of the article on cross-composition by myself, Hans Bodlaender, and Stefan Kratsch [7].

2.2   Intuition for Polynomial Kernelization Lower Bounds

To show the impossibility (under suitable complexity-theoretic conjectures) of a parameterized problem Q having small kernels, one must prove that the existence of a small kernel would imply algorithmic consequences that are "too good to be true" and therefore violate established beliefs such as $P \neq NP$. Before presenting the formal details, let me try to convey some intuition behind such proofs, elaborating on what these consequences are. Consider the NP-hard VERTEX COVER problem, whose inputs consist of a graph G and integer k. The question is whether G has a set S of at most k vertices, such that each edge of G contains at least one vertex from S. Using the Nemhauser-Trotter theorem [31], an instance (G,k) can efficiently be reduced to an equivalent instance (G',k') where G' has at most $2k$ vertices and $k' \leq k$. The kernelized instance can be encoded in $\mathcal{O}(k^2)$ bits, by writing down the adjacency matrix of G' and the integer k' in unary. How could one prove that it is impossible to always obtain a kernel of, say, $\mathcal{O}(k^{1.9})$ bits?

Let us consider algorithmic consequences which are "too good to be true". First of all, suppose there would be a polynomial-time algorithm $\mathcal{A}$ that, given a sequence of t instances $(G_1, k_1), \ldots, (G_t, k_t)$ of VERTEX COVER, would be able to distinguish between the case that all input instances have answer NO, and the case that there is at least one YES-instance among the inputs. Then using $\mathcal{A}$, we could solve the VERTEX COVER decision problem in polynomial time: to determine the answer to (G, k), just ask $\mathcal{A}$ whether a sequence consisting of some arbitrarily chosen NO-instances, together with the instance (G, k), contains at least one YES-instance. Hence such an algorithm $\mathcal{A}$ is too good to be true.

Now suppose that there is an algorithm $\mathcal{B}$ that, given a sequence of $t \in N^{\mathcal{O}(1)}$ instances of VERTEX COVER, each on N bits, outputs a single VERTEX COVER instance (G^*, k^*) on $(t-1) \cdot N$ bits whose answer is YES if and only if there was a YES-instance in the input sequence. Even though such an algorithm $\mathcal{B}$ does not directly give a way to solve VERTEX COVER in polynomial time, I will argue it is still too good to be true. To achieve an output size of $(t-1) \cdot N$ bits, intuitively it has to omit one of the input instances when building the output (G^*, k^*). But to ensure that the output (G^*, k^*) accurately represents the logical OR of the input sequence, it seems that $\mathcal{B}$ must solve an input instance before it can be sure that it is safe to omit it. After all, if the omitted instance was the only one in the sequence whose answer was YES, then omitting it changes the value of the logical OR. Since we believe VERTEX COVER cannot be solved in polynomial time, it becomes intuitively clear that algorithm $\mathcal{B}$ is also too good to be true.

We can therefore prove the impossibility of having kernels of bitsize $\mathcal{O}(k^{1.9})$ by developing a so-called cross-composition algorithm $\mathcal{C}$ which, together with such a kernel, would yield algorithm $\mathcal{B}$. Suppose $\mathcal{C}$ can take any sequence of t instances of VERTEX COVER, each on n vertices and therefore $N \in \mathcal{O}(n^2)$ bits, and composes these into a single instance (G', k') whose answer is the logical OR of the answers to the inputs, on $\mathcal{O}(\sqrt{t} \cdot n)$ vertices such that $k' \in \mathcal{O}(\sqrt{t} \cdot n)$. Suppose further that $\mathcal{D}$ is a kernelization for VERTEX COVER of bitsize $\mathcal{O}(k^{1.9})$. Then by pipe-lining algorithms $\mathcal{C}$ and $\mathcal{D}$, we obtain an algorithm such as $\mathcal{B}$, thereby showing that $\mathcal{D}$ should not exist. Indeed, if we take a sequence of $t = n^{50}$ instances of VERTEX COVER, each on n vertices, then the composition $\mathcal{C}$ merges these into a single instance with parameter value $k' \in \mathcal{O}(\sqrt{t} \cdot n) = \mathcal{O}(n^{26})$. Applying the kernelization $\mathcal{D}$ to this composed instance, reduces it to size $\mathcal{O}((n^{26})^{1.9}) \leq \mathcal{O}(n^{49.4}) < (t-1) \cdot N \approx n^{52}$, therefore forming an algorithm of type $\mathcal{B}$ that is too good to be true; hence if we can find such a cross-composition $\mathcal{C}$, then such a kernel $\mathcal{D}$ should not exist. By increasing the number of input instances from n^{50} to larger powers of n, the same intuitive reasoning rules out the existence of a kernel of bitsize $\mathcal{O}(k^{2-\varepsilon})$ for any $\varepsilon > 0$.

Note that the key property of $\mathcal{C}$ that makes this work, is that it manages to compress the information of t instances on n vertices each, into a single instance on $\mathcal{O}(\sqrt{t} \cdot n)$ vertices. While this may seem far-fetched at first, there are elementary reductions achieving this. They exploit the fact that t instances on n

vertices carry $t \cdot n^2$ bits of information (for each of the t graphs, which of the n^2 potential edges exist?) while a single instance on $\mathcal{O}(\sqrt{t} \cdot n)$ vertices has $\mathcal{O}(t \cdot n^2)$ potential edges, and can therefore encode the same amount of information if it is packed efficiently. Now let me make this intuition precise.

2.3   The Formal Cross-Composition Framework

The fact that algorithms such as $\mathcal{B}$ are too good to be true[4] was proven by Dell and van Melkebeek [16, §6], building on work by Fortnow and Santhanam [23], which in turn was triggered by the seminal work on kernelization lower bounds by Hans with Downey, Fellows, and Hermelin [4]. The existence of algorithms like $\mathcal{B}$ does not directly lead to the consequence that $\mathsf{P} = \mathsf{NP}$, but implies the complexity-theoretic containment $\mathsf{NP} \subseteq \mathsf{coNP/poly}$, which is still considered very unlikely. To prove kernelization lower bounds, we therefore formalize what the composition algorithm, referred to above as $\mathcal{C}$, has to achieve.

In many cases, it turns out to be easier to build composition algorithms for sequences of "similarly-sized" input instances $(G_1, k_1), \ldots, (G_t, k_t)$, for example when the input graphs all have the same number of vertices, edges, and target vertex cover size k. Since there are exponentially many different instances of a given number of vertices, edges, and target cover size, such a restriction does not make the algorithmic task much easier. Indeed, even if the composition algorithm $\mathcal{C}$ described above is only applied to sequences of similarly-sized inputs, the combination of $\mathcal{C}$ and $\mathcal{D}$ still leads to algorithms that are too good to be true. For that reason, the cross-composition framework [7] allows one to choose an equivalence relation on inputs, efficiently grouping inputs of bitsize N into $N^{\mathcal{O}(1)}$ different classes, and only requires a composition algorithm to be able to merge inputs coming from the same class.

Definition 2 (Polynomial equivalence relation). *An equivalence relation $\mathcal{R}$ on Σ^* is called a* polynomial equivalence relation *if the following conditions hold.*

1. *There is an algorithm that, given two strings $x, y \in \Sigma^*$, decides whether x and y belong to the same equivalence class in time polynomial in $|x| + |y|$.*
2. *For any finite set $S \subseteq \Sigma^*$ the equivalence relation $\mathcal{R}$ partitions the elements of S into a number of classes that is polynomially bounded in the size of the largest element of S.*

Definition 2 allows one to circumvent padding arguments that were frequently used in earlier proofs. Using this notion, we can now formalize cross-composition for proving polynomial lower bounds on kernelization. For this overview article, rather than defining bounded-cost cross-composition in general (cf. [7, §3.2]), I restrict myself to the less technical degree-2 cross-compositions like the ones described above, which are used to rule out kernels of subquadratic size. In the following definition, I use the shorthand $[n] = \{1, \ldots, n\}$ for $n \in \mathbb{N}$.

---

[4] The formal details differ slightly from my intuitive interpretation above.

Definition 3 (Degree-2 cross-composition). *Let $L \subseteq \Sigma^*$ be a language, let $\mathcal{R}$ be a polynomial equivalence relation on Σ^*, and let $\mathcal{Q} \subseteq \Sigma^* \times \mathbb{N}$ be a parameterized problem. A* degree-2 OR-cross-composition *of L into $\mathcal{Q}$ with respect to $\mathcal{R}$ is an algorithm that, given t instances $x_1, x_2, \ldots, x_t \in \Sigma^*$ of L belonging to the same equivalence class of $\mathcal{R}$, takes time polynomial in $\sum_{i=1}^{t} |x_i|$ and outputs an instance $(x^*, k^*) \in \Sigma^* \times \mathbb{N}$ such that:*

1. *the parameter k^* is bounded by $\mathcal{O}(\sqrt{t} \cdot (\max_i |x_i|)^c)$, where c is some constant independent of t, and*
2. *$(x^*, k^*) \in \mathcal{Q}$ if and only if there is an $i \in [t]$ such that $x_i \in L$.*

The adjective *cross* in the name cross-composition comes from the fact that the reduction crosses over from inputs of problem L, into an input of parameterized problem $\mathcal{Q}$. This contrasts the earlier plain *composition* framework of Bodlaender, Downey, Fellows, and Hermelin [4] which required problems to be composed into themselves. As we will see, crossing over from one problem to another makes it easier to prove lower bounds in several settings.

When building a degree-2 OR-cross-composition, it will be convenient if the number of input instances t is a square: then $\sqrt{t} \in \mathbb{N}$, which allows the input instances to be enumerated as $x_{i,j}$ for $i, j \in [\sqrt{t}]$. This assumption can be made without loss of generality. Suppose we have a cross-composition $\mathcal{C}$ that works if t is a square, and we want to obtain a cross-composition $\mathcal{C}'$ for an arbitrary number of inputs. Then $\mathcal{C}'$ can be obtained as follows. Given an input sequence $x_1, \ldots, x_t$ to $\mathcal{C}'$, let $t' \geq t$ be the smallest square that is larger than t. Note that $t' \leq 2t$ because the nearest power of two suffices. Then build a new sequence of t' inputs by appending $t' - t$ copies of x_1 to $x_1, \ldots, x_t$, and apply $\mathcal{C}$ to this sequence. Clearly the logical OR of the old and new sequences have the same value, and it is easy to verify that $\mathcal{C}'$ satisfies all conditions of Definition 3. We will therefore assume without loss of generality that t is a square.

The following theorem shows that degree-2 OR-cross-compositions indeed rule out subquadratic kernels, under the assumption that NP $\not\subseteq$ coNP/poly.

Theorem 1 ([7, Thm. 3.8, Prop. 2.3]). *Let $L \subseteq \Sigma^*$ be a language that is NP-hard under Karp reductions, let $\mathcal{Q} \subseteq \Sigma^* \times \mathbb{N}$ be a parameterized problem, and let $\varepsilon > 0$ be a real number. If L is NP-hard under Karp reductions and has a degree-2 OR-cross-composition into $\mathcal{Q}$, and $\mathcal{Q}$ parameterized by k has a polynomial (generalized) kernelization of bitsize $\mathcal{O}(k^{2-\varepsilon})$, then NP $\subseteq$ coNP/poly.*

In the upcoming case studies, I will consider a number of graph problems parameterized by the number of vertices n. Hence the parameter value, denoted by k in the definitions of the framework, will be the number of vertices of the input graph that is commonly denoted as n. Any pair (G, k) consisting of an n-vertex graph, together with a target value k in the range $\{0, \ldots, n\}$, can trivially be encoded in $\mathcal{O}(n^2)$ bits. Graph problems whose input is of the form (G, k) therefore have trivial polynomial kernels of bitsize $\mathcal{O}(n^2)$ when parameterized by the number of vertices n. The lower bounds will prove that this cannot be significantly improved. So intuitively, the lower bounds will rule out that there is an efficient *sparsification* algorithm that reduces a dense n-vertex instance to an equivalent one with a subquadratic number of edges. Note that such a sparsification bound directly implies that, for any problem parameter ℓ whose value on n-vertex graphs is $\mathcal{O}(n)$, there cannot be a kernel of bitsize $\mathcal{O}(\ell^{2-\varepsilon})$.

3   Vertex Cover

The first case study concerns the VERTEX COVER problem. The problem admits a simple degree-based kernelization due to Sam Buss [11] that reduces inputs (G, k) to $\mathcal{O}(k^2)$ vertices and edges. The same bounds can be obtained via the sunflower lemma [20, §9.1]. The linear-programming based kernelization based on the Nemhauser-Trotter theorem [31] achieves a better bound of $2k$ vertices, but may still have $\Omega(k^2)$ edges. In a breakthrough paper [15,16], Dell and van Melkebeek proved that this is optimal up to $k^{o(1)}$ factors: they proved that VERTEX COVER has no kernel of bitsize $\mathcal{O}(k^{2-\varepsilon})$ for any $\varepsilon > 0$, unless NP $\subseteq$ coNP/poly. Their proof is based on a nontrivial number-theoretic construction called the *Packing Lemma* [16, Lemma 2], which shows how to construct graphs whose edges partition into many large cliques, in such a way that no large cliques exist other than in the packing. Later, Dell and Marx [14, Thm. C.1] showed how to obtain the same lower bound using an elementary gadget construction, by composing instances of MULTICOLORED BICLIQUE based on a table layout (cf. [21, §20.2]).

To illustrate the technique of degree-2 OR-cross-composition, I will present an alternative elementary lower-bound construction for VERTEX COVER, which avoids the intermediate problem of MULTICOLORED BICLIQUE. It will be useful to use a restricted version of VERTEX COVER as a starting point for the composition, though, which is formalized in the following way.

VERTEX COVER ON SUBDIVIDED GRAPHS
Input: A graph G, an integer k, and a partition of $V(G)$ into $A \cup B$ such that $G[A]$ is an independent set and each connected component of $G[B]$ consists of a single edge.
Question: Does G contain a set $S \subseteq V(G)$ of size at most k, such that $S \cap \{u, v\} \neq \emptyset$ for each edge $\{u, v\} \in E(G)$?

The partition of $V(G)$ into sets A and B that induce subgraphs of a specific form will be useful when merging a series of inputs into one. Intuitively, it will allow us to merge the sets A of various distinct inputs into a single set, while preserving the adjacency information of the original inputs.

Lemma 1. VERTEX COVER ON SUBDIVIDED GRAPHS *is NP-hard under Karp reductions.*

Proof. Consider an instance (G, k) of VERTEX COVER, and pick an arbitrary edge $\{x, y\}$. Let G' be the graph obtained from G by removing the edge $\{x, y\}$,

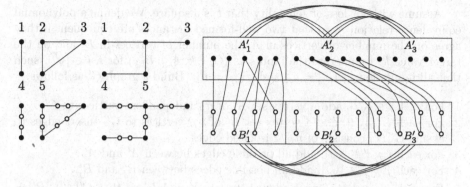

Fig. 1. Left: A 5-vertex graph at the top; below it, the graph obtained by subdividing every edge twice, which is used as input $G_{1,1}$ to the cross-composition. In the subdivided instance, the set A of original vertices is in black, the set B of subdividers is in white. Middle: Another 5-vertex graph with its double-subdivision, used as instance $G_{2,3}$. Right: Illustration of the cross-composition of Theorem 2 for $t = 3 \cdot 3$ inputs of the subdivided problem with 5 vertices in A and 8 vertices in B. Edges between different sets $A'_i, A'_{i'}$ are visualized schematically, similarly for $B'_j, B'_{j'}$. Only the edges inserted on account of instances $G_{1,1}$ and $G_{2,3}$ are shown.

inserting two new vertices x', y' and the edges $\{x, x'\}, \{x', y'\}, \{y', y\}$. Intuitively, G' is obtained by subdividing the edge $\{x, y\}$ twice. It is easy to verify that G has a vertex cover of size k if and only if G' has a vertex cover of size $k+1$, which was first observed by Poljak [32] for the complementary problem INDEPENDENT SET (cf. [9, Lemma 7]).

To prove the lemma, we use the following reduction from the NP-complete VERTEX COVER problem. Given an instance (G, k) on $m = |E(G)|$ edges, let G' be obtained by replacing each edge of G by a three-edge path as above, and let $k' := k+m$. By the observation above, G has a vertex cover of size k if and only if G' has a vertex cover of size k'. Letting $A = V(G)$ denote the original vertices in G', and letting B denote the inserted subdivider vertices, we have that $G'[A]$ is an independent set and $G'[B]$ consists of isolated edges. Hence (G', k', A, B) is a valid equivalent instance of VERTEX COVER ON SUBDIVIDED GRAPHS. $\square$

Using this starting problem, I now present the degree-2 OR-cross-composition that rules out subquadratic kernels for VERTEX COVER. Refer to Fig. 1 for an illustration.

Theorem 2. *For any $\varepsilon > 0$, VERTEX COVER parameterized by the number of vertices n does not admit a generalized kernelization of bitsize $\mathcal{O}(n^{2-\varepsilon})$ unless* NP $\subseteq$ coNP/poly.

Proof. By Theorem 1 and Lemma 1 it suffices to give a cross-composition of VERTEX COVER ON SUBDIVIDED GRAPHS into VERTEX COVER, such that any sequence of t inputs of bitsize at most N each, is composed into a single instance (G', k') on $n \in \mathcal{O}(\sqrt{t} \cdot N^{\mathcal{O}(1)})$ vertices.

Assume without loss of generality that t is a square. We define a polynomial equivalence relation $\mathcal{R}$ so that two well-formed instances are equivalent if they agree on the number of vertices in A, the number of vertices in B, and on the target value k. Enumerate the inputs as $(G_{i,j}, k, A_{i,j}, B_{i,j})$ for $i, j \in [\sqrt{t}]$, such that all inputs have $|A_{i,j}| = n_A$ and $|B_{i,j}| = n_B$. Build a graph G' as follows.

1. For each $i \in [\sqrt{t}]$, add a vertex set A'_i of n_A independent vertices to G'.
2. For each $i \in [\sqrt{t}]$, add a vertex set B'_i of n_B vertices to G'. Insert edges to ensure $G'[B'_i]$ consists of $n_B/2$ isolated edges.
3. For each $i \neq i' \in [\sqrt{t}]$, add all possible edges between A'_i and $A'_{i'}$.
4. For each $j \neq j' \in [\sqrt{t}]$, add all possible edges between B'_j and $B'_{j'}$.
5. For each $i, j \in [\sqrt{t}]$, insert edges between A'_i and B'_j so that $G'[A'_i \cup B'_j]$ is isomorphic to $G_{i,j}$.

To complete the cross-composition, we set $k' := (\sqrt{t} - 1)(n_A + n_B) + k$. Observe that G' has $\sqrt{t} \cdot (n_A + n_B)$ vertices, which is suitably bounded for a degree-2 OR-cross-composition since input instances have $N \geq n_A + n_B$ bits. The construction can easily be performed in polynomial time. It remains to verify that G' has a vertex cover of size k' if and only if some input instance $G_{i,j}$ has a vertex cover of size k.

Suppose first that there is a YES-instance G_{i^*,j^*} among the inputs that has a vertex cover of size at most k. Since $G'[A'_{i^*}, B'_{j^*}]$ is isomorphic to G_{i^*,j^*} by construction, it has a vertex cover S'_{i^*,j^*} of size at most k. Combined with all $(\sqrt{t} - 1)(n_A + n_B)$ remaining vertices of G', this yields a vertex cover of size at most k' of G', proving that the result of the composition is a YES-instance.

For the other direction, suppose S' is a vertex cover of size at most k' in G'. Since all possible edges are present between distinct sets A'_i and $A'_{i'}$, there is at most one set A'_{i^*} from which S' does not contain all vertices. Similarly, there is at most one set B'_{j^*} from which S' does not contain all vertices. Since $|S'| \leq k' = (\sqrt{t} - 1)(n_A + n_B) + k$, while S' contains all $(\sqrt{t} - 1)(n_A + n_B)$ vertices of $(\bigcup_{i \neq i^*} A'_i) \cup (\bigcup_{j \neq j^*} B'_j)$, it follows that $S' \cap (A'_{i^*} \cup B'_{j^*}) \leq k$. Since $G'[A'_{i^*} \cup B'_{j^*}]$ is isomorphic to G_{i^*,j^*}, this proves that G_{i^*,j^*} has a vertex cover of size at most k, so that there is a YES-instance among the inputs.   □

Let me point out two crucial features of the cross-composition above. First, note that in Step 5 of the construction we heavily exploit the fact that all graphs $G_{i,j}[A_{i,j}]$ are isomorphic, and similarly that all graphs $G_{i,j}[B_{i,j}]$ are isomorphic. The fact that the vertex sets of the input graphs can be partitioned into two parts that induce very uniformly structured subgraphs, will also be exploited in the upcoming lower bounds. Second, we used some problem-specific gadgeteering to ensure that solutions to G' must contain all but one of the groups A'_i in their entirety, and similarly for the groups B'_j. The step of inserting all possible edges between the groups ensures that effectively, a good solution in a single instance $G_{i^*,j^*} = G'[A'_{i^*} \cup B'_{j^*}]$ is sufficient to guarantee the existence of a good solution in G'. More involved gadgeteering will be needed to achieve a similar OR behavior in future constructions.

4 Feedback Vertex Set

We move on to another classic vertex-deletion problem, FEEDBACK VERTEX SET. We consider the problem on *undirected* graphs; in fact, I will consider only undirected graphs throughout this article. An instance (G, k) therefore consists of an undirected graph G and integer k, and asks whether G has a subset S of at most k vertices, whose removal from G leaves an acyclic graph. Several polynomial kernels were developed for the problem [2,10,25,33], one of which is famously due to Hans [1]. The current-best kernel [25] has $\mathcal{O}(k^2)$ vertices and edges, and can be encoded in $\mathcal{O}(k^2 \log k)$ bits. Dell and van Melkebeek [16] show that FEEDBACK VERTEX SET does not have kernels of bitsize $\mathcal{O}(k^{2-\varepsilon})$ unless NP $\subseteq$ coNP/poly. However, their proof does not say anything about the possibility of sparsifying n-vertex instances to $\mathcal{O}(n^{2-\varepsilon})$ bits. I will show that the latter is also impossible, assuming NP $\not\subseteq$ coNP/poly, by adapting the construction of Theorem 2. We will again need a version of the problem whose vertex set partitions into two parts that induce uniformly structured subgraphs. Since FEEDBACK VERTEX SET remains NP-complete [27] on bipartite graphs (subdividing every edge preserves the answer to the problem, and yields a bipartite graph), the following NP-complete problem will be used as the source problem for the cross-composition.

FEEDBACK VERTEX SET ON BIPARTITE GRAPHS
Input: An undirected graph G, an integer k, and a partition of $V(G)$ into $A \cup B$ such that $G[A]$ and $G[B]$ are both independent sets.
Question: Does G contain a vertex set $S \subseteq V(G)$ of size at most k, such that $G - S$ is acyclic?

Theorem 3. *For any $\varepsilon > 0$, FEEDBACK VERTEX SET parameterized by the number of vertices n does not admit a generalized kernelization of bitsize $\mathcal{O}(n^{2-\varepsilon})$ unless NP $\subseteq$ coNP/poly.*

Proof. I present a degree-2 OR-cross-composition. Let $(G_{i,j}, k, A_{i,j}, B_{i,j})$ for $i, j \in [\sqrt{t}]$ be a sequence of t input instances of FEEDBACK VERTEX SET ON BIPARTITE GRAPHS that all share the same target value k, the same number n_A of vertices in the A-set, and the same number n_B of vertices in the B-set, which may be assumed by a suitable choice of $\mathcal{R}$. Build an instance (G', k') as follows.

1. For each $i \in [\sqrt{t}]$, add a set A'_i of n_A independent vertices to G' and number these from 1 to n_A.
2. For each $i \in [\sqrt{t}]$, add a set B'_i of n_B independent vertices to G' and number these from 1 to n_B.
3. For each $i, j \in [\sqrt{t}]$, insert edges between A'_i and B'_j so that $G'[A'_i \cup B'_j]$ is isomorphic to $G_{i,j}$.
4. For each $i \in [\sqrt{t}]$, add a vertex set $A^*_i = \{a^*_{i,x,y,c} \mid x, y \in [n_A], c \in [2]\}$ to G'. For each $i < i' \leq \sqrt{t}$, for each $x, y \in [n_A]$, for each $c \in [2]$, make $a^*_{i,x,y,c}$ adjacent to the xth vertex of A'_i and the yth vertex of $A'_{i'}$.

5. For each $j \in [\sqrt{t}]$, add a vertex set $B_j^* = \{b_{j,x,y,c}^* \mid x, y \in [n_B], c \in [2]\}$ to G'. For each $j < j' \leq \sqrt{t}$, for each $x, y \in [n_B]$, for each $c \in [2]$, make $b_{j,x,y,c}^*$ adjacent to the xth vertex of B_j' and the yth vertex of $B_{j'}'$.

Define $A^* := \bigcup_i A_i^*$ and $B^* := \bigcup_j B_j^*$. By the last two steps, for each $i \in [\sqrt{t}]$, each vertex of A^* is adjacent to at most one vertex in A_i', and symmetrically for adjacencies of B^* into sets B_j'. For every pair of vertices $x \in A_i', y \in A_{i'}'$ for $i < i'$, there are two vertices $a_{i,x,y,1}^*$ and $a_{i,x,y,2}^*$ adjacent to both x and y. Effectively, these form a cycle with x and y, prompting the following observation.

Observation 1. *If S' is a feedback vertex set in G', and $i < i' \in [\sqrt{t}]$ such that there exist $x \in A_i' \setminus S'$ and $y \in A_{i'}' \setminus S'$, then S' contains a vertex of $\{a_{i,x,y,1}^*, a_{i,x,y,2}^*\}$. The analogous statement for B' also holds.*

To complete the cross-composition, we set $k' := (\sqrt{t} - 1)(n_A + n_B) + k$. Observe that G' has $\mathcal{O}(\sqrt{t} \cdot (n_A + n_B)^2)$ vertices, suitably bounded for a degree-2 OR-cross-composition. It remains to verify that G' has a feedback vertex set of size k' if and only if some input instance $G_{i,j}$ has one of size k.

Suppose first that G_{i^*,j^*} has a feedback vertex set of size k. Since $G'[A_{i^*}' \cup B_{j^*}']$ is isomorphic to G_{i^*,j^*}, it has a feedback vertex set S_{i^*,j^*} of size k. Consider $S' := S_{i^*,j^*} \cup (\bigcup_{i \neq i^*} A_i') \cup (\bigcup_{j \neq j^*} B_j')$, which has size at most k'. Now observe that $G' - S'$ can be obtained from the acyclic graph $G'[A_{i^*}' \cup B_{j^*}'] - S_{i^*,j^*}$ by inserting the vertices $A^* \cup B^*$ along with their edges into $(A_{i^*}' \cup B_{j^*}') \setminus S_{i^*,j^*}$. As each vertex of A^* is adjacent to at most one vertex of A_{i^*}', and each vertex of B^* is adjacent to at most one vertex of B_{j^*}', the graph $G' - S'$ is obtained from an acyclic graph by inserting vertices of degree at most one, which does not introduce any cycles. Hence S' is a feedback vertex set of size at most k' in G'.

For the reverse direction, suppose that G' has a feedback vertex set S' of size at most k'. If there are distinct indices $i < i' \in [\sqrt{t}]$ for which $A_i' \setminus S'$ and $A_{i'}' \setminus S'$ are both nonempty, then normalize S' as follows. Let i^* be the largest index for which $A_{i^*}' \setminus S' \neq \emptyset$, and define $S'' := (S' \setminus A^*) \cup (\bigcup_{i \neq i^*} A_i')$. Since all vertices of A^* have at most one neighbor in A_{i^*}', they have degree at most one in $G' - S''$ and therefore $G' - S''$ is also acyclic. Let me show that $|S''| \leq |S'|$. For each $i < i^*$, for each vertex $x \in A_i' \setminus S'$, vertex x belongs to S'' but not to S'. To show S'' is not larger than S', we charge each such x to a unique vertex that is contained in S' but not in S''. Fix an arbitrary $y \in A_{i^*}' \setminus S'$. By Observation 1 the solution S' contains a vertex of $\{a_{i,x,y,1}^*, a_{i,x,y,2}^*\}$ to which we can charge $x \in A_i' \setminus S'$. In this way we can charge each $x \in S'' \setminus S'$ to a unique pair, implying that $|S''| \leq |S'|$. Hence this normalization process yields a feedback vertex set S^* of size at most k' that contains all vertices of $\bigcup_{i \neq i^*} A_i'$, for a suitable choice of $i^* \in [\sqrt{t}]$. By a second analogous and independent normalization step for B', we may assume there is an index j^* such that S^* contains all vertices of $\bigcup_{j \neq j^*} B_j'$. Since $|S^*| \leq k' = (\sqrt{t} - 1)(n_A + n_B) + k$, the set S^* contains at most k vertices from $G'[A_{i^*}' \cup B_{j^*}']$, which is isomorphic to G_{i^*,j^*} by construction. Hence G_{i^*,j^*} has a feedback vertex set of size at most k. $\square$

The type of construction of Theorem 3 can be used to prove analogous lower bounds for many other vertex-deletion problems to nontrivial hereditary graph classes: all one has to do is change the source problem of the composition, and change the gadgets in the sets A^*, B^* which ensure that there is an optimal solution that avoids vertices from at most one set A'_{i^*} and at most one set B'_{j^*}. Such gadgets have been developed by Lewis and Yannakakis in their generic NP-completeness proof [30]. As their description is somewhat technical, I will not treat them here.

5 H-Factor

I devote the last case study of this article to generalizations of the MATCHING problem in graphs. For (undirected, simple) graphs G and H, an H-packing in G is a collection $H_1, \ldots, H_k$ of vertex-disjoint subgraphs of G, each of which is iso-morphic to H. An H-factor in G is an H-packing $H_1, \ldots, H_k$ whose vertex sets partition $V(G)$. The corresponding decision problem H-FACTOR asks if an input graph G has an H-factor. Kirkpatrick and Hell proved [28] that H-FACTOR is NP-complete when H contains a connected component of three or more ver-tices, and is polynomial-time solvable otherwise by a reduction to MAXIMUM MATCHING. Dell and Marx proved [14, Thm. 1.4] under the standard assump-tion NP $\not\subseteq$ coNP/poly that for connected graphs H on at least three vertices, the H-FACTOR problem parameterized by the number of vertices n does not admit a generalized kernel of bitsize $\mathcal{O}(n^{2-\varepsilon})$ for any $\varepsilon > 0$. Their proof relies on the Packing Lemma. In this section, I will give an elementary proof of the same theorem. The proof uses the following gadgets by Kirkpatrick and Hell.

Lemma 2 ([28, Lemma 3.5], cf. [14, Lemma 4.2]). *For each connected graph H on at least three vertices, there is a graph H' called the local H-coordinator gadget, which contains $|V(H)|$ distinct connector vertices $C \subseteq V(H')$ as an independent set, and has $V(H') \setminus C$ as its interior vertices, such that:*

- *There is an H-factor of H' and there is an H-factor of $H' - C$.*
- *For each $\emptyset \subsetneq C' \subsetneq C$ there is no H-factor of $H' - C'$.*
- *The graph $H' - C$ is connected.*
- *If a graph G contains H' as an induced subgraph, such that no interior vertex of H' is adjacent to a vertex outside of H', then in any H-factor $H_1, \ldots, H_k$ of G, the following holds: for every interior vertex $v \in V(H') \setminus C$, if $v \in V(H_i)$ then $V(H_i) \subseteq V(H')$.*

See Fig. 2 for an example. The last *coherence* property of the gadget effec-tively ensures that the H-subgraphs covering interior vertices of H', cannot cover any vertices not belonging to H'.

Consider a fixed graph H on h vertices. Given a graph G, the operation of *attaching a local H-coordination gadget* onto a set $S = \{v_1, \ldots, v_h\}$ of h vertices in G is defined as follows: insert a new disjoint copy of the local coordination gadget H', and denote its connector vertices by $c_1, \ldots, c_h$. For each $j \in [h]$, identify v_j with c_j, and use v_j as the identity of the merged vertex. I will use this operation in several constructions.

Fig. 2. Left: Local coordination gadget H' for $H = K_3$, whose connector vertices C are visualized by squares. Middle: a K_3-factor of H'. Right: a K_3-factor of $H' - C$.

Lemma 2 yields the following useful property. If a graph G is obtained by attaching a local H-coordination gadget H' onto an independent set S in an existing graph, and possibly inserting other vertices and edges that are not incident to the interior vertices of H' in such a way that S remains an independent set, then in any H-factor of G the following holds: either all connector vertices are covered by copies of H contained entirely within H', or no connector vertex is covered by a copy of H that contains other vertices of H'. In the former case, I say that the gadget H' *absorbs* all its connector vertices; in the latter, that the gadget absorbs none of its connector vertices.

As in the earlier sections, a more structured NP-hard version of H-PACKING is needed as the source problem for the cross-composition. For fixed connected graphs H and F, it is defined as follows.

H-FACTOR WITH F-PARTITION
Input: A graph G and a partition of $V(G)$ into $A \cup B$ such that $G[A]$ is an independent set, each connected component of $G[B]$ is isomorphic to F, and $|A|$ and $|B|$ are both multiples of $|V(H)|$.
Question: Does G have an H-factor?

For each connected graph H for which H-FACTOR is NP-hard, there is a connected graph F for which the above problem is NP-hard; F can be chosen as the local H-coordination gadget without its connector vertices. This follows from the construction of Kirkpatrick and Hell [28, Lemma 4.1]. I sketch a proof below, to highlight how the local coordination gadget can be exploited.

Lemma 3. *Let H be a connected graph on at least three vertices, let H' be the local H-coordination gadget with connector vertices C as in Lemma 2, and let $F := H' - C$. Then H-FACTOR WITH F-PARTITION is NP-hard under Karp reductions.*

Proof (sketch). Recall that for every integer $d \geq 3$, the PERFECT d-SET PACKING problem is defined as follows: given a collection of $S_1, \ldots, S_m$ of subsets of size d of a universe U, decide whether there exist $|U|/d$ *pairwise disjoint* sets in the collection (whose union therefore contains every element of U). The PERFECT d-SET PACKING problem is NP-complete for each $d \geq 3$ [24, SP1]. To prove the lemma, I show that for $h := |V(H)|$ there is a Karp reduction from PERFECT h-SET PACKING to H-FACTOR WITH F-PARTITION.

Consider an input $S_1, \ldots, S_m$ over a universe U for PERFECT h-SET PACKING. If $|U|$ is not a multiple of h, then clearly the answer is NO and we may output a fixed NO instance. In the remainder, assume $|U|$ is a multiple of h. Construct a graph G as follows. Initialize G as the edgeless graph on vertex set U, and attach a local H-coordination gadget onto the vertices of S_i for each $i \in [m]$. Since Lemma 2 guarantees that the connector vertices form an independent set in H', this preserves the fact that $G[U]$ is edgeless. As $G - U$ consists of copies of the connected graph $H' - C = F$, each connected component of $G' - U$ is isomorphic to F. Hence $(G, A := U, B := V(G) \setminus U)$ is a valid instance of H-FACTOR WITH F-PARTITION. Since $|A| = |U|$, it is a multiple of $h = |V(H)|$. To see that $|B|$ is a multiple of h, it suffices to note that $G[B]$ consists of disjoint copies of $H' - C$, each of which has an H-factor by Lemma 2 and therefore has an integer multiple of h many vertices. As Lemma 2 guarantees that, for each gadget attached onto a set S_i, the gadget either absorbs all attached vertices or none, while the interior vertices of gadgets attached for unused sets can be also covered by an H-factor, it is easy to verify that (G, A, B) is equivalent to the set packing instance we started from. □

Lemma 3 provides us with a starting problem for the cross-composition. Before presenting that composition, some more gadgeteering is required. While the local gadget of Lemma 2 synchronizes the behavior of $|V(H)|$ vertices at a time, in the construction we will need to synchronize the behavior of arbitrarily large vertex sets. For that reason we need a global coordination gadget, which will be described in Lemma 4. The following proposition is needed for its construction.

Proposition 1. *There is a polynomial-time algorithm that, given an integer $h \geq 3$ and an integer $m \geq 1$, outputs a connected bipartite multigraph F with partite sets A and B. Set A has $m \cdot h$ vertices, each of degree $h-1$, and set B has $m \cdot (h-1)$ vertices, each of degree h.*

Proof. Initialize F as a cycle on $2m(h-1)$ vertices, half of which belong to A and the other half to B. Then insert m additional vertices into A, each of which is connected by an edge to a distinct vertex of B. Clearly, F is a connected bipartite graph with partite sets of the right size. No vertex exceeds its intended degree bound since $h \geq 3$. Greedily extend F to the desired regular bipartite multigraph: as long as there is a vertex in one partite set whose degree is still too small, there is an accompanying vertex in the other partite set whose degree is also too small: insert an edge between them into the multigraph. □

Now I present the global coordination gadget.

Lemma 4. *For each fixed connected graph H on $h \geq 3$ vertices, there is a polynomial-time algorithm that, given an integer n that is a multiple of h, constructs a graph H^* on $\mathcal{O}(n)$ vertices together with an independent set C^* of n connector vertices in H^*, such that:*

1. *There is an H-factor of H^* and there is an H-factor of $H^* - C^*$.*
2. *For each $\emptyset \subsetneq C' \subsetneq C^*$ there is no H-factor of $H^* - C'$.*

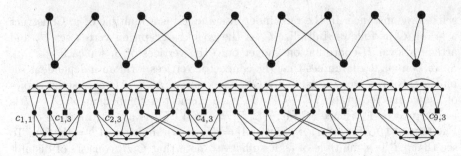

Fig. 3. Bottom: A global coordination gadget H^* for $H = K_3$, whose $n = 3 \cdot 3 = 9$ connector vertices $C^* = \{c_{1,3}, c_{2,3}, \ldots, c_{9,3}\}$ are visualized as squares. Top: The connected bipartite multigraph F whose use in Step 3 of Lemma 4 leads to the bottom gadget H^*. Observe that F can be obtained from H^* by taking one vertex for every local coordination gadget that was inserted, and adding an edge for every pair of gadgets that share a vertex.

3. If a graph G contains H^* as an induced subgraph, such that none of the interior vertices $V(H^*) \setminus C^*$ are adjacent to vertices outside of H^*, then for every H-factor $H_1, \ldots, H_k$ of G, the following holds: if v is an interior vertex of H^* and $v \in H_i$, then $V(H_i) \subseteq V(H^*)$.

Proof. Let $n = m \cdot h$. We build H^* as follows. (See Fig. 3 for an illustration.)

1. Initialize H^* as an independent set on vertex set $C = \{c_{i,j} \mid i \in [n], j \in [h]\}$. Define $C^* := \{c_{i,h} \mid i \in [n]\}$ to be the n connector vertices.
2. For each $i \in [n]$, insert a local H-coordinator gadget $\mathcal{A}_i$ into H^* and attach $\mathcal{A}_i$ onto $\{c_{i,j} \mid j \in [h]\}$. The gadgets $\mathcal{A}_1, \ldots, \mathcal{A}_n$ added in this step are referred to as the *top gadgets*, reflecting their visualization in Fig. 3.
3. Invoke Proposition 1 to construct a connected bipartite multigraph F with one $(h-1)$-regular partite set $A = \{a_1, \ldots, a_{m \cdot h}\}$, and one h-regular partite set $B = \{b_1, \ldots, b_{m \cdot (h-1)}\}$. Order the edges incident on each vertex a_i arbitrarily from 1 to $h-1$. Associate to each vertex b_k a private set of h vertices from C: for each edge e connecting b_k to a neighbor a_i, if e is the ℓ-th incident edge of a_i in the ordering, then associate $c_{i,\ell}$ to b_k. For each $b_k \in B$, insert a local H-coordination gadget $\mathcal{B}_k$ into H^* and attach it onto the h vertices associated to b_k. This leads to the bottom row of coordinator gadgets in Fig. 3. The regularity conditions of F ensure that we attach exactly one bottom-row gadget onto each vertex of $\{c_{i,j} \mid i \in [n], 1 \leq j \leq h-1\}$. Moreover, for vertices $b_k, b_{k'}$ with a common neighbor a_i in F, the vertex sets onto which gadgets $\mathcal{B}_k$ and $\mathcal{B}_{k'}$ are attached both include a vertex of $\{c_{i,j} \mid 1 \leq j \leq h-1\}$.

It is easy to see that the construction can be carried out in polynomial time and results in a graph on $\mathcal{O}(n)$ vertices. Note that since H is fixed, the size of a local coordinator gadget is constant. Let us verify the claimed properties.

(1) To get an H-factor of H^*, we combine the H-factors of the local coordinator gadgets whose existence is guaranteed by Lemma 2, as follows. For each

top-row gadget inserted in Step 2, use an H-factor of the gadget that absorbs all connector vertices. For each bottom-row gadget inserted in Step 3, use an H-factor that absorbs no connector vertices.

To get an H-factor of $H^* - C^*$, we do the opposite: top-row gadgets inserted in Step 2 absorb no connector vertices, but bottom-row gadgets inserted in Step 3 absorb all connector vertices.

(2) Consider an H-factor of $H^* - C'$ for some nonempty $C' \subseteq C^*$; I will show that $C' = C^*$. Consider a connector vertex $c_{i,h} \in C'$ that is not used in the H-factor of the subgraph. Then the vertices $X = \{c_{i,j} \mid 1 \leq j \leq h - 1\}$ are not absorbed by the corresponding top-row gadget $\mathcal{A}_i$ that was attached to $X \cup \{c_{i,h}\}$, since Lemma 2 guarantees that $\mathcal{A}_i$ absorbs either all or none of its local connector vertices. Since graph $H^* - C'$ contains X and $C \supseteq X$ is an independent set in H^*, the vertices from X must therefore be absorbed by bottom-row gadgets in the H-factor. Let $\mathcal{B}_k$ be a bottom-row gadget that was attached onto a vertex of X. Since $\mathcal{A}_i$ does not absorb its local connector vertices, the vertex shared between $\mathcal{A}_i$ and $\mathcal{B}_k$ must be absorbed by $\mathcal{B}_k$, which therefore absorbs all its connector vertices. This means that no top-row gadget that shares a vertex with $\mathcal{B}_k$ can absorb any of its connector vertices. If vertices $b_k, b_{k'}$ have a common neighbor $a_{i'}$ in F, this implies that $\mathcal{A}_{i'}$ absorbs no connector vertices, so that $\mathcal{B}_{k'}$ absorbs all its connector vertices. Since F is connected, repeating this argument shows that all bottom-row gadgets absorb all their local connector vertices, while no top-row gadgets absorb any of their local connector vertices. Consequently, the H-factor of $H^* - C'$ does not contain any vertex of C^* and therefore $C' = C^*$.

(3) Suppose that H^* is contained as an induced subgraph in a larger graph G, such that no interior vertex is adjacent to a vertex outside H^*. Consider an H-factor $H_1, \ldots, H_k$ of G, and fix an interior vertex v of H^*. If v is an interior vertex of some local coordination gadget, then Lemma 2 ensures that the H-subgraph H_i containing v is contained within the local coordination gadget, and therefore $V(H_i) \subseteq V(H^*)$. Otherwise, v is of the form $c_{i,j}$ for $i \in [n]$ and $j \in [h - 1]$. But then the only neighbors of v in G are interior vertices of local coordination gadgets inserted into H^*. Since H is connected and has at least three vertices, vertex v is contained in some H-subgraph H_i together with an internal vertex of a local coordination gadget, ensuring $V(H_i) \subseteq V(H^*)$ by the previous argument. This concludes the proof. $\square$

Using the properties of Lemma 4, the terminology of attaching coordination gadgets and absorbing connector vertices extends to global coordination gadgets in the natural way. The sparsification lower bound for H-FACTOR now follows cleanly by combining the two ingredients developed so far: the "bipartite" NP-hard source problem of Lemma 3 and the global coordination gadget of Lemma 4.

Theorem 4. *For any $\varepsilon > 0$, for any connected graph H on at least three vertices, H-FACTOR parameterized by the number of vertices n does not admit a generalized kernelization of bitsize $\mathcal{O}(n^{2-\varepsilon})$ unless* NP $\subseteq$ coNP/poly.

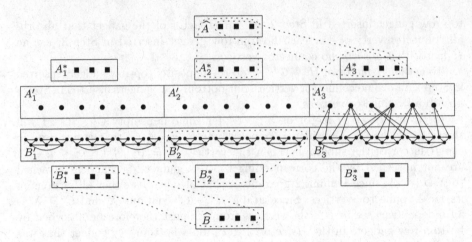

Fig. 4. Schematic visualization of the result of the cross-composition of Theorem 4 for $H = K_3$, applied to $t = 9$ inputs of H-FACTOR WITH F-PARTITION. Of the edges between different sets A_i' and B_j', only those corresponding to the YES-instance induced by A_3' and B_3' have been drawn. Vertices of inserted global coordination gadgets are not drawn. For some global coordination gadgets, the vertices they have been attached onto have been highlighted by a dotted curve.

Proof. I present a degree-2 OR-cross-composition. Fix a graph F such that the source problem H-FACTOR WITH F-PARTITION is NP-hard. Let $(G_{i,j}, A_{i,j}, B_{i,j})$ for $i, j \in [\sqrt{t}]$ be a sequence of t input instances that all share the same number n_A of vertices in the A-set and the same number n_B of vertices in the B-set, which may be assumed by a suitable choice of $\mathcal{R}$. By definition of the source problem, both n_A and n_B are multiples of $h = |V(H)|$. Build an instance G' of H-FACTOR; refer to Fig. 4 for an illustration.

1. For each $i \in [\sqrt{t}]$, add a set A_i' of n_A independent vertices to G'.
2. For each $i \in [\sqrt{t}]$, add a set B_i' of n_B vertices to G'. Insert edges so that $G'[B_i']$ forms $n_B/|V(F)|$ vertex-disjoint copies of F.
3. For each $i, j \in [\sqrt{t}]$, insert edges between A_i' and B_j' so that $G'[A_i' \cup B_j']$ is isomorphic to $G_{i,j}$. This is simultaneously possible for all i, j since all graphs $(G_{i,j}[A_{i,j}])_{i,j \in [\sqrt{t}]}$ are isomorphic to each other, and similarly all graphs $(G_{i,j}[B_{i,j}])_{i,j \in [\sqrt{t}]}$ are isomorphic to each other. Since each connected component of $G_{i,j}[B_{i,j}]$ is isomorphic to the fixed graph F, this step can be performed in polynomial time.
4. For each $i \in [\sqrt{t}]$, add a vertex set A_i^* of size h to G', and attach a new global coordination gadget $\mathcal{A}_i$ with $n_A + h$ connector vertices onto $A_i' \cup A_i^*$.
5. For each $j \in [\sqrt{t}]$, add a vertex set B_j^* of size h to G', and attach a new global coordination gadget $\mathcal{B}_j$ with $n_B + h$ connector vertices onto $B_j' \cup B_j^*$.
6. Add a vertex set $\widehat{A}$ of size h to G'. For each $i \in [\sqrt{t}]$, insert a global coordination gadget $\widehat{\mathcal{A}}_i$ with $2h$ connector vertices, and attach it onto $\widehat{A} \cup A_i^*$.

7. Add a vertex set $\widehat{B}$ of size h to G'. For each $j \in [\sqrt{t}]$, insert a global coordination gadget $\widehat{\mathcal{B}}_j$ with $2h$ connector vertices, and attach it onto $\widehat{B} \cup B_j^*$.

This concludes the description of graph G'. It is easy to see that the construction can be performed in polynomial time. Let us analyze the number of vertices in G'. It is easy to verify that, apart from the vertices of the inserted global coordination gadgets, the graph has $\mathcal{O}(\sqrt{t}(n_A + n_B))$ vertices, treating $|V(H)|$ as a constant. In Steps 4–5 we insert $\mathcal{O}(\sqrt{t})$ global coordination gadgets for $\mathcal{O}(n_A + n_B)$ connector vertices each, which therefore contribute $\mathcal{O}(\sqrt{t}(n_A + n_B))$ vertices to G'. Finally, the last two steps contribute $\mathcal{O}(\sqrt{t})$ additional vertices. It follows that $|V(G)'| \in \mathcal{O}(\sqrt{t}(n_A + n_B))$, which is suitably bounded.

To complete the cross-composition, we verify that G' has an H-factor if and only if some input instance $G_{i,j}$ has an H-factor.

Suppose first that G_{i^*,j^*} has an H-factor. Since $G'[A'_{i^*} \cup B'_{j^*}]$ is isomorphic to G_{i^*,j^*}, it has an H-factor. To extend it to an H-factor of all of G', we do the following. For each $i \in [\sqrt{t}] \setminus \{i^*\}$, use an H-factor of the gadget $\mathcal{A}_i$ together with its connector vertices, absorbing $A'_i \cup A_i^*$. Similarly, for each $j \in [\sqrt{t}] \setminus \{j^*\}$, use an H-factor of $\mathcal{B}_j$ together with its connector vertices, absorbing $B'_j \cup B_j^*$. Use an H-factor of gadget $\widehat{\mathcal{A}}_{i^*}$ with its connector vertices, absorbing $A_{i^*}^* \cup \widehat{A}$. Similarly, use an H-factor of gadget $\widehat{\mathcal{B}}_{j^*}$ with its connector vertices, absorbing $B_{j^*}^* \cup \widehat{B}$. For the remaining global coordination gadgets, use H-factors of only the interior of the gadget without absorbing connector vertices. This yields an H-factor of G'.

For the other direction, suppose that G' has an H-factor. Since the vertices of $\widehat{A}$ and $\widehat{B}$ form independent sets, whose only interaction with the rest of the graph is through the attachment of coordination gadgets, using Lemma 4 it follows there is a global coordination gadget $\widehat{\mathcal{A}}_{i^*}$ that absorbs all its connection vertices $A_{i^*}^* \cup \widehat{A}$, and analogously a gadget $\widehat{\mathcal{B}}_{j^*}$ absorbing $B_{j^*}^* \cup \widehat{B}$. Consequently, for each $i \neq i^*$ the vertices of A_i^* are absorbed by the global coordination gadget $\mathcal{A}_i$, which therefore also absorbs A'_i. Similarly, for each $j \neq j^*$ the vertices of B_j^* are absorbed by $\mathcal{B}_j$, thereby also absorbing B'_j. This implies that in the H-factor of G', vertices of $A'_{i^*} \cup B'_{j^*}$ are not contained in H-subgraphs together with vertices outside of $A'_{i^*} \cup B'_{j^*}$. Consequently, the H-factor of G' restricted to $A'_{i^*} \cup B'_{j^*}$ is an H-factor of $G'[A'_{i^*} \cup B'_{j^*}]$, which is isomorphic to G_{i^*,j^*}. Hence G_{i^*,j^*} is a YES-instance. This concludes the proof of Theorem 4. □

Theorem 4 shows that for *connected* graphs H on at least three vertices, the H-FACTOR problem admits no nontrivial polynomial-time sparsification unless $\mathsf{NP} \subseteq \mathsf{coNP/poly}$. It is not difficult to extend this to disconnected graphs H that have a connected component of at least three vertices, thereby extending the lower bound to all graphs H for which H-FACTOR is NP-hard.

Corollary 1. *For any $\varepsilon > 0$, for any graph H that contains a connected component on at least three vertices, H-FACTOR parameterized by the number of vertices n does not admit a generalized kernelization of bitsize $\mathcal{O}(n^{2-\varepsilon})$ unless $\mathsf{NP} \subseteq \mathsf{coNP/poly}$.*

Proof. Consider such a graph H, and let H' be a connected component of H on at least three vertices that maximizes the number of edges. Using a reduction of Kirkpatrick and Hell [28, Lemma 2.1], I give a polynomial-time reduction from H'-FACTOR instances on n vertices to H-FACTOR instances on $\mathcal{O}(n)$ vertices. This reduction, together with a subquadratic generalized kernelization for H-FACTOR, would yield a subquadratic generalized kernelization for H'-FACTOR, which is ruled out by Theorem 4. Hence it suffices to give the reduction.

Given a graph G' for which we want to determine the existence of an H'-factor, we may assume without loss of generality that $|V(G')|$ is a multiple of $|V(H')|$ (otherwise the answer is trivially NO). If $|V(G')| = d \cdot |V(H')|$, then for the graph G obtained as the disjoint union of G together with d copies of $H - H'$, it is easy to verify (cf. [28, Lemma 2.1]) that G' has an H'-factor if and only if G' has an H'-factor. Since H is fixed, $|V(G')| \in \mathcal{O}(|V(G)|)$, which concludes the proof. $\square$

6   Conclusion

In this article, I showed how the technique of cross-composition [7] that was developed together with Hans Bodlaender and Stefan Kratsch can be used to give elementary proofs that VERTEX COVER, FEEDBACK VERTEX SET, and H-PACKING do not admit generalized kernels of bitsize $\mathcal{O}(n^{2-\varepsilon})$. The constructions all boil down to the appropriate combination of two key ingredients: a suitable NP-hard starting problem whose inputs can be partitioned into two regularly structured induced subgraphs, and a gadget that allows some parts of the input to be disabled in order to achieve the desired logical OR.

Following the case-by-case investigation of sparsification lower bounds in this article, one is naturally led to ask whether it is possible to prove sparsification lower bounds on a larger scale, capturing entire classes of problems at the same time. In recent years, I have pursued this direction together with my PhD student Astrid Pieterse (Hans' academic granddaughter!) by investigating the sparsifiability of CONSTRAINT SATISFACTION problems. A key challenge for the future consist of the following: for which constraint languages (defining the types of constraints that one is allowed to use in the problem) do CSPs over the Boolean domain allow for a linear sparsification? I refer the interested reader to recent papers [12, 29] for further information.

An entirely different, but equally exciting, direction is the study of the positive toolkit: sparsification algorithms. While nontrivial sparsification algorithms exist for several satisfiability problems, such as 3-NOT ALL EQUAL SATISFIABILITY parameterized by the number of variables n, which can be sparsified [26] to instances with $\mathcal{O}(n^2)$ clauses, to this day we do not have any good examples of NP-hard *graph* problems that admit nontrivial polynomial-time sparsification. Have we not been looking in the right places, or could there be a reason for their nonexistence?

I conclude this article by thanking Hans; for saving me from becoming a computer-game programmer; for inspiring me to do research; and for his good-humored companionship, discussions, and boardgames spanning over a decade and multiple continents. Happy 60th birthday, Hans!

References

1. Bodlaender, H.L.: A cubic kernel for feedback vertex set. In: Thomas, W., Weil, P. (eds.) STACS 2007. LNCS, vol. 4393, pp. 320–331. Springer, Heidelberg (2007). https://doi.org/10.1007/978-3-540-70918-3_28
2. Bodlaender, H.L., van Dijk, T.C.: A cubic kernel for feedback vertex set and loop cutset. Theory Comput. Syst. **46**(3), 566–597 (2010). https://doi.org/10.1007/s00224-009-9234-2
3. Bodlaender, H.L., Downey, R.G., Fellows, M.R., Hermelin, D.: On problems without polynomial kernels (extended abstract). In: Aceto, L., Damgård, I., Goldberg, L.A., Halldórsson, M.M., Ingólfsdóttir, A., Walukiewicz, I. (eds.) ICALP 2008. LNCS, vol. 5125, pp. 563–574. Springer, Heidelberg (2008). https://doi.org/10.1007/978-3-540-70575-8_46
4. Bodlaender, H.L., Downey, R.G., Fellows, M.R., Hermelin, D.: On problems without polynomial kernels. J. Comput. Syst. Sci. **75**(8), 423–434 (2009). https://doi.org/10.1016/j.jcss.2009.04.001
5. Bodlaender, H.L., Jansen, B.M.P., Kratsch, S.: Cross-composition: a new technique for kernelization lower bounds. In: Schwentick, T., Dürr, C. (eds.) Proceedings of the 28th STACS. LIPIcs, vol. 9, pp. 165–176. Schloss Dagstuhl - Leibniz-Zentrum fuer Informatik (2011). https://doi.org/10.4230/LIPIcs.STACS.2011.165
6. Bodlaender, H.L., Jansen, B.M.P., Kratsch, S.: Kernel bounds for path and cycle problems. Theor. Comput. Sci. **511**, 117–136 (2013). https://doi.org/10.1016/j.tcs.2012.09.006
7. Bodlaender, H.L., Jansen, B.M.P., Kratsch, S.: Kernelization lower bounds by cross-composition. SIAM J. Discrete Math. **28**(1), 277–305 (2014). https://doi.org/10.1137/120880240
8. Bodlaender, H.L., Thomassé, S., Yeo, A.: Kernel bounds for disjoint cycles and disjoint paths. Theor. Comput. Sci. **412**(35), 4570–4578 (2011). https://doi.org/10.1016/j.tcs.2011.04.039
9. Brandt, S.: Computing the independence number of dense triangle-free graphs. In: Möhring, R.H. (ed.) WG 1997. LNCS, vol. 1335, pp. 100–108. Springer, Heidelberg (1997). https://doi.org/10.1007/BFb0024491
10. Burrage, K., Estivill-Castro, V., Fellows, M., Langston, M., Mac, S., Rosamond, F.: The undirected feedback vertex set problem has a poly(k) kernel. In: Bodlaender, H.L., Langston, M.A. (eds.) IWPEC 2006. LNCS, vol. 4169, pp. 192–202. Springer, Heidelberg (2006). https://doi.org/10.1007/11847250_18
11. Buss, J.F., Goldsmith, J.: Nondeterminism within P. SIAM J. Comput. **22**(3), 560–572 (1993). https://doi.org/10.1137/0222038
12. Chen, H., Jansen, B.M.P., Pieterse, A.: Best-case and worst-case sparsifiability of Boolean CSPs. In: Paul, C., Pilipczuk, M. (eds.) Proceedings of 13th IPEC. LIPIcs, vol. 115, pp. 15:1–15:13. Schloss Dagstuhl - Leibniz-Zentrum fuer Informatik (2018). https://doi.org/10.4230/LIPIcs.IPEC.2018.15
13. Cygan, M., et al.: Parameterized Algorithms. Springer, Cham (2015). https://doi.org/10.1007/978-3-319-21275-3
14. Dell, H., Marx, D.: Kernelization of packing problems. In: Rabani, Y. (ed.) Proceedings of the 23rd SODA, pp. 68–81. SIAM (2012). https://doi.org/10.1137/1.9781611973099.6

15. Dell, H., van Melkebeek, D.: Satisfiability allows no nontrivial sparsification unless the polynomial-time hierarchy collapses. In: Schulman, L.J. (ed.) Proceedings of the 42nd STOC, pp. 251–260. ACM (2010). https://doi.org/10.1145/1806689.1806725

16. Dell, H., van Melkebeek, D.: Satisfiability allows no nontrivial sparsification unless the polynomial-time hierarchy collapses. J. ACM **61**(4), 23:1–23:27 (2014). https://doi.org/10.1145/2629620

17. Dom, M., Lokshtanov, D., Saurabh, S.: Kernelization lower bounds through colors and IDs. ACM Trans. Algorithms **11**(2), 13 (2014). https://doi.org/10.1145/2650261

18. Downey, R., Fellows, M.R.: Parameterized Complexity. Monographs in Computer Science. Springer, New York (1999). https://doi.org/10.1007/978-1-4612-0515-9

19. Downey, R.G., Fellows, M.R.: Fundamentals of Parameterized Complexity. Texts in Computer Science. Springer, London (2013). https://doi.org/10.1007/978-1-4471-5559-1

20. Flum, J., Grohe, M.: Parameterized Complexity Theory. Springer, Heidelberg (2006). https://doi.org/10.1007/3-540-29953-X

21. Fomin, F.V., Lokshtanov, D., Saurabh, S., Zehavi, M.: Kernelization: Theory of Parameterized Preprocessing. Cambridge University Press, Cambridge (2019). https://doi.org/10.1017/9781107415157

22. Fortnow, L., Santhanam, R.: Infeasibility of instance compression and succinct PCPs for NP. In: Dwork, C. (ed.) Proceedings of the 40th STOC, pp. 133–142. ACM (2008). https://doi.org/10.1145/1374376.1374398

23. Fortnow, L., Santhanam, R.: Infeasibility of instance compression and succinct PCPs for NP. J. Comput. Syst. Sci. **77**(1), 91–106 (2011). https://doi.org/10.1016/j.jcss.2010.06.007

24. Garey, M.R., Johnson, D.S.: Computers and Intractability, A Guide to the Theory of NP-Completeness. W.H. Freeman and Company, New York (1979)

25. Iwata, Y.: Linear-time kernelization for feedback vertex set. In: Chatzigiannakis, I., Indyk, P., Kuhn, F., Muscholl, A. (eds.) Proceedings of the 44th ICALP. LIPIcs, vol. 80, pp. 68:1–68:14. Schloss Dagstuhl - Leibniz-Zentrum fuer Informatik (2017). https://doi.org/10.4230/LIPIcs.ICALP.2017.68

26. Jansen, B.M.P., Pieterse, A.: Optimal sparsification for some binary CSPs using low-degree polynomials. TOCT **11**(4), 28:1–28:26 (2019). https://doi.org/10.1145/3349618

27. Karp, R.M.: Reducibility among combinatorial problems. In: Miller, R.E., Thatcher, J.W., Bohlinger, J.D. (eds.) Complexity of Computer Computations. The IBM Research Symposia Series, pp. 85–103. Springer, Boston (1972). https://doi.org/10.1007/978-1-4684-2001-2_9

28. Kirkpatrick, D.G., Hell, P.: On the complexity of general graph factor problems. SIAM J. Comput. **12**(3), 601–609 (1983). https://doi.org/10.1137/0212040

29. Lagerkvist, V., Wahlström, M.: Kernelization of constraint satisfaction problems: a study through universal algebra. In: Beck, J.C. (ed.) CP 2017. LNCS, vol. 10416, pp. 157–171. Springer, Cham (2017). https://doi.org/10.1007/978-3-319-66158-2_11

30. Lewis, J.M., Yannakakis, M.: The node-deletion problem for hereditary properties is NP-complete. J. Comput. Syst. Sci. **20**(2), 219–230 (1980). https://doi.org/10.1016/0022-0000(80)90060-4

31. Nemhauser, G., Trotter, L.: Vertex packings: structural properties and algorithms. Math. Program. **8**, 232–248 (1975). https://doi.org/10.1007/BF01580444

32. Poljak, S.: A note on stable sets and colorings of graphs. Commentationes Mathematicae Universitatis Carolinae **015**(2), 307–309 (1974). http://eudml.org/doc/16622
33. Thomassé, S.: A $4k^2$ kernel for feedback vertex set. ACM Trans. Algorithms **6**(2) (2010). https://doi.org/10.1145/1721837.1721848

Open Access This chapter is licensed under the terms of the Creative Commons Attribution 4.0 International License (http://creativecommons.org/licenses/by/4.0/), which permits use, sharing, adaptation, distribution and reproduction in any medium or format, as long as you give appropriate credit to the original author(s) and the source, provide a link to the Creative Commons license and indicate if changes were made.

The images or other third party material in this chapter are included in the chapter's Creative Commons license, unless indicated otherwise in a credit line to the material. If material is not included in the chapter's Creative Commons license and your intended use is not permitted by statutory regulation or exceeds the permitted use, you will need to obtain permission directly from the copyright holder.

Efficient Graph Minors Theory
and Parameterized Algorithms
for (Planar) Disjoint Paths

Daniel Lokshtanov[1], Saket Saurabh[2,3], and Meirav Zehavi[4(✉)]

[1] University of California, Santa Barbara, USA
daniello@ucsb.edu
[2] IRL 2000 ReLaX, Indian Institute of Mathematical Sciences, Chennai, India
saket@imsc.res.in
[3] University of Bergen, Bergen, Norway
[4] Ben-Gurion University of the Negev, Beersheba, Israel
meiravze@bgu.ac.il

Abstract. In the DISJOINT PATHS problem, the input consists of an n-vertex graph G and a collection of k vertex pairs, $\{(s_i, t_i)\}_{i=1}^k$, and the objective is to determine whether there exists a collection $\{P_i\}_{i=1}^k$ of k pairwise vertex-disjoint paths in G where the end-vertices of P_i are s_i and t_i. This problem was shown to admit an $f(k)n^3$-time algorithm by Robertson and Seymour *Graph Minors XIII, The Disjoint Paths Problem, JCTB*. In modern terminology, this means that DISJOINT PATHS is fixed parameter tractable (FPT) with respect to k. Remarkably, the above algorithm for DISJOINT PATHS is a cornerstone of the entire Graph Minors Theory, and conceptually vital to the $g(k)n^3$-time algorithm for MINOR TESTING (given two undirected graphs, G and H on n and k vertices, respectively, determine whether G contains H as a minor).

In this semi-survey, we will first give an exposition of the Graph Minors Theory with emphasis on efficiency from the viewpoint of Parameterized Complexity. Secondly, we will review the state of the art with respect to the DISJOINT PATHS and PLANAR DISJOINT PATHS problems. Lastly, we will discuss the main ideas behind a new algorithm that combines treewidth reduction and an algebraic approach to solve PLANAR DISJOINT PATHS in time $2^{k^{\mathcal{O}(1)}} n^{\mathcal{O}(1)}$ (for undirected graphs).

Keywords: Disjoint paths · Planar disjoint paths · Graph minors · Treewidth

This project has received funding from the European Research Council (ERC) under the European Union's Horizon 2020 research and innovation programme (grant agreement no. 819416 and no. 715744). The second author also acknowledges the support of Swarnajayanti Fellowship grant DST/SJF/MSA-01/2017-18. The third author acknowledges the support of ISF grant no. 1176/18. The first and third authors also acknowledge the support of BSF grant no. 2018302.

© Springer Nature Switzerland AG 2020
F. V. Fomin et al. (Eds.): Bodlaender Festschrift, LNCS 12160, pp. 112–128, 2020.
https://doi.org/10.1007/978-3-030-42071-0_9

1   Background on Graph Minors Theory with Emphasis on Efficiency

Arguably, the origin of Parameterized Complexity is the *graph minors project of Robertson and Seymour*. Recollecting the birth of Parameterized Complexity, Downey [21] stated not only that "a real inspiration was the theorem of Robertson and Seymour", but also that in the early years of the field, "many listeners thought that what we were doing was basically applying Robertson-Seymour." The concept of a minor originated already in the early 20th century. Formally, for any two graphs G and H, we say that H is a *minor* of G if there exists a series of edge deletions, edge contractions and vertex deletions in G that yields H. One of the most famous results in Graph Theory is Kuratowski's theorem [42], which states that a graph is planar if and only if it does not contain the graphs $K_{3,3}$ and K_5 as minors. That is, the class of planar graphs is characterized by a set of two forbidden minors. Robertson and Seymour set out to prove a' vast generalization of Kuratowski's theorem, namely, Wagner's conjecture [60]: *Any infinite sequence of graphs contains two graphs such that one is a minor of the other (that is, the class of all graphs is well-quasi ordered by the minor relation).* Equivalently, Wagner's conjecture states that *any* minor-closed family of graphs can be characterized by a finite set of forbidden minors.

In perhaps one of the most amazing feats of modern mathematics, Robertson and Seymour managed to prove Wagner's conjecture. The endeavour of Robertson and Seymour to prove this conjecture spans a series of over 23 papers, published from 1983 to 2004. One of the main reasons why this theory has had such a great impact is the sheer number of novel algorithms and algorithmic techniques that were developed as a part of it. A few notable algorithmic highlights are their parameterized algorithms for MINOR TESTING (given two graphs, G on n vertices and H on k vertices, decide whether H is a minor of G), DISJOINT PATHS (given a graph G and a collection of k terminal pairs, $\{(s_i, t_i)\}|_{i=1}^{k}$, decide whether G has k pairwise vertex-disjoint paths, $\{P_i\}|_{i=1}^{k}$, where for every $i \in \{1, 2, \ldots, k\}$, the endpoints of P_i are s_i and t_i) and a constant-factor approximation parameterized algorithm to compute the treewidth of a given graph. Their project introduced key definitions and concepts such as those of an *excluded grid* and a *tree decomposition* (a decomposition of a graph into a tree-like structure), along with key structural results such as duality theorems (e.g., the characterization of treewidth in the terms of a family of connected graphs called a *bramble*). Additionally, their project presented new methods such as the so-called *irrelevant vertex technique*. Notably, *all* of these intermediate results have found applications and implications across a wide range of research domains.

Unfortunately, the hidden constants in the results above, both in terms of time complexities and in the structural theorems themselves, are *really, really bad*. In fact, the immense parameter dependence of algorithms based on the graph minors project earned them their own name—"galactic algorithms" [43]. As phrased by Johnson [35], "for any instance $G = (V, E)$ that one could fit into the known universe, one would easily prefer $|V|^{70}$ to even constant time, if that constant had to be one of Robertson and Seymour's". In light of this, Downey [21] stated that "in retrospect, it might have been a bit unfortunate to tie the FPT material to the Robertson-Seymour material when we spoke". Indeed, keeping in mind that the primary objective of the paradigm of Parameterized Complexity is to cope with computational intractability, we are facing a blatant discrepancy:

While Parameterized Complexity does provide an extremely rich toolkit to design efficient parameterized algorithms, one of its foundations and still most powerful tools yields algorithms that are wildly impractical.

In 1989, Fellows [24] noted that "it is likely to be many years before the practical significance of Robertson-Seymour theorems is fully understood". Nevertheless, for some of the algorithms, substantial advances have been made. In particular, Grohe et al. [34] gave an algorithm for computing *Robertson and Seymour's structural decomposition* (stating that all graphs excluding some fixed graph as a minor have a tree decomposition with bags that are almost embeddable in a fixed surface), that runs in time $f(k)n^2$ (for some function f of k), improving over the $f(k)n^3$-time algorithm of Robertson and Seymour. Prior to this result, Kawarabayashi and Wollan [38] gave a simplified proof of correctness of the graph minors algorithm. This proof yields a parametrized algorithm for MINOR TESTING with the best currently known parameter dependence. Here, $f(k)$ is a "tower of powers of 2 of height at most 5, with k^{1000} on top" [61]. From the work of Kawarabayashi et al. [37] on DISJOINT PATHS, we also know that MINOR TESTING is solvable in time $f(k)n^2$. Chuzhoy [10], building upon the seminal work of Chekuri and Chuzhoy [9], gave an improved algorithm for a weaker variant of Robertson and Seymour's structural decomposition. However, here the improvement is in the quality of the output decomposition, not in the time it takes to compute it.

Both the algorithm for MINOR TESTING and the structural theorem of Robertson and Seymour were discovered as consequences of the entire graph minors theory that they built. In retrospect, however, the "converse" also holds true: almost the entire graph minors theory would have had to be built in order to achieve either one of these goals, that is, the algorithm for MINOR TESTING as well as the structural theorem of Robertson and Seymour! In plain words, the design of an efficient algorithm for MINOR TESTING as a goal *necessitates* to devise efficient versions of large parts of the whole graph minors theory. A problem that is tightly linked to MINOR TESTING, yet seemingly more difficult than it, is TOPOLOGICAL MINOR TESTING: given two graphs, G on n vertices and H on k vertices, decide whether H is a *topological minor* of G. That is, the objective is to determine whether there exists a series of operations that delete an edge, *dissolve* an edge (i.e. delete a degree-2 vertex and make its two neighbors adjacent) and delete a vertex in G that yields H. We remark that Kuratowski's theorem [42] was, in fact, originally phrased in the terms of topological minors rather than minors. Unlike MINOR TESTING and DISJOINT PATHS, the question of the parameterized complexity of TOPOLOGICAL MINOR TESTING was not resolved by Robertson and Seymour, and was first stated explicitly as an open problem by Downey and Fellows in [22]. Since then, this question was restated as an open problem many times, until it was positively resolved by Grohe et al. [33], who designed an $f(k)n^3$-time (galactic) algorithm.

Some Central Applications

Graph minors in general, and MINOR TESTING in particular, have enjoyed numerous applications over the past 30 years. These applications span a wide range of areas, including (but not limited to) Approximation Algorithms, Exact Exponential and Polynomial-Time Algorithms, Parameterized Complexity, Logic, Computational Geometry and Property Testing. It is highly conceivable that algorithmically (and structurally) efficient Graph Minors Theory, being a core engine behind all of these applications, will have great impact on all of these areas simultaneously. For the sake of illustration, we briefly discuss three important discoveries (with emphasis on Parameterized Complexity) that build upon graph minors.

Classification. By Robertson and Seymour's theorem, every minor-closed family of graphs can be characterized by a finite set of forbidden minors. In particular, for any minor-closed family of graphs G, the family of graphs obtained from the graphs in G by adding at most k vertices is also minor-closed. As MINOR TESTING is solvable in time $f(k)n^2$ [37], this observation immediately shows that a vast range of parametrized problems, such as FEEDBACK VERTEX SET, PLANAR VERTEX DELETION and GRAPH GENUS, are *non-uniformly FPT*: for every integer k, the set of forbidden minors might be different. Furthermore, the result is non-constructive as long as we do not know how to compute the finite set of forbidden minors. Nevertheless, this result provides a very useful classification tool (for whether a problem is FPT or not), known since the early days of Parameterized Complexity.

Bidimensionality. The Graph Minors Theory laid the foundation for studying how computational problems that are hard on general graphs behave when restricted to minor-free graphs. The theory of *bidimensionality* [19] builds upon this knowledge, particularly on the relationship between grids and treewidth. While most NP-hard graph problems remain NP-hard even on planar graphs [30], many problems that are fixed-parameter intractable on general graphs are FPT on planar graphs, and even on graph classes excluding a fixed H as a minor. Bidimensionality simultaneously yields linear-time parameterized algorithms with subexponential parameter dependence [19], polynomial-time approximation schemes [20,26], and linear kernels [28] for many problems on minor-free graphs, with applications even in computational geometry [18,27]. Nevertheless, there are fundamental graph problems on planar and minor-free graphs for which bidimensionality seems insufficient. Examples of such problems include the LONGEST PATH problem on directed graphs, STEINER TREE, ODD CYCLE TRANSVERSAL, and many others [25,46,51]. More information can be found in two other chapters in this volume (one by D. Marx, and the other by Ma. Pilipczuk).

Irrelevant Vertices. The irrelevant vertex technique originated from Robertson and Seymour's algorithm for the DISJOINT PATHS problem [54]. Since then, this technique has found several other applications [4,17,32,33,48]. Roughly speaking, as long as the treewidth of the graph is large, the technique is applied by repeatedly finding an *irrelevant vertex*—a vertex whose deletion does not change the answer to the problem. For an illustrative application of this technique in textbook level of detail, we refer to Chapter 7.8 in [16].

2   (Planar) Disjoint Paths: State of the Art

Conceptually vital to the algorithm for MINOR TESTING of Robertson and Seymour, and the source of the irrelevant vertex technique, is their $f(k)n^3$-time algorithm for the DISJOINT PATHS problem [54]. The current state-of-the-art is the algorithm developed by Kawarabayashi et al. [37] in 2012, which runs in time $f(k)n^2$. Just like the case of MINOR TESTING, the parameter dependence of this algorithm on k renders it a "galactic algorithm". The DISJOINT PATHS problem is important on its own right due to its applications in the contexts of transportation networks, VLSI layout and virtual circuit routing [29,50,58,59]. It was shown to be NP-complete by Karp in [36], being one of Karp's original NP-complete problems. In fact, it remains NP-complete even if the input graph is restricted to be a grid [41].

2.1   Minor Testing and Disjoint Paths on Planar Graphs

We first remark that the MINOR TESTING, DISJOINT PATHS and TOPOLOGICAL MINOR TESTING problems are well known to be NP-hard also when restricted to the class of planar graphs [47]. Moreover, it is easy to see that if (TOPOLOGICAL) MINOR TESTING is solvable in time $2^{k^{\mathcal{O}(1)}}n^c$ (for some $c > 0$) on H-minor free graphs, then DISJOINT PATHS is also solvable in time $2^{k^{\mathcal{O}(1)}}n^c$ (for the same c) on this class of graphs. To see this, consider an instance $(G, \{(s_i, t_i)\}|_{i=1}^k, k)$ of DISJOINT PATHS on H-minor free graphs. Briefly, the idea is to attach, to each terminal s_i or t_i, a "large enough" clique (on $\mathcal{O}(|V(H)|k)$ vertices) of unique size, and define the graph to be sought as a minor as an appropriate combination of these cliques. Unfortunately, this reduction idea is clearly tailored specifically to H-minor free graphs—in particular, it is inapplicable to planar graphs and general graphs.

Apart from being important problems on their own right, the design of algorithms for MINOR TESTING, DISJOINT PATHS and TOPOLOGICAL MINOR TESTING on planar graphs also serves as a critical building block for the design of algorithms for these problems on general graphs in view of the way Robertson and Seymour's Graph Minors Theory is structured. Without delving into technical details, we note that all known algorithms for (TOPOLOGICAL) MINOR TESTING and DISJOINT PATHS (on general graphs) are based on the distinction between the case where the input graph G contains a large clique as a minor, and the case where it does not. Already at this stage, we see that the resolution of (TOPOLOGICAL) MINOR TESTING and DISJOINT PATHS on H-minor-free graphs is a building block towards the resolution of these problems on general graphs. Moreover, when the input graph does not contain a large clique as a minor, known algorithms distinguish between the case where the treewidth of G is small, and the case where the treewidth of G is large. In the latter case, G contains a so-called *flat wall* that further motivates, or even necessitates, the study of these problems on planar and "almost planar" graph classes.

With respect to known algorithms, for the MINOR TESTING problem on planar graphs, the design of a $2^{k^{O(1)}}n$-time algorithm is folklore: if the input graph G has treewidth larger than some function linear in k, then it necessarily contains the sought graph as a minor (this is a property holds only for planar graphs!), and otherwise it is possible to solve the problem in time $2^{k^{O(1)}}n$ via dynamic programming over tree decompositions (e.g., using [2]). We also remark that, for MINOR TESTING on planar graphs, Adler et al. [3] developed an algorithm that runs in time $O(2^{O(k)}n + n^2 \log n)$.

For the DISJOINT PATHS problem on planar graphs [52,53], and even on graphs of bounded genus [23,40,52], there already exist algorithms with running times whose dependency on n in linear, but whose dependency on k is prohibitive. Additionally, for the DISJOINT PATHS problem on planar graphs, Adler et al. [5] developed a $2^{2^{O(k)}}n^2$-time algorithm; in particular, towards that end, they presented a so-called *unique linkage theorem* that states that, in every instance of DISJOINT PATHS on planar graphs where the treewidth is larger than 2^{ck} (for some $c > 0$), there exists an irrelevant vertex and it is computable in linear time. Such a relation with single exponential dependency of the treewidth bound on k also holds for graphs of bounded genus [49]. However, for the more general H-minor-free graphs, the dependency becomes a tower of exponents [31,39] (prior to these works—that is, from the project of Robertson and Seymour—not even the computability of the bound was not known).

The PLANAR DISJOINT PATHS problem is intensively studied also from the perspective of approximation algorithms, with a burst of activity in recent years [11–15]. Highlights of this work include an approximation algorithm with approximation factor $O(n^{9/19} \log^{O(1)} n)$ [12] and, under reasonable complexity-theoretic assumptions, the proof of hardness of approximating the problem within a factor of $2^{\Omega(\frac{1}{(\log \log n)^2})}$ [14]. For the DIRECTED DISJOINT PATHS problem on planar graphs, Schrijver [57] gave an algorithm with running time $n^{O(k)}$, in contrast to the NP-hardness for $k = 2$ on general directed graphs. Almost 20 years later, Cygan et al. [17] improved over the algorithm of Schrijver and showed that DIRECTED DISJOINT PATHS on planar graphs is FPT by giving an algorithm with running time $2^{2^{O(k^2)}}n^{O(1)}$.

2.2 General Structure of Algorithms for (Planar) Disjoint Paths

All known algorithms for both DISJOINT PATHS and PLANAR DISJOINT PATHS have the same high level structure. In particular, given a graph G we distinguish between the cases of G having "small" or "large" treewidth. In case the treewidth is large, we distinguish between two further cases: either G contains a "large" clique minor or it does not. This results in the following case distinctions.

1. **Treewidth is small.** Let the treewidth of G be w. Then, we use the known dynamic programming algorithm with running time $2^{\mathcal{O}(w \log w)} n^{\mathcal{O}(1)}$ [56] to solve the problem. It is important to note that, assuming the Exponential Time Hypothesis (ETH), there is neither an algorithm for DISJOINT PATHS running in time $2^{o(w \log w)} n^{\mathcal{O}(1)}$ [44], nor an algorithm for PLANAR DISJOINT PATHS running in time $2^{o(w)} n^{\mathcal{O}(1)}$ [7].

2. **Treewidth is large and G has a large clique minor.** In this case, we use the good routing property of the clique to find an irrelevant vertex and delete it without changing the answer to the problem. Since this case does not arise for graphs embedded on a surface or for planar graphs, we do not discuss it in more detail.

3. **Treewidth is large and G has no large clique minor.** Using a fundamental structure theorem for minors called the flat wall theorem, we can conclude that G contains a large planar piece and a vertex v that is sufficiently insulated in the middle of it. Applying the unique linkage theorem [55] to this vertex, we conclude that it is irrelevant and remove it. For planar graphs, one can use the unique linkage theorem of Adler et al. [5]:

   Any instance of DISJOINT PATHS consisting of a planar graph with treewidth at least $82k^{3/2}2^k$ and k terminal pairs contains a vertex v such that every solution to DISJOINT PATHS can be replaced by an equivalent one whose paths avoid v.

   This result says that if the treewidth of the input planar graph is (roughly) $\Omega(2^k)$, then we can find an irrelevant vertex and remove it. A natural question is whether we can guarantee an irrelevant vertex even if the treewidth is $\Omega(\mathsf{poly}(k))$. Adler and Krause [6] exhibited a planar graph G with $k + 1$ terminal pairs such that G contains a $(2^k + 1) \times (2^k + 1)$ grid as a subgraph, DISJOINT PATHS on this input has a unique solution, and the solution uses all vertices of G; in particular, *no vertex of G is irrelevant*. This implies that the irrelevant vertex technique can only guarantee a treewidth of $\Omega(2^k)$, even if the input graph is planar.

Combining items (1) and (3), observe that the known methodology for DISJOINT PATHS can only guarantee an algorithm with running time $2^{2^{\mathcal{O}(k)}} n^2$ even when restricted to planar graphs. Thus, a $2^{k^{\mathcal{O}(1)}} n^{\mathcal{O}(1)}$-time algorithm for PLANAR DISJOINT PATHS appears to require entirely new ideas. As this obstacle was known to Adler et al. [1], it is likely to be the main motivation for Adler to pose the existence of a $2^{k^{\mathcal{O}(1)}} n^{\mathcal{O}(1)}$ time algorithm for PLANAR DISJOINT PATHS as an open problem.

3   Recent Development: Combination of Treewidth Reduction and an Algebraic Approach

Recent joint work of the authors of this paper with Misra and Pilipczuk [45] led to the development of an FPT algorithm for PLANAR DISJOINT PATHS whose parameter dependency on k is bounded by $2^{\mathcal{O}(k^2)}$. Specifically, we proved the following theorem.

Theorem 1 ([45]). *The* PLANAR DISJOINT PATHS *problem is solvable in time* $2^{\mathcal{O}(k^2)} n^{\mathcal{O}(1)}$.

Our algorithm is based on a novel combination of two techniques that do not seem to give the desired outcome when used on their own. The first ingredient is the treewidth reduction theorem of Adler et al. [5] that proves that given an instance of PLANAR DISJOINT PATHS, the treewidth can be brought down to $2^{\mathcal{O}(k)}$ (see item (3) in Sect. 2.2). This by itself is sufficient for an FPT algorithm (this is what Adler et al. [5] do), but as explained above, it seems hopeless that it will bring a $2^{k^{\mathcal{O}(1)}} n^{\mathcal{O}(1)}$-time algorithm.

We circumvent the obstacle by using an algorithm for a more difficult problem with a worse running time, namely, Schrijver's $n^{\mathcal{O}(k)}$-time algorithm for DISJOINT PATHS on directed planar graphs [57]. Schrijver's algorithm has two steps: a "guessing" step where one (essentially) guesses the homology class of the solution paths, and then a surprising homology-based algorithm that, given a homology class, finds a solution in that class (if one exists) in polynomial time. Our key insight is that for PLANAR DISJOINT PATHS, if the instance that we are considering has been reduced according to the procedure of Adler et al. [5], then we only need to iterate over $2^{\mathcal{O}(k^2)}$ homology classes in order to find the homology class of a solution, if one exists. The proof of this key insight is highly non-trivial, and builds on a cornerstone ingredient of the FPT algorithm of Cygan et al. [17] for DIRECTED DISJOINT PATHS on planar graphs. To the best of our knowledge, this is the first algorithm that finds the exact solution to a problem that exploits that the treewidth of the input graph is small in a way that is different from doing dynamic programming.

In what follows, we begin with an explanation of the statement of the main technical result of Schrijver [57]. Then, we will introduce a special Steiner tree that is one of the key components in the design of our algorithm, in order to explain one of the arguments where treewidth reduction is critical.

3.1   Schrijver's Main Technical Result

The starting point of our algorithm is Schrijver's view [57] of a collection of "non-crossing" (but possibly not vertex- or even edge-disjoint) sets of walks as flows. To work with flows (defined immediately), we deal with directed graphs. (In this context, undirected graphs are treated as directed graphs by replacing each edge by two parallel arcs of opposite directions.) Specifically, we denote an instance of DIRECTED PLANAR DISJOINT PATHS as a tuple (D, S, T, g, k) where D is a directed plane graph, $S, T \subseteq V(D)$, $k = |S|$ and $g : S \to T$ is bijective. Then, a *solution* is a set $\mathcal{P}$ of pairwise vertex-disjoint directed paths in D containing, for each vertex $s \in S$, a path directed from s to $g(s)$. In the language of flows, each arc of D is assigned a word with letters in $T \cup T^{-1}$ (that is, we treat the set of vertices T also as an alphabet), where $T^{-1} = \{t^{-1} : t \in T\}$. A word is *reduced* if, for all $t \in T$, the letters t and t^{-1} do not appear consecutively. Then, a *flow* is an assignment of reduced words to arcs that satisfies two constraints. First, when we concatenate the words assigned to the arcs incident to a vertex $v \notin S \cup T$ in clockwise order, where words assigned to ingoing arcs are reversed and their letters negated, the result (when reduced) is the empty word (see Fig. 1). This is an algebraic interpretation of the standard flow-conservation constraint. Second, when we do the same operation with respect to a vertex $v \in S \cup T$, then when the vertex is in S, the result is $g(s)$ (rather than the empty word), and when it is in T, the result is t. There is a natural association of flows to solutions: for every $t \in T$, assign the letter t to all arcs used by the path from $g^{-1}(t)$ to t.

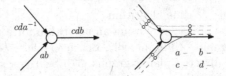

Fig. 1. Flow at a vertex and its reduction.

Roughly speaking, Schrijver proved that if a flow ϕ is given along with the instance (D, S, T, g, k), then in *polynomial time* we can either find a solution or determine that there is no solution "similar to ϕ". Specifically, two flows are *homologous* (which is the notion of similarity) if one can be obtained from the other by a *set* of "face operations" defined as follows.

Definition 2. *Let D be a directed plane graph with outer face f, and denote the set of faces of D by $\mathcal{F}$. Two flows ϕ and ψ are* homologous *if there exists a function $h : \mathcal{F} \to (T \cup T^{-1})^*$ such that (i) $h(f) = 1$, and (ii) for every arc $e \in A(D)$, $h(f_1)^{-1} \cdot \phi(e) \cdot h(f_2) = \psi(e)$ where f_1 and f_2 are the faces at the left-hand side and the right-hand side of e, respectively.*

Then, a slight modification of Schrijver's theorem [57] yields the following corollary.

Corollary 3. *There is a polynomial-time algorithm that, given an instance (D, S, T, g, k) of* DIRECTED PLANAR DISJOINT PATHS, *a flow ϕ, and a subset $X \subseteq A(D)$, either finds a solution of $(D - X, S, T, g, k)$ or decides that there is no solution of it such that the "flow associated with it" and ϕ are homologous in D.*

Discrete Homotopy and Our Objective. The language of flows brings several technicalities such as having different sets of non-crossing walks corresponding to the same flow (see Fig. 2). Instead, we may define a notion of *discrete*

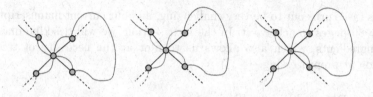

Fig. 2. Two different ways of extracting a walk from a flow.

homotopy, which is an equivalence relation that consists of three face operations. Then, we deal only with collections of non-crossing edge-disjoint walks, called *weak linkages*. Roughly speaking, two weak linkages are *discretely homotopic* if one can be obtained from the other by using "face operations" that push/stretch its walks across faces and keep them non-crossing and edge-disjoint (see Fig. 3). We note that the order in which face operations are applied is important in discrete homotopy (unlike homology)—e.g., we cannot stretch a walk across a face if no walk passes its boundary, but we can execute operations that will move a walk to that face, and then stretch it. We can translate Corollary 3 to discrete homotopy (and undirected graphs) to derive the following result.

Lemma 4. *There is a polynomial-time algorithm that, given an instance* (G, S, T, g, k) *of* PLANAR DISJOINT PATHS, *a weak linkage* $\mathcal{W}$ *in* G *and a subset* $X \subseteq E(G)$, *either finds a solution of* $(G - X, S, T, g, k)$ *or decides that no solution of it is discretely homotopic to* $\mathcal{W}$ *in* G.

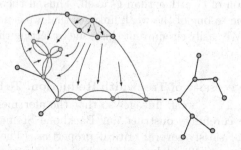

Fig. 3. Moving a weak linkage (having one walk) with "face operations".

In light of this result, our objective is reduced to the following task.

> Compute a collection of weak linkages such that if there exists a solution, then there also exists a solution (*possibly a different one!*) that is discretely homotopic to one of the weak linkages in our collection. To prove Theorem 1, the size of the collection should be upper bounded by $2^{\mathcal{O}(k^2)}$.

This task turns out to be very challenging, and our current manuscript spans roughly 80 pages to achieve it. In the next section, we will describe one of the main ingredients, which also allows us to hint at the necessity of treewidth reduction at preprocessing.

3.2 Key Player: Steiner Tree

A key to the proof of our theorem is a very careful construction (done in three steps) of a so-called *Backbone Steiner tree R*. We use the term Steiner tree to refer to any tree in the *radial completion* of G (the graph obtained by placing a vertex on each face and making it adjacent to all vertices incident to the face) whose set of leaves is precisely $S \cup T$. Having the aforementioned Backbone Steiner tree R at hand, we have a more focused goal: we will zoom into weak linkages that are "pushed onto R", and we will only generate such weak linkages to construct our collection. Informally, a weak linkage is *pushed onto R* if all of the edges used by all of its walks are *parallel to* edges of R. We do not demand that the edges belong to R itself, because then the aforementioned goal cannot be achieved— some edges will have to be used several times, which prevents satisfying the edge disjointness requirement. Instead, we make $\Theta(n)$ parallel copies of each edge in the radial completion (the precise number arises from considerations in the "pushing process"), and then impose the weaker demand of being parallel. Now, our goal is to show the following statement: If there exists a solution, then there also exists one that can be pushed onto R by applying face operations (in discrete homotopy) so that it becomes *identical* to one of the weak linkages in our collection (see Fig. 3).

At this point, one remark is in place. Our Steiner tree R is a subtree of the radial completion of G rather than G itself. Thus, if there exists a solution discretely homotopic to one of the weak linkages that we generate, it might not be a solution in G. We easily circumvent this issue by letting the set X in Lemma 4 contain all "fake" edges.

Example of the Necessity of Treewidth Reduction. To be able to generate a collection of only $2^{k^{O(1)}}$ weak linkages so that the aforementioned statement can be proven, we carefully construct our Backbone Steiner tree R in three steps so that it will satisfy several critical properties. (The third step is the most technical one, and will not be presented here). The first step is merely an initialization step, where we consider an arbitrary Steiner tree as R. Then, in the second step we modify R as follows. For each "long" maximal degree-2 path P of R with endpoints u and v, we will compute two minimum-size vertex sets, S_u and S_v, such that S_u separates (i.e., intersects all paths between) the following two subgraphs in the radial completion of G: *(i)* the subtree of R that contains u after the removal of a vertex u_1 of P that is "very close" to u, and *(ii)* the subtree of R that contains v after the removal of a vertex u_2 that is "close" to u. The condition satisfied by S_v is symmetric (see Fig. 4). Here, "very close" refers to distance $2^{c_1 k}$ and "close" refers to distance $2^{c_2 k}$ for some constants $c_1 < c_2$. (The selection of u' not to be u itself is of use in the third modification of R.)

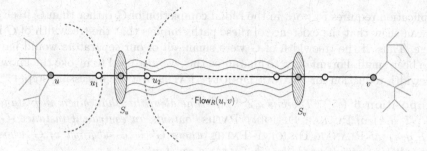

Fig. 4. Separators and flows for a long maximal degree-2 path P in R.

To utilize these separators, we need their sizes to be upper bounded by $2^{\mathcal{O}(k)}$. For our initial Steiner tree R, such small separators may not exist. However, the modification we will present now can be shown to guarantee their existence. Specifically, we will ensure that R does not have any *detour*, which roughly means that each of its maximal degree-2 paths is a shortest path connecting the two subtrees obtained once it is removed. More formally, we define a detour as follows (see Fig. 5).

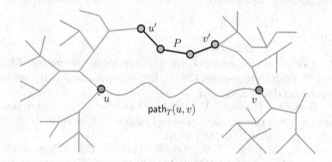

Fig. 5. A detour in a Steiner tree T.

Definition 5. *A detour in R is a pair of vertices $u, v \in V_{\geq 3}(R) \cup V_{=1}(R)$ that are endpoints of a maximal degree-2 path L of R, and a path P in the radial completion of G, such that (i) P is shorter than L, (ii) one endpoint of P belongs to the component of $R - V(L) \setminus \{u, v\}$ containing u, and (iii) one endpoint of P belongs to the component of $R - V(L) \setminus \{u, v\}$ containing v.*

By repeatedly "short-cutting" R, a process that terminates in a linear number of steps, we obtain a new Steiner tree R with no detour. Now, if the separator S_u is large, then there is a large number of vertex-disjoint paths that connect the two subtrees separated by S_u, and all of these paths are "long", namely, of length at least $2^{c_2 k} - 2^{c_1 k}$. Based on a result by Bodlaender et al. [8] (whose application requires to work in the radial completion of G rather than G itself), we can show that the existence of these paths implies that the treewidth of G is large. Thus, if the treewidth of G were small, all of our separators would have also been small. Fortunately, to guarantee this, we just need to invoke the known treewidth reduction for PLANAR DISJOINT PATHS in a preprocessing step:

Proposition 6 ([5]). *There is a $2^{\mathcal{O}(k)}n^2$-time algorithm that, given an instance (G, ST, g, k) of* PLANAR DISJOINT PATHS, *outputs an equivalent instance $(G',$ $S, T, g, k)$ of* PLANAR DISJOINT PATHS *where G' is a subgraph of G whose treewidth is upper bounded by 2^{ck} for some constant c.*

Intuitively, having separators of size $2^{\mathcal{O}(k)}$ is useful with respect to the proof of sufficiency of generating a collection of only $2^{k^{\mathcal{O}(1)}}$ weak linkages as follows. When we consider a solution $\mathcal{P}$ (if one exists), then from the fact that the paths in $\mathcal{P}$ are vertex disjoint, we immediately have the property that the paths in $\mathcal{P}$ can "go across" different maximal degree-2 paths of R only $2^{\mathcal{O}(k)}$ many times (because when going across different maximal degree-2 paths, either the path needs to intersect a $2^{\mathcal{O}(k)}$-sized separator or a "short" maximal degree-2 path). Having any solution $\mathcal{P}$ satisfy this property helps in proving that when we push any solution $\mathcal{P}$ onto R, at least those parts of the paths in $\mathcal{P}$ that go across different maximal degree-2 paths are only $2^{\mathcal{O}(k)}$ in number, and therefore they can be accommodated using only $2^{\mathcal{O}(k)}$ parallel edges for each edge of R.

References

1. Adler, I.: List of open problems. In: 6th Workshop on Graph Classes, Optimization, and Width Parameters (2013). http://www.cs.upc.edu/~sedthilk/grow/Open_Problems_GROW_2013.pdf
2. Adler, I., Dorn, F., Fomin, F.V., Sau, I., Thilikos, D.M.: Faster parameterized algorithms for minor containment. Theor. Comput. Sci. **412**(50), 7018–7028 (2011). https://doi.org/10.1016/j.tcs.2011.09.015
3. Adler, I., Dorn, F., Fomin, F.V., Sau, I., Thilikos, D.M.: Fast minor testing in planar graphs. Algorithmica **64**(1), 69–84 (2012). https://doi.org/10.1007/s00453-011-9563-9
4. Adler, I., Kolliopoulos, S.G., Krause, P.K., Lokshtanov, D., Saurabh, S., Thilikos, D.: Tight bounds for linkages in planar graphs. In: Aceto, L., Henzinger, M., Sgall, J. (eds.) ICALP 2011. LNCS, vol. 6755, pp. 110–121. Springer, Heidelberg (2011). https://doi.org/10.1007/978-3-642-22006-7_10
5. Adler, I., Kolliopoulos, S.G., Krause, P.K., Lokshtanov, D., Saurabh, S., Thilikos, D.M.: Irrelevant vertices for the planar disjoint paths problem. J. Comb. Theory Ser. B **122**, 815–843 (2017). https://doi.org/10.1016/j.jctb.2016.10.001
6. Adler, I., Krause, P.K.: A lower bound for the tree-width of planar graphs with vital linkages. CoRR abs/1011.2136 (2010). http://arxiv.org/abs/1011.2136
7. Baste, J., Sau, I.: The role of planarity in connectivity problems parameterized by treewidth. Theor. Comput. Sci. **570**, 1–14 (2015). https://doi.org/10.1016/j.tcs.2014.12.010
8. Bodlaender, H.L., Fomin, F.V., Lokshtanov, D., Penninkx, E., Saurabh, S., Thilikos, D.M.: (Meta) kernelization. J. ACM **63**(5), 44:1–44:69 (2016). https://doi.org/10.1145/2973749

9. Chekuri, C., Chuzhoy, J.: Polynomial bounds for the grid-minor theorem. In: Symposium on Theory of Computing, STOC 2014, New York, NY, USA, 31 May–03 June 2014, pp. 60–69 (2014)

10. Chuzhoy, J.: Improved bounds for the flat wall theorem. In: Proceedings of the Twenty-Sixth Annual ACM-SIAM Symposium on Discrete Algorithms, SODA 2015, San Diego, CA, USA, pp. 256–275 (2015)

11. Chuzhoy, J., Kim, D.H.K.: On approximating node-disjoint paths in grids. In: Approximation, Randomization, and Combinatorial Optimization. Algorithms and Techniques, APPROX/RANDOM 2015, Princeton, NJ, USA, 24–26 August 2015. LIPIcs, vol. 40, pp. 187–211. Schloss Dagstuhl - Leibniz-Zentrum fuer Informatik (2015)

12. Chuzhoy, J., Kim, D.H.K., Li, S.: Improved approximation for node-disjoint paths in planar graphs. In: Proceedings of the 48th Annual ACM SIGACT Symposium on Theory of Computing, STOC 2016, Cambridge, MA, USA, 18–21 June 2016, pp. 556–569. ACM (2016)

13. Chuzhoy, J., Kim, D.H.K., Nimavat, R.: New hardness results for routing on disjoint paths. In: Proceedings of the 49th Annual ACM SIGACT Symposium on Theory of Computing, STOC 2017, Montreal, QC, Canada, 19–23 June 2017, pp. 86–99. ACM (2017)

14. Chuzhoy, J., Kim, D.H.K., Nimavat, R.: Almost polynomial hardness of node-disjoint paths in grids. In: Proceedings of the 50th Annual ACM SIGACT Symposium on Theory of Computing, STOC 2018, Los Angeles, CA, USA, 25–29 June 2018, pp. 1220–1233. ACM (2018)

15. Chuzhoy, J., Kim, D.H.K., Nimavat, R.: Improved approximation for node-disjoint paths in grids with sources on the boundary. In: 45th International Colloquium on Automata, Languages, and Programming, ICALP 2018, Prague, Czech Republic, 9–13 July 2018. LIPIcs, vol. 107, pp. 38:1–38:14 (2018)

16. Cygan, M., et al.: Parameterized Algorithms. Springer, Cham (2015). https://doi.org/10.1007/978-3-319-21275-3

17. Cygan, M., Marx, D., Pilipczuk, M., Pilipczuk, M.: The planar directed k-vertex-disjoint paths problem is fixed-parameter tractable. In: 54th Annual IEEE Symposium on Foundations of Computer Science, FOCS 2013, Berkeley, CA, USA, 26–29 October 2013, pp. 197–206 (2013). https://doi.org/10.1109/FOCS.2013.29

18. Demaine, E.D., Fomin, F.V., Hajiaghayi, M.T., Thilikos, D.M.: Fixed-parameter algorithms for (k, r)-center in planar graphs and map graphs. ACM Trans. Algorithms **1**(1), 33–47 (2005)

19. Demaine, E.D., Fomin, F.V., Hajiaghayi, M., Thilikos, D.M.: Subexponential parameterized algorithms on bounded-genus graphs and H-minor-free graphs. J. ACM **52**(6), 866–893 (2005)

20. Demaine, E.D., Hajiaghayi, M.T.: Bidimensionality: new connections between FPT algorithms and PTASs. In: Proceedings of the Sixteenth Annual ACM-SIAM Symposium on Discrete Algorithms, SODA 2005, Canada, pp. 590–601 (2005)

21. Downey, R.: The birth and early years of parameterized complexity. In: Bodlaender, H.L., Downey, R., Fomin, F.V., Marx, D. (eds.) The Multivariate Algorithmic Revolution and Beyond. LNCS, vol. 7370, pp. 17–38. Springer, Heidelberg (2012). https://doi.org/10.1007/978-3-642-30891-8_2

22. Downey, R.G., Fellows, M.R.: Fixed-parameter intractability. In: Proceedings of the Seventh Annual Structure in Complexity Theory Conference, Boston, Massachusetts, USA, 22–25 June 1992, pp. 36–49 (1992). https://doi.org/10.1109/SCT.1992.215379

23. Dvorak, Z., Král, D., Thomas, R.: Coloring triangle-free graphs on surfaces. In: Proceedings of the Twentieth Annual ACM-SIAM Symposium on Discrete Algorithms, SODA 2009, New York, NY, USA, 4–6 January 2009, pp. 120–129 (2009). http://dl.acm.org/citation.cfm?id=1496770.1496784

24. Fellows, M.R.: The Robertson-Seymour theorems: a survey of applications. In: Graphs and Algorithms: Proceedings of the AMS-IMS-SIAM Joint Summer Research Conference Held 28 June–4 July 1987 with Support from the National Science Foundation, pp. 1–18 (1989)

25. Fellows, M.R., Fomin, F.V., Lokshtanov, D., Rosamond, F.A., Saurabh, S., Villanger, Y.: Local search: is brute-force avoidable? J. Comput. Syst. Sci. **78**(3), 707–719 (2012)

26. Fomin, F.V., Lokshtanov, D., Raman, V., Saurabh, S.: Bidimensionality and EPTAS. In: Proceedings of the Twenty-Second Annual ACM-SIAM Symposium on Discrete Algorithms, SODA 2011, USA, pp. 748–759 (2011)

27. Fomin, F.V., Lokshtanov, D., Saurabh, S.: Bidimensionality and geometric graphs. In: Proceedings of the Twenty-Third Annual ACM-SIAM Symposium on Discrete Algorithms, SODA 2012, Kyoto, Japan, 17–19 January 2012, pp. 1563–1575 (2012)

28. Fomin, F.V., Lokshtanov, D., Saurabh, S., Thilikos, D.M.: Bidimensionality and kernels. In: Proceedings of the Twenty-First Annual ACM-SIAM Symposium on Discrete Algorithms, SODA 2010, USA, pp. 503–510 (2010)

29. Frank, A.: Packing paths, cuts, and circuits-a survey. Paths, Flows and VLSI-Layout, pp. 49–100 (1990)

30. Garey, M.R., Johnson, D.S.: Computers and Intractability: A Guide to the Theory of NP-Completeness. W. H. Freeman, New York (1979)

31. Geelen, J., Huynh, T., Richter, R.B.: Explicit bounds for graph minors. J. Comb. Theory Ser. B **132**, 80–106 (2018). https://doi.org/10.1016/j.jctb.2018.03.004

32. Golovach, P.A., van't Hof, P.: Obtaining planarity bycontracting few edges. Theor. Comput. Sci. **476**, 38–46 (2013). https://doi.org/10.1016/j.tcs.2012.12.041

33. Grohe, M., Kawarabayashi, K., Marx, D., Wollan, P.: Finding topological subgraphs is fixed-parameter tractable. In: Proceedings of the 43rd ACM Symposium on Theory of Computing, STOC 2011, San Jose, CA, USA, 6–8 June 2011, pp. 479–488 (2011). https://doi.org/10.1145/1993636.1993700

34. Grohe, M., Kawarabayashi, K., Reed, B.A.: A simple algorithm for the graph minor decomposition - logic meets structural graph theory. In: Proceedings of the Twenty-Fourth Annual ACM-SIAM Symposium on Discrete Algorithms, SODA 2013, New Orleans, Louisiana, USA, pp. 414–431 (2013)

35. Johnson, D.S.: The NP-completeness column: an ongoing guide. J. Algorithms **8**(2), 285–303 (1987). https://doi.org/10.1016/0196-6774(87)90043-5

36. Karp, R.M.: On the computational complexity of combinatorial problems. Networks **5**(1), 45–68 (1975). https://doi.org/10.1002/net.1975.5.1.45

37. Kawarabayashi, K., Kobayashi, Y., Reed, B.A.: The disjoint paths problem in quadratic time. J. Comb. Theory Ser. B **102**(2), 424–435 (2012). https://doi.org/10.1016/j.jctb.2011.07.004

38. Kawarabayashi, K., Wollan, P.: A shorter proof of the graph minor algorithm: the unique linkage theorem. In: Proceedings of the 42nd ACM Symposium on Theory of Computing, STOC 2010, Cambridge, Massachusetts, USA, 5–8 June 2010, pp. 687–694 (2010)

39. Kawarabayashi, K., Wollan, P.: A simpler algorithm and shorter proof for the graph minor decomposition. In: Proceedings of the 43rd ACM Symposium on Theory of Computing, STOC 2011, San Jose, CA, USA, 6–8 June 2011, pp. 451–458 (2011)

40. Kobayashi, Y., Kawarabayashi, K.: Algorithms for finding an induced cycle in planar graphs and bounded genus graphs. In: Proceedings of the Twentieth Annual ACM-SIAM Symposium on Discrete Algorithms, SODA 2009, New York, NY, USA, 4–6 January 2009, pp. 1146–1155 (2009). http://dl.acm.org/citation.cfm?id=1496770.1496894

41. Kramer, M.R., van Leeuwen, J.: The complexity of wirerouting and finding minimum area layouts for arbitrary VLSI circuits. Adv. Comput. Res. **2**, 129–146 (1984)

42. Kuratowski, K.: Sur le problème des courbes gauches en topologie. Fund. Math. **15**, 271–283 (1930). (in French)

43. Lipton, R.J., Regan, K.W.: People, Problems, and Proofs: Essays from Gödel's Lost Letter: 2010. Springer, Heidelberg (2013). https://doi.org/10.1007/978-3-642-41422-0

44. Lokshtanov, D., Marx, D., Saurabh, S.: Slightly superexponential parameterized problems. SIAM J. Comput. **47**(3), 675–702 (2018). https://doi.org/10.1137/16M1104834

45. Lokshtanov, D., Misra, P., Pilipczuk, M., Saurabh, S., Zehavi, M.: An exponential time parameterized algorithm for planar disjoint paths. Manuscript in Preparation (2019)

46. Lokshtanov, D., Saurabh, S., Wahlström, M.: Subexponential parameterized odd cycle transversal on planar graphs. In: IARCS Annual Conference on Foundations of Software Technology and Theoretical Computer Science, FSTTCS 2012, India, pp. 424–434 (2012)

47. Lynch, J.F.: The equivalence of theorem proving and the interconnection problem. ACM SIGDA Newslett. **5**(3), 31–36 (1975)

48. Marx, D.: Chordal deletion is fixed-parameter tractable. Algorithmica **57**(4), 747–768 (2010). https://doi.org/10.1007/s00453-008-9233-8

49. Mazoit, F.: A single exponential bound for the redundant vertex theorem on surfaces. arXiv preprint arXiv:1309.7820 (2013)

50. Ogier, R.G., Rutenburg, V., Shacham, N.: Distributed algorithms for computing shortest pairs of disjoint paths. IEEE Trans. Inf. Theory **39**(2), 443–455 (1993). https://doi.org/10.1109/18.212275

51. Pilipczuk, M., Pilipczuk, M., Sankowski, P., van Leeuwen, E.J.: Network sparsification for Steiner problems on planar and bounded-genus graphs. In: 55th IEEE Annual Symposium on Foundations of Computer Science, FOCS 2014, USA, pp. 276–285 (2014)

52. Reed, B.A.: Rooted routing in the plane. Discrete Appl. Math. **57**(2–3), 213–227 (1995). https://doi.org/10.1016/0166-218X(94)00104-L

53. Reed, B.A., Robertson, N., Schrijver, A., Seymour, P.D.: Finding disjoint trees in planar graphs in linear time. In: Graph Structure Theory, Proceedings of a AMS-IMS-SIAM Joint Summer Research Conference on Graph Minors held June 22 to July 5, 1991, at the University of Washington, Seattle, USA, pp. 295–301 (1991)

54. Robertson, N., Seymour, P.D.: Graph minors. XIII. The disjoint paths problem. J. Comb. Theory Ser. B **63**(1), 65–110 (1995). https://doi.org/10.1006/jctb.1995.1006

55. Robertson, N., Seymour, P.D.: Graph minors. XXII. Irrelevant vertices in linkage problems. J. Comb. Theory Ser. B **102**(2), 530–563 (2012). https://doi.org/10.1016/j.jctb.2007.12.007

56. Scheffler, P.: A practical linear time algorithm for disjoint paths in graphs with bounded tree-width. TU, Fachbereich 3 (1994)

57. Schrijver, A.: Finding k disjoint paths in a directed planar graph. SIAM J. Comput. **23**(4), 780–788 (1994). https://doi.org/10.1137/S0097539792224061

58. Schrijver, A.: Combinatorial Optimization: Polyhedra and Efficiency, vol. 24. Springer, Heidelberg (2003)

59. Srinivas, A., Modiano, E.: Finding minimum energy disjoint paths in wireless ad-hoc networks. Wirel. Netw. **11**(4), 401–417 (2005). https://doi.org/10.1007/s11276-005-1765-0

60. Wagner, K.: Über eine eigenschaft der ebenen komplexe. Math. Ann **114**(1), 570–590 (1937)

61. Wollan, P.: Personal communication, January 2015

Four Shorts Stories on Surprising Algorithmic Uses of Treewidth

Dániel Marx[(⊠)]

Max Planck Institute for Informatics, Saarland Informatics Campus,
Saarbrücken, Germany
dmarx@mpi-inf.mpg.de

Dedicated to Hans L. Bodlaender on the occasion of his 60th birthday.

Abstract. This article briefly describes four algorithmic problems where the notion of treewidth is very useful. Even though the problems themselves have nothing to do with treewidth, it turns out that combining known results on treewidth allows us to easily describe very clean and high-level algorithms.

Keywords: Treewidth · Parameterized complexity · Fixed-parameter tractability · Bidimensionality

1 Introduction

While the definition of treewidth may seem very technical at first sight, the naturality of treewidth is witnessed by the fact that it was introduced independently at least three times with equivalent definitions by different authors [7,50,69]. One may arrive to the study of treewidth from various directions and justify its importance with different arguments. One can, for example, argue that graphs of low treewidth (or some generalization of it) appear naturally in certain applications [14,38,60,73], hence algorithms for such graphs could be of practical interest. Or one could say that algorithms on bounded-treewidth graphs are based on the fundamental idea of recursively splitting the problem along small separators, and the study of treewidth is a good formalization of the study of this basic principle. But perhaps the nicest and most surprising reason for arriving at this notion is when the original goal has nothing to do with treewidth, but suddenly treewidth appears as the right theoretical tool for handling the problem. This article contains four such "war stories," where the notion of treewidth and algorithms for bounded-treewidth graphs give very elegant solutions, which are sometimes in fact more efficient than those that were obtained earlier by involved and problem-specific techniques.

This research is a part of a project that has received funding from the European Research Council (ERC) under the European Union's Horizon 2020 research and innovation programme under grant agreement SYSTEMATICGRAPH (No. 725978).

© Springer Nature Switzerland AG 2020
F. V. Fomin et al. (Eds.): Bodlaender Festschrift, LNCS 12160, pp. 129–144, 2020.
https://doi.org/10.1007/978-3-030-42071-0_10

The four stories below are intentionally kept very brief in order to highlight the conceptual simplicity of the arguments. The aim is to show how certain high-level results can be combined in a clean way to achieve our goals. The detailed discussions or proofs of the results we are building on are beyond the scope of this article. Later in this volume, the article of Marcin Pilipczuk contains more advanced examples of algorithmic use of treewidth bounds [63].

2   Bidimensionality

Restricting an algorithmic problem to a certain family of graphs can make it easier than trying to solve it in general on every possible graph. A large part of the literature on algorithmic graph theory concerns algorithms for restricted classes of graphs that are of practical or theoretical significance. Restriction to planar graphs are studied both because of their interesting mathematical properties and as a starting point for modelling, e.g., road networks or 2D geometric problems.

From the viewpoint of polynomial-time solvability vs. NP-hardness, the restriction to planarity does not seem to make the problem significantly easier. Most of the classic NP-hard problems (e.g., 3-COLORING, MAXIMUM INDEPENDENT SET, HAMILTONIAN CYCLE, etc.) remain NP-hard on planar graphs. The situation is very different from the viewpoint of parameterized complexity. Many of the basic problems that are W[1]-hard on general graphs turn out to be FPT on planar graphs. In fact, it took some time to arrive to the first relatively simple and natural problems that are W[1]-hard on planar graphs [13, 19].

The restriction to planarity can help even for problems that are already FPT for general graphs. One of the main goals of the area of parameterized algorithms is to design algorithms with running time $f(k)n^{O(1)}$ such that the dependence $f(k)$ on the parameter is a function that grows as slowly as possible. For many of the fundamental problems studied in parameterized algorithms (e.g., VERTEX COVER, FEEDBACK VERTEX SET, k-PATH, ODD CYCLE TRANSVERSAL), algorithms with running time $2^{O(k)}n^{O(1)}$ are known. Furthermore, it is very likely that this form of running time is optimal: it is known that, under the Exponential Time Hypothesis (ETH) [51, 52], no algorithm with running time $2^{o(k)}n^{O(1)}$ exists for these problems. When restricted to planar graphs, significantly better algorithms are known for many of these problems, typically with running times of the form $2^{O(\sqrt{k})}n^{O(1)}$ or $2^{O(\sqrt{k}\log k)}n^{O(1)}$. Below we show how a very clean argument based on treewidth delivers such agorithms for certain basic problems; for others, more involved problem-specific ideas are needed [1, 35, 44, 54, 58, 64, 65]. The main argument we present here was described first by Fomin and Thilikos [46] (for the DOMINATING SET problem) and was further developed under the name "bidimensionality" (see, e.g., [28–31]).

Let us consider the k-PATH problem as our running example: given a (planar) graph G and an integer k, we have to decide if G contains a simple path on k vertices. Let us first note that k-PATH is FPT parameterized by the treewidth

w of the input graph G. More precisely, standard dynamic programming techniques give $2^{O(w \log w)} n^{O(1)}$ running time, while more sophisticated arguments are needed to obtain $2^{O(w)} n^{O(1)}$ time [11, 25, 33, 34, 36, 37, 45] (note that some of these algorithms are randomized and some of these algorithms work only on planar graphs).

Theorem 1. k-PATH *can be solved in time* $2^{O(w)} n^{O(1)}$ *if a tree decomposition of width w is given in the input.*

The second ingredient that we need is the Planar Excluded Grid Theorem [48, 68]. A *minor* of a graph G is a graph H that is obtained by a sequence of vertex deletions, edge deletions, and edge contractions. A $k \times k$ *grid* is a graph with vertex set $[k] \times [k]$, where vertices (x, y) and (x', y') are adjacent if and only if $|x - x'| + |y - y'| = 1$. The following theorem states that, in a very tight sense, the existence of a grid minor is the canonical reason why a planar graph has large treewidth:

Theorem 2 (Planar Excluded Grid Theorem). *Every planar graph with treewidth at least $4.5k$ has a $k \times k$ grid minor.*

In particular, Theorem 2 implies that an n-vertex planar graph has treewidth $O(\sqrt{n})$: it certainly cannot contain a grid minor larger than $\sqrt{n} \times \sqrt{n}$.

Finally, we have to make two simple observations about the k-PATH problem:

(1) The $k \times k$ grid contains a path on k^2 vertices: imagine a "snake" that visits the rows one after the other.
(2) If H is a minor of G, then the length of the longest path in H is not larger than in G. This can be proved by verifying that none of vertex deletion, edge deletion, or edge contraction can increase the length of the longest path.

Now the claimed algorithm can be obtained by putting together these ingredients using a win/win approach. For simplicity, we describe an algorithm for the decision version of the problem where only a YES/NO answer has to be returned.

Theorem 3. k-PATH *on planar graphs can be solved in time* $2^{O(\sqrt{k})} n^{O(1)}$.

Proof. Let $w := 4.5\lceil\sqrt{k}\rceil$. If G is a graph with treewidth at least w, then Theorem 2 implies that G contains a $\lceil\sqrt{k}\rceil \times \lceil\sqrt{k}\rceil$ grid minor H. Then the first observation above shows that H contains a path on k vertices and the second observation shows that G also contains a path on k vertices. Therefore, we can conclude that if the input graph G has treewidth at least w, then it is a YES-instance: it surely contains a path on k vertices.

The algorithm proceeds as follows. First, we compute an (approximate) tree decomposition of G. For this purpose, it is convenient to use the algorithm of Bodlaender et al. [12], which, given an integer w and a graph G, in time $2^{O(w)} \cdot n = 2^{O(\sqrt{k})} \cdot n$ either correctly states that treewidth of G is larger than w, or gives a tree decomposition of width at most $5w + 4$. We can complete the computation in both cases:

- If the algorithm states that G has treewidth larger than w, then, as we have seen above, the answer is YES.
- If the algorithm returns a tree decomposition of width at most $5w + 4 = O(\sqrt{k})$, then we can invoke Theorem 1 to decide the existence of a path on k vertices and return YES or NO accordingly. The running time is $2^{O(w)}n^{O(1)} = 2^{O(\sqrt{k})}n^{O(1)}$, as required.

Thus we have an algorithm that returns a correct YES/NO-answer in time $2^{O(\sqrt{k})} \cdot n^{O(1)}$.

The same argument works for FEEDBACK VERTEX SET and VERTEX COVER. Only the analogs of the two observations (1) and (2) need to be verified: the optimum value is $\Omega(k^2)$ on the $k \times k$ grid and that the minor operation cannot increase the optimum value. A variant of the argument, based on contractions instead of minors, can give algorithms for INDEPENDENT SET and DOMINATING SET. There are also less straighforward uses of Theorem 2, where it is invoked not on the input graph itself, but on some auxilliary graph defined in a nonobvious way; see the article of Marcin Pilipczuk later in this volume for some examples [63].

3   Exponential-Time Algorithms for Graphs of Maximum Degree 3

If the task is to find a subset of vertices satisfying certain properties, then we can typically solve the problem in time $2^n \cdot n^{O(1)}$ on graphs with n vertices by enumerating every subset. For many problems, it is easy to improve on this brute force algorithm. For example, in the case of the MAXIMUM INDEPENDENT SET problem (for graphs with arbitrarily large degree), there is a simple textbook example of an improved branching algorithm that beats the $2^n \cdot n^{O(1)}$ running time. As long as there is a vertex v of degree at least 3, branch into two directions: either the solution avoids v (in which case we can remove v, decreasing the size of the graph by 1) or it contains v (in which case we can remove v and its neighbors from the problem, decreasing the size of the graph by at least 4 vertices). The problem can be solved in polynomial time if every vertex has degree at most 2. Analyzing the algorithm shows that its running time is $1.3803^n \cdot n^{O(1)}$. Further improvements are possible with more and more involved techniques [18, 41, 53, 55, 70, 72] with the current best algorithm having running time $1.1996^n \cdot n^{O(1)}$ [77]. Similar "races" for the best exponential-time algorithm are known for many other problems [43]. Let us remark that for some problems just beating the trivial $2^n \cdot n^{O(1)}$ running time is already highly nontrivial [10, 26, 66].

For the MAXIMUM INDEPENDENT SET problem on graphs of maximum degree 3, the current best algorithm has running time $1.0836^n \cdot n^{O(1)}$ [76]. Here we would like to highlight an earlier, less efficient algorithm that can be explained using the notion of treewidth very easily. Fomin and Høie [42] proved, using an earlier result of Monien and Preis [62], that the pathwidth (and hence the treewidth) of an n-vertex graph with maximum degree 3 is essentially at most $n/6$. More precisely:

Theorem 4. (Fomin and Høie [42]). *For any $\epsilon > 0$, there is an integer n_ϵ such that the pathwidth of any graph on $n > n_\epsilon$ vertices and maximum degree at most 3 is at most $(1/6 + \epsilon)n$.*

Together with the fact that a MAXIMUM INDEPENDENT SET on an n-vertex graph can be solved in time $2^w \cdot n^{O(1)}$ if a tree decomposition of width w is given, it follows that the problem can be solved in time $2^{n/6} \cdot n^{O(1)} = 1.1225^n \cdot n^{O(1)}$. The running time obtained as a simple consequence of this pathwidth bound was better than some earlier work at that time [5,20], but since then improved algorithms with more complicated and problem-specific arguments were found for this problem [17,18,67,76]. In a similar way, algorithms for MINIMUM DOMINATING SET and MAX CUT follow immediately from Theorem 4, which were better than some of the algorithms found by earlier problem specific techniques [42].

4   Finding and Counting Permutation Patterns

Interesting combinatorial and algorithmic problems can be defined on permutations and on the patterns they contain or avoid. A *permutation* of length n is a bijection $\pi : [n] \rightarrow [n]$; typically we describe permutations by the sequence $(\pi(1), \pi(2), \ldots, \pi(n))$. We say that a permutation σ of length n *contains* a permutation π of length k if there is a mapping $f : [k] \rightarrow [n]$ such that $f(1) < f(2) < \cdots < f(k)$ and $\pi(i) < \pi(j)$ if and only if $\sigma(f(i)) < \sigma(f(j))$. That is, σ contains π if the sequence $(\pi(1), \ldots, \pi(k))$ can be mapped to a subsequence of $(\sigma(1), \ldots, \sigma(n))$ in a way that preserves the relative order of the values. As an example, the permutation $(3, 4, 5, 2, 1, 7, 8, 6)$ contains the permutation $(2, 1, 3, 4)$ (e.g., by the mapping $(f(1), f(2), f(3), f(4)) = (1, 4, 6, 7)$), but it does not contain the permutation $(4, 3, 2, 1)$. Observe that the permutations *not* containing $(1, 2)$ are exactly the decreasing sequences, while the permutations *not* containing $(2, 1)$ are exactly the increasing sequences. As shown by Knuth [56, § 2.2.1], the permutations avoiding $(2, 3, 1)$ are exactly the permutations sortable by a single stack. From the extremal combinatorics point of view, a very natural question is to bound the number of permutations of length n avoiding a fixed permutation π. Marcus and Tardos [61] proved a long-standing conjecture of Stanley and Wilf[1] by showing that for every fixed permutation π, there is a constant $c(\pi)$ such that the number of permutations of length n avoiding π is at most $2^{c(\pi) \cdot n}$. This has to be contrasted with the fact that the total number of permutations of length n is $n! = 2^{O(n \log n)}$.

From the algorithmic point of view, perhaps the most fundamental question is testing for containment: given a permutation σ of length n and a permutation

---

[1] Marcus and Tardos [61] mentions that the conjecture was formulated around 1992 (but it is hard to find a citable source) and the PhD thesis of Julian West is an even earlier source [75].

π of length k, does σ contain π? The problem is often called PERMUTATION PATTERN MATCHING and is known to be NP-hard [16], but of course can be solved in time $O(n^k)$ by brute force. Albert et al. [3] improved this to $O(n^{2/3k+1})$ time, Ahal and Rabinovich [2] further improved it to $n^{0.47k+o(k)}$ time, and Berendsohn et al. [6] gave an $n^{0.25k+o(k)}$ time algorithm. Guillemot and Marx [49] showed that PERMUTATION PATTERN MATCHING can be solved in time $2^{O(k^2 \log k)} \cdot n$, that is, it is fixed-parameter tractable (FPT) parameterized by the length of π.

Even though the problem is FPT, algorithms with running time n^{ck} can be still interesting for two reasons. First, if k is fairly large, say, $\Omega(\log n)$, then $2^{O(k^2 \log k)} \cdot n$ is actually worse than $n^{O(k)}$. Thus unless we have $2^{O(k)} \cdot n^{O(1)}$ FPT algorithms for the problem, we need different type of algorithms to understand the complexity of the problem in the regime where k is large. Second, the n^{ck} time algorithms [2,3,6] can be easily modified to count the total number of solutions, while the FPT algorithm of Guillemot and Marx [49] returns only a single solution. This is not just a shortcoming of the presentation [49]: the FPT algorithm contains a step where a certain structure is discovered that guarantees that every permutation of length k appears in σ. Then the algorithm stops and does not look for any further occurences of π. Furthermore, it is unlikely that the algorithm can be extended to a counting version: Berendsohn et al. [6] proved that the counting problem is #W[1]-hard.

The n^{ck} algorithms for PERMUTATION PATTERN MATCHING [2,3,6] are implicitly or explicitly based on dynamic programming on a certain tree decomposition. Here we follow the presentation of Berendsohn et al. [6], where it is shown how high-level arguments and previous results on treewidth can be combined to obtain an $n^{k/3+o(k)}$ time in a very clean way (a further improvement, based on a technical idea of Cygan et al. [24], reduces the running time to $n^{0.25k+o(k)}$ [6]).

A permutation $\pi : [k] \rightarrow [k]$ can be seen as a k-element point set $S_\pi = \{(i, \pi(i)) \mid i \in [k]\}$ (see Fig. 1). With this interpretation, σ contains π if S_π can be mapped to a subset of S_σ in a way that the mapping preserves the relative ordering of any two points along both the horizontal axis and the vertical axis. For a point $p \in S_\pi$, we will denote by $p.x$ and $p.y$ the first and second coordinates of p, respectively. For each point $(x, y) \in S_\pi$, we define the four neighbors of (x, y) as follows:

$$N^R((x,y)) = (x + 1, \ \pi(x + 1)),$$
$$N^L((x,y)) = (x - 1, \ \pi(x - 1)),$$
$$N^U((x,y)) = (\pi^{-1}(y + 1), \ y + 1),$$
$$N^D((x,y)) = (\pi^{-1}(y - 1), \ y - 1).$$

The superscripts R, L, U, D are meant to evoke the directions *right, left, up, down*, when plotting S_σ in the plane. That is, if we start sweeping the vertical line going through (x, y) to the *R*ight, then $N^R((x, y))$ is the next point that we meet, and similarly with the other directions. Note that some neighbors of a point may coincide.

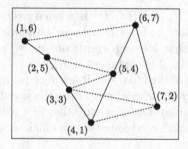

Fig. 1. Permutation $\pi = (6, 5, 3, 1, 4, 7, 2)$ and its incidence graph G_π. Solid lines indicate neighbors by index (L-R), dashed lines indicate neighbors by value (U-D). Indices plotted on x-coordinate, values plotted on y-coordinate.

The *incidence graph* G_π of π is a graph on S_π where each point is connected to its four neighbors (when defined). It is easy to see that G_π is the union of two Hamiltonian paths on the same set S_π of vertices, with one path going in the left-right direction in the plane, while the other path going in the top-bottom direction.

The key lemma that allows a clean abstraction of the problem is the following characterization of solutions.

Lemma 1. *Let $\sigma : [n] \to [n]$ and $\pi : [k] \to [k]$ be two permutations. Then σ contains π if and only if there is a function $f : S_\pi \to S_\sigma$ such that for every $p \in S_\pi$*

$$f(N^L(p)).x < \quad f(p).x < \quad f(N^R(p)).x, \quad and \tag{1}$$
$$f(N^D(p)).y < \quad f(p).y < \quad f(N^U(p)).y, \tag{2}$$

whenever the corresponding neighbor of p is defined.

It is not very difficult to prove Lemma 1 using the definitions and we can also see that the functions f satisfying the requirements of Lemma 1 are in one to one correspondence with the occurrences of π in σ. The inequalities in the first line ensure that the mapping of points represent the left-to-right ordering, while the inequalities in the second line handle the top-to-bottom ordering. The key observation is that even though we require these inequalities only between neighbors in G_π, it follows as consequence that every pairwise inequality in the definition of containment holds. For example, if $\pi(i) < \pi(j)$, then $(j, \pi(j))$ can be reached from $(i, \pi(i))$ by going through a sequence of U-neighbors, hence a sequence of inequalities ensure that the second coordinate of $f((i, \pi(i))$ is less than the second coordiante of $f((j, \pi(j)))$.

Readers familar with the notion of Constraint Satisfaction Problems (CSPs) may recognize that Lemma 1 cleanly transforms the problem into a binary constraint satisfaction problem. A *binary CSP* instance is a triplet (V, D, C), where V is a set of variables, D is a set of admissible values (the *domain*), and C is a set of constraints $C = \{c_1, \ldots, c_m\}$, where each constraint c_i is of the form

$((x,y), R)$, where $x, y \in V$, and $R \subseteq D^2$ is a binary relation. A solution of the CSP instance is a function $f : V \to D$ (i.e., an assignment of admissible values to the variables), such that for each constraint $c_i = ((x_i, y_i), R_i)$, the pair of assigned values $(f(x_i), f(y_i))$ is contained in R_i.

The *constraint graph* of the binary CSP instance (also known as *primal graph* or *Gaifman graph*) is a graph whose vertices are the variables V and whose edges connect all pairs of variables that occur together in a constraint. Low treewidth of the constraint graph can be exploited for an efficient solution of the problem:

Theorem 5 [27,47]. *A binary CSP instance (V, D, C) can be solved in time $O(|D|^{t+1})$ where t is the treewidth of the constraint graph.*

To view the PERMUTATION PATTERN MATCHING problem as a binary CSP instance, let $V = S_\pi$ be the set of variables and let $D = S_\sigma$ be the domain. Then we want to find a function f that satisfies the inequalities in Lemma 1. Each inequality is a binary constraint between p and $N^\alpha(p)$ for some $\alpha \in \{L, R, D, U\}$, restricting the possible combination of values that $f(p)$ and $f(N^\alpha(p))$ can take. Thus we end up with a CSP instance on k variables, domain size n, and whose constraint graph is exactly G_π.

In order to invoke Theorem 5 on this instance, we need to bound the treewidth of G_π. Recall that G_π has k vertices and maximum degree 4. By splitting each degree-4 vertex into two degree-3 vertices connected by an edge, we can create a graph G'_π that has at most $2k$ vertices, maximum degree 3, and G_π is a minor of G'_π. Then Theorem 4 shows that G'_π has treewidth $2k/6 + o(k) = k/3 + o(k)$ and G_π being a minor of G'_π shows that the same bound holds for G_π as well. Therefore, we can conclude that Theorem 5 solves the instance in time $n^{k/3+o(k)}$. It is not difficult to modify the algorithm to count the number of solutions. Therefore, the combination of an easy observation (Lemma 1), a combinatorial treewidth bound (Theorem 4), and a known general algorithm (Theorem 5) solves the problem in a very clean way.

In Lemma 1, the functions f satisfying the requirements are in one to one correspondence with the occurences of π in σ and Theorem 5 can be extended to a counting version. Theorem 4 is purely combinatorial, thus it is of course irrelevant if we are using it for the decision or the counting problem. Thus the same algorithmic idea goes through.

Theorem 6 (Berendsohn et al. [6]). *Given a length-k permutation π and length-n permutation σ, the number of occurrences of π in σ can be counted in time $n^{k/3+o(k)}$.*

5   Counting Subgraphs

It is a well-known phenomenon in theoretical computer science that in many cases finding a solution is easier than counting the number of all solutions. For example, it can be checked in polynomial time if a bipartite graph contains a perfect matching, but the seminal result of Valiant shows that counting the number of perfect matchings is #P-hard and hence unlikely to be polynomial-time solvable [74]. By now, many other examples of hard counting problems are known.

Flum and Grohe [39] started the investigation of the complexity of counting in the setting of parameterized complexity. They introduced the notion of #W[1]-hardness to give evidence that certain parameterized counting problems are unlikely to be FPT. As a highly nontrivial example, they considered the k-PATH problem: the decision version is known to be FPT by various techniques [4,45], but they showed that the counting version of the problem is #W[1]-hard. In the same paper, they asked as an open question whether the counting version of the polynomial-time solvable k-MATCHING problem is FPT. This question was resolved in the negative by the #W[1]-hardness proof of Curticapean [21], which used heavy algebraic machinery, and by the later simpler proof given by Curticapean and Marx [23]. More recently, Dell et al. [22] described and exploited a connection beween subgraph counting and homomorphism counting problems. This connection can be useful in two different ways: it gives new subgraph-counting algorithms by reducing it to homomorphism-counting problems, and gives hardness results for subgraph counting (including new and clean #W[1]-hardness proofs of k-MATCHING and k-PATH) based on our understanding of the complexity of counting homomorphisms. Below we give an example of the algorithmic use of this connection.

Given the #W[1]-hardness of k-PATH, we cannot hope for an FPT algorithm solving the problem. But it is still an interesting question whether we can improve on the trivial $n^{k+O(1)}$ time brute force algorithm. The "meet in the middle" approach can be used to improve this to $n^{k/2+O(1)}$ time [8,57], which was further improved by Björklund et al. [9] to $n^{0.455k+O(1)}$. Here we describe an algorithm with running time $k^{O(k)} \cdot n^{0.174k+o(k)}$, which has a much smaller exponent for a fixed k and at the same time conceptually much simpler.

Let us first review some basic background on homomorphisms. A *homomorphism* from graph H to graph G is a mapping $f : V(H) \to V(G)$ such that for every edge $uv \in E(H)$, we have $f(u)f(v) \in E(G)$. We will denote by #Hom$(H \to G)$ the number of homomorphisms from H to G. Given a tree decomposition of H, standard dynamic programming techniques can be used to compute the number of homomorphisms from H to a given graph G.

Theorem 7 (Díaz et al. [32]). *Given graphs H and G, #Hom$(H \to G)$ can be computed in time $(|V(H)| + |V(G)|)^{w+O(1)}$, where w is the treewidth of H.*

Note that the algorithm of Theorem 7 does not need a decomposition of H, as it can be found in time $|V(H)|^{c+O(1)}$.

A homomorphism $f : V(H) \to V(G)$ is *injective* if $f(u) \neq f(v)$ for any two distinct $u, v \in V(H)$; let #Emb$(H \to G)$ denote the number of such homomorphisms. Let us denote by #Sub$(H \to G)$ the number of subgraphs of G that are isomorphic to H. It is well known and easy to see that #Emb$(H \to G) = $ #Sub$(H \to G) \cdot $#Aut$(H)$, where #Aut$(H) = $ #Emb$(H \to H)$ is the number of automorphisms of the graph H. Therefore, for a fixed H, computing #Sub$(H \to G)$ is essentially equivalent to computing #Emb$(H \to G)$, the number of injective homomorphisms. In order to explain the connection between counting homomorphisms and subgraphs, it will be more convenient to work with $\#\mathsf{Emb}(H \to G)$ than with $\#\mathsf{Sub}(H \to G)$, as the former is already defined in terms of homomorphisms.

Of course, not every homomorpism from H to G is injective, the images of some vertices may coincide. For example, if H is the 4-cycle on vertices $1, 2, 3, 4$, then a homomorphism from H to a loopless graph G either (1) is injective, (2) identifies 1 with 3, (3) identifies 2 with 4, (4) identifies 1 with 3, and 2 with 4. In case (1), the image of H is a 4-cycle; in cases (2) and (3), the image of H is the path P_3 on three vertices; and in case (4), the image of H is the path P_2 on two vertices. This shows that the following formula holds for the number of homomorphisms:

$$\#\mathsf{Hom}(C_4 \to G) = \#\mathsf{Emb}(C_4 \to G) + 2 \cdot \#\mathsf{Emb}(P_3 \to G) + \#\mathsf{Emb}(P_2 \to G).$$

More generally, we can classify the homomorphisms according to which sets of vertices they identify. To each homomorphism $h : V(G) \to V(H)$, we can associate a partition ρ_h of $V(H)$ with the meaning that, for every $u, v \in V(H)$, we have $h(u) = h(v)$ if and only u and v are in the same block of ρ. For a partition ρ of $V(H)$, let H/ρ be the *quotient graph* obtained by consolidating each block of ρ into a single vertex. The key observation is that the homomorphisms from H to G having type ρ are in one-to-one correspondence with the injective homomorphisms from H/ρ to G. Therefore, we can express the number of homomorphisms from H to G as

$$\#\mathsf{Hom}(H \to G) = \sum_{\rho} \#\mathsf{Emb}(H/\rho \to G), \qquad (3)$$

where the sum ranges over every partition ρ of $V(H)$.

Why is this useful for us? Observe that $H = H/\rho$ holds only for the partition ρ_0 where every block has size exactly one and H/ρ has strictly fewer vertices for every other ρ. Therefore, Eq. (3) can be written as

$$\#\mathsf{Hom}(H \to G) = \#\mathsf{Emb}(H \to G) + \sum_{\rho \neq \rho_0} \#\mathsf{Emb}(H/\rho \to G),$$

and hence

$$\#\mathsf{Emb}(H \to G) = \#\mathsf{Hom}(H \to G) - \sum_{\rho \neq \rho_0} \#\mathsf{Emb}(H/\rho \to G). \qquad (4)$$

That is, Eq. (4) reduces the problem of computing $\#\mathsf{Emb}(H \to G)$ to the problem of computing $\#\mathsf{Hom}(H \to G)$ and to computing some number of $\#\mathsf{Emb}(H/\rho \to G)$ values, where H/ρ has strictly fewer vertices than $|V(H)|$. Therefore, we can repeat the same argument and recursively replace each term $\#\mathsf{Emb}(H/\rho \to G)$ with a $\#\mathsf{Hom}$ term and some number of $\#\mathsf{Emb}$ terms. As the replacement strictly decreases the number of vertices in the $\#\mathsf{Emb}$ terms, eventually all these terms disappear, and we can express $\#\mathsf{Emb}(H \to G)$ as the linear combination of $\#\mathsf{Hom}(H' \to G)$ values for various graphs H'. This means that we can reduce the problem of computing $\#\mathsf{Emb}(H \to G)$ to computing certain homomorphism values.

Which graphs H' can appear in the $\#\mathsf{Hom}(H' \to G)$ terms when we express $\#\mathsf{Emb}(H \to G)$ this way? It is easy to see that the quotient graph of a quotient of H is also a quotient graph of H. This means that every graph H' appearing in this linear combination is a quotient graph of H. Thus we can express $\#\mathsf{Emb}(H \to G)$ as

$$\#\mathsf{Emb}(H \to G) = \sum_\rho \beta_{\rho,H} \cdot \#\mathsf{Hom}(H/\rho \to G), \qquad (5)$$

where $\beta_{\rho,H}$ is a constant depending only on ρ and H. The argument described above gives an algorithm for writing $\#\mathsf{Emb}(H \to G)$ in this form and for computing the constants $\beta_{\rho,H}$ (and the work of Lovász et al. [15,59] gives more explicit formulas for these constants). Given this expression, we can reduce the problem of computing $\#\mathsf{Emb}(H \to G)$ to computing the values $\#\mathsf{Hom}(H/\rho \to G)$. If H has k vertices, then the sum ranges over $k^{O(k)}$ different partitions ρ. Therefore, if every H/ρ has treewidth bounded by c, then invoking Theorem 7 for the computation of each $\#\mathsf{Hom}(H/\rho \to G)$ results in an algorithm with running time $k^{O(k)} \cdot n^{c+O(1)}$ for the computation of $\#\mathsf{Emb}(H \to G)$ (and hence of $\#\mathsf{Sub}(H \to G)$).

These considerations show that bounding the running time of our algorithm essentially boils down to a bound on the maximum treewidth of H/ρ. The treewidth of H/ρ can be much larger than the treewidth of H. For example, it is not difficult to see that if H is a matching with k independent edges, then we can obtain any connected graph with k edges as H/ρ for an appropriate partition ρ. However, this operation cannot increase the number of edges: if H has k edges, then H/ρ has at most k edges. We can use the following bound on the treewidth of graphs with at most k edges:

Theorem 8 [40,71]. *Every graph with at most k edges has treewidth $0.174k + o(k)$.*

This immediately gives an upper bound on the running time needed if H has at most k edges.

Theorem 9 (Dell et al. [22]). *If H has at most k edges, then $\#\mathsf{Emb}(H \to G)$ and $\#\mathsf{Sub}(H \to G)$ can be computed in time $k^{O(k)} \cdot n^{0.174k+o(k)}$.*

In particular, we obtain algorithms with running time $k^{O(k)} \cdot n^{0.174k+o(k)}$ if H is a path with k edges (the k-PATH problem) or a matching with k edges (the k-MATCHING problem). We want to emphasize that for a fixed H, the algorithm is very simple: it consists of invoking Theorem 7 for various graphs $H' = H/\rho$ and then taking a linear combination of these values. All the real work is done by the computation of the fixed constants $\beta_{\rho,H}$ and by the algorithm of Theorem 7 exploiting low treewidth and tree decompositions.

References

1. Aboulker, P., Brettell, N., Havet, F., Marx, D., Trotignon, N.: Coloring graphs with constraints on connectivity. J. Graph Theory **85**(4), 814–838 (2017)
2. Ahal, S., Rabinovich, Y.: On complexity of the subpattern problem. SIAM J. Discret. Math. **22**(2), 629–649 (2008). https://doi.org/10.1137/S0895480104444776
3. Albert, M.H., Aldred, R.E.L., Atkinson, M.D., Holton, D.A.: Algorithms for pattern involvement in permutations. In: Eades, P., Takaoka, T. (eds.) ISAAC 2001. LNCS, vol. 2223, pp. 355–367. Springer, Heidelberg (2001). https://doi.org/10. 1007/3-540-45678-3_31. http://dl.acm.org/citation.cfm?id=646344.689586
4. Alon, N., Yuster, R., Zwick, U.: Color-coding. J. ACM **42**(4), 844–856 (1995). https://doi.org/10.1145/210332.210337
5. Beigel, R.: Finding maximum independent sets in sparse and general graphs. In: Proceedings of the Tenth Annual ACM-SIAM Symposium on Discrete Algorithms, Baltimore, Maryland, USA, 17–19 January 1999, pp. 856–857 (1999). http://dl. acm.org/citation.cfm?id=314500.314969
6. Berendsohn, B.A., Kozma, L., Marx, D.: Finding and counting permutations via CSPs, accepted to IPEC (2019)
7. Bertelè, U., Brioschi, F.: On non-serial dynamic programming. J. Comb. Theory Ser. A **14**(2), 137–148 (1973). https://doi.org/10.1016/0097-3165(73)90016-2
8. Björklund, A., Husfeldt, T., Kaski, P., Koivisto, M.: Counting paths and packings in halves. In: Fiat, A., Sanders, P. (eds.) ESA 2009. LNCS, vol. 5757, pp. 578–586. Springer, Heidelberg (2009). https://doi.org/10.1007/978-3-642-04128-0_52
9. Björklund, A., Kaski, P., Kowalik, L.: Counting thin subgraphs via packings faster than meet-in-the-middle time. In: Proceedings of the 25th Annual Symposium on Discrete Algorithms (SODA), pp. 594–603 (2014). https://doi.org/10.1137/1. 9781611973402.45
10. Bliznets, I., Fomin, F.V., Pilipczuk, M., Villanger, Y.: Largest chordal and interval subgraphs faster than 2^n. Algorithmica **76**(2), 569–594 (2016). https://doi.org/10. 1007/s00453-015-0054-2
11. Bodlaender, H.L., Cygan, M., Kratsch, S., Nederlof, J.: Deterministic single exponential time algorithms for connectivity problems parameterized by treewidth. Inf. Comput. **243**, 86–111 (2015). https://doi.org/10.1016/j.ic.2014.12.008
12. Bodlaender, H.L., Drange, P.G., Dregi, M.S., Fomin, F.V., Lokshtanov, D., Pilipczuk, M.: A $c^k \cdot n$ 5-approximation algorithm for treewidth. SIAM J. Comput. **45**(2), 317–378 (2016). https://doi.org/10.1137/130947374
13. Bodlaender, H.L., Lokshtanov, D., Penninkx, E.: Planar capacitated dominating set is $W[1]$-Hard. In: Chen, J., Fomin, F.V. (eds.) IWPEC 2009. LNCS, vol. 5917, pp. 50–60. Springer, Heidelberg (2009). https://doi.org/10.1007/978-3-642-11269-0_4
14. Bonifati, A., Martens, W., Timm, T.: An analytical study of large SPARQL query logs. PVLDB **11**(2), 149–161 (2017). https://doi.org/10.14778/3149193.3149196. http://www.vldb.org/pvldb/vol11/p149-bonifati.pdf
15. Borgs, C., Chayes, J., Lovász, L., Sós, V.T., Vesztergombi, K.: Counting graph homomorphisms. Top. Discret. Math. **26**, 315–371 (2006). https://doi.org/10. 1007/3-540-33700-8_18
16. Bose, P., Buss, J.F., Lubiw, A.: Pattern matching for permutations. Inf. Process. Lett. **65**(5), 277–283 (1998). https://doi.org/10.1016/S0020-0190(97)00209-3

17. Bourgeois, N., Escoffier, B., Paschos, V.T.: An $O^*(1.0977^n)$ exact algorithm for MAX INDEPENDENT SET in sparse graphs. In: Grohe, M., Niedermeier, R. (eds.) IWPEC 2008. LNCS, vol. 5018, pp. 55–65. Springer, Heidelberg (2008). https://doi.org/10.1007/978-3-540-79723-4_7

18. Bourgeois, N., Escoffier, B., Paschos, V.T., van Rooij, J.M.M.: Fast algorithmsfor max independent set. Algorithmica **62**(1–2), 382–415 (2012). https://doi.org/10.1007/s00453-010-9460-7

19. Cai, L., Fellows, M.R., Juedes, D.W., Rosamond, F.A.: The complexity of polynomial-time approximation. Theory Comput. Syst. **41**(3), 459–477 (2007). https://doi.org/10.1007/s00224-007-1346-y

20. Chen, J., Kanj, I.A., Xia, G.: Labeled search trees and amortized analysis: Improved upper bounds for np-hard problems. Algorithmica **43**(4), 245–273 (2005). https://doi.org/10.1007/s00453-004-1145-7

21. Curticapean, R.: Counting matchings of size k Is ♯W[1]-Hard. In: Fomin, F.V., Freivalds, R., Kwiatkowska, M., Peleg, D. (eds.) ICALP 2013. LNCS, vol. 7965, pp. 352–363. Springer, Heidelberg (2013). https://doi.org/10.1007/978-3-642-39206-1_30

22. Curticapean, R., Dell, H., Marx, D.: Homomorphisms are a good basis for counting small subgraphs. In: Proceedings of the 49th Annual ACM SIGACT Symposium on Theory of Computing, STOC 2017, Montreal, QC, Canada, 19–23 June 2017, pp. 210–223 (2017). https://doi.org/10.1145/3055399.3055502

23. Curticapean, R., Marx, D.: Complexity of counting subgraphs: only the bounded-ness of the vertex-cover number counts. In: 55th IEEE Annual Symposium on Foundations of Computer Science, FOCS 2014, Philadelphia, PA, USA, 18–21 October 2014, pp. 130–139 (2014). https://doi.org/10.1109/FOCS.2014.22

24. Cygan, M., Kowalik, L., Socala, A.: Improving TSP tours using dynamic programming over tree decompositions. In: 25th Annual European Symposium on Algorithms, ESA 2017, 4–6 September 2017, Vienna, Austria, pp. 30:1–30:14 (2017). https://doi.org/10.4230/LIPIcs.ESA.2017.30

25. Cygan, M., Nederlof, J., Pilipczuk, M., Pilipczuk, M., van Rooij, J.M.M., Wojtaszczyk, J.O.: Solving connectivity problems parameterized by treewidth in single exponential time. In: Ostrovsky, R. (ed.) IEEE 52nd Annual Symposium on Foundations of Computer Science, FOCS 2011, Palm Springs, CA, USA, 22–25 October 2011, pp. 150–159. IEEE Computer Society (2011). https://doi.org/10.1109/FOCS.2011.23

26. Cygan, M., Pilipczuk, M., Pilipczuk, M., Wojtaszczyk, J.O.: Solving the 2-disjoint connected subgraphs problem faster than 2 n. Algorithmica **70**(2), 195–207 (2014). https://doi.org/10.1007/s00453-013-9796-x

27. Dechter, R., Pearl, J.: Tree clustering for constraint networks. Artif. Intell. **38**(3), 353–366 (1989). https://doi.org/10.1016/0004-3702(89)90037-4. http://www.sciencedirect.com/science/article/pii/0004370289900374

28. Demaine, E.D., Fomin, F.V., Hajiaghayi, M.T., Thilikos, D.M.: Bidimensional parameters and local treewidth. SIAM J. Discret. Math. **18**(3), 501–511 (2004). https://doi.org/10.1137/S0895480103433410

29. Demaine, E.D., Fomin, F.V., Hajiaghayi, M.T., Thilikos, D.M.: Fixed-parameteralgorithms for (k, r)-center in planar graphs and map graphs. ACM Trans. Algorithms **1**(1), 33–47 (2005). https://doi.org/10.1145/1077464.1077468

30. Demaine, E.D., Fomin, F.V., Hajiaghayi, M.T., Thilikos, D.M.: Subexponential parameterized algorithms on bounded-genus graphs and H-minor-freegraphs. J. ACM **52**(6), 866–893 (2005). https://doi.org/10.1145/1101821.1101823

31. Demaine, E.D., Hajiaghayi, M.T., Thilikos, D.M.: Exponential speedup offixed-parameter algorithms for classes of graphs excluding single-crossinggraphs as minors. Algorithmica **41**(4), 245–267 (2005). https://doi.org/10.1007/s00453-004-1125-y

32. Díaz, J., Serna, M.J., Thilikos, D.M.: Counting H-colorings of partial k-trees. Theor. Comput. Sci. **281**(1–2), 291–309 (2002). https://doi.org/10.1016/S0304-3975(02)00017-8

33. Dorn, F.: Dynamic programming and planarity: Improved tree-decomposition basedalgorithms. Discret. Appl. Math. **158**(7), 800–808 (2010). https://doi.org/10.1016/j.dam.2009.10.011

34. Dorn, F.: Planar subgraph isomorphism revisited. In: 27th International Symposium on Theoretical Aspects of Computer Science, STACS 2010, 4–6 March 2010, Nancy, France, pp. 263–274 (2010). https://doi.org/10.4230/LIPIcs.STACS.2010.2460

35. Dorn, F., Fomin, F.V., Lokshtanov, D., Raman, V., Saurabh, S.: Beyond bidimensionality: parameterized subexponential algorithms on directed graphs. Inf. Comput. **233**, 60–70 (2013). https://doi.org/10.1016/j.ic.2013.11.006

36. Dorn, F., Fomin, F.V., Thilikos, D.M.: Catalan structures and dynamic programming in h-minor-free graphs. J. Comput. Syst. Sci. **78**(5), 1606–1622 (2012). https://doi.org/10.1016/j.jcss.2012.02.004

37. Dorn, F., Penninkx, E., Bodlaender, H.L., Fomin, F.V.: Efficient exact algorithms on planar graphs: exploiting sphere cut decompositions. Algorithmica **58**(3), 790–810 (2010). https://doi.org/10.1007/s00453-009-9296-1

38. Fischl, W., Gottlob, G., Longo, D.M., Pichler, R.: Hyperbench: a benchmark and tool for hypergraphs and empirical findings. In: Proceedings of the 13th Alberto Mendelzon International Workshop on Foundations of Data Management, Asunción, Paraguay, 3–7 June 2019 (2019). http://ceur-ws.org/Vol-2369/short02.pdf

39. Flum, J., Grohe, M.: The parameterized complexity of counting problems. SIAM J. Comput. **33**(4), 892–922 (2004). https://doi.org/10.1137/S0097539703427203

40. Fomin, F.V., Gaspers, S., Saurabh, S., Stepanov, A.A.: On two techniques ofcombining branching and treewidth. Algorithmica **54**(2), 181–207 (2009). https://doi.org/10.1007/s00453-007-9133-3

41. Fomin, F.V., Grandoni, F., Kratsch, D.: A measure & conquer approach for the analysis of exact algorithms. J. ACM **56**(5), 25:1–25:32 (2009). https://doi.org/10.1145/1552285.1552286

42. Fomin, F.V., Høie, K.: Pathwidth of cubic graphs and exact algorithms. Inf. Process. Lett. **97**(5), 191–196 (2006). https://doi.org/10.1016/j.ipl.2005.10.012

43. Fomin, F.V., Kratsch, D.: Exact Exponential Algorithms. TTCSAES. Springer, Heidelberg (2010). https://doi.org/10.1007/978-3-642-16533-7

44. Fomin, F.V., Lokshtanov, D., Marx, D., Pilipczuk, M., Pilipczuk, M., Saurabh, S.: Subexponential parameterized algorithms for planar and apex-minor-free graphs via low treewidth pattern covering. In: FOCS 2016, pp. 515–524. IEEE Computer Society (2016)

45. Fomin, F.V., Lokshtanov, D., Panolan, F., Saurabh, S.: Efficient computation of representative families with applications in parameterized and exact algorithms. J. ACM **63**(4), 29:1–29:60 (2016). https://doi.org/10.1145/2886094

46. Fomin, F.V., Thilikos, D.M.: Dominating sets in planar graphs: branch-width andexponential speed-up. SIAM J. Comput. **36**(2), 281–309 (2006). https://doi.org/10.1137/S0097539702419649

47. Freuder, E.C.: Complexity of k-tree structured constraint satisfaction problems. In: Proceedings of the Eighth National Conference on Artificial Intelligence, AAAI 1990, vol. 1, pp. 4–9. AAAI Press (1990), http://dl.acm.org/citation.cfm?id=1865499.1865500

48. Gu, Q., Tamaki, H.: Improved bounds on the planar branchwidth with respect tothe largest grid minor size. Algorithmica **64**(3), 416–453 (2012). https://doi.org/10.1007/s00453-012-9627-5

49. Guillemot, S., Marx, D.: Finding small patterns in permutations in linear time. In: Proceedings of the Twenty-Fifth Annual ACM-SIAM Symposium on Discrete Algorithms, SODA 2014, Portland, Oregon, USA, 5–7 January 2014, pp. 82–101 (2014). https://doi.org/10.1137/1.9781611973402.7

50. Halin, R.: S-functions for graphs. J. Geom. **8**(1–2), 171–186 (1976)

51. Impagliazzo, R., Paturi, R.: On the complexity of k-SAT. J. Comput. Syst. Sci. **62**(2), 367–375 (2001)

52. Impagliazzo, R., Paturi, R., Zane, F.: Which problems have strongly exponential complexity? J. Comput. Syst. Sci. **63**(4), 512–530 (2001)

53. Jian, T.: $O(2^{0.304n})$ algorithm for solving maximum independent set problem. IEEE Trans. Comput. **35**(9), 847–851 (1986). https://www.scopus.com/inward/record.uri?eid=2-s2.0-0022787854&partnerID=40&md5=c723ea6d9074acfa3d6f6c73e3439007

54. Klein, P.N., Marx, D.: A subexponential parameterized algorithm for subset TSP on planar graphs. In: SODA 2014, pp. 1812–1830. SIAM (2014)

55. Kneis, J., Langer, A., Rossmanith, P.: A fine-grained analysis of a simple independent set algorithm. In: IARCS Annual Conference on Foundations of Software Technology and Theoretical Computer Science, FSTTCS 2009, 15–17 December 2009, IIT Kanpur, India, pp. 287–298 (2009). https://doi.org/10.4230/LIPIcs.FSTTCS.2009.2326

56. Knuth, D.E.: The Art of Computer Programming, Volume I: Fundamental Algorithms. Addison-Wesley, Boston (1968)

57. Koutis, I., Williams, R.: LIMITS and applications of group algebras for parameterized problems. ACM Trans. Algorithms **12**(3), 31:1–31:18 (2016). https://doi.org/10.1145/2885499

58. Lokshtanov, D., Saurabh, S., Wahlström, M.: Subexponential parameterized odd cycle transversal on planar graphs. In: FSTTCS 2012. LIPIcs, vol. 18, pp. 424–434. Schloss Dagstuhl – Leibniz-Zentrum für Informatik (2012)

59. Lovász, L.: Operations with structures. Acta Math. Hungarica **18**(3–4), 321–328 (1967)

60. Maniu, S., Senellart, P., Jog, S.: An experimental study of the treewidth of real-world graph data. In: 22nd International Conference on Database Theory, ICDT 2019, 26–28 March 2019, Lisbon, Portugal, pp. 12:1–12:18 (2019). https://doi.org/10.4230/LIPIcs.ICDT.2019.12

61. Marcus, A., Tardos, G.: Excluded permutation matrices and the Stanley-Wilf conjecture. J. Comb. Theory Ser. A **107**(1), 153–160 (2004). https://doi.org/10.1016/j.jcta.2004.04.002

62. Monien, B., Preis, R.: Upper bounds on the bisection width of 3- and 4-regular graphs. In: Mathematical Foundations of Computer Science 2001, 26th International Symposium, MFCS 2001 Marianske Lazne, Czech Republic, 27–31 August 2001, Proceedings, pp. 524–536 (2001). https://doi.org/10.1007/3-540-44683-4_46

63. Pilipczuk, M.: Surprising applications of treewidth bounds for planar graphs. In: Fomin, F.V., et al. (eds.) Bodlaender Festschrift. LNCS, vol. 12160, pp. 173–188. Springer, Heidelberg (2020). https://doi.org/10.1007/978-3-030-42071-0_13

64. Pilipczuk, M., Pilipczuk, M., Sankowski, P., van Leeuwen, E.J.: Subexponential-time parameterized algorithm for Steiner tree on planar graphs. In: STACS 2013. LIPIcs, vol. 20, pp. 353–364. Schloss Dagstuhl – Leibniz-Zentrum für Informatik (2013)

65. Pilipczuk, M., Pilipczuk, M., Sankowski, P., van Leeuwen, E.J.: Network sparsification for steiner problems on planar and bounded-genus graphs. In: FOCS 2014, pp. 276–285. IEEE Computer Society (2014)

66. Razgon, I.: Computing minimum directed feedback vertex set in $O(1.9977^n)$. In: Proceedings on Theoretical Computer Science, 10th Italian Conference, ICTCS 2007, Rome, Italy, 3–5 October 2007, pp. 70–81 (2007)

67. Razgon, I.: Faster computation of maximum independent set and parameterized vertex cover for graphs with maximum degree 3. J. Discret. Algorithms $7(2)$, 191–212 (2009). https://doi.org/10.1016/j.jda.2008.09.004

68. Robertson, N., Seymour, P., Thomas, R.: Quickly excluding a planar graph. J. Comb. Theory Ser. B $62(2)$, 323–348 (1994). https://doi.org/10.1006/jctb.1994.1073

69. Robertson, N., Seymour, P.D.: Graph minors. III. planar tree-width. J. Comb. Theory Ser. B $36(1)$, 49–64 (1984). https://doi.org/10.1016/0095-8956(84)90013-3

70. Robson, J.M.: Algorithms for maximum independent sets. J. Algorithms $7(3)$, 425–440 (1986). https://doi.org/10.1016/0196-6774(86)90032-5

71. Scott, A.D., Sorkin, G.B.: Linear-programming design and analysis of fast algorithms for Max 2-CSP. Discret. Optim. $4(3-4)$, 260–287 (2007). https://doi.org/10.1016/j.disopt.2007.08.001

72. Tarjan, R.E., Trojanowski, A.E.: Finding a maximum independent set. SIAM J. Comput. $6(3)$, 537–546 (1977). https://doi.org/10.1137/0206038

73. Thorup, M.: All structured programs have small tree-width and good register allocation. Inf. Comput. $142(2)$, 159–181 (1998). https://doi.org/10.1006/inco.1997.2697

74. Valiant, L.G.: The complexity of computing the permanent. Theor. Comput. Sci. 8, 189–201 (1979). https://doi.org/10.1016/0304-3975(79)90044-6

75. West, J.: Permutations with restricted subsequences and stack-sortable permutations. Ph.D. thesis, MIT, Cambridge, MA (1990)

76. Xiao, M., Nagamochi, H.: Confining sets and avoiding bottleneck cases: a simple maximum independent set algorithm in degree-3 graphs. Theor. Comput. Sci. 469, 92–104 (2013). https://doi.org/10.1016/j.tcs.2012.09.022

77. Xiao, M., Nagamochi, H.: Exact algorithms for maximum independent set. Inf. Comput. 255, 126–146 (2017). https://doi.org/10.1016/j.ic.2017.06.001

Algorithms for NP-Hard Problems via Rank-Related Parameters of Matrices

Jesper Nederlof[(✉)] [iD]

Eindhoven University of Technology, Utrecht University, Utrecht, The Netherlands
j.nederlof@uu.nl

Abstract. We survey a number of recent results that relate the fine-grained complexity of several NP-Hard problems with the rank of certain matrices. The main technical theme is that for a wide variety of Divide & Conquer algorithms, structural insights on associated *partial solutions matrices* may directly lead to speedups.

1 Introduction

Rank is a fundamental concept in linear algebra to express algebraic dependence in relations described by matrices. It has numerous applications in theoretical computer science and mathematics, ranging from algebraic complexity [BCS97], communication complexity [LS88], to extremal combinatorics [Mat10].

A common phenomenon in these areas is that low rank often helps in proving combinatorial upper bounds or in designing algorithms, e.g., through representative sets [BCKN15,FLPS16,KW14] or the polynomial method [Wil14].

In particular, rank has recently found applications in *fine-grained complexity* and the closely related area of *parameterized complexity*. For example, influential results such as algorithms for kernelization [KW14], the longest path problem [Mon85], and connectivity problems parameterized by treewidth [CNP+11, CKN13,BCKN15], rely crucially on certain low-rank factorizations.

Low-rank factorizations especially arise very naturally when applying the general Divide & Conquer and the closely related Dynamic Programming technique. Recall that these techniques (conceptually) partition a solution into *partial solutions*. Typically, lists of candidates for these partial solutions are maintained by an algorithm that gradually filters and extends these partial solutions to a complete solution.

The dominating term in the runtime of such an algorithm is the *number* of such partial solutions. But sometimes, there is no need to keep track of all partial solutions because of *group domination*: For example, suppose that partial solutions $s_0, s_1, \ldots, s_l$ are such that for any partial solution t that forms a complete solution with s_0 there is also an $i > 0$ such s_i forms a complete solution with t. Then of course, s_0 can be safely disregarded as partial solution.

Supported by the Netherlands Organization for Scientific Research under project no. 024.002.003 and the European Research Council under project no. 617951.

© Springer Nature Switzerland AG 2020
F. V. Fomin et al. (Eds.): Bodlaender Festschrift, LNCS 12160, pp. 145–164, 2020.
https://doi.org/10.1007/978-3-030-42071-0_11

In this survey we study several standard Divide & Conquer algorithms from the field of fine-grained complexity for NP-hard problems and explore how group domination helps to improve them. A crucial tool in these are *partial solution matrices*. Given two groups of partial solutions R and C a partial solution matrix $\mathbf{A} \in \{0,1\}^{R \times C}$ is a matrix such that $\mathbf{A}[p,q] = 1$ if and only if partial solutions p and q combine to a complete solution. We will see that insights on rank-related parameters of the partial solution matrices can be used to speed up the associated Divide & Conquer algorithms in a variety of settings.

Organization. This survey is organized as follows: In Sect. 2 we introduce used notation In Sect. 3 we introduce some matrices along with their various parameters, which will be used in Sect. 4 to provide algorithms for various NP-complete problems. Finally, in Sect. 5 we mention some other directions that we do not fully touch.

2   Preliminaries

We let A^B denote the set of vectors or functions indexed by B with values in A. The symbol ε denotes the empty string, vector or partition. If b is a Boolean, we denote $[b]$ for 1 if b if true and 0 otherwise. On the other hand, if $[b]$ is an integer we use $[b]$ to denote $\{1, \ldots, b\}$.

In this survey all matrices will be written in bold font. If $\mathbf{M} \in \{0,1\}^{R \times C}$ is a matrix and $X \subseteq R$ and $Y \subseteq C$ we denote $\mathbf{M}[X, Y]$ for the matrix induced by rows X and columns Y. If either X or Y is replaced with a $\cdot$ this means no restriction is placed on the rows or columns, respectively. We let $\equiv_2$ denote equivalence modulo 2. If $Y, Y' \subseteq R$, we denote $\mathbf{M}[X, Y \circ Y']$ for the matrix obtained by horizontally concatenating the matrices $\mathbf{M}[X, Y]$ and $\mathbf{M}[X, Y']$.

Partitions and the Partition Lattice. Given a set U, we use $\Pi(U)$ for the set of all partitions of U, i.e. a family of subsets of U that are pairwise disjoint and whose union equals U. It is known that, together with the coarsening relation $\sqsubseteq$, $\Pi(U)$ gives a lattice, with the maximum element being $\{U\}$ and the minimum element being the partition into singletons. We denote $\sqcup$ for the join operation and $\sqcap$ for the meet operation in this lattice; these operators are associative and commutative. I.e., for two partitions p and q, $p \sqcup q$ is obtained as follows: let $\sim$ be the relation on the elements with $v \sim w$, if and only if v and w belong to the same set in p or v and w belong to the same set in q. Now, $p \sqcup q$ is the partition of U into the equivalence classes of the transitive closure of $\sim$. (In simple graph terms: build a graph H with a vertex set U, by turning each set in p and each set in q into a clique. Now, $p \sqcup q$ is the partition of U into the connected components of H.) $p \sqcap q$ precisely consists of all sets that are the nonempty intersection of a set from p and a set from q. We use $\Pi_m(U) \subset \Pi(U)$ to denote the set of all partitions of U in blocks of size 2, or equivalently, the set of perfect matchings over U. Moreover, $\Pi_2(U)$ denotes the set of all partitions with two blocks, i.e.

cuts. Thus there are partitions $\{X, Y\}$ where $X \cap Y = \emptyset$, $X \cup Y = U$ and X or Y may equal the empty set. Given $p \in \Pi(U)$ we let #blocks(p) denote the number of blocks of p. We sometimes formally interpret a partition as a family of disjoint subsets in the natural way. If $p = \{P_1, \ldots, P_l\} \in \Pi(U)$ and $X \subseteq U$ we define $p_{|X} = \{P_1 \setminus X, \ldots, P_l \setminus X\}$ as the restriction of p onto X. Also, if $A \subseteq U$, we let $\{A\}$ denote the partition with the single non-trivial block A.

3   Some Matrices and Their Rank-Related Parameters

In this section we introduce and study a number of families of matrices that will serve as partial solution matrices in the next section. In order to use them as such the following terminology will be useful:

Definition 3.1. *A family of matrices* $\{\mathbf{A}_t\}_t$ *is explicit if the following holds for every t: If* $\mathbf{A}_t$ *is an* $n \times n$ *matrix, then given t and* $1 \leq r, c \leq n$, *the entry* $\mathbf{A}_t[i, j]$ *can be computed in* polylog(n) *time. A factorization* $\mathbf{A}_t = \mathbf{L}_t \mathbf{R}_t$ *is explicit if* $\{\mathbf{L}_t\}_t$ *is explicit.*

This section is organized as follows: In Subsect. 3.1 we define a number of rank-related parameters. In the subsequent subsections we present case studies of matrices where the different rank-related parameters are useful: In Subsect. 3.2 we study the field rank, in Subsect. 3.3 the Boolean rank, and in Subsect. 3.4 the support rank.

3.1   Some Rank-Related Parameters of Matrices

We study several parameters that express various sorts of (algebraic) dependence between rows of a matrix. Let $\mathbf{A}_t$ be a binary matrix, for a field $\mathbb{F}$, we denote $\mathrm{rk}_{\mathbb{F}}(\mathbf{A}_t)$ for the rank of $\mathbf{A}_t$ over $\mathbb{F}$. We define the *field rank* of $\mathbf{A}_t$ as the minimum of $\mathrm{rk}_{\mathbb{F}}(\mathbf{A}_t)$ over all reasonable[1] fields $\mathbb{F}$.

We define the *support rank* supRank$(\mathbf{A}_t)$ of a matrix $\mathbf{A}_t$ to be the minimum rank of a matrix $\mathbf{A}_t'$ over a finite field $\mathbb{F}$ with the property that $\mathbf{A}_t[p, q]$ is non-zero if and only if $\mathbf{A}_t'[p, q]$ is non-zero for every p, q. This parameter goes by several names, such as the 'non-deterministic rank' [Wol03], and its computation has received significant attention by researchers working on linear algebra.[2]

We let boolRank$(\mathbf{A}_t)$ denote the *Boolean rank* of matrix $\mathbf{A}_t$. This is the minimum size of a family $\mathcal{F}$ of submatrices of $\mathbf{A}$ with value 1 in each cell with the following property: every matrix cell with of $\mathbf{A}$ with value 1 is contained in at least one submatrix in $\mathcal{F}$. Such a family $\mathcal{F}$ is often called a *rectangle cover*. Boolean rank can also be defined as the rank of $\mathbf{A}_t$ over the Boolean semiring $(\{0, 1\}, \wedge, \vee)$: A matrix $\mathbf{A}_t$ has Boolean rank at most r if there exist Boolean matrices $\mathbf{L}_t$ and $\mathbf{R}_t$ such that $\mathbf{A}_t[p, q] = \vee_{i=1}^{r}(\mathbf{L}_t[p, i] \wedge \mathbf{R}_t[i, q])$.

---

[1] Since things get a bit tricky formally here, let's just say we restrict attention to the fields $\mathbb{R}$ and $\mathbb{F}_p$ for finite p.

[2] https://aimath.org/pastworkshops/matrixspectrum.html.

The Boolean rank can also be interpreted as the minimum 'biclique cover' of the bipartite graph of which $\mathbf{A}$ is the incidence matrix. It is worthwhile noticing that $\texttt{boolRank}(\mathbf{A}_t)$ is equal to the logarithm of the non-deterministic communication complexity [KN97].

We let $\texttt{indMatch}(\mathbf{A}_t)$ denote the maximum size of a permutation matrix (i.e. exactly one cell with value 1 per row and column) that is a submatrix of $\mathbf{A}_t$. We use this notation since $\texttt{indMatch}(\mathbf{A}_t)$ can be seen to be equal to the largest *induced matching* of the bipartite graph that has $\mathbf{A}_t$ as its incidence matrix.

Definition 3.2. *Given a matrix $\mathbf{A}_t \in \{0,1\}^{R \times C}$ and a subset $X \subseteq R$, a subset $X' \subseteq X$ is a* representative set *of X with respect to $\mathbf{A}_t$ if for every $c \in C$, there exists an $r \in X$ such that $\mathbf{A}_t[r,c] = 1$ only if there exists $r' \in X'$ such that $\mathbf{A}_t[r',c] = 1$.*

It is easy to see that representation is *transitive*: If X is a representative set of Y and Y is a representative set of Z, then X is a representative set of Z.

We observe that a set of rows X has no representative set of X as a strict subset if and only if every element $r \in X$ has a 'reason' to be included, i.e. a column c such that $\mathbf{A}_t[r,c] = 1$ and $\mathbf{A}_t[r',c] = 0$ for every $r' \in X_t \setminus r$.

Observation 3.1. *Let $\mathbf{A}_t \in \{0,1\}^{R \times C}$. A set $X \subseteq R$, is an inclusion-wise minimal representative set of itself if and only if there exists $Y \subseteq C$ such that $\mathbf{A}_t[X,Y]$ is a permutation matrix.*

We are interested in computing small representative sets for any (worst-case) set of rows. Observation 3.1 implies that the minimum size representative set of any set of rows is at most $\texttt{indMatch}(\mathbf{A}_t)$, and that there exists some set of rows for which the minimum size of a representative set equals $\texttt{indMatch}(\mathbf{A}_t)$. Thus to understand the exact efficiency of computing representative sets, the quantity $\texttt{indMatch}(\mathbf{A}_t)$ is of relevance.

Unfortunately it is NP-complete to compute $\texttt{indMatch}(\mathbf{A}_t)$ even in special cases such as matrices with at most 3 non-zero values per row and column [Loz02]. Moreover, even if there would be a polynomial time algorithm, in many cases we would like to avoid to construct the matrix $\mathbf{A}_t$ explicitly. Fortunately, the following lemma shows that often we can compute representative sets in time *sublinear* in terms of the dimensions of the matrix if it has a small factorization.

Lemma 3.1. *Suppose $\mathbf{A}_t \in \{0,1\}^{R \times C}$ has field, support or Boolean rank r and the associated factorization is explicit. Then, a representative set of a given subset of rows $X \subseteq R$ can be found in $|X| r^{\omega-1} \text{polylog}(r)$ time, where $\omega < 2.371$ is the smallest number such that two $(n \times n)$-matrices can be multiplied in $n^{\omega+o(1)}$ time. Moreover, if $\mathbf{A}_t$ has Boolean rank r, the runtime can be reduced to $|X| r \cdot \text{polylog}(r)$ time.*

Proof. Let $\mathbf{A}'_t = \mathbf{L}_t \mathbf{R}_t$ be the explicit factorization of rank r, where $\mathbf{A}'_t$ is a matrix such that $\mathbf{A}'_t \neq 0$ if and only if $\mathbf{A}_t \neq 0$. So $\mathbf{L}_t$ is an $(|R| \times r)$-matrix and $\mathbf{R}_t$ is an $(r \times |C|)$-matrix. We first focus on the field and support rank.

Construct the matrix $\mathbf{L}_t[X, \cdot]$ explicitly. Note this is possible within $|X| \cdot r \cdot$ polylog(r) time: This matrix has $|X| \times r$ entries, and that each entry of it can be computed in polylog(r) time since the factorization $\mathbf{A}'_t = \mathbf{L}_t \mathbf{R}_t$ is assumed to be explicit. Now use fast Gaussian elimination algorithm based on fast matrix multiplication algorithms [BCKN15, Lemma 3.15] to compute a row basis X' of $\mathbf{L}_t[X, \cdot]$ in time $|X| r^{\omega-1}$, where $\omega < 2.373$ is a number such that two $n \times n$ matrices can be multiplied within $n^{\omega+o(1)}$ time.

It remains to show that X' is a representative set of X. Let c be a column and let r be a row such that $\mathbf{A}[r, c] \neq 0$. This implies that $\mathbf{A}'[r, c] \neq 0$. Since X' is a row-basis of $\mathbf{L}'_t$, there exist $r_1, \ldots, r_\ell \in R$ and $\lambda_1, \ldots, \lambda_l \in \mathbb{F}$ such that

$$\sum_{i=1}^{\ell} \lambda_i \mathbf{L}'_t[r_i, \cdot] = \mathbf{L}_t[r, \cdot], \quad \text{which implies} \quad \sum_{i=1}^{\ell} \lambda_i \mathbf{A}'_t[r_i, \cdot] = \mathbf{A}'_t[r, \cdot].$$

Note that the implication follows from post-multiplying both sides of the first equation with $\mathbf{R}_t$. In particular, the latter implies that $\sum_{i=1}^{\ell} \lambda_i \mathbf{A}'_t[r_i, c] \neq 0$. Thus $\mathbf{A}'_t[r_i, c] \neq 0$ for some i, as required.

For the Boolean rank factorization, let $\mathbf{L}_t$ and $\mathbf{R}_t$ be the matrices of the explicit factorization (note that now the factorization is over the $\wedge - \vee$ semiring). Construct the matrix $\mathbf{L}_t[X, \cdot]$ explicitly and let $X' \subseteq X$ be all elements $r \in X$ for which there is a c such that $\mathbf{L}_t[r, c] = 1$ and $\mathbf{L}_t[r', c] = 0$ for every $r' \in X \setminus r$. It is clear that X' is a representative set of X since the set of columns with a cell with value 1 is by construction the same in $\mathbf{L}_t[X, \cdot]$ and in $\mathbf{L}_t[X', \cdot]$. This computes a representative set in $|X| \cdot r \cdot$ polylog(r) time. $\square$

3.2 Field Rank: Partitions and Matchings

We now introduce two matrices that express connectivity of subgraphs.

Partitions Connectivity Matrix. The following matrix was instrumental for the derandomization of the Cut&Count approach [CNP+11] from [BCKN15].

Definition 3.3. *For $t \geq 0$, define matrix* $\mathbf{P}_t \in \{0, 1\}^{\Pi([t]) \times \Pi([t])}$ *as*

$$\mathbf{P}_t[p, q] = \begin{cases} 1, & \text{if } p \sqcup q = \{[t]\}, \\ 0, & \text{otherwise.} \end{cases}$$

Suppose that t is odd and let $P, Q \subseteq \Pi([t])$ be all partitions with one block of size $(t-1)/2 + 1$ that contains the element 1 and all other blocks singleton. It is easy to see that for $p \in P$ and $q \in Q$ we have $p \sqcup q = \{[t]\}$ if and only if the non-singleton blocks of p and q are X and $([t] \setminus X) \cup 1$, for some $X \subseteq [t] \setminus 1$. This shows that $\text{indMatch}(\mathbf{P}_t)$ roughly 2^t. We continue with showing that the rank of $\mathbf{P}_t$ over $\mathbb{F}_2$ is only slightly higher. To do so we first define the factorizing matrices:

Definition 3.4. *For* $t \geq 0$, *define matrix* $\mathbf{F}_t \in \{0,1\}^{\Pi([t]) \times \Pi_2([t]))}$ *that has rows index by partitions and columns indexed by cuts as*

$$\mathbf{F}_t[p, \{X, Y\}] = \begin{cases} 1, & if\ p\ refines\ \{X, Y\}, \\ 0, & otherwise. \end{cases}$$

Since there are at most 2^t cuts the following implies the promised rank upper bound:

Lemma 3.2 (Cut&Count factorization). $\mathbf{P}_t \equiv_2 \mathbf{F}_t \cdot \mathbf{F}_t^T$.

Proof. Let $p, q \in \Pi([t])$. By expanding the definition of matrix multiplication, we have that

$$(\mathbf{F}_t \cdot \mathbf{F}_t^T)[p, q] = \sum_{\{X,Y\} \in \Pi_2([t])} [p \sqsubseteq \{X, Y\}] \cdot [q \sqsubseteq \{X, Y\}].$$

Since $(\Pi([t]), \sqsubseteq)$ is a lattice, $p, q \sqsubseteq \{X, Y\}$ is equivalent with $p \sqcup q \sqsubseteq \{X, Y\}$ and we can rewrite into

$$= \sum_{\{X,Y\} \in \Pi_2([t])} [p \sqcup q \sqsubseteq \{X, Y\}]$$

The number of cuts that coarsen a partition is exactly its number of blocks minus 1 since for each component we can choose a side and divide by 2 because of a cut is an unordered pair.

$$= 2^{\#\texttt{blocks}(p \sqcup q)-1}$$
$$\equiv_2 [\#\texttt{blocks}(p \sqcup q) = 1] \quad = \quad [p \sqcup q = \{[t]\}] \quad = \quad \mathbf{P}_t[p, q]. \qquad \square$$

It can also be shown that $\mathrm{rk}_\mathbb{R}(\mathbf{P}_t) \leq 4^t$ using the 'squared determinant approach' from [BCKN15].

Matchings Connectivity Matrix. Note that the aforementioned construction of an induced matching of $\mathbf{P}_t$ crucially relies on partitions with many singleton blocks. A natural question is how large induced matchings exist in the submatrix of $\mathbf{P}_t$ induced by all partitions without singleton blocks. While the answer to this question is not known,[3] significant progress was made on the following even smaller submatrix of $\mathbf{P}_t$:

Definition 3.5. *For* $t \geq 0$, *define matrix* $\mathbf{H}_t \in \{0,1\}^{\Pi_m([t]) \times \Pi_m([t])}$ *as*

$$\mathbf{H}_t[P, Q] = \begin{cases} 1, & if\ P \cup Q\ is\ a\ Hamiltonian\ Cycle, \\ 0, & otherwise. \end{cases}$$

We now define a family of matchings of $\mathbf{H}_t$ that are crucial to understand the structure of $\mathbf{H}_t$. See Fig. 1 for an illustration.

---

[3] At least, to the author.

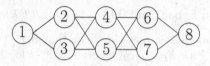

Fig. 1. The graph Z_8.

Definition 3.6 (Basis matchings). *Let $t \geq 2$ be an even integer, and let $Z_t = ([t], E)$ be a graph with vertices $[t]$ and edges $E = \{\{i, j\} : \lfloor j/2 \rfloor = \lfloor i/2 \rfloor + 1\}$. Define $\mathcal{X}_t$ to be the set of perfect matchings of Z_t.*

It can be shown that for every perfect matching M of Z_t there is a unique different perfect matching $\overline{M}$ such that $M \cup \overline{M}$ is a Hamiltonian cycle of Z_t. This proves that $\text{indMatch}_t(\mathbf{H}_t) \geq |\mathcal{X}_t|$. This bound turns out to be tight. Even stronger, it turns out that $\mathcal{X}_t$ is a row-basis of $\mathbf{H}_t$, and thus $\text{rk}_{\mathbb{F}_2}(\mathbf{H}_t) = |\mathcal{X}_t| = 2^{t/2-1}$, by virtue of the following factorization:

Lemma 3.3 ([CKN13]). *If P, Q are two perfect matchings of K_t, then*

$$\mathbf{H}_t[P, Q] \equiv_2 \sum_{M \in \mathcal{X}_t} [P \cup M \text{ is an Ham. Cycle }] \cdot [Q \cup \overline{M} \text{ is an Ham. Cycle}].$$

Let us remark that other variants of the rank of $\mathbf{H}_t$ over the reals also have been studied. In [RS95], the authors showed that if $\mathbf{H}_t$ is restricted to all perfect matchings on the complete balanced bipartite graph on t vertices, then its rank is $\binom{t-2}{t/2-1}$. Their motivation was to disprove the original formulation of the 'log-rank conjecture' in communication complexity. They achieved this by relating their rank bound to a second bound: The non-deterministic communication complexity of the same submatrix of $\mathbf{H}_t$ is $\Omega(n \log \log n)$. In [CLN18] the authors showed that $\text{rk}_{\mathbb{R}}(\mathbf{A}_t)$ equals 4^t, modulo some poly(t) factors.

3.3   Boolean Rank: Disjointness Matrix

We now define one of the most well-studied families of matrices in the field of communication complexity:

Definition 3.7. *For $t \geq p, q \geq 1$, define matrix $\mathbf{D}_{t,p,q} \in \{0,1\}^{\binom{[t]}{p} \times \binom{[t]}{q}}$ as*

$$\mathbf{D}_{t,p,q}[P, Q] = \begin{cases} 1, & \text{if } P \cap Q = \emptyset, \\ 0, & \text{otherwise.} \end{cases}$$

This time, we focus on the Boolean rank:

Lemma 3.4. *For even* k, $\texttt{boolRank}(\mathbf{D}_{t,k/2,k/2}) = O(2^k k \log t)$.

Proof. We use the probabilistic method. Note that if $P \cap Q = \emptyset$, then $\Pr[P \subseteq S \wedge S \cap Q = \emptyset] = 2^{-k}$. Pick $S_1, \ldots, S_l$, where $l = 2^k \cdot k \log 20t$. If $P \cap Q = \emptyset$, the probability that there is no i such that $P \subseteq S_i$ and $S_i \cap Q = \emptyset$ is

$$(1 - 2^{-k})^l \leq \exp(-l/2^k) = \exp(-20k \log t) \leq 1/t^k.$$

By a union bound, with positive probability there exists an i such that $P \subseteq S_i$ and $S_i \cap Q = \emptyset$ for each of the $\binom{t}{k/2}^2 \leq t^k$ possible disjoint pairs P, Q. In particular, a family $S_1, \ldots, S_l$ with this property exists, and this can be used as the rectangle cover. $\qquad\square$

Note the above proof is standard in Communication Complexity[4]. The above argument can be generalized to upper bounds on the Boolean rank of $\mathbf{D}_{t,p,q}$ by choosing a different distribution of the S_i's, and also can be made explicit by employing techniques reminiscent to [AYZ95] to get the following result:

Lemma 3.5 ([FLPS16]). $\texttt{boolRank}(\mathbf{D}_{t,p,q}[P,Q]) = O(\binom{p+q}{p} 2^{o(p+q)} \log t)$, *and the associated factorization is semi-explicit, in the sense that it could be computed in* $O(\binom{p+q}{p} 2^{o(p+q)} t \log t)$ *time.*

3.4 Support Rank: Linear Independence and Bipartite Colorings

Sometimes, in order to compute small representative sets quickly, it may be needed to consider the rank of different matrices with the same support. Consider the following example: Let $\mathbf{A}$ be the complement of a $t \times t$ identity matrix. It is easily seen that $\texttt{indMatch}(\mathbf{A}) = 2$, but there is a large gap with the rank of $\mathbf{A}$ which typically is t or $t - 1$. We resolve this gap by studying the rank of a different matrix with same support.

Linear Independence. The following matrix expresses when two linear independent sets again form an linear independent set. It arises frequently especially due to connections with matroid theory.

Definition 3.8. *Let* $\mathbb{F}$ *be a field and let* $\mathbf{M} \in \mathbb{F}^{R \times C}$ *be a matrix. Define a matrix* $\mathbf{L_M} \in \{0,1\}^{\binom{C}{p} \times \binom{C}{q}}$ *as*

$$\mathbf{L_M}[P,Q] = \begin{cases} 1, & \text{if } \mathrm{rk}_{\mathbb{F}}(\mathbf{M}[\cdot, P \cup Q]) = p + q, \\ 0, & \text{otherwise.} \end{cases}$$

Note that, even if $p = q = 1$ and $\mathbf{M}$ is an identity matrix, then the matrix $\mathbf{L_M}$ is the complement of the $|C| \times |C|$ identity matrix which has high rank (as mentioned above). Therefore, indeed resorting to support rank is needed here to get a low rank factorization. Define $\bar{I} := [p+q] \setminus I$ to be the complement of I, and define $\Sigma I = \sum_{i \in I} i$.

---

[4] See e.g. http://www.tcs.tifr.res.in/~prahladh/teaching/2011-12/comm/lectures/l03.pdf.

Lemma 3.6 (Generalized Laplace Expansion, Lemma [CFK+15]). *Let* $\mathbf{M} \in \{0,1\}^{(p+q)\times(p+q)}$ *and let* $P, Q \subseteq [p+q]$ *with* $|P| = p$ *and* $|Q| = q$. *Then*

$$\det(\mathbf{M}) = (-1)^{\lceil p/2 \rceil} \sum_{I \subseteq [p+q], |I|=p} (-1)^{\Sigma I} \det(\mathbf{M}[I, P]) \cdot \det(\mathbf{M}[\bar{I}, Q])$$

We start with employing generalized Laplace expansions to factorize $\mathbf{L}$ in a natural special case:

Lemma 3.7. *If* $p + q = |R|$, $\mathrm{supRank}(\mathbf{L_M}[P,Q]) = \binom{p+q}{p}$ *and the associated factorization is explicit.*

Proof. Define $\mathbf{L'_M}[P,Q] = \det(\mathbf{M}[\cdot, P \circ Q])$, where the $\circ$ operator denotes concatenation (see Sect. 2). We will show that $\mathbf{L'_M}$ has the same support as $\mathbf{L'_M}$ and low rank over $\mathbb{F}$. As the determinant of a square matrix is non-zero if and only if it is of full rank, we have that $\det(\mathbf{M}[\cdot, P \circ Q])$ is non-zero if and only if $\mathrm{rk}_\mathbb{F}(\mathbf{M}[\cdot, P \cup Q]) = p + q$, as required. The lemma now is a consequence of the following factorization implied by Lemma 3.6.

$$\mathbf{L'_M}[P,Q] = (-1)^{\lceil p/2 \rceil} \sum_{I \subseteq [p+q], |I|=p} (-1)^{\Sigma I} \det(\mathbf{M}[I, P]) \cdot \det(\mathbf{M}[\bar{I}, Q]). \qquad \square$$

We continue with focusing on the case $p + q \ll |R|$. A natural idea is to pre-multiply $\mathbf{M}$ with a random $(p + q) \times n$ matrix. This indeed works if we allow for randomized algorithms, but with constant probability the sought independent set may become dependent. A derandomized version of this 'truncation' operation was presented in [LMPS18], leading to the following result:

Lemma 3.8 ([LMPS18]). $\mathrm{supRank}(\mathbf{L_M}) = \binom{p+q}{p}$, *and the associated factorization is explicit.*

This bound has quite diverse applications: For example, it generalizes and refines the rank bound from Lemma 3.2, and it even strengthens this bound in the special case that the partitions are 'unbalanced'. See [LMPS18] for more details.

Colorings Matrix. We now introduce a matrix that naturally arises in graph coloring problems. It was defined for this purpose in [JN18], but somewhat surprisingly also found an application in the area of online algorithms [BEKN18].

Definition 3.9. *For an integer* $c \geq 1$ *and bipartite graph* H *with parts* $X = \{x_1, \ldots, x_t\}$ *and* $Y = \{y_1, \ldots, y_{t'}\}$ *and ordered edges in* $X \times Y$, *define matrix* $\mathbf{C}_{c,H} \in \{0,1\}^{[c]^X \times [c]^Y}$ *as*

$$\mathbf{C}_{c,H}[p,q] = \begin{cases} 1, & \text{if } p_i \neq q_j \text{ for every } (i,j) \in E(H)), \\ 0, & \text{otherwise.} \end{cases}$$

Note that even if H is a single edge, $\mathbf{C}_{c,H}$ is the complement of the $(c \times c)$ identity matrix. Therefore, indeed resorting to support rank is needed here to get a low rank factorization.

Lemma 3.9. $\mathrm{supRank}(\mathbf{C}_{c,H}) = 2^t$, and the associated factorization is explicit.

Proof. Define a matrix $\mathbf{C}'_{c,H}$ as follows

$$\mathbf{C}'_{c,H}[p,q] = \prod_{(i,j)\in E(H)} (p_i - q_j).$$

Since the product vanishes whenever $p_i = q_j$ for some $(i,j) \in E(H)$ and it is the product of positive numbers otherwise, we see that indeed $\mathbf{C}'_{c,H}[p,q] \neq 0$ if and only if $\mathbf{C}_{c,H}[p,q] \neq 0$. Moreover, this matrix has a low rank factorization that follows directly from expanding the parentheses to state the polynomial in its standard form: In particular, we have that $\mathbf{C}'_{c,H}[p,q]$ equals

$$\prod_{(i,j)\in E(H)} (p_i - q_j)$$

$$= \sum_{W \subseteq E(H)} \left(\prod_{i\in X} p_i^{d_W(i)} \right) \left(\prod_{j\in Y} (-q_j)^{d_{E(H)\setminus W}(j)} \right)$$

$$= \sum_{(d_i\in\{0,\ldots,d_{E(H)}(i)\})_{i\in X}} \left(\prod_{i\in X} p_i^{d_i} \right) \left(\sum_{\substack{W \subseteq E(H) \\ \forall i\in X: d_W(i)=d_i}} \prod_{j\in Y} (-q_j)^{d_{E(H)\setminus W}(j)} \right), \quad (1)$$

where the second equality follows by expanding the product and the third equality follows by grouping the summands on the number of edges incident to vertices in W included in X. It is easily seen that (1) gives a factorization of $\mathbf{C}'_{c,H}$ of rank at most the maximum number of the possibilities for d, since the inner dimension of the implied factorization is indexed by the possible vectors d. These are vectors d with $|X|$ coordinates where each $d_i \in \{0,\ldots,d_{|E(H)|}(i)\}$. The vector $d_{E(H)}$ that maximizes the number of such possible vectors while satisfying $\sum_{i\in X} d_{E(H)}(i) = k$ is the vector with k coordinates being equal to 1 by convexity (i.e., H is a matching) in which case the number of possibilities for $d_i \in \{0,1\}$ for all k vertices in X. $\square$

4   Using Low Rank Matrix Factorizations to Speed up Dynamic Programming

In this section we will use the insights from the previous section to speed up several natural dynamic programming algorithms. In Subsect. 4.1 this is a natural $O^*(q^k)$ time algorithm to decide whether a given graph with given permutation of cutwidth k has a proper q-coloring (see the section for definitions). We improve the runtime to $O^*(c^k)$ time where c is a constant independent of q.

In Subsect. 4.2 we study two connectivity problems parameterized by pathwidth and show they can be solved in $O^*(c^{pw})$ time by building on natural $O^*(pw^{pw})$ time dynamic programming algorithms.

Finally, in Subsect. 4.3 we present one of the first uses of representative sets to solve k-path in $O^*(c^k)$ by speeding up a natural $n^{O(k)}$ time algorithm.

4.1 Cutwidth

In this subsection we demonstrate the methods based on low rank factorizations on the graph coloring problem. Recall that in the graph coloring problem one is given an undirected graph $G = (V, E)$ and an integer q, and one is asked whether there exists a *proper coloring*, which is a vector $x \in [q]^V$ such that $x_v \neq x_w$ for every $\{v, w\} \in E$. Let $\{v_1, \dots, v_n\} = V(G)$ be a linear ordering of its vertices.

We denote all edges as directed pairs (v_i, v_j) with $i < j$. For $i = 1, \dots, n$, define V_i as the i'th prefix of this ordering, C_i as the i'th cut in this ordering, and X_i and Y_i as the left and respectively right endpoints of the edges in this cut, i.e.

$$
\begin{aligned}
V_i &= \{v_1, \dots, v_i\}, \\
C_i &= \{(v_l, v_r) \in E(G) : l \leq i < r\}, \\
X_i &= \{v_l \in V(G) : \exists (v_l, v_r) \in C_i \wedge l < r\}, \\
Y_i &= \{v_r \in V(G) : \exists (v_l, v_r) \in C_i \wedge l < r\}.
\end{aligned}
$$

Note that $X_i \subseteq X_{i-1} \cup \{v_i\}$ and $Y_{i-1} \subseteq Y_i \cup \{v_i\}$. We let H_i denote the bipartite graph with parts X_i, Y_i and edge set C_i. We study the graph coloring problem in the setting where one is given a permutation of low cutwidth, which is defined as follows:

Definition 4.1. *The cutwidth of the linear order $\{v_1, \dots, v_n\}$ is the maximum value of $|C_i|$ taken over all i.*

We use the following notation: A vector $x \in V^I$ is *an extension* of a vector $x' \in V^{I'}$ if $I' \subseteq I$ and $x'_i = x_i$ for every $i \in I'$. If $x \in V^I$ and $P \subseteq I$ then the projection $x_{|P}$ is defined as the unique vector in V^P of which x is an extension. For $i = 1, \dots, n$, we define $T[i] \subseteq [q]^{X_i}$ to be the set of all q-colorings of the vertices in X_i that can be extended to a proper q-coloring of $G[V_i]$. The following lemma allows to compute representative sets of $T[i]$.

Lemma 4.1 ([JN18]). *If $T'[i-1]$ is a representative set of $T[i-1]$ with respect to $\mathbf{C}_{q,H_{i-1}}$, then $T'[i]$ is a representative set of $T[i]$ with respect to $\mathbf{C}_{q,H_i}$, where*

$$
T'[i] = \left\{ (x \cup (v_i, c))_{|X_i} : x \in T'[i-1], c \in [q], (\forall v \in N(v_i) \cap X_{i-1} : x_v \neq c) \right\}.
$$

We remark that the lemma is very similar to the recurrence underlying a standard $O^*(q^k)$ time dynamic programming algorithm for the task at hand, but it is formulated in the language of this survey in order to allow for a speed up via representative sets as we now outline:

Theorem 4.1 ([JN18]). *The graph coloring problem can be solved in $O^*(2^{\omega \cdot k})$ time, assuming a linear order of cutwidth at most k is given.*

Proof. Compute $T'[0] = T[0]$ to be the singleton set with only the unique zero-dimensional vector. The for each $i = 1, \ldots, n$ to the following: First use Lemma 4.1 to compute $T'[i]$ from $T'[i-1]$. After each such step, use Lemma 3.1 with the explicit factorization of Lemma 3.9 to compute a subset $T''[i]$ of $T'[i]$ that represents $T'[i]$ with respect to $\mathbf{C}_{q,H_i}$. By transitivity it also represents $T[i]$ and we can set $T'[i] := T''[i]$ and continue with computing $T'[i+1]$. In the end we can check whether a proper q-coloring exists since it does if and only if $T[n]$ (and thus $T'[n]$) is non-empty by definition of $T[n]$. The run time follows since the number of partial solutions is at most 2^k at every step and the bottleneck is due to the application of Lemma 3.1. □

4.2 Pathwidth

A *path decomposition* of a graph $G = (V, E)$ is a path $\mathbb{P}$ in which each node x has an associated set of vertices $B_x \subseteq V$ (called a *bag*) such that $\bigcup B_x = V$ and the following properties hold:

1. For each edge $\{u, v\} \in E(G)$ there is a node x in $\mathbb{P}$ such that $u, v \in B_x$.
2. If $v \in B_x \cap B_y$ then $v \in B_z$ for all nodes z on the (unique) path from x to y in $\mathbb{P}$.

The *pathwidth* of $\mathbb{P}$ is the size of the largest bag minus one, and the pathwidth of a graph G is the minimum pathwidth over all possible path decompositions of G. We define *nice path decompositions* as follows.

Definition 4.2 (Nice Path Decomposition). *A nice path decomposition is a path decomposition where the underlying path of nodes is ordered from left to right (the predecessor of any node is its left neighbor) and in which each bag is of one of the following types:*

First bag: the bag associated with the leftmost node x is empty, $B_x = \emptyset$.

Introduce vertex bag: an internal node x of $\mathbb{P}$ with predecessor y such that $B_{x'} = B_y \cup \{v\}$ for some $v \notin B_y$. This bag is said to introduce v.

Introduce edge bag: an internal node x of $\mathbb{P}$ labeled with an edge $\{u, v\} \in E(G)$ with one predecessor y for which $u, v \in B_x = B_y$. This bag is said to introduce $\{u, v\}$, and every edge is introduced by exactly one bag.

Forget bag: an internal node x of $\mathbb{P}$ with one predecessor y for which $B_x = B_y \setminus \{v\}$ for some $v \in B_y$. This bag is said to forget v.

Last bag: the bag associated with the rightmost node x is empty, $B_x = \emptyset$.

It is easy to verify that any given path decomposition of pathwidth pw can be transformed in time $|V(G)|pw^{O(1)}$ into a nice path decomposition without increasing the width. For a bag B_i, we define the $G_i = (\cup_{j=1}^i B_j, E_i)$ where E_i are all edges introduced in bags $B_1, \ldots, B_i$.

Steiner Tree. In the Steiner Tree problem[5] one is given an undirected graph G, a vertex subset $K \subseteq V(G)$, and an integer s. The goal is to determine if there exists a subset $K \subseteq Y \subseteq V(G)$ such that $|Y| \leq s$ and $G[Y]$ is connected. As in the previous case studies, we first present a recurrence that allows to gradually build partial solutions. To facilitate this we use the following notation:

Definition 4.3. *Given a graph* G', *a subset* $Y \subseteq V(G')$ *and a partition* $p \in \Pi(X)$ *where* $X \subseteq Y$, *we say that* Y *connects* p *in* G' *if for every two vertices* $u, v \in X$ *the following holds: u and v are connected in $G'[Y]$ if and only if u and v are in the same block in p.*

For a bag B_i, a subset X and an integer $\dot{s}$ we define $T[i, X, s]$ to be the set of partitions $p \in \Pi(X)$ such that there exists a subset $Y \subseteq V(G_i)$ that connects p in G_i and satisfies $K \subseteq Y$, $|Y| \leq s$ and

$$\forall u \in Y \; \exists v \in X \; : u \text{ and } v \text{ are connected in } G_i[Y].$$

We now show how to compute entries $T[i, \cdot, \cdot]$ given the appropriate entries $T[i-1, \cdot, \cdot]$, by distinguishing on what kind of bag X_i is:

First Bag. If i is the first bag, $T[i, X, s] = \{\varepsilon\}$, where ε is the empty partition.

Introduce Vertex Bag. If B_i introduces a vertex v, note that G_i contains v as an isolated vertex (as we did not introduce any of its incident edges). If $v \in K$ it needs to be included in X. Hence, if $v \notin X$ we have that

$$T[i, X, s] = \begin{cases} \emptyset & \text{if } v \in K \text{ and } v \notin X, \\ T[i-1, X, s], & \text{if } v \notin K \text{ and } v \notin X. \end{cases}$$

Moreover if v is included in the solution, it should also be included in the partitions as a singleton:

$$T[i, X \cup \{v\}, s] = \{p \cup \{\{v\}\} \mid p \in T[i-1, X, s-1]\}.$$

Introduce Edge Bag. If an edge $\{u, v\}$ is introduced in B_i we have that $T[i, X, s] = T[i-1, X, s]$ if $\{u, v\} \not\subseteq X$, and otherwise

$$T[i, X, s] = \{p \sqcup \{\{u, v\}\} \mid p \in T[i-1, X, s]\}.$$

Note that here $\{u, v\}$ denoted the partition of X with $\{u, v\}$ as only non-trivial block.

Forget Vertex Bag. If a vertex v is forgotten in B_i, all partitions in $T[i-1, X, s]$ remain in $T[i, X, s]$ and all partitions in

$$T[i-1, X \cup \{v\}, s]$$

remain in $T[i, X, s]$ if they do not include v as a singleton:

$$T[i, X, s] = T[i-1, X, s] \cup \{p_{|X} \mid p \in T[i, X \cup \{v\}], \{v\} \notin p\}.$$

---

[5] For ease of exposition, we discuss a less general variant of the Steiner tree problem. The same methods can also solve more general versions within time that only depends linearly on the number of vertices, see [BCKN15] or the exposition in [CFK+15].

With all recurrences in place, we are ready to sketch the algorithm for Steiner tree:

Theorem 4.2. *Given a graph G and a path decomposition of G of width pw, any instance of Steiner tree on G can be solved in $O^*((1 + 2^\omega)^{pw})$ time.*

Proof Sketch. Let $\{B_i\}_{i=1}^l$ be the path decomposition. For every X and s, we compute a family of partitions $T'[i, X, s]$ that is a representative set for $T[i, X, s]$ with respect to $\mathbf{P}_{|X|}$, based on representative sets $T'[i-1, X', s]$ of $T[i-1, X', s]$ with respect to $\mathbf{P}_{|X'|}$. By following the above recurrence (but with all occurrences of $T[\cdot, \cdot, \cdot]$ with $T'[\cdot, \cdot, \cdot]$). It can be shown that in all cases indeed the resulting set $T'[i, X, s]$ is representative of $T[i, X, s]$. Alternating this computation with the table reduction procedure from Lemma 3.1 ensures $|T'[i, X, s]| \leq 2^{|X|} \text{poly}(n)$ and it runs in $2^{\omega|X|}$ time. Summing over all possibilities for X per bag, the run time becomes $n^{O(1)} \sum_{X \subseteq B_i} 2^{\omega|X|} = n^{O(1)}(1 + 2^\omega)^{pw}$ time.   □

For completeness, we remark that the Steiner tree problem can be solved in $O^*(3^{pw})$ time by a randomized algorithm [CNP+11].

Hamiltonian Cycle. In the Hamiltonian cycle problem one is given an undirected graph G, and is asked whether there exists a simple cycle $C \subseteq E(G)$ with $|C| = n$.

Definition 4.4. *Given a graph G', a subset $Y \subseteq E(G')$ and a partition $p \in \Pi(X)$ where $X \subseteq V(G')$, we say that Y connects p if for every two vertices $u, v \in X$ the following holds: u and v are connected in $(\cup_{e \in Y} e, Y)$ if and only if u and v are in the same block in p.*

For a bag B_i, a vector $d \in \{0, 1, 2\}^{B_i}$ we define $T[i, d]$ to be

$$\Big\{ M \in \Pi_m(d^{-1}(1)) \Big| \exists Y \subseteq E(G_i) : Y \text{ connects } p \wedge \forall v \in B_i : d_Y(v) = d_v$$

$$\wedge \forall v \in V(G_i) \setminus B_i : d_Y(v) = 2 \Big\},$$

where we let $d_Y(v)$ denote the number of edges in Y that is incident to v.

Similar to the algorithm for Steiner tree, a recurrence for $T[i, d]$ in terms of $T[i-1, d]$ can be formulated, and the same recurrence can be used to compute a set $T'[i, d]$ that is a representative set of $T[i, d]$ with respect to $\mathbf{H}_{|d^{-1}(1)|}$ from entries of the type $T'[i-1, d]$ that are representative sets of $T[i-1, d]$ with respect to $\mathbf{H}_{|d^{-1}(1)|}$. We refer to [BCKN15] for details.

By interleaving these computations with an algorithm implied by Lemma 3.1 with the matrix $\mathbf{H}_t$ and its factorization from Lemma 3.3 we can obtain the following theorem in a way similar to the previous sections:

Theorem 4.3 ([BCKN15]). *Given a graph G with path decomposition of width pw, it can be determined in $O^*((2 + 2^{\omega/2})^{pw})$ time whether G has a Hamiltonian cycle.*

In fact, the same running time can be obtained for the weighted version of the problem (the Traveling Salesman Problem). We would also like to mention that the problem can be solved in $O^*((2 + \sqrt{2})^{pw})$ time with a randomized algorithm [CKN13].

4.3 k-Path

In the k-path problem one is given a graph $G = ([n], E)$ and an integer k. The task is to determine whether G has a path on at least k vertices. Recall a path is a sequence of vertices such that consecutive vertices are adjacent and each vertex occurs at most once in the sequence. We outline an approach that was originally described in the paper that introduced representative sets [Mon85][6] For every $i = 1, \ldots, k$ and $v \in V$ we define

$$T[i, v] = \left\{ X \in \binom{[n]}{i} \Big| \exists \text{ path that ends at } v \text{ and visits } X \right\}.$$

By trying all possibilities for the penultimate vertex v' in the path the following recurrence can be obtained:

$$T[i, v] = \{X \cup \{v\} : X \in T[i-1, v'], v \in N(v')\}.$$

Similarly we have that

Lemma 4.2. *If $T'[i-1, v']$ is a representative set of $T[i-1, v']$ with respect to $\mathbf{D}_{n,i-1,k-(i-1)}$, then $T'[i, v]$ is a representative set of $T[i, v]$ with respect to $\mathbf{D}_{n,i,k-i}$ where*

$$T'[i, v] = \{X \cup \{v\} : X \in T'[i-1, v'], v \in N(v)\}.$$

Similarly as before, we use Lemma 3.1 in combination with Lemma 4.2 to obtain the following result:

Theorem 4.4. *Given a graph G and an integer k, it can be determined in $O^*(4^k)$ time whether G has a path on at least k vertices.*

Proof. Compute $T'[1, \{v\}] = T[1, \{v\}] = \{\{v\}\}$. For $i = 2, \ldots, k$ do the following: Compute $T'[i, v]$ as defined in Lemma 4.2 for every $v \in V$. Afterwards use Lemma 3.1 to compute a set $T''[i, v]$ that is a representative set of $T'[i, v]$ with respect to $\mathbf{D}_{n,i,k-i}$. By Lemma 3.1, $|T''[i, v]| \leq \binom{k}{i}$ and the time required to compute the set is at most 4^k. By transitivity, it will also be a representative set of $T[i, v]$ and we can set $T'[i, v] = T''[i, v]$ and use it in the next iteration to compute a family that is a representative for $T[i+1, v]$.

Afterwards, we can return whether G has a path on at least k vertices since it does if and only if $T'[k, v]$ is non-empty for some vertex v. □

---

[6] Indeed, the idea of representing partial solutions with a strict subset is natural, but to the author's knowledge [Mon85] was the first paper (in parameterized complexity) to use a generalization of this concept beyond equivalence classes.

Let us remark for completeness that the currently fastest deterministic algorithm for k-path refines the above approach and solves the problem in $O^*(2.597^k)$ time [Zeh15]. In the randomized setting, a beautiful algorithm from [BHKK17] solves the problem in $O^*(1.66^k)$ time.

5   Other Relevant Directions

This survey focused on only few applications in the area of parameterized complexity. We list a few of the most relevant directions not yet discussed.

5.1   Pair Problems

For a fixed family of explicit matrices $\{\mathbf{A}_t\}_t$, we may study the following problem PAIR($\mathbf{A}$): Given an integer t and sets P, Q such that $\mathbf{A}_t \in \{0,1\}^{R \times C}$ and $P \subseteq R$ and $Q \subseteq C$, the goal is to detect whether there exists $p \in P$ and $q \in Q$ such that $\mathbf{A}_t[p,q] = 1$. Freivalds [Fre77] famous matrix multiplication algorithm can be used to obtain the following result by computing $\mathbf{A}_t = (r\mathbf{L}_t)(\mathbf{R}_t r')$ with random vectors $r \in \{0,1\}^P$ and $r' \in \{0,1\}^Q$:

Observation 5.1. *If $\mathbf{A}_t$ has an explicit[7] field, support or Boolean rank r_t factorization, then an instance (t, P, Q) of PAIR($\mathbf{A}$) can be solved with a randomized algorithm in $r_t(|P| + |Q|) \cdot \mathrm{polylog}(t)$ time.*

An interesting special case is PAIR($\mathbf{D}_{t,p,q}$), also known as the *orthogonal vectors* problem. Several algorithms for this problem have been developed that rely on interesting rank parameters of $\mathbf{D}_{t,p,q}$ such as the rank over the reals [BHKK09] and an intriguing variant of 'probabilistic rank' [AWY15, AW17].

An especially interesting theme is that of *sparse factorizations*. That is, a factorization $\mathbf{A}_t = \mathbf{L}_t \mathbf{R}_t$ such that both $\mathbf{L}$ and $\mathbf{R}$ are relatively sparse.

Sparse factorizations for PAIR($\mathbf{D}_{t,p,q}$) are used in for example in [FLPS16]. In an unpublished note [Ned17], the author observed that if a natural algorithm that relies on a sparse Boolean rank decomposition of Lemma 3.5 can be improved slightly, a classic algorithm for the Subset Sum problem can be improved.

In a very recent work [Ned19] on improving the Bellman-Held-Karp algorithm for the Traveling Salesman Problem, the author studied the problem PAIR($\mathbf{H}$). That is, given two families of perfect matchings $P, Q \subseteq \Pi_m([t])$, determine whether there exist perfect matchings $p \in P$ and $q \in Q$ that form a Hamiltonian cycle of the complete graph K_t on t vertices. By combining the rank bound $\mathrm{rk}_{\mathbb{F}_2}(\mathbf{H}_t) = 2^{t/2-1}$ with Observation 5.1 this problem can be solved in $O((|P| + |Q|)2^{t/2}t^{O(1)})$ time with a randomized algorithm. In [Ned19] an $O(((|P|+|Q|)2^{3t/10} + 3^{t/2})t^{O(1)})$ time randomized algorithm was given that was instrumental to obtain a new result on TSP. Curiously this faster algorithm for PAIR($\mathbf{H}$) uses a factorization of $\mathbf{H}$ of *higher* rank, but since it is much *sparser* it is nevertheless more useful to solve PAIR($\mathbf{H}$). We refer to [Ned19] for more discussion and details.

---

[7] As a minor technical caveat, both $\mathbf{L}_t$ and $\mathbf{R}_t$ need to be explicit.

5.2   Matrix Multiplication

It should be noted that often the use of Lemma 3.1 as described in Sect. 4 does not yield the fastest algorithms, as these rely on more algebraic ideas such as Observation 5.1. At a high level, these algorithms are obtained by applying the low rank factorization at a more general level. Slightly more formal, one could see many dynamic programming algorithms as evaluating a chain of matrix multiplications $\mathbf{A}_1 \mathbf{A}_2 \cdots \mathbf{A}_l$ where $\mathbf{A}_1$ is a row vector and $\mathbf{A}_l$ is a column vector. Given low-rank factorization of these matrices, their product can be evaluated quickly if their products are evaluated in a clever order. If the factorizations are over finite fields or in terms of support rank, typically standard tools in complexity theory such as polynomial identity testing or the isolation lemma can be used to solve the (unweighted variants) of the decision problems by introducing randomization.

Notably, the algorithms obtained via this method are often known to be optimal under the Strong Exponential Time Hypothesis[8] (SETH). For example, this gives rise to an $O^*(3^{pw})$ time algorithm for Steiner Tree, an $O^*((2 + \sqrt{2})^{pw})$ algorithm for Hamiltonian cycle, and an $O^*(2^k)$ time algorithm for graph coloring (where k is the cutwidth of a given permutation). Furthermore, these algorithms cannot be improved to $O^*((3 - \varepsilon)^{pw})$ time, $O^*((2 + \sqrt{2} - \varepsilon)^{pw})$ time, and $O^*((2 - \varepsilon)^k)$ time for any $\varepsilon > 0$, unless the SETH fails.

5.3   Counting Algorithms

If the number solutions needs to be counted instead of detecting only one, only the rank over the reals can be applied in general. Instead, if one needs to count the number of solutions modulo a prime p, the rank over $\mathbb{Z}_p$ can be used.

A particularly non-trivial case is that of counting Hamiltonian cycles parameterized by the pathwidth. In [CLN18] a general connection between the complexity of the problem and the rank of $\mathbf{H}$ was shown:

Theorem 5.1. *Let* $r = \lim_{t \to \infty} \log_2(\text{rk}(\mathbf{H}_t))/t$. *Assuming SETH, there is no* $\varepsilon > 0$ *such that the number of Hamiltonian cycles can be computed in* $O^*((2 + r - \varepsilon)^{pw})$ *time on graphs with a given path decomposition of width pw. For prime numbers* p, *the same applies to counting Hamiltonian cycles modulo* p *when replacing* r *by* r_p, *which is defined analogously to* r *by taking the rank over* $\mathbb{Z}_p$.

Determining the rank of $\mathbf{H}_t$ over various fields turns out a challenging job. Over the reals, it was shown in [CLN18] that the rank of $\mathbf{H}$ is (up to factors polynomial in t) equal to 4^t. Thus, by Theorem 5.1 the existing $O^*(6^{pw})$ algorithm from [BCKN15] cannot be significantly improved, assuming SETH.

---

[8] This hypothesis postulates that for every $\varepsilon > 0$ there is an integer k such k-CNF satisfiability on n variables cannot be solved in $O^*((2 - \varepsilon)^n)$ time.

5.4   Further Results

This survey is far from exhaustive and biased towards the familiarity of the author. Other interesting connections between fine-grained complexity can be found in papers on the *probabilistic rank* [AW17], Waring rank [Pra18]. Since many algorithms on fine-grained complexity of hard problems rely on *fast matrix multiplication*, the rich theory underlying these fast algorithms that features a plethora of variants (tensor) rank can also be considered to be in the same category.

Let us conclude by remarking that studying problem specific tensors arising from divide and conquer algorithms that merge triples of partial solutions into a complete solution may be a good source of further research opportunities.

Acknowledgements. The author would like to thank Johan van Rooij and Stefan Kratsch for their valuable feedback on a previous version of this survey.

References

[AW17] Alman, J., Williams, R.R.: Probabilistic rank and matrix rigidity. In: Proceedings of the 49th Annual ACM SIGACT Symposium on Theory of Computing, STOC 2017, Montreal, QC, Canada, 19–23 June 2017, pp. 641–652 (2017)

[AWY15] Abboud, A., Williams, R.R., Yu, H.: More applications of the polynomial method to algorithm design. In: Proceedings of the Twenty-Sixth Annual ACM-SIAM Symposium on Discrete Algorithms, SODA 2015, San Diego, CA, USA, 4–6 January 2015, pp. 218–230 (2015)

[AYZ95] Alon, N., Yuster, R., Zwick, U.: Color-coding. J. ACM **42**(4), 844–856 (1995)

[BCKN15] Bodlaender, H.L., Cygan, M., Kratsch, S., Nederlof, J.: Deterministic single exponential time algorithms for connectivity problems parameterized by treewidth. Inf. Comput. **243**, 86–111 (2015)

[BCS97] Bürgisser, P., Clausen, M., Shokrollahi, M.A.: Algebraic Complexity Theory. Grundlehren der mathematischen Wissenschaften, vol. 315. Springer, Heidelberg (1997). https://doi.org/10.1007/978-3-662-03338-8

[BEKN18] Bansal, N., Eliás, M., Koumoutsos, G., Nederlof, J.: Competitive algorithms for generalized k-server in uniform metrics. In: Czumaj, A., (ed.), Proceedings of the Twenty-Ninth Annual ACM-SIAM Symposium on Discrete Algorithms, SODA 2018, New Orleans, LA, USA, 7–10 January 2018, pp. 992–1001. SIAM (2018)

[BHKK09] Björklund, A., Husfeldt, T., Kaski, P., Koivisto, M.: Counting paths and packings in halves. In: Fiat, A., Sanders, P. (eds.) ESA 2009. LNCS, vol. 5757, pp. 578–586. Springer, Heidelberg (2009). https://doi.org/10.1007/978-3-642-04128-0_52

[BHKK17] Björklund, A., Husfeldt, T., Kaski, P., Koivisto, M.: Narrow sieves for parameterized paths and packings. J. Comput. Syst. Sci. **87**, 119–139 (2017)

[CFK+15] Cygan, M., et al.: Parameterized Algorithms. Springer, Heidelberg (2015). https://doi.org/10.1007/978-3-319-21275-3

[CKN13]   Cygan, M., Kratsch, S., Nederlof, J.: Fast Hamiltonicity checking via bases of perfect matchings. In: Symposium on Theory of Computing Conference, STOC 2013, Palo Alto, CA, USA, 1–4 June 2013, pp. 301–310 (2013)

[CLN18]   Curticapean, R., Lindzey, N., Nederlof, J.: A tight lower bound for counting Hamiltonian cycles via matrix rank. In: Proceedings of the Twenty-Ninth Annual ACM-SIAM Symposium on Discrete Algorithms, SODA 2018, New Orleans, LA, USA, 7–10 January 2018, pp. 1080–1099 (2018)

[CNP+11] Cygan, M., Nederlof, J., Pilipczuk, M., Pilipczuk, M., van Rooij, J.M.M., Wojtaszczyk, J.O.: Solving connectivity problems parameterized by treewidth in single exponential time. In: Ostrovsky, R. (ed.), IEEE 52nd Annual Symposium on Foundations of Computer Science, FOCS 2011, Palm Springs, CA, USA, 22–25 October 2011, pp. 150–159. IEEE Computer Society (2011)

[FLPS16] Fomin, F.V., Lokshtanov, D., Panolan, F., Saurabh, S.: Efficient computation of representative families with applications in parameterized and exact algorithms. J. ACM **63**(4), 29:1–29:60 (2016)

[Fre77]   Freivalds, R.: Probabilistic machines can use less running time. In: Information Processing, Proceedings of the 7th IFIP Congress 1977, Toronto, Canada, 8–12 August 1977, pp. 839–842 (1977)

[JN18]   Jansen, B.M.P., Nederlof, J.: Computing the chromatic number using graph decompositions via matrix rank. In: Azar, Y., Bast, H., Herman, G. (eds.) 26th Annual European Symposium on Algorithms, ESA 2018, 20–22 August 2018, Helsinki, Finland. LIPIcs, vol. 112, pp. 47:1–47:15. Schloss Dagstuhl - Leibniz-Zentrum fuer Informatik (2018)

[KN97]   Kushilevitz, E., Nisan, N.: Communication Complexity. Cambridge University Press, Cambridge (1997)

[KW14]   Kratsch, S., Wahlström, M.: Compression via matroids: a randomized polynomial kernel for odd cycle transversal. ACM Trans. Algorithms **10**(4), 20:1–20:15 (2014)

[LMPS18] Lokshtanov, D., Misra, P., Panolan, F., Saurabh, S.: Deterministic truncation of linear matroids. ACM Trans. Algorithms **14**(2), 14:1–14:20 (2018)

[Loz02]   Lozin, V.V.: On maximum induced matchings in bipartite graphs. Inf. Process. Lett. **81**(1), 7–11 (2002)

[LS88]   Lovász, L., Saks, M.E.: Lattices, Möbius functions and communication complexity. In: 29th Annual Symposium on Foundations of Computer Science, White Plains, New York, USA, 24–26 October 1988, pp. 81–90. IEEE Computer Society (1988)

[Mat10]   Matoušek, J.: Thirty-Three Miniatures: Mathematical and Algorithmic Applications of Linear Algebra, vol. 53. American Mathematical Society, Providence (2010)

[Mon85]   Monien, B.: How to find long paths efficiently. North-Holland Math. Stud. **109**, 239–254 (1985). Analysis and Design of Algorithms for Combinatorial Problems

[Ned17]   Nederlof, J.: Faster subset sum via improved orthogonal vectors (2017). http://www.win.tue.nl/~jnederlo/problem.pdf

[Ned19]   Nederlof, J.: Bipartite TSP in $O(1.9999^n)$ time, assuming quadratic time matrix multiplication (2019). Unpublished

[Pra18]   Pratt, K.: Faster algorithms via waring decompositions. In: The Proceedings of FOCS 2018 (2018, to appear)

[RS95]   Raz, R., Spieker, B.: On the "log rank"-conjecture in communication complexity. Combinatorica **15**(4), 567–588 (1995)

[Wil14] Williams, R.: The polynomial method in circuit complexity applied to algorithm design (invited talk). In: Raman, V., Suresh, S.P. (eds.), 34th International Conference on Foundation of Software Technology and Theoretical Computer Science, FSTTCS 2014, 15–17 December 2014, New Delhi, India. LIPIcs, vol. 29, pp. 47–60. Schloss Dagstuhl-Leibniz-Zentrum fuer Informatik (2014)

[Wol03] de Wolf, R.: Nondeterministic quantum query and communication complexities. SIAM J. Comput. **32**(3), 681–699 (2003)

[Zeh15] Zehavi, M.: Mixing color coding-related techniques. In: Bansal, N., Finocchi, I. (eds.) ESA 2015. LNCS, vol. 9294, pp. 1037–1049. Springer, Heidelberg (2015). https://doi.org/10.1007/978-3-662-48350-3_86

A Survey on Spanning Tree Congestion

Yota Otachi[✉]

Kumamoto University, Kumamoto, Japan
otachi@cs.kumamoto-u.ac.jp

Abstract. In this short survey dedicated to Hans L. Bodlaender on the occasion of his 60th birthday, we review known results and open problems about the spanning tree congestion problem. We focus mostly on the algorithmic results, where his contribution was precious.

Keywords: Spanning tree congestion · Parameterized complexity · Exact algorithm · Approximation algorithm

1 Introduction

The spanning tree congestion of a connected graph is the minimum congestion over all its spanning trees, formally defined as follows. Let $G = (V, E)$ be a connected graph and T a spanning tree of G. For each $e \in E$, the *congestion* of e, denoted $\mathsf{cng}_{G,T}(e)$, is the number of edges in E that connect the two components in $T - e$. The *(edge) congestion* of T is $\mathsf{ec}(G; T) = \max_{e \in E} \mathsf{cng}_{G,T}(e)$. Then, the *spanning tree congestion* of G is $\mathsf{stc}(G) = \min_T \mathsf{cng}_G(T)$, where the minimum is taken over all spanning trees T of G.

The *spanning tree congestion problem* is the problem of finding a spanning tree with the minimum congestion. The concept of this problem was described intuitively as follows [28]:

> Imagine that edges of G are roads and edges of T are those roads that are cleared of snow following a snowstorm. If we assume that for each edge in G there is a flow of traffic between its ends; these flows are the same for each edge, and that after a snowstorm each driver takes the corresponding (unique) detour in T, then $\mathsf{ec}(G; T)$ describes the *traffic congestion* at the most congested road of T. We are interested in spanning trees that minimize the congestion.

In the example above, one may find that minimizing the length of a longest detour is important as well. Such a variant is rather well studied and known as the *tree spanner problem* [9]. Although there are some strong relations between the two problems (e.g., on planar graphs [27]), we will focus only on the spanning tree congestion problem in this survey.

Partially supported by JSPS KAKENHI Grant Numbers JP18H04091, JP18K11168, JP18K11169.

© Springer Nature Switzerland AG 2020

F. V. Fomin et al. (Eds.): Bodlaender Festschrift, LNCS 12160, pp. 165–172, 2020.
https://doi.org/10.1007/978-3-030-42071-0_12

The concept of spanning tree congestion was formally introduced by Ostrovskii [34], but prior to that, Shimonson [38] studied it as an auxiliary tool for approximating cutwidth of graphs. After the introduction by Ostrovskii [34], the problem was studied intensively and many combinatorial results are obtained [10, 19–22, 26–29, 32, 35, 36]. Also, some student projects on the spanning tree congestion of special graphs were done [1]. The algorithmic studies started slightly later [5, 6, 23–25, 33], and as we will see, there still remain some basic questions unanswered. Recently, Chandran et al. [11] have studied upper bounds on the spanning tree congestion of a graph and gave both combinatorial and algorithmic results.

2 Known Results and Open Problems

All graphs considered in this survey are simple, finite, undirected, and connected. We assume that the reader is familiar with graph classes [7,16], width parameters of graphs [4,18], and parameterized complexity [14].

2.1 Extremal Cases

Here we discuss how large the spanning tree congestion of a graph can be. Let $\mu(n) = \max\{\mathsf{stc}(G) \mid G = (V,E), |V| = n\}$ and $\mu(m,n) = \max\{\mathsf{stc}(G) \mid G = (V,E), |E| = m, |V| = n\}$. One can see that $\mu(n) \geq n-1$ as $\mathsf{stc}(K_n) = n-1$ [34], where K_n is the complete graph on n vertices. Also, $\mu(n) \leq \binom{n}{2}$ and $\mu(m,n) \leq m$ trivially hold. It is known that these trivial lower and upper bounds can be much improved.

Ostrovskii [34] constructed a graph with $n = 3\ell - 2\sqrt{\ell}$ vertices and spanning tree congestion at least $\frac{1}{4}\ell^{3/2}$, where the graph consists of three cliques C_1, C_2, C_2 of size ℓ where $|C_1 \cap C_2| = |C_2 \cap C_3| = \sqrt{\ell}$ and $C_1 \cap C_3 = \emptyset$. This implies the following lower bound of $\mu(n)$.

Theorem 2.1 (Ostrovskii [34][1]). $\mu(n) \in \Omega(n^{3/2})$.

Löwenstein et al. [32] showed that this bound is best possible using the Győri-Lovász theorem [17,30].

Theorem 2.2 (Löwenstein et al. [32]). $\mu(n) \leq n^{3/2}$.

Chandran et al. [11] improved the bound $\mu(n) \in \Theta(n^{3/2})$ by replacing the $\Theta(n)$ factor with a $\Theta(\sqrt{m})$ factor. To show the new bound, they used the generalized Győri-Lovász theorem [12] for which they give the first constructive proof.

Theorem 2.3 (Chandran et al. [11]). $\mu(m,n) \in \Theta(\sqrt{mn})$.

---

[1] Ostrovskii [36] later showed that this bound holds even for bipartite graphs.

They presented an $O^*(2^{O(n \log n/\sqrt{m/n})})$-time[2] algorithm for computing a spanning tree with congestion $O(\sqrt{mn})$ and a polynomial-time algorithm for computing a spanning tree with congestion $O(\sqrt{mn} \log n)$.

Problem 2.4. *Is it possible to find a spanning tree with congestion $O(\sqrt{mn})$ in polynomial time?*

For planar graphs, almost tight bounds are known. Let $\mu_p(n) = \max\{\text{stc}(G) \mid \text{planar graph} G = (V, E), |V| = n\}$.

Theorem 2.5 (Law et al. [27]). *For $n \geq 5$,*

$$\mu_p(n) \in \begin{cases} \{n, n+1\} & \text{if } n \text{ is even,} \\ \{n-1, n\} & \text{if } n \text{ is odd.} \end{cases}$$

Problem 2.6. *Close the ± 1 gaps in Theorem 2.5.*

For k-outer planar graphs, Shimonson [38] conjectured that $\text{stc}(G) \leq k \cdot \Delta(G)$. This bound is later proved by Bodlaender et al. [6].

2.2   Graph Classes

We discuss here the problem of finding a spanning tree with the minimum congestion. We call the problem STC. The problem STC is known to be NP-hard in general [31, Section 5.6] and even for planar graphs [37], but not so much is known about polynomial-time solvable cases. We list all known and some unsettled cases.

Shimonson [38] showed (in our terminology) that $\text{stc}(G) \leq \Delta(G) + 1$ for outerplanar graphs G and gave a linear-time algorithm for computing a spanning tree achieving this bound. Bodlaender et al. [6] gave a characterization of the spanning tree congestion using a technique provided by Ostrovskii [35] and presented a linear-time algorithm for computing $\text{stc}(G)$ for outerplanar graphs.

Theorem 2.7 (Bodlaender et al. [6]). *For outerplanar graphs, STC is linear-time solvable.*

They asked whether the result can be generalized for k-outerplanar graphs.

Problem 2.8. *Is STC fixed-parameter tractable for k-outerplanar graphs, when parameterized by k?*

Since k-outerplanar graphs have treewidth at most $3k - 1$ [4], this problem is related to the fixed-parameter tractability question parameterized by treewidth. See Problem 2.16.

We often study graph problems on chordal graphs or on chordal bipartite graphs. It turned out that the problem is hard even on small subclasses.

---

[2] The O^* notation suppresses the factors polynomial in the input size. See [15].

Theorem 2.9 (Okamoto et al. [33]). *STC is NP-hard on chain graphs and on split graphs.*

The hardness on split graphs is not surprising since we know many such examples. However, the hardness on chain graphs is rather surprising (at least to the author). Recall that a bipartite graph $B = (U, V; E)$ is a *chain graph* if it contains no induced $2K_2$; that is, it cannot have two independent edges. To the best of our knowledge, there is no other graph parameter that is difficult to determine on (unweighted) chain graphs. It is known that every chain graph has clique-width at most 3 [8]. It would be natural to ask whether the problem is polynomial-time solvable for graphs of clique-width at most 2; that is, for cographs.

Problem 2.10. *Is STC polynomial-time solvable on cographs?*

Kubo et al. [23–25] studied STC on graphs of small diameter. Among other things, they showed that STC on co-chain graphs is tractable.

Theorem 2.11 (Kubo et al. [23, 25]). *STC is polynomial-time solvable on co-chain graphs.*

One can see that every co-chain graph is a proper interval graph. To get a better understanding of the problem, it would be useful to know the complexity of the problem on proper interval graphs and interval graphs. Note that on trivially perfect graphs (another well-known subclass of interval graphs), the problem becomes trivial as a connected trivially perfect graph has a universal vertex of degree $n - 1$ and a star-shaped spanning tree centered at the universal vertex minimizes the congestion (see [33]).

Problem 2.12. *Is STC polynomial-time solvable on interval graphs or on proper interval graphs?*

2.3 Parameterized Complexity

By k-STC, we denote the parameterized version of STC; that is, given a graph G, k-STC is the problem of deciding whether $\text{stc}(G) \leq k$ parameterized by k. Bodlaender et al. [5] studied the complexity of k-STC and showed some sharp contrasts between tractable and intractable cases. There main results can be summarized as follows.

Theorem 2.13 (Bodlaender et al. [5]). *k-STC is linear-time solvable if one of the following conditions is satisfied:*

- $k \leq 3$,
- *the input graph has bounded degree, or*
- *the input graph is an apex-minor-free graph.*

Theorem 2.14 (Bodlaender et al. [5]). *For $k \geq 8$, k-STC is NP-complete even if all the following conditions are satisfied:*

– the input graph is K_6-minor-free, and
– in the input graph, all but one vertex has degree at most k.

Let $\mathsf{tw}(G)$ denote the *treewidth* of G. To show their positive results (Theorem 2.13), they showed that k-STC can be expressed in MSO_2. This implies, by Courcelles's theorem [13] and Bodlaender's algorithm [3], that if a graph class $\mathcal{G}$ has a constant $c_\mathcal{G}$ such that $\mathsf{stc}(G) \geq c_\mathcal{G} \cdot \mathsf{tw}(G)$ for each $G \in \mathcal{G}$, then k-STC is linear-time solvable on $\mathcal{G}$. For example, we can see that for a chordal graph G, $\mathsf{stc}(G) \geq \omega(G) - 1 = \mathsf{tw}(G)$, where $\omega(G)$ is the maximum clique size of G.

Corollary 2.15. *k-STC is linear-time solvable on chordal graphs.*

The MSO_2 expressibility implies that k-STC is fixed-parameter tractable when parameterized by both k and treewidth. Since the problem is NP-complete when parameterized by k only, it is natural to ask the following question.

Problem 2.16. *Is STC parameterized by treewidth fixed-parameter tractable?*

An obvious gap between Theorems 2.13 and 2.14 is the case where $k \in \{4, ..., 7\}$.

Problem 2.17. *Is k-STC polynomial-time solvable for $4 \leq k \leq 7$?*

The case $k \leq 3$ was solved by showing that $\mathsf{stc}(G) \leq 3$ only if G is planar and that $\mathsf{stc}(G) \geq (\mathsf{tw}(G) - 49)/24$ if G is planar [5]. Such an argument cannot be used for $k \geq 4$ since there are graphs G of unbounded treewidth that still have $\mathsf{stc}(G) = 4$. For example, the graph obtained from a wall by adding a universal vertex satisfies these conditions.

Another related topic is the exact (exponential) complexity. A straightforward approach of trying all edge subsets gives an $O^*(2^m)$-time algorithm. Okamoto et al. [33] presented a faster algorithm by enumerating all possible combinations of cuts and then carefully applying the fast subset convolution method [2].

Theorem 2.18 (Okamoto et al. [33]). *STC can be solved in time $O^*(2^n)$.*

Problem 2.19. *Is the running time $O^*(2^n)$ optimal under some reasonable assumptions?*

2.4 Approximation

There are no non-trivial results known on approximation of spanning tree congestion for general graphs. The hardness of 8-STC can be used to show the following hardness of approximation.

Corollary 2.20 (Bodlaender et al. [5]). *It is NP-hard to approximate the spanning tree congestion within a factor better than $9/8$.*

Observe that $ec(G; T) \leq m - n + 2$ for any spanning tree T of G and that $stc(G) \geq 2m/(n-1) - 1$. These facts imply that every spanning tree gives an $n/2$-approximation. A nontrivial improvement on the approximation ratio would be a nice challenge.

Problem 2.21. *Are there polynomial-time algorithms approximating the spanning tree congestion within factor $o(n)$ or $O(\log n)$?*

For some graph classes of diameter at most 3, Kubo et al. [23–25] presented constant-factor approximation algorithms.

Theorem 2.22 (Kubo et al. [23–25]). *There are polynomial-time approximation algorithms with factor 2 for cographs, and with factor 3 for chain graphs. There is a randomized polynomial-time 2-approximation algorithm for split graphs.*

3   A Personal Story as a Conclusion

The spanning tree congestion problem was the first topic I worked with Hans. In 2009, I visited Hans as a visiting student for six weeks (from the mid-September to the late-October). It was my first research stay abroad and I had never met him before. I was a bit nervous on the first day, but after the first conversation with him I quickly understood that the stay was going to be great. We had meetings almost every day. It was magical for me that whenever I made some progress, he always had a next nice step for me. The stay was quite productive and we ended up with two papers on spanning tree congestion [6,37]. Since then, he has been my great teacher.

Happy birthday Hans! I really hope I will have the next opportunity to work with you soon.

References

1. REU project: California State University, San Bernardino. https://www.math.csusb.edu/reu/studentwork.html. Accessed 13 Sept 2019
2. Björklund, A., Husfeldt, T., Kaski, P., Koivisto, M.: Fourier meets Möbius: fast subset convolution. STOC **2007**, 67–74 (2007). https://doi.org/10.1145/1250790.1250801
3. Bodlaender, H.L.: A linear-time algorithm for finding tree-decompositions of small treewidth. SIAM J. Comput. **25**(6), 1305–1317 (1996). https://doi.org/10.1137/S0097539793251219
4. Bodlaender, H.L.: A partial k-arboretum of graphs with bounded treewidth. Theor. Comput. Sci. **209**(1–2), 1–45 (1998). https://doi.org/10.1016/S0304-3975(97)00228-4
5. Bodlaender, H.L., Fomin, F.V., Golovach, P.A., Otachi, Y., van Leeuwen, E.J.: Parameterized complexity of the spanning tree congestion problem. Algorithmica **64**(1), 85–111 (2012)

6. Bodlaender, H.L., Kozawa, K., Matsushima, T., Otachi, Y.: Spanning tree congestion of k-outerplanar graphs. Discrete Math. **311**(12), 1040–1045 (2011). https://doi.org/10.1016/j.disc.2011.03.002

7. Brandstädt, A., Le, V.B., Spinrad, J.P.: Graph Classes: A Survey. SIAM (1999)

8. Brandstädt, A., Lozin, V.V.: On the linear structure and clique-width of bipartite permutation graphs. Ars Comb. **67**, 273–281 (2003)

9. Cai, L., Corneil, D.G.: Tree spanners. SIAM J. Discrete Math. **8**(3), 359–387 (1995). https://doi.org/10.1137/S0895480192237403

10. Castejón, A., Ostrovskii, M.I.: Minimum congestion spanning trees of grids and discrete toruses. Discuss. Math. Graph Theory **29**(3), 511–519 (2009). https://doi.org/10.7151/dmgt.1461

11. Chandran, L.S., Cheung, Y.K., Issac, D.: Spanning tree congestion and computation of generalized Győri-Lovász partition. In: ICALP 2018. LIPIcs, vol. 107, pp. 32:1–32:14 (2018). https://doi.org/10.4230/LIPIcs.ICALP.2018.32

12. Chen, J., Kleinberg, R.D., Lovász, L., Rajaraman, R., Sundaram, R., Vetta, A.: (Almost) tight bounds and existence theorems for single-commodity confluent flows. J. ACM **54**(4), 16 (2007). https://doi.org/10.1145/1255443.1255444

13. Courcelle, B.: The monadic second-order logic of graphs III: tree-decompositions, minor and complexity issues. Theor. Inform. Appl. **26**, 257–286 (1992). https://doi.org/10.1051/ita/1992260302571

14. Cygan, M., Fomin, F.V., Kowalik, L., Lokshtanov, D., Marx, D., Pilipczuk, M., Pilipczuk, M., Saurabh, S.: Parameterized Algorithms. Springer, Heidelberg (2015). https://doi.org/10.1007/978-3-319-21275-3

15. Fomin, F.V., Kratsch, D.: Exact Exponential Algorithms. Springer, Heidelberg (2010). https://doi.org/10.1007/978-3-642-16533-7

16. Golumbic, M.C.: Algorithmic Graph Theory and Perfect Graphs. Annals of Discrete Mathematics, 2nd edn., vol. 57, North Holland (2004)

17. Győri, E.: On division of graphs to connected subgraphs. Combinatorics, pp. 485–494 (1978). Proceedings of Fifth Hungarian Combinatorial Coll., 1976, Keszthely

18. Hliněný, P., Oum, S.-I., Seese, D., Gottlob, G.: Width parameters beyond treewidth and their applications. Comput. J. **51**(3), 326–362 (2008). https://doi.org/10.1093/comjnl/bxm052

19. Hruska, S.W.: On tree congestion of graphs. Discrete Math. **308**(10), 1801–1809 (2008). https://doi.org/10.1016/j.disc.2007.04.030

20. Kozawa, K., Otachi, Y.: On spanning tree congestion of hamming graphs. CoRR, abs/1110.1304 (2011). arXiv:1110.1304

21. Kozawa, K., Otachi, Y.: Spanning tree congestion of Rook's graphs. Discuss. Math. Graph Theory **31**(4), 753–761 (2011). https://doi.org/10.7151/dmgt.1577

22. Kozawa, K., Otachi, Y., Yamazaki, K.: On spanning tree congestion of graphs. Discrete Math. **309**(13), 4215–4224 (2009). https://doi.org/10.1016/j.disc.2008.12.021

23. Kohei, K.: Spanning tree congestion problem on graphs of small diameter. Master's thesis, Kyushu University (2015). (in Japanese)

24. Kubo, K., Yamauchi, Y., Kijima, S., Yamashita, M.: Approximating the spanning tree congestion for split graphs with iterative rounding. IPSJ SIG Technical Report, 2014-AL-150(17), 1–8 (2014). http://id.nii.ac.jp/1001/00106885/. (in Japanese)

25. Kubo, K., Yamauchi, Y., Kijima, S., Yamashita, M.: Spanning tree congestion problem on graphs of small diameter. RIMS Kôkyûroku **1941**, 17–21 (2015). http://www.kurims.kyoto-u.ac.jp/~kyodo/kokyuroku/contents/pdf/1941-03.pdf. (in Japanese)

26. Law, H.-F.: Spanning tree congestion of the hypercube. Discrete Math. **309**(23–24), 6644–6648 (2009). https://doi.org/10.1016/j.disc.2009.07.007

27. Law, H.-F., Leung, S.L., Ostrovskii, M.I.: Spanning tree congestion of planar graphs. Involve. A J. Math. **7**(2), 205–226 (2014). https://doi.org/10.2140/involve.2014.7.205

28. Law, H.-F., Ostrovskii, M.I.: Spanning tree congestion: duality and isoperimetry; with an application to multipartite graphs. Graph Theory Notes N. Y. **58**, 18–26 (2010). http://gtn.kazlow.info/GTN58.pdf

29. Law, H.-F., Ostrovskii, M.I.: Spanning tree congestion of some product graphs. Indian J. Math. **52**(Suppl.), 103–111 (2010)

30. Lovász, L.: A homology theory for spanning tress of a graph. Acta Math. Acad. Sci. Hung. **30**(3–4), 241–251 (1977). https://doi.org/10.1007/BF01896190

31. Löwenstein, C.: In the complement of a dominating set. Ph.D. thesis, Technische Universität Ilmenau (2010). https://nbn-resolving.org/urn:nbn:de:gbv:ilm1-2010000233

32. Löwenstein, C., Rautenbach, D., Regen, F.: On spanning tree congestion. Discrete Math. **309**(13), 4653–4655 (2009). https://doi.org/10.1016/j.disc.2009.01.012

33. Okamoto, Y., Otachi, Y., Uehara, R., Uno, T.: Hardness results and an exact exponential algorithm for the spanning tree congestion problem. J. Graph Algorithms Appl. **15**(6), 727–751 (2011). https://doi.org/10.7155/jgaa.00246

34. Ostrovskii, M.I.: Minimal congestion trees. Discrete Math. **285**(1–3), 219–226 (2004). https://doi.org/10.1016/j.disc.2004.02.009

35. Ostrovskii, M.I.: Minimum congestion spanning trees in planar graphs. Discrete Math. **310**(6–7), 1204–1209 (2010). https://doi.org/10.1016/j.disc.2009.11.016

36. Ostrovskii, M.I.: Minimum congestion spanning trees in bipartite and random graphs. Acta Mathematica Scientia **31B**(2), 634–640 (2011). https://doi.org/10.1016/S0252-9602(11)60263-4

37. Otachi, Y., Bodlaender, H.L., van Leeuwen, E.J.: Complexity results for the spanning tree congestion problem. In: Thilikos, D.M. (ed.) WG 2010. LNCS, vol. 6410, pp. 3–14. Springer, Heidelberg (2010). https://doi.org/10.1007/978-3-642-16926-7_3

38. Simonson, S.: A variation on the min cut linear arrangement problem. Math. Syst. Theory **20**(4), 235–252 (1987). https://doi.org/10.1007/BF01692067

Surprising Applications of Treewidth Bounds for Planar Graphs

Marcin Pilipczuk[✉]

Institute of Informatics, University of Warsaw, Warsaw, Poland
`malcin@mimuw.edu.pl`

Abstract. In this chapter we continue a story told from in Dániel Marx's Chapter and present three examples of surprising use of treewidth. In all cases, we present a state-of-the-art and provably optimal (assuming the Exponential Time Hypothesis) algorithm exploiting a sublinear bound on the treewidth of an auxiliary graph taken out of the blue.

Keywords: Planar graphs · Subexponential algorithm

1 Introduction

In this chapter we will show three unexpected examples of the use of the following cornerstone statement.

Theorem 1. *An n-vertex planar graph has treewidth $\mathcal{O}(\sqrt{n})$.*

In other words, this chapter continues the story presented in [5] on surprising usages of treewidth. In all our examples, the graph we apply the sublinear bound of Theorem 1 to is not visible at a first glance, but comes out of the blue.

In fact, all three applications that we present here make use of some extra topological properties that can be inferred about the tree decomposition of width $\mathcal{O}(\sqrt{n})$ promised by Theorem 1.

To describe these topological properties, we need some basic terminology. Consider a plane graph G, that is, a graph G embedded in the Euclidean plane $\mathbb{R}^2$; we will rely here on the intuitive understanding of the notions of an embedding, faces, etc. A *noose* is a closed Jordan curve without self-intersections that intersects the drawing of G only in vertices. A noose γ is *simple* if it visits every face of G at most once. Observe that every noose separates the plane into two regions, one inside the noose and one outside of the noose. Consequently, a noose γ separates $V(G)$ into three parts: vertices lying inside γ, outside γ, and on the noose γ.

The property of planar graphs that is absolutely crucial for this chapter is the following incarnation of the Cycle Separator Theorem of Miller [9].

© Springer Nature Switzerland AG 2020
F. V. Fomin et al. (Eds.): Bodlaender Festschrift, LNCS 12160, pp. 173–188, 2020.
https://doi.org/10.1007/978-3-030-42071-0_13

Theorem 2 (Theorem 1 of [9] applied to the face-vertex graph of G).
If G is an n-vertex plane graph with ℓ faces, then there exists a noose γ in G that visits $\mathcal{O}(\sqrt{n})$ vertices of G and at most $2\ell/3$ faces of G lie entirely inside γ and at most $2\ell/3$ faces of G lie entirely outside γ. Furthermore, if G is 2-edge-connected, then γ can be assumed to be simple.

We will call such a curve γ a *balanced separator*.

One can leverage Theorem 2 to prove Theorem 1 with the extra property that every bag in the tree decomposition is associated with a noose γ such that the bag is exactly the set of vertices traversed by γ. Such a decomposition, called a *sphere-cut decomposition* in the literature, turned out to be extremely useful in algorithmic design, see e.g. [3,4].

In this chapter, the main focus is on the choice of the graph G to apply Theorem 2 to, as it is surprising and highly nonobvious in all three examples. To hide tedious details of the internals of dynamic programming algorithms on tree decompositions, in the first two examples we will rely on Theorem 2 and present the algorithms in the divide-and-conquer paradigm.

Finally, we remark that in all three presented cases, matching lower bounds based on Exponential Time Hypothesis are known. That is, the presented algorithms are not only some improvements obtained by the same technique, but probably "the best" ways of solving the problems in questions.

2   Scattered Set and Voronoi Diagrams

Our first example considers the SCATTERED SET problem. Here, we are given an undirected graph G together with edge weights $\mathbf{w} : E(G) \to \mathbb{Z}_+$ (that is, the edge weights are positive integers) and two integers k and d. The goal is to check if G contains a d-scattered set of size k: a set $X \subseteq V(G)$ of k vertices such that every two distinct vertices from X are within distance strictly larger than d.

We sketch the proof of the following result of Marx and Pilipczuk [8].

Theorem 3. *On n-vertex planar graphs* SCATTERED SET *can be solved in time $n^{\mathcal{O}(\sqrt{k})}$.*

Here, we assume that basic operations on edge weights (addition, subtraction, comparison) can be done in constant time. Consequently, there is no dependence on d in the running time bound.

We start with some simplification and preprocessing steps. First, we can assume that the input graph G is simple: loops can be safely deleted and among parallel edges only the one with the one with the smallest weight counts. Second, by adding edges of weight $d+1$, we can turn G into a triangulation: a connected plane graph where every face is a triangle (a cycle with three edges). Third, by multiplying edge weights (and the parameter d) by a large factor and adding small perturbations to edge weights, we can assume that all shortest paths are unique and of unique weight (i.e., for every distinct $u, v \in V(G)$, there is a unique shortest path from u to v and, furthermore, the weight of this path uniquely

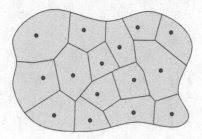

Fig. 1. An example of a Voronoi diagram.

determines the pair $\{u, v\}$). Note that a straightforward way to ensure the above is to replace the weight of the i-th edge e_i with $\mathbf{w}(e_i) \cdot 2^{|E(G)|} + 2^{i-1}$ (and d by $d \cdot (2^{|E(G)|} + 1) - 1$) and this will incurr a polynomial-in-n multiplicative factor in the running time bound due to the arithmetic operations on the increased edge weights. Finally, we will be solving a slightly more general problem where we are additionally given a set $A \subseteq V(G)$ of *allowed vertices* and we seek a d-scattered set $X \subseteq A$ of maximum possible size, but not larger than k (i.e., if there is no d-scattered set of size k contained in A, we want a d-scattered set of the largest possible cardinality).

In what follows, the crucial notions are of *Voronoi cell* and *Voronoi diagram*. A set $X \subseteq V(G)$ induces a partition of $V(G)$ into *Voronoi cells* $(\mathrm{Cell}(x))_{x \in X}$ defined as follows: for $x \in X$ the cell $\mathrm{Cell}(x) \subseteq V(G)$ consists of those vertices $y \in V(G)$ for which x is the closest to y among all vertices from X. Since the distances in G are unique, there are no ties in the above definition and $x \in \mathrm{Cell}(x)$ for every $x \in X$.

The partition of $V(G)$ into Voronoi cells induces a *Voronoi diagram* in the dual graph G^*. Recall that a dual of a plane graph contains a vertex for every face of the primal graph and every edge of the primal graph projects to an edge in the dual graph connecting the two incident faces. Since G is a triangulation, the dual G^* is a 3-regular plane graph.

The Voronoi diagram Diag_X° is the subgraph of G^* consisting of those edges of G^* that in the primal graph connect vertices from two distinct Voronoi cells. If G is a triangulation, then Diag_X° is a subdivision of a 3-regular plane graph, where vertices of degree 3 correspond to faces of G where three distinct Voronoi cells meet, i.e., where all three vertices belong to three distinct Voronoi cells. Let Diag_X be the graph Diag_X° with all vertices of degree 2 suppressed; Diag_X is a 3-regular plane graph. See Fig. 1 for an example.

Note that both Diag_X° and Diag_X may be disconnected, contain parallel edges or loops, but every connected component of Diag_X is 2-connected. In what follows, we ignore the nuisance of Diag_X being disconnected and assume that it is a 2-connected 3-regular plane graph.

The graph Diag_X° (and thus Diag_X) has exactly $|X|$ faces, one for each Voronoi cell. Consequently, Diag_X° has $\mathcal{O}(|X|)$ vertices of degree 3 and Diag_X has $\mathcal{O}(|X|)$ vertices and edges. Hence, both these graphs have treewidth $\mathcal{O}(\sqrt{|X|})$

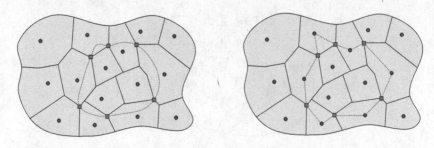

Fig. 2. A balanced separator γ (left) and the corresponding walk W_γ (right).

and a balanced separator γ (in the sense of Theorem 2) of size $\mathcal{O}(\sqrt{|X|})$; this will be the crucial property we use later.

Consider now an optimum solution X to the input (G, A, k) (that is, $X \subseteq A$ is a d-scattered set in G of maximum possible size, but not larger than k) and let us inspect the graph Diag_X. By Theorem 2, there exists a noose γ for Diag_X that visits $\mathcal{O}(\sqrt{|X|}) = \mathcal{O}(\sqrt{k})$ vertices of Diag_X and that leaves at most $2|X|/3$ faces corresponding to the vertices of X inside γ and at most $2|X|/3$ faces corresponding to the vertices of X outside of γ. Let X_γ be the set of vertices of X corresponding to the faces visited by γ. See the left panel of Fig. 2 for an illustration.

With the noose γ we can associate a walk W_γ in G defined as follows: whenever γ traverses a vertex $f \in V(\mathrm{Diag}_X)$ (which is a triangular face of G) from a Voronoi cell of a vertex $x \in X_\gamma$ to a Voronoi cell of a vertex $y \in X_\gamma$, we denote by v_x and v_y the vertices incident with f that belong to the Voronoi cells of x and y, respectively, and lead the walk W_γ via the shortest path from x to v_x, via the edge $v_x v_y$, and via the shortest path from v_y to y. See the right panel of Fig. 2 for an illustration.[1]

Since we ensured the unique shortest paths property in G, the knowledge of the tuple (x, v_x, v_y, y) fully determines the part of the walk W_γ from x to y. Consequently, the knowledge of all tuples (x, v_x, v_y, y) for all vertices of Diag_X visited by W_γ fully determines the walk W_γ. Hence, there are at most $n^{\mathcal{O}(\sqrt{k})}$ options for the walk W_γ. More formally, we have the following statement.

Claim. One can enumerate in $n^{\mathcal{O}(\sqrt{k})}$ time a family $\mathcal{W}$ of $n^{\mathcal{O}(\sqrt{k})}$ pairs (X_W, W) where W is a walk in G and $X_W \subseteq V(G)$ such that $\mathcal{W}$ contains the pair (X_γ, W_γ).

We invoke the algorithm of Claim 2 and branch into $|\mathcal{W}| = n^{\mathcal{O}(\sqrt{k})}$ options, considering each $(X_W, W) \in \mathcal{W}$ as potential (X_γ, W_γ). Thus, in what follows, we can assume that the pair (X_γ, W_γ) is known to the algorithm.

Observe that W_γ is "almost" a cycle: since we assumed that Diag_X is 2-connected, γ is simple, and the only anomalies W_γ may have is when it visits its $x \in X$ via two shortest paths that overlap for a while. However, such an

---

[1] Both Figs. 1 and 2 are due to Dániel Marx and Michał Pilipczuk, from their slides on the work [8], reproduced with authors' permission.

"almost"cycle W_γ allows us to define the part of the graph G inside and outside W_γ (with the edges and vertices of W_γ belonging to both parts) with the following properties: for every face of Diag_X that is inside (outside) γ, the corresponding cell in G is in the part of G inside (outside) W_γ. We denote these parts by G^1 and G^2, respectively.

Let $X^i := V(G^i) \cap (X \setminus X_\gamma)$, so that $X^1 \cup X^2 \cup X^\gamma$ is a partition of X. By the properties of γ from Theorem 2, we have that $|X^1|, |X^2| \leq 2k/3$.

We would like now to recurse on two instances, $(G, A \cap V(G^i), \lfloor 2k/3 \rfloor)$ for $i = 1, 2$. However, we need to ensure that while merging the solutions found for the subinstances, the vertices chosen in the first subinstance are within distance more than d from the vertices chosen in the second subinstance, as well as from the vertices of X_γ. Luckily, we can easily ensure this using the way W_γ is constructed.

For every $v \in V(W_\gamma)$, let $x_v \in X_\gamma$ be the vertex of X_γ closest to v. Observe that $v \in \text{Cell}(x_v)$ and that:

Claim. Let $i \in \{1, 2\}$, $y \in X^i$, and $v \in V(W_\gamma)$. Then,

1. $\text{dist}_G(y, X_\gamma) > d$,
2. $\text{dist}_G(v, y) > \text{dist}(v, X_\gamma)$, and
3. $\text{dist}_G(v, y) > d/2$.

Proof. The first claim follows from the fact that X is a d-scattered set. The second claim follows from the fact that $v \in \text{Cell}(x_v)$ and $x_v \in X_\gamma$. Finally, the last claim follows from the first two and the triangle inequality.

Claim 2 motivates the following definition for $i = 1, 2$:

$$A^i = \{u \in A \cap V(G^i) \mid \text{dist}_G(u, V(W_\gamma)) > d/2 \wedge \text{dist}_G(u, X_\gamma) > d\}.$$

The algorithm invokes itself recursively to $(G, A^i, \lfloor 2k/3 \rfloor)$, obtaining a d-scattered set Y^i for $i = 1, 2$, and returns the set $Y^1 \cup Y^2 \cup X_\gamma$.[2] To show the correctness of the algorithm, we observe the following two statements.

Claim. $|Y^i| \geq |X^i|$ for $i = 1, 2$.

Proof. Claim 2 ensures that $X^i \subseteq A^i$, making X^i a feasible output for the recursive call. Furthermore, Theorem 2 ensures that $|X^i| \leq 2k/3$.

Claim. $Y^1 \cup Y^2 \cup X_\gamma$ is a d-scattered set.

Proof. The claim follows from the definition of A^i, as every vertex of $A^1 \cup A^2$ is within distance more than d from X_γ and any path from a vertex in A^1 to a vertex in A^2 crosses W_γ, and thus consists of two parts of length more than $d/2$ each by Claim 2.

---

[2] Technically speaking, the largest of the sets $Y^1 \cup Y^2 \cup X_\gamma$ among all guesses of which element of W is the pair (X_γ, W_γ), and a negative answer if no found set is of size at most k.

We conclude with complexity analysis. A recursive call on (G, A, k) results in $n^{\mathcal{O}(\sqrt{k})}$ recursive subcalls on $(G, A', \lfloor 2k/3 \rfloor)$. Thus, the size of the recursion tree is bounded by

$$n^{c\sqrt{k}} \cdot n^{c\sqrt{(2/3) \cdot k}} \cdot n^{c\sqrt{(2/3)^2 \cdot k}} \cdot \ldots \leq n^{\frac{c}{1 - \sqrt{2/3}} \cdot \sqrt{k}}$$

for some universal constant c. Beside the recursion, each recursive call performs polynomial-time computations. The running time bound follows.

This concludes the proof of Theorem 3.

3   Geometric Halfspace Cover and Uncovered Polytope

The second example is the most geometric one and considers HALFSPACE COVER: given a set U of points in $\mathbb{R}^3$ and a set of halfspaces $\mathcal{A}$, find the minimum size of a set $\mathcal{B} \subseteq \mathcal{A}$ that covers U, that is, $U \subseteq \bigcup_{A \in \mathcal{B}} A$. We are interested in the parameterized decision version of the problem where we are additionally given a parameter k and ask for the existence of a solution $\mathcal{B}$ of cardinality at most k.

We sketch the proof of the following result of Bringmann, Kisfaludi-Bak, Pilipczuk, and van Leeuwen [1]:

Theorem 4. HALFSPACE COVER *can be solved in time* $|\mathcal{A}|^{\mathcal{O}(\sqrt{k})} \cdot |U|^{\mathcal{O}(1)}$.

We assume here that all halfspaces are closed, that is, formally a halfspace A is defined by a vector $v_A \in \mathbb{R}^3$ and a threshold $b_A \in \mathbb{R}$ and consists of all points $x \in \mathbb{R}^3$ such that $v_A \cdot x \geq b_A$, where $(\cdot)$ denotes here the scalar product. We say that a halfspace A *covers* a point p if $p \in A$. We will also freely alternate between the definition of a halfspace A as a subset of $\mathbb{R}^3$ and a pair (v_A, b_A). Furthermore, we assume that all basic operations on real numbers and real vectors can be done in constant time, in particular we ignore all representation issues.

As in the previous example, we will design a divide-and-conquer algorithm that at every step uses Theorem 2 to find a separator that splits the instance into a number of subinstances with the parameter k decreased multiplicatively. However, here the input to the problem does not even contain a graph, so finding the correct graph to apply Theorem 2 is even more challenging.

We start with describing the base of the recursion. We will use the following result of Danzer, Grünbaum, and Klee [2] that asserts that one cannot nontrivially cover the entire space $\mathbb{R}^3$ with halfspaces.

Theorem 5. *Let $\mathcal{B}$ be a set of halfspaces that covers the entire space, that is, $\mathbb{R}^3 = \bigcup_{A \in \mathcal{B}} A$. Then there exists a subset $\mathcal{B}' \subseteq \mathcal{B}$ of cardinality at most 4 such that already $\mathbb{R}^3 = \bigcup_{A \in \mathcal{B}'} A$.*

Our algorithm starts with checking by brute-force in time $\mathcal{O}(|\mathcal{A}|^4 \cdot |U|)$ if there is a solution of cardinality at most 4. If this is the case, then we return the smallest such solution. Also, if no solution is found and $k \leq 4$, we return a negative answer.

In what follows we can assume that $k > 4$ and that there is no solution of cardinality at most 4. From Theorem 5 we infer that any solution $\mathcal{B}$ satisfies $\mathbb{R}^3 \neq \bigcup_{A \in \mathcal{B}} A$, that is,

$$\mathbf{P} := \mathbb{R}^3 \setminus \bigcup_{A \in \mathcal{B}} A$$

is a nonempty (open) polytope.

We handwave two standardization assumptions. First, by adding some guard halfspaces to $\mathcal{B}$, we may assume that $\mathbf{P}$ is bounded. Second, by slightly perturbing each halfspace, we can assume that no two halfspaces have parallel boundaries (have parallel vectors v_A) and there are no four halfspaces in $\mathcal{A}$ whose boundaries intersect in a common point. We omit here the technical details of these standardization steps.

Assume that we are given a yes-instance and let $\mathcal{B}$ be a minimum-size solution. The minimality of $\mathcal{B}$ implies a one-to-one correspondence between faces of $\mathbf{P}$ and the elements of $\mathcal{B}$. Furthermore, the standardization steps ensure that every vertex of $\mathbf{P}$ is incident to exactly three faces of $\mathbf{P}$.

Observe that $\mathbf{P}$ induces a graph $G_{\mathbf{P}}$ with vertices being the vertices of $\mathbf{P}$ and edges being the intersections of boundaries of two halfspaces of $\mathcal{B}$ that meet in a vertex. The graph $G_{\mathbf{P}}$ is 3-regular, 2-connected, plane, simple, and has $|\mathcal{B}| \leq k$ faces. Consequently, Theorem 2 implies that there is a noose γ in $G_{\mathbf{P}}$ that visits a set X_γ of $\mathcal{O}(\sqrt{|\mathcal{B}|}) = \mathcal{O}(\sqrt{k})$ vertices of $G_{\mathbf{P}}$. Let $\mathcal{B}_\gamma \subseteq \mathcal{B}$ be the set of halfspaces corresponding to the faces visited by γ and incident with all vertices visited by γ. Finally, let $U_\gamma \subseteq U$ be the set of points covered by $\mathcal{B}_\gamma$, that is, $U_\gamma = U \cap \bigcup_{A \in \mathcal{B}_\gamma} A$.

Since $|X_\gamma|, |\mathcal{B}_\gamma| = \mathcal{O}(\sqrt{k})$, there are $|\mathcal{A}|^{\mathcal{O}(\sqrt{k})}$ options for these two sets (note that every $x \in X_\gamma$ can be defined by three halfspaces in $\mathcal{A}$ that contain x in their boundaries). By iterating over all options, the algorithm guesses X_γ and $\mathcal{B}_\gamma$. Thus, henceforth we assume that they are known to the algorithm. In particular, the set U_γ (infered from $\mathcal{B}_\gamma$) is covered by halfspaces from $\mathcal{B}_\gamma$ and no longer needs to be covered.

The curve γ partitions G_γ into two parts and, consequently, partitions $\mathcal{B} \setminus \mathcal{B}_\gamma$ into two sets, $\mathcal{B}^1$ and $\mathcal{B}^2$, corresponding to the faces of $G_{\mathbf{P}}$ on the two sides of γ. The balancedness property of Theorem 2 ensures that $|\mathcal{B}^1|, |\mathcal{B}^2| \leq 2k/3$. We would like to partition $U \setminus U_\gamma$ into two sets corresponding to the points covered by halfspaces in $\mathcal{B}^1$ and $\mathcal{B}^2$, respectively. We could then recurse on these two sets with parameter $2k/3$, giving the desired divide-and-conquer step. To this end, we need to deepen our understanding on how γ partitioned the input instance.

First, we observe that the partition $\mathcal{B} = \mathcal{B}_\gamma \uplus \mathcal{B}^1 \uplus \mathcal{B}^2$ induces also a partition of U.

Claim. If a point $p \in U$ is covered by a halfspace from $\mathcal{B}^1$ and a halfspace from $\mathcal{B}^2$, then it is also covered by a halfspace from $\mathcal{B}_\gamma$, that is, $p \in U_\gamma$.

Proof. Let A^1 and A^2 be the halfspaces covering p from $\mathcal{B}^1$ and $\mathcal{B}^2$. For $i = 1, 2$, let q^i be any internal point on the face of $\mathbf{P}$ corresponding to A^i. The convexity of

P ensures that q^1 is not covered by $\mathcal{B}^2$ and q^2 is not covered by $\mathcal{B}^1$. In particular, p, q^1, and q^2 are three distinct points.

Consider a plane π passing through the points p, q^1, and q^2. Since q^2 is not covered by $\mathcal{B}^1$, the intersection of π with the boundary of A^1 is a line ℓ^1 containing q^1. Similarly, the intersection of π with the boundary of A^2 is a line ℓ^2 containing q^2.

Let **Q** be the convex polygon being the intersection of **P** and π. For $i = 1, 2$, the line ℓ^i contains a side of **Q** that contains q^i in its interior. Furthermore, since p is covered by both $\mathcal{B}^1$ and $\mathcal{B}^2$, it lies on the other side than **Q** of both ℓ^1 and ℓ^2; see Fig. 3. Hence, one can walk around **Q** from the side contained in ℓ^1 to the side contained in ℓ^2 in such a way that for every visited side ℓ corresponding to a halfspace A, the point p is on the other side of ℓ than the polygon **Q**. Since $A^1 \in \mathcal{B}^1$ and $A^2 \in \mathcal{B}^2$ while no two faces corresponding to a halfspace in $\mathcal{B}^1$ and in $\mathcal{B}^2$ meet in **P**, there is a visited side ℓ_0 and a corresponding halfspace A_0 satisfying $A_0 \in \mathcal{B}_\gamma$. But then p is covered by A_0, as desired.

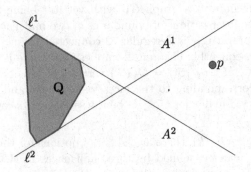

Fig. 3. Proof of Claim 3.

Claim 3 implies that $U \setminus U_\gamma$ can be partitioned into two parts U^1 and U^2: points covered only by halfspaces from $\mathcal{B}^1$ and points covered only by halfspaces from $\mathcal{B}^2$. However, the knowledge of X_γ and $\mathcal{B}_\gamma$ does not immediately give us the partition of $U \setminus U_\gamma$ into U^1 and U^2, and we need to work a bit more.

The knowledge of the points X_γ allow us to define an open polytope $\mathbf{P}_\gamma = \mathrm{int}\,\mathrm{conv}(X_\gamma)$, which is a subset of the polytope **P**. To make the interior operation meaningful, we need $\mathbf{P}_\gamma$ to be a three-dimensional object, that is, not contained in any two-dimensional subspace of $\mathbb{R}^3$. If this is not the case, we guess an additional set Y_γ of at most four vertices of **P**, such that $\mathbf{P}_\gamma = \mathrm{int}\,\mathrm{conv}(X_\gamma \cup Y_\gamma)$ is a three-dimensional (open) polytope.

Consider two points $p^1, p^2 \in U \setminus U_\gamma$. The following observation is straightforward.

Claim. If p^1 and p^2 are two points of $U \setminus U_\gamma$ covered by the same halfspace $A \in \mathcal{B}$, then the straight line segment from p^1 to p^2 does not intersect the interior of **P**, so in particular it does not intersect the interior of $\mathbf{P}_\gamma$.

The crucial insight is that if $p^1 \in U^1$ and $p^2 \in U^2$, then the segment from p^1 to p^2 necessarily crosses the interior of $\mathbf{P}_\gamma$. We prove it in a series of two claims.

Claim. For every $p^1 \in U^1$ and $p^2 \in U^2$, the straight line segment from p^1 to p^2 crosses the interior of $\mathbf{P}$.

Proof. Assume the contrary. Since $\mathbf{P}$ is convex and open, there exists a vector $v \in \mathbb{R}^3$ and a threshold $b \in \mathbb{R}$ such that

$$v \cdot p^1 \geq b \qquad \text{and} \qquad v \cdot p^2 \geq b,$$

but

$$v \cdot q < b \qquad \text{for every } q \in \mathbf{P}.$$

We can chose b so that there exists a vertex q_0 of $\mathbf{P}$ with

$$v \cdot q_0 = b.$$

Let A_1, A_2, and A_3 be the three halfspaces of $\mathcal{B}$ corresponding to the faces meeting at q. For $i = 1, 2, 3$, let A_i be defined as $\{p \in \mathbb{R}^3 \mid p \cdot v_i \geq b_i\}$. Then, there exist nonnegative coefficients α_1, α_2, and α_3 such that

$$v = \alpha_1 v_1 + \alpha_2 v_2 + \alpha_3 v_3.$$

Since q_0 lies on the boundary of all A_i and $v \cdot q_0 = b$, we have

$$b = \alpha_1 b_1 + \alpha_2 b_2 + \alpha_3 b_3.$$

Since $v \cdot p^1 \geq b$ and $v \cdot p^2 \geq b$, there exists $i, j \in \{1, 2, 3\}$ such that

$$v_i \cdot p^1 \geq b_i \qquad \text{and} \qquad v_j \cdot p^2 \geq b_j.$$

However, this implies that $A_i \in \mathcal{B}^1$ while $A_j \in \mathcal{B}^2$, a contradiction as no two faces corresponding to halfspaces from $\mathcal{B}^1$ and $\mathcal{B}^2$ meet in one vertex.

Claim. For every $p^1 \in U^1$ and $p^2 \in U^2$, the straight line segment from p^1 to p^2 crosses the interior of $\mathbf{P}_\gamma$.

Proof. From Claim 3 we already know that the segment from p^1 to p^2 crosses the interior of $\mathbf{P}$; let this intersection be an open segment from q^1 to q^2 where q^1 lies closer to p^1 than q^2.

Note that q^1 may lie on an edge of $\mathbf{P}$ or even be a vertex of $\mathbf{P}$. We claim that *all* (at most three) faces containing q^1 correspond to halfspaces of $\mathcal{B}^1$. Recall that the segment from p^1 to p^2 crosses the interior of $\mathbf{P}$, so there exists a point q on this segment such that $v \cdot q < b$ for every $A = (v, b) \in \mathcal{B}$; note that q lies strictly between q^1 and q^2. On the other hand, for every $A = (v, b) \in \mathcal{B}$ that contains q^1 in its corresponding face, we have $v \cdot q^1 = b$. Consequently, we have $v \cdot p' > b$ for every p' between p^1 (inclusive) and q^1 (exclusive). In particular, A covers p^1. Since $p^1 \in U^1$, we have $A \in \mathcal{B}^1$. Symmetrically, all faces containing q^2 correspond to halfspaces of $\mathcal{B}^2$. Consequently, a small neighborhood of q^1 on the segment between q^1 and q^2 is contained in a different connected component of $\mathbf{P} \setminus \mathbf{P}_\gamma$ than an analogous small neighborhood of q^2. This concludes the proof.

Claim 3 motivates the following step in our algorithm. Construct an auxiliary graph H with vertex set $U \setminus U_\gamma$ and two points p^1, p^2 connected by an edge if the straight line segment from p^1 to p^2 does not intersect the interior of $\mathbf{P}_\gamma$. Claim 3 ensures that the points of U^1 and of U^2 belong to different connected components of H. On the other hand, Claim 3 implies that the points covered by a single halfspace form a clique in H, in particular, they stay in the same connected component. We expect at most $|\mathcal{B} \setminus \mathcal{B}_\gamma| \leq k$ connected components of H and every vertex set U^0 of a connected component of H to be covered by at most $2k/3$ elements of $\mathcal{B}$.

We recurse on instances $(\mathcal{A}, U^0, \lfloor 2k/3 \rfloor)$ for all vertex sets U^0 of connected components of H and output the union of U_γ and all solutions output by the recursive calls. We return a negative answer if any of the recursive calls returns a negative answer or if the final union is of cardinality larger than k.[3] Since a recursive call branches into $|\mathcal{A}|^{\mathcal{O}(\sqrt{k})} \cdot k$ subcases with k decreased to $\lfloor 2k/3 \rfloor$ and we can assume $k \leq |\mathcal{A}|$ (as otherwise the problem can be solved trivially), for some universal constant c the running time of the algorithm can be bounded by

$$|\mathcal{A}|^{\mathcal{O}(1)} \cdot |U|^{\mathcal{O}(1)} \cdot \left(|\mathcal{A}|^{c\sqrt{k}} + |\mathcal{A}|^{c\sqrt{(2/3) \cdot k}} + |\mathcal{A}|^{c\sqrt{(2/3)^2 \cdot k}} \cdots \right)$$

$$= |\mathcal{A}|^{\frac{c}{1 - \sqrt{2/3}} \cdot \sqrt{k} + \mathcal{O}(1)} |U|^{\mathcal{O}(1)}.$$

This concludes the proof of Theorem 4.

4   Steiner Tree and a Local Search Optimum

In our last example we look at the STEINER TREE problem: given an undirected graph G with positive edge weights $\mathbf{w} : E(G) \rightarrow \mathbb{Z}_+$ and a set of terminals $T \subseteq V(G)$, find a connected subgraph H of G that contains T and is of minimum possible total weight. We will sketch the proof of the following result of Marx, Pilipczuk, and Pilipczuk [7].

Theorem 6. STEINER TREE *can be solved in time* $n^{\mathcal{O}(\sqrt{|T|})}$, *assuming that the input is an n-vertex planar graph and the edge weights are bounded polynomially in n.*

In Theorem 6, the assumption on polynomially bounded edge weights is a technical condition that ensures convergence of a local search routine. Let $(G, \mathbf{w}, T)$ be a STEINER TREE instance. Without loss of generality assume that G is connected. We need the following definitions. A subgraph H of G is a *solution* if it is connected and contains T, and *minimal* if one cannot delete an edge of H while keeping it a solution. Furthermore, in a graph H, a *2-path* is a path whose all internal vertices are nonterminals of degree 2 in H; a *maximal 2-path* is a 2-path with inclusion-wise maximal set of edges.

---

[3] Again, since in fact the algorithm guesses X_γ and $\mathcal{B}_\gamma$ by iterating over all options, we return the best solution found for all cases.

Note that every minimal solution is a tree whose every leaf is a terminal. In every minimal solution H, a vertex is *important* if it is a terminal or is of degree at least three. Note that there are less than $2|T|$ important vertices in any minimal solution. Observe that a path P in H is a maximal 2-path if and only if it is a path connecting two important vertices without any important vertex as an internal vertex.

A *k-step local search algorithm* for STEINER TREE maintains a minimal solution H and iteratively tries to decrease its weight by deleting from H the edges of at most k maximal 2-paths and reconnecting the obtained connected components in an optimal way. It is straightforward to implement one iteration of the k-step local search algorithm in time $n^{\mathcal{O}(k)}$. The algorithm stops when there is no possibility to decrease the weight of H in the aforementioned way; if this is the case, the current minimal solution H is called a *k-step local search minimum*.

In our algorithm we shall use a 2-step local search minimum H. By the above discussion and the assumption that the weights of G are positive integers bounded polynomially in n, such an H can be computed in polynomial time. Note that this is the only moment where we use the assumption of the weights of G being polynomially bounded; if one can ensure the convergence of the local search algorithm in a different way, this assumption can be omitted.

The main engine of the approach is the following combinatorial statement.

Lemma 1. *Let $(G, \mathbf{w}, T)$ be a STEINER TREE instance where G is planar, let H^2 be a 2-step local search optimum, and let H^* be a minimum-weight solution that minimizes the number of edges of $E(H^*) \setminus E(H^2)$. Then the graph $H_0 := H^2 \cup H^*$ has at most $105|T|$ vertices of degree at least three and, consequently, has treewidth $\mathcal{O}(\sqrt{|T|})$.*

In this exposition, we limit ourselves to the proof of Lemma 1 only. That is, we omit how to algorithmically leverage on Lemma 1 to get Theorem 6: this is a relatively standard application of dynamic programming that unfortunately cannot be easily phrased as a divide-and-conquer algorithm as in the previous examples. We refer to the arXiv version [6] of the original work [7] for full details.

Let H be the graph H_0 where every maximal 2-path is replaced with a single edge. To prove Lemma 1 it suffices to show that H has size $\mathcal{O}(|T|)$.

Recall that both H^* and H^2 are minimal solutions. A vertex $v \in V(H_0)$ is *important* if it is an important vertex of either H^* or of H^2. Since a vertex $v \in H_0$ that is a nonterminal vertex of degree 2 is not important, the set of important vertices is present also in H.

Let v be a nonterminal vertex of degree 2 in H_0. Since H^* and H^2 are minimal solutions, the edges incident to v in H_0 either both belong to $E(H^2)$ or both do not belong to $E(H^2)$. Consequently, the edge set of every maximal 2-path in H_0 is either completely contained in $E(H^2)$ or disjoint with $E(H^2)$. A symmetrical statement holds for H^*. Consequently, we can partition $E(H)$ into three sets:

- E^{both}, the edges corresponding to maximal 2-paths that are both contained in H^2 and in H^*;
- E^2, the edges corresponding to maximal 2-paths that are contained in H^2 but not in H^*; and

– E^*, the edges corresponding to maximal 2-paths that are contained in H^* but not in H^2.

Fix an embedding of G, which induces embeddings of H^2, H^*, and H.

The following claim is the crucial observation that allows us to bound the size of H. It relies on the fact that H^2 is a 2-step local search minimum in an essential way. Intuitively, it says that every face that is not incident to an important vertex or a bridge of H needs to be large: more precisely, a bit larger than what Euler's formula would allow. As we will see later, the number (or even the sum of the degrees in H) of important vertices in H and the number of bridges in H can be easily bounded by $\mathcal{O}(|T|)$. Hence, Claim 4 gives the crucial property that allows us later to perform a discharging argument bounding the size of H.

Claim. Every face of H is incident to an important vertex, to a bridge of H, or to at least six edges of $E^2 \cup E^*$.

Proof. Assume the contrary and let f be an offending face, not incident to any bridge nor important vertex and incident to at most five edges of $E^2 \cup E^*$.

In particular, since f is not incident to any bridge, all edges in the walk W around f are pairwise distinct (but a vertex can appear multiple times). Since H^2 and H^* are trees, f is incident to at least one edge of E^2 and at least one edge of E^*.

Orient the walk W clockwise around f. A subwalk W' of W is E^2-*maximal* if it starts with an edge of E^2, contains only edges of $E^2 \cup E^{\text{both}}$, and is maximal with this property. Similarly we define an E^*-*maximal* subwalks. Since W contains at least one edge of E^2 and at least one edge of E^*, the family of E^2-maximal and E^*-maximal subwalks partitions W into walks $W_1, W_2, \ldots, W_\ell$ such that W_i is E^2-maximal for odd i and E^*-maximal for even i. In particular, the number of walks ℓ is even.

Since f is not incident with any important vertex of neither H^2 nor H^*, every path W_i corresponds to a 2-path P_i in H^2 (for odd i) or H^* (for even i).

Since f is incident to at most 5 edges of $E^2 \cup E^*$, we infer that $\ell = 2$ or $\ell = 4$. We consider these two cases separately.

Case $\ell = 2$. Here, P_1 and P_2 are two paths between the same endpoints. Since f is not incident with any terminal, the minimality of H^2 ensures that P_1 is a shortest path between its endpoints. However, then the minimality of H^* implies that P_2 is also a shortest path between the same two endpoints and, furthermore, the minimality of $E(H^*) \setminus E(H^2)$ implies $P_2 = P_1$, a contradiction.

Case $\ell = 4$. In this more involved case we distinguish two subcases, depending on the topology of the picture; see Fig. 4. Recall that both P_1 and P_3 are 2-paths in H^2. The *interface endpoint* of P_1 is the endpoint v_1 of P_1 that is closer to P_3 in the tree H^2. Similarly we define the interface endpoint v_3 of P_3 as the one closer to P_1 on H^2 and the interface endpoints v_2 of P_2 and v_4 of P_4, referring to H^* instead of H^2.

In the first case, there exists an endpoint v of one of the paths P_i that is not an interface endpoint of neither of the incident two paths. Without loss of generality, assume that v is a common endpoint of P_1 and P_2 and $v \notin \{v_1, v_2\}$. Then, observe that if we replace P_1 with P_2 in the solution H^2, we obtain a solution as well. Thus, the (2-step local search) minimality of H^2 implies that the weight of P_1 is not larger than the weight of P_2. However, then we can replace P_2 with P_1 in the minimum-weight solution H^*, obtaining a new minimum-weight solution with smaller number of edges not in H^2. This contradicts the choice of H^*.

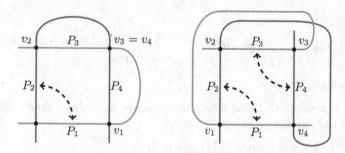

Fig. 4. Two subcases of the replacement argument of the case $\ell = 4$ in the proof of Claim 4. The solution H^2 is green and the solution H^* is red. Figure taken from [6], reproduced with authors' permission. (Color figure online)

In the second case, by symmetry we can assume that v_i is a common endpoint of P_i and P_{i+1} (with $P_1 = P_5$). The crucial observation is that now we obtain a new solution by either:

– taking H^2 and swapping both P_1 with P_2 and P_3 with P_4, or
– taking H^* and swapping both P_2 with P_1 and P_4 with P_3.

The minimality of both H^2 and H^* implies that in both swaps we do not change the total weight of the corresponding solution. However, then the second swap decreases the number of edges of H^* that are not in H^2, a contradiction to the choice of H^*. This finishes the proof of the claim.

We now move to the promised bounds on the important vertices and bridges in H. The bound on the number of important vertices is straightforward.

Claim. There are less than $3|T|$ important vertices in H.

Proof. There are $|T|$ terminals, less than $|T|$ vertices of degree at least 3 in H^2, and less than $|T|$ vertices of degree at least 3 in H^*.

The statement of Claim 4 can be improved to bound the total degree of important vertices. Let S be the set of important vertices in H.

Claim.

$$\sum_{v \in S} \deg_H(v) < 12|T|.$$

Proof. Since every leaf of H^2 and H^* is a terminal, we have that

$$\sum_{v \in S \cap V(H^2)} (\deg_{H^2}(v) - 2) = \sum_{v \in V(H^2)} (\deg_{H^2}(v) - 2) = 2|E(H^2)| - 2|V(H^2)| = -2,$$

$$\sum_{v \in S \cap V(H^2)} (\deg_{H^*}(v) - 2) = \sum_{v \in V(H^*)} (\deg_{H^*}(v) - 2) = 2|E(H^*)| - 2|V(H^*)| = -2.$$

Hence,

$$\sum_{v \in S} \deg_H(v) \leq \sum_{v \in S \cap V(H^2)} \deg_{H^2}(v) + \sum_{v \in S \cap V(H^*)} \deg_{H^*}(v) \leq 4|S| - 4 < 12|T|.$$

This finishes the proof.

Clearly, H is connected. Using a similar argument as in the previous claims, one can bound the number of bridges in H.

Claim. There are less than $2|T|$ bridges in H.

Proof. First, observe that every bridge of H lies in E^{both}, that is, corresponds to a maximal 2-path belonging both to H^2 and H^*.

Let B be the set of bridges in H and consider a connected component C of $H - B$. Assume that C contains no important vertex of H^2. Then, C is incident to exactly two bridges of H (as otherwise it would need to contain an important vertex of H^* and an important vertex of H^2 since all bridges lie both in H^* and H^2) and contains no terminals. The (2-step local search) minimality of H^2 implies that H^2 contains a shortest path between the two vertices of $V(C) \cap V(B)$. The minimality of H^* implies that H^* contains such a shortest path as well and, furthermore, the minimality of $E(H^*) \setminus E(H^2)$ implies that this is the same shortest path as in H^2. This is a contradiction with the definition of H.

Hence, every connected component of $H - B$ contains an important vertex of H^2. Since there are less than $2|T|$ such vertices, the claim follows.

Claim 4, together with the bounds above, allow us to conclude the proof of Lemma 1 with a discharging argument.

Set Up. We set up initial charges in two phases. In the first phase, every vertex and face of H receives a charge of -6 and every edge receives a charge of $+6$. Euler's formula ensures that the total charge assigned in the first phase is $-12 < 0$.

In the second phase, every important vertex v receives a charge of $+6\deg_H(v) + 7$ and every bridge receives a charge of $+6$. By Claims 4, 4, and 4, the total charge assigned in the second phase is bounded by

$$6 \cdot 12|T| + 7 \cdot 3|T| + 6 \cdot 2|T| = 105|T|.$$

Discharging Rules. We distribute the charge as follows.

1. **Every important vertex v sends a charge of $+6$ to every incident face.** After this step, every face incident with an important vertex has a nonnegative charge, as it started with a charge of -6, while every important vertex still keeps a charge of $+1$ (-6 from the first phase of the set up and $+7$ from the second phase).
2. **Every bridge e sends a charge of $+6$ to the unique incident face.** After this step, every face incident with a bridge of H has a nonnegative charge.
3. **Every edge $e \in E^{\text{both}}$ sends a charge of $+3$ to each of the two incident vertices.** After this step, every edge $e \in E^{\text{both}}$ has no charge.
4. **Every edge $e \in E^2 \cup E^*$ sends a charge of $+2$ to each of the two incident vertices and a charge of $+1$ to each of its two incident faces.** Note that the two incident faces are distinct as every bridge of H lies in E^{both}. After this step, again every edge $e \in E^2 \cup E^*$ has no charge. Furthermore, every face incident with at least six edges of $E^2 \cup E^*$ has a nonnegative charge, as it received a charge of $+1$ from at least six edges.

Final Analysis. As discussed above, every edge has no charge while every face has nonnegative charge thanks to Claim 4. To conclude the proof of Lemma 1, we show that every vertex ends up with charge of at least $+1$. This has already been discussed for important vertices.

Let v be an unimportant vertex of H. The initial charge of v is -6. By the last two discharging rules, v received a charge of $+2$ or $+3$ from each of its incident edges. By the construction of H and since v is not a terminal, the degree of v in H is at least 3. Thus, v ends up with a positive charge unless it is of degree exactly 3 and received only a charge of $+2$ from each of its three incident edges. However, this can only happen if the three incident edges with v belong to $E^2 \cup E^*$, which is impossible. This is the final contradiction that concludes the proof of Lemma 1.

References

1. Bringmann, K., Kisfaludi-Bak, S., Pilipczuk, M., van Leeuwen, E.J.: On geometric set cover for orthants. In: Bender, M.A., Svensson, O., Herman, G. (eds.) 27th Annual European Symposium on Algorithms, ESA 2019, 9–11 September 2019, Munich/Garching, Germany. LIPIcs, vol. 144, pp. 26:1–26:18. Schloss Dagstuhl - Leibniz-Zentrum für Informatik (2019). https://doi.org/10.4230/LIPIcs.ESA.2019.26
2. Danzer, L., Grünbaum, B., Klee, V.: Helly's theorem and its relatives. Proc. Symp. Pure Math. **7**, 101–180 (1963)
3. Dorn, F., Penninkx, E., Bodlaender, H.L., Fomin, F.V.: Efficient exact algorithms on planar graphs: exploiting sphere cut decompositions. Algorithmica **58**(3), 790–810 (2010)
4. Fomin, F.V., Thilikos, D.M.: New upper bounds on the decomposability of planar graphs. J. Graph Theor. **51**(1), 53–81 (2006)

5. Marx, D.: Four shorts stories on surprising algorithmic uses of treewidth. In: Fomin, F.V., Kratch, S., Leeuwen, E.J. (eds.) Treewidth, kernels and algorithms – Essays and Surveys Dedicated to Hans L. Bodlaender on the Occasion of His 60th Birthday. LNCS, vol. 12160. Springer, Heidelberg (2020)

6. Marx, D., Pilipczuk, M., Pilipczuk, M.: On subexponential parameterized algorithms for Steiner tree and directed subset TSP on planar graphs. CoRR abs/1707.02190 (2017). http://arxiv.org/abs/1707.02190

7. Marx, D., Pilipczuk, M., Pilipczuk, M.: On subexponential parameterized algorithms for Steiner tree and directed subset TSP on planar graphs. In: Thorup, M. (ed.) 59th IEEE Annual Symposium on Foundations of Computer Science, FOCS 2018, Paris, France, 7–9 October 2018, pp. 474–484. IEEE Computer Society (2018). https://doi.org/10.1109/FOCS.2018.00052

8. Marx, D., Pilipczuk, M.: Optimal parameterized algorithms for planar facility location problems using Voronoi diagrams. In: Bansal, N., Finocchi, I. (eds.) ESA 2015. LNCS, vol. 9294, pp. 865–877. Springer, Heidelberg (2015). https://doi.org/10.1007/978-3-662-48350-3_72

9. Miller, G.L.: Finding small simple cycle separators for 2-connected planar graphs. J. Comput. Syst. Sci. **32**(3), 265–279 (1986). https://doi.org/10.1016/0022-0000(86)90030-9

Computing Tree Decompositions

Michał Pilipczuk$^{(\boxtimes)}$

Institute of Informatics, University of Warsaw, Warsaw, Poland
michal.pilipczuk@mimuw.edu.pl

Abstract. In this chapter we review the most important algorithmic approaches to the following problem: given a graph G, compute a tree decomposition of G of (nearly) optimum width. We present the 4-approximation algorithm running in time $\mathcal{O}(27^k \cdot k^2 \cdot n^2)$, which was first proposed by Robertson and Seymour in the Graph Minors series, and we discuss the main ideas behind the exact algorithm of Bodlaender that runs in linear fixed-parameter time [2].

1 Introduction

If for a given graph G we can find a tree decomposition $\mathcal{T}$ of small width, say at most k, then many problems that are hard in general can be solved efficiently in G. More precisely, by employing bottom-up dynamic programming over $\mathcal{T}$ we can obtain algorithms with a typical running time of the form $f(k) \cdot n$ for some function f, which is *linear-time fixed parameter tractable* when parameterized by the width k. See e.g. [10, Chapter 7] for a comprehensive introduction to dynamic programming on tree decompositions. However, in order to apply such an approach we need to get the decomposition $\mathcal{T}$ from somewhere. This leads to a natural question: given the graph G alone, how efficiently can we find a tree decomposition of G of optimum or near-optimum width?

It turns out that the problem of determining the treewidth of a given graph is NP-hard. The first proof of this fact was given by Arnborg, Corneil, and Proskurowski [1] and its idea, inspired by the proof of the NP-hardness of the MINIMUM FILL-IN problem due to Yannakakis [32], is based on restricting attention to the class of co-bipartite graphs. Here, a graph is *co-bipartite* if it is a complement of a bipartite graph, or equivalently its vertex set can be partitioned into two cliques. It turns out that one can describe optimum-width tree decompositions of co-bipartite graphs in a very concrete form: every co-bipartite graph G, say with vertices partitioned into cliques A and B, admits an optimum-width tree decomposition that is actually a path decomposition where the bags

The work of Mi. Pilipczuk on this article is supported by the project TOTAL that has received funding from the European Research Council (ERC) under the European Union's Horizon 2020 research and innovation programme (grant agreement No. 677651).

© Springer Nature Switzerland AG 2020

F. V. Fomin et al. (Eds.): Bodlaender Festschrift, LNCS 12160, pp. 189–213, 2020.
https://doi.org/10.1007/978-3-030-42071-0_14

associated with the endpoints of the path respectively consist of all the vertices of A and of all the vertices of B. Thus, traversing this path decomposition from the A-endpoint to the B-endpoint can be imagined as sweeping the graph while gradually exchanging vertices of A with vertices of B so that the maximum total number of vertices kept at any point is minimized. This view leads to a transparent reduction from the problem of determining the *cutwidth* of a graph, which is known to be NP-hard, see [15, GT44]. All in all, a corollary is that it is even NP-hard to compute the treewidth, equivalently the pathwidth, of a co-bipartite graph.

A natural approach would be to resort to *approximation*: instead of asking for a decomposition of the smallest possible width, we would like to efficiently compute a decomposition whose width is provably not much larger than the optimum. Note that posing the question in this way is justified by applications: having computed a decomposition we probably would like to run some dynamic programming algorithm on it, so obtaining a slightly larger width should result in a moderate increase in the running time of this follow-up procedure. Unfortunately, even in this relaxed setting the problem remains hard: Wu et al. [31] proved that assuming the Small Set Expansion conjecture of Raghavendra and Steurer [24], achieving any constant factor approximation is NP-hard. The best known approximation algorithm running in polynomial time, due to Feige et al. [13], achieves an approximation factor of $\mathcal{O}(\sqrt{\log \mathrm{OPT}})$; that is, on a graph of treewidth k it outputs a tree decomposition of width $\mathcal{O}(k\sqrt{\log k})$.

However, observe the following: If having computed a tree decomposition of width k we anyway apply some dynamic programming algorithm with running time of the form $f(k) \cdot n$, that is, FPT for the parameter k, then we can afford also a running time of this form when constructing the decomposition itself. It turns out that if we allow FPT running time, then there is a wide range of interesting algorithmic techniques that lead to fixed-parameter algorithms for constructing tree decompositions of (near) optimum width. These typically fall into two groups:

- *Parameterized exact algorithms* that given a graph G and parameter k, either certify that the treewidth of G is larger than k or construct a decomposition of width at most k.
- *Parameterized approximation algorithms* that work in the setting as above, but we allow that the output decomposition has width at most $g(k)$ for some (hopefully slowly growing) function g.

In this chapter we present two such algorithms that arguably had the largest impact: the 4-approximation algorithm of Robertson and Seymour and the exact algorithm of Bodlaender.

The approximation algorithm of Robertson and Seymour came as a byproduct of their work on the theory of graph minors. The result can be formally stated as follows.

Theorem 1 (Robertson and Seymour, [26]). *There is an algorithm that given a graph G on n vertices and an integer k, works in time $\mathcal{O}(27^k \cdot k^2 \cdot n^2)$*

and either correctly concludes that the treewidth of G is larger than k, or outputs a tree decomposition of G of width at most $4k + 4$.

Different expositions of the algorithm of Robertson and Seymour can be found in a number of sources, for instance in the textbooks of Kleinberg and Tardos [19] and of Cygan et al. [10]. In Sect. 2 we take the liberty of presenting a yet another interpretation, close to the one from [10], where we focus on the duality connection between the treewidth of a graph and the sizes of so-called *well-linked sets* that can be found in it. Informally speaking, this connection explains that the problem of approximating the treewidth of a graph is essentially equivalent to the problem of finding balanced separators in it, as well as it highlights objects dual to tree decompositions, which witness that the treewidth of a graph is large.

The basic idea behind the algorithm of Robertson and Seymour proved to be very influential throughout the years. This is because in fact it provides a robust framework for approximating not only the treewidth of a graph, but also other width parameters based on hierarchical decompositions, such as the branchwidth of a matroid for example. Different parts of this framework can be conveniently replaced and modified, leading to algorithms achieving various properties of the constructed decompositions.

Theorem 1 provides a fast parameterized approximation algorithm for treewidth, but looking at the result one can ask two natural questions. First, can we efficiently construct a tree decomposition of really optimum width? Second, can we reduce the dependency of the running time on the size of the graph to linear (i.e. $\mathcal{O}(n)$, as a graph of treewidth k has at most kn edges)? Addressing these question seems to require ideas beyond simple variations of the approximation algorithm of Robertson and Seymour, and a positive answer to both was given by Bodlaender in [2].

Theorem 2 (Bodlaender, [2]). *There is an algorithm that given a graph G on n vertices and an integer k, works in time $2^{\mathcal{O}(k^3)} \cdot n$ and either correctly concludes that the treewidth of G is larger than k, or outputs a tree decomposition of G of width at most k.*

Note that compared to the approximation algorithm of Robertson and Seymour, the running time of Bodlaender's algorithm has a far higher parametric factor (i.e., dependency on the parameter k). This makes it a less favorable choice in the settings where optimizing the parametric factor is of importance. While it is known that a constant-factor approximation of treewidth can be computed in time $2^{\mathcal{O}(k)} \cdot n$ [3], whether the treewidth can be computed *exactly* in a similar running time, or even in time $2^{o(k^3)} \cdot n^{\mathcal{O}(1)}$, remains open.

Bodlaender's algorithm is far more complex than the approximation algorithm of Robertson and Seymour. We sketch the crucial ideas behind it in Sect. 3.

2   Approximating Treewidth: The Algorithm of Robertson and Seymour

A tree decomposition of a graph G of width k, after rooting it in an arbitrary bag, can be roughly imagined as a hierarchical decomposition of G using separators of size at most $k + 1$. We start with the whole graph G, then we break it using the root bag into two or more smaller pieces, corresponding to subtrees of the decomposition rooted at the children of the root, then we break each of these pieces using the bag at the root of the corresponding subtree, and so on. If we are given a graph G alone and we would like to compute an approximately optimum tree decomposition of G, then the idea is that we would like to implement such a decomposition process using separators chosen somehow greedily. The assumption that the graph G at hand admits some (unknown to the algorithm) tree decomposition of width k should ensure us that we can indeed find such greedy separators in G.

2.1   Balanced Separators and Well-Linked Sets

Hence, our first goal is to understand separator properties of bounded treewidth graphs. The idea is that if G has treewidth k, then it admits a small—of size at most $k+1$—separator that splits it in a balanced way. This applies not only to splitting the whole vertex set of G, but also to splitting any its subsets, which can be conveniently phrased using weights on vertices.

We need a few simple definitions to facilitate our discussion. A *separation* of a graph G is a pair (A, B) of subsets of vertices of G such that $A \cup B = V(G)$ and there is no edge between $A \setminus B$ and $B \setminus A$. The *order* of separation (A, B) is the cardinality of its *separator* $A \cap B$. If $\omega \colon V(G) \to \mathbb{R}_{\geqslant 0}$ is a nonnegative weight function on vertices of G, then for $X \subseteq V(G)$ we write $\omega(X) = \sum_{u \in X} \omega(u)$ and for a subgraph H of G we write $\omega(H) = \omega(V(H))$.

Lemma 1. *Let G be a graph of treewidth at most k and $\omega \colon V(G) \to \mathbb{R}_{\geqslant 0}$ be a nonnegative weight function on vertices of G. Then there exists a set of vertices $X \subseteq V(G)$ such that $|X| \leqslant k + 1$ and*

$$\omega(C) \leqslant \frac{1}{2}\omega(G) \qquad \text{for every connected component } C \text{ of } G - X.$$

Proof. Let (T, β) be a tree decomposition of G of width at most k. With every edge $e = xy$ of T and its orientation (x, y) we can associate a separation $(A_{x,y}, B_{x,y})$ defined as follows: if T_x and T_y are the connected components of $T - xy$ that contain x and y, respectively, then

$$A_{x,y} = \bigcup_{z \in V(T_x)} \beta(z) \quad \text{and} \quad B_{x,y} = \bigcup_{z \in V(T_y)} \beta(z).$$

It is easy to verify using the definition of a tree decomposition that $(A_{x,y}, B_{x,y})$ is a separation of G with separator $A_{x,y} \cap B_{x,y} = \beta(x) \cap \beta(y)$. Note that the reversed orientation of xy is assigned the reversed separation: $(A_{y,x}, B_{y,x}) = (B_{x,y}, A_{x,y})$.

Now, for every edge xy of T let us choose an orientation (x, y) so that

$$\omega(A_{x,y}) \leqslant \omega(B_{x,y});$$

clearly, one of the two orientations has this property. In other words, every edge is oriented towards the "heavier" side. Observe that thus T becomes a tree with $|V(T)| - 1$ oriented edges, which therefore have $|V(T)| - 1$ tails in total. Therefore, there exists a node z of T that is not the tail of any edge of T.

We claim that $X = \beta(z)$ satisfies the required property. As (T, β) has width at most k, we have $|X| \leqslant k + 1$. Take any connected component C of $G - X$. By the connectedness of C and the properties of tree decompositions, there exists a neighbor w of z in T such that all nodes of T whose bags contain some vertex of C belong to the connected component of $T - wz$ that contains w. Hence $V(C) \subseteq A_{w,z}$. Since C is disjoint with $\beta(z)$, which in turn contains $A_{w,z} \cap B_{w,z} = \beta(w) \cap \beta(z)$, we in fact have $V(C) \subseteq A_{w,z} \setminus B_{w,z}$. By the choice of z the edge wz is oriented towards w, which means that $\omega(A_{w,z}) \leqslant \omega(B_{w,z})$. This entails $\omega(A_{w,z} \setminus B_{w,z}) \leqslant \omega(B_{w,z} \setminus A_{w,z})$ and, as $A_{w,z} \setminus B_{w,z}$ and $B_{w,z} \setminus A_{w,z}$ are disjoint, also $\omega(A_{w,z} \setminus B_{w,z}) \leqslant \frac{1}{2}\omega(G)$. We conclude that

$$\omega(C) \leqslant \omega(A_{w,z} \setminus B_{w,z}) \leqslant \frac{1}{2}\omega(G),$$

as required.   □

Suppose S is a subset of vertices of a graph G of treewidth at most k. Letting $\mathbb{1}_S$ be the indicator function of S (i.e. $\mathbb{1}_S$ assigns 1 to vertices of S and 0 to other vertices of G), Lemma 1 applied to $\mathbb{1}_S$ provides a separator of size at most $k + 1$ whose removal splits S into connected components in a balanced way. Intuitively, this means that in a graph of treewidth at most k we cannot find a large set of vertices that is "concentrated" in the sense that one cannot break it in a balanced way using a separation of small order. This is formalized in the next definition.

Definition 1. *A set S of vertices of a graph G is* well-linked *in G if there is no separation (A, B) of G such that*

$$|A \cap S| > |A \cap B| \qquad and \qquad |B \cap S| > |A \cap B|.$$

Lemma 2. *If G is a graph of treewidth at most k, then every set of vertices of G of size at least $3k + 4$ is not well-linked. Moreover, this is witnessed by a separation of G of order at most $k + 1$.*

Proof. Let S be a set of size at least $3k + 4$ in G. By Lemma 1 applied to the indicator function $\mathbb{1}_S$, we conclude that there exists $X \subseteq V(G)$ such that $|X| \leqslant k + 1$ and

$$|V(C) \cap S| \leqslant \frac{1}{2}|S| \qquad \text{for every connected component } C \text{ of } G - X.$$

The idea now is to partition the connected components of $G - X$ into two groups so that each of them contains at most $\frac{2}{3}|S|$ vertices of S.

Suppose for a moment that we have achieved this goal. We then define a separation (A, B) of G as follows: A consists of X and the vertex sets of the connected components of one of the groups, and similarly B consists of X and the components of the second group. By the condition imposed on the second group, we have $|(B \setminus A) \cap S| \leqslant \frac{2}{3}|S|$, which implies that

$$|A \cap S| \geqslant \frac{1}{3}|S| > k + 1 \geqslant |X| = |A \cap B|.$$

Symmetrically, we conclude that $|B \cap S| > |A \cap B|$. Thus, (A, B) witnesses that S is not well-linked and the order of (A, B) is $|X| \leqslant k + 1$.

We are left with constructing the groups. We start with partitioning the connected components of $G - X$ into singleton groups: each consisting of one component. Then we iteratively merge the groups until only two groups are left using the following rule: take the two groups that contain the least number number of vertices of S and merge them into one group. Observe that as long as there are at least four groups left, the two with the least number of vertices of S in total contain at most $\frac{1}{2}|S|$ vertices of S. Thus, up to the point when there are only three groups left we maintain the following invariant: each group contains at most $\frac{1}{2}|S|$ vertices of S. Finally, in the last step out of three groups we merge the two with the least number of vertices of S. This means that the merged groups together contain at most $\frac{2}{3}|S|$ vertices of S and we are done. □

In our approximation algorithm we will iteratively break sets of vertices using separations witnessing that they are not well-linked. Therefore, we need an efficient way of testing that a set of vertices is well-linked. While this seems hard in general, we only need to treat sets of small size, so we can use a fixed-parameter algorithm for the parameterization by the cardinality of the tested set.

Lemma 3. *Given a graph G on n vertices and m edges and a set of its vertices S, one can in time $\mathcal{O}(3^{|S|} \cdot |S|(n + m))$ verify whether S is well-linked in G. In case of a negative answer, the algorithm provides a minimum-order separation of G that witnesses that S is not well-linked.*

Proof. The algorithm proceeds as follows: consider all pairs of subsets (S_A, S_B) of S satisfying $S_A \cup S_B = S$ and for each of them apply the Ford-Fulkerson max-flow algorithm to find a minimum-order separation (A, B) of G satisfying $S_A \subseteq A$ and $S_B \subseteq B$. On one hand, if we find that $\min(|S_A|, |S_B|) > |A \cap B|$, then (A, B) witnesses that S is not well-linked. On the other hand, if there exists some separation (A_0, B_0) witnessing that S is not well-linked, then for the choice $(S_A, S_B) = (S \cap A_0, S \cap B_0)$ some witness of non-well-linkedness—which is possibly different from (A_0, B_0), but not of higher order—will be found. Hence, if the algorithm finds some separations witnessing that S is not well-linked, then it can output any such separation of the smallest order. Otherwise, in the absence of witnesses, we can safely conclude that S is well-linked.

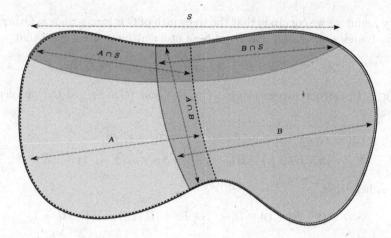

Fig. 1. Splitting the graph using a separation (A, B) that breaks S in a balanced way.

For the time complexity, there are $3^{|S|}$ pairs (S_A, S_B) to consider. For each of them we apply the Ford-Fulkerson algorithm, which runs as many steps of flow augmentation as the maximum number of vertex-disjoint paths from S_A to S_B, which is upper bounded by $|S|$. As each iteration can be implemented in time $\mathcal{O}(n + m)$, the claimed time complexity of $\mathcal{O}(3^{|S|} \cdot |S|(n + m))$ follows.   □

2.2 Recursive Construction of a Decomposition

With all the tools prepared, we are ready to present the algorithm. As usual with recursive procedures, we will solve a more general problem defined as follows. We are given a graph G, integer k, and a set of vertices S satisfying $|S| \leqslant 3k + 4$. The goal is to reach one of the following outcomes:

- conclusions that the treewidth of G is larger than k; or
- construction of a tree decomposition of G of width at most $4k + 4$ such that one of its bags contains the whole set S.

As the running time of the algorithm will be $\mathcal{O}(27^k \cdot k^2 \cdot n^2)$, Theorem 1 will follow by applying this procedure to empty S. The parameter k will be fixed throughout the description, so instances of this general problem will be written as pairs of the form (G, S).

The algorithm works as follows. First, if $S = V(G)$, that is, there are no vertices outside of S, then we can output a trivial decomposition consisting of one bag with the whole vertex set inside. Second, if $|S| < 3k + 4$, then we can take an arbitrary vertex $u \notin S$ and add it to S. Indeed, a valid outcome of the algorithm for the instance $(G, S \cup \{s\})$ is also valid for (G, S).

Hence, from now on we may assume that $|S| = 3k + 4$. We invoke the algorithm of Lemma 3 on S. If it turns out that S is well-linked, or that a minimum-order separation witnessing that it is not well-linked has order larger than $k + 1$, then by Lemma 2 we conclude that the treewidth of G is larger than k. Otherwise, we have found a separation (A, B) of G of order at most $k + 1$ such that

$$|A \cap S| > |A \cap B| \quad \text{and} \quad |B \cap S| > |A \cap B|;$$

see Fig. 1. Construct two instances (G_A, S_A) and (G_B, S_B) of the problem as follows:

$$G_A = G[A], \qquad\qquad G_B = G[B],$$
$$S_A = (S \setminus B) \cup (A \cap B), \qquad S_B = (S \setminus A) \cup (A \cap B).$$

Observe here that

$$|S_A| = |S \setminus B| + |A \cap B| < |S \setminus B| + |B \cap S| = |S| = 3k + 4,$$

and similarly $|S_B| < 3k + 4$.

Hence, we can apply the algorithm recursively to the instances (G_A, S_A) and (G_B, S_B). If this results in the conclusion that either G_A or G_B has treewidth larger than k, then the same holds also for G. Otherwise, we obtain tree decompositions (T_A, β_A) and (T_B, β_B) of G_A and G_B such that there are nodes x_A and x_B of T_A and T_B, respectively, satisfying $S_A \subseteq \beta_A(x_A)$ and $S_B \subseteq \beta_B(x_B)$. Now, we construct a tree decomposition (T, β) of G as follows:

- T is constructed from the disjoint union of T_A and T_B by adding a new node x and making it adjacent to x_A and x_B;
- β is obtained by taking the union of mappings β_A and β_B and additionally setting $\beta(x) = S \cup (A \cap B)$.

It is straightforward to verify that (T, β) is a tree decomposition of G. Moreover, we have

$$|\beta(x)| \leqslant |S| + |A \cap B| \leqslant (3k + 4) + (k + 1) = 4k + 5.$$

This, together with the decompositions (T_A, β_A) and (T_B, β_B) having width at most $4k + 4$ due to being obtained from recursion, implies that (T, β) has width at most $4k + 4$. Finally, we have $S \subseteq \beta(x)$, hence the decomposition (T, β) is a valid outcome of the algorithm.

We are left with estimating the running time of the algorithm. First, recall that a graph on n vertices of treewidth at most k has at most kn edges, hence we can assume this about the input graph G; otherwise, the conclusion that its treewidth is larger than k can be immediately reached. Observe now that the internal computations in each recursive call boil down to one application of the algorithm of Lemma 3 to a set of size $3k + 4$, hence they take time $\mathcal{O}(27^k \cdot k^2 \cdot n)$. We conclude that the whole algorithm runs in time $\mathcal{O}(27^k \cdot k^2 \cdot n^2)$, it suffices to argue that the algorithm executes at most $\mathcal{O}(n)$ recursive calls in total.

For this, observe that in fact in each call of the algorithm we always have $|S| < 3k + 4$, not only $|S| \leqslant 3k + 4$. Indeed, this is true both in the first call when $S = \emptyset$ and in each further recursive call. Thus, each recursive call, apart from the calls where $S = V(G)$, starts with adding one additional vertex to S. It is not hard to see that those added vertices are pairwise different for different calls, hence the total number of calls where $S \neq V(G)$ is at most n. On the other hand, every call where $S = V(G)$ is invoked inside a call where $S \neq V(G)$. Since each call invokes at most two subcalls, we conclude that the total number of recursive calls invoked by the algorithm is at most $3n$.

Having presented the algorithm, let us pause for a moment and elaborate on the intuition on what actually happened. The first, naive approach would be to recursively decompose the graph by always taking any its balanced separator and recursing into connected components obtained by removing it from the vertex set. However, when translating this to a construction of a tree decomposition, the balanced separators used in the consecutive levels of the recursion would need to be added to the current "interface" towards the rest of the graph, represented by the set S. As this interface needs to be incorporated in the bags, we would not obtain any reasonable bound on the sizes of bags. The idea for solving this problem is not to split the graph in the balanced way, but the interface S itself. Thus, in each recursive subcall we cease to "see" a significant portion of S: just enough to amortize adding the separator to the interface. Note that the condition expressed in the definition of well-linkedness (Definition 1) is *precisely* the condition sufficient for this amortization trick to work.

An important conclusion that can be drawn from the presented algorithm is that treewidth is closely related, in the dual way, to well-linked sets. Precisely, if by $\mathrm{wl}(G)$ we denote the size of the largest well-linked set in G, then we have the following. On one hand, Lemma 2 directly implies that $\mathrm{wl}(G) \leqslant 3k + 3$ for every graph G of treewidth at most k. On the other hand, by a straightforward adaptation of the algorithm, we can conclude that, when applied on a graph without well-linked sets of size ℓ, it produces a tree decomposition of width smaller than $\frac{3}{2}\ell$. These two observations together imply the following.

Corollary 1. *For every graph G we have*

$$\frac{2\mathrm{tw}(G) - 1}{3} \leqslant \mathrm{wl}(G) \leqslant 3\mathrm{tw}(G) + 3.$$

Conceptually, this means that tree decompositions and well-linked sets are dual to each other: if a graph G has high treewidth, then this is witnessed by the existence of a large well-linked set in G. Then such a large well-linked set can be a good starting point for finding more concrete combinatorial obstacles for treewidth, such as grid minors or embedded expanders.

2.3 Futher Reading

The obvious question is to what extent the running time of the presented algorithm is the best possible. For the parametric factor, the base of the exponent can be improved, even by just slightly restructuring the argument; for instance, Cygan et al. [10] present the algorithm somewhat differently so that they achieve a running time of $\mathcal{O}(8^k \cdot k^2 \cdot n^2)$. However, it seems difficult to imagine improving the parametric to subexponential in k, even though there are no known lower bounds against it.

A perhaps more important question is to reduce the dependency on n to from quadratic to linear or near-linear. This is because the computation of a low-width tree decomposition is usually succeeded by running a dynamic programming algorithm on it, which typically works in time linear in the size of the graph. Thus, quadratic running time of the construction of a tree decomposition becomes the dominant factor in the running time. If the parametric factor of the running time of the dynamic programming is high anyway, e.g., it is obtained using Courcelle's theorem, then one can instead compute the treewidth exactly in time $2^{\mathcal{O}(k^3)} \cdot n$ using the algorithm of Theorem 2. However, if we would like to optimize the parametric factor and the dependency on n at the same time, then the algorithms of Theorems 1 and 2 have actually incomparable running times: one has high parametric factor while achieving linear depency on n, and the second has moderate parametric factor while being quadratic in n.

An improvement of this state of the art came from Reed [25], who gave a constant factor approximation algorithm working in time $k^{\mathcal{O}(k)} \cdot n \log n$. The main insight here is that the classic approximation algorithm, as presented here, achieves quadratic running time because it may run a linear-time procedure on each of $\mathcal{O}(n)$ levels of recursion. However, if we inspect any fixed recursion level, then the instances solved at this level are essentially disjoint: they can only share parts of the set S, which is small anyway. Therefore, the total amount of work used by the algorithm on each level of recursion is linear in total, so to achieve dependency on the graph size of the order $n \log n$ it suffices to limit the depth of the recursion to $\mathcal{O}(\log n)$. This can be done by, in each recursive subcall, breaking not only the set S in a balanced way but also the graph that we are currently decomposing. This idea was pushed further by Bodlaender et al. [3], who, using much more technical work, gave a 5-approximation algorithm with running time $2^{\mathcal{O}(k)} \cdot n$. We note that in [25], Reed specifies neither the approximation ratio nor the parametric factor of the running time of his algorithm, but based on his ideas Bodlaender et al. describe in [3] a relatively simple 3-approximation algorithm with running time $2^{\mathcal{O}(k)} \cdot n \log n$.

One of the main advantages of the presented approximation algorithm of Robertson and Seymour is that it is rather a "template" of the algorithm, where the step of finding a balanced separator for the set S (which we essentially encapsulated in Lemma 3) can be replaced by different subprocedures, which leads to achieving different trade-offs between approximation and running time guarantees. For instance, using a polynomial-time $\mathcal{O}(\sqrt{\log \mathrm{OPT}})$-approximation algorithm for balanced separators based on semi-definite programming (SDP), Feige et al. [13] gave a polynomial-time $\mathcal{O}(\sqrt{\log \mathrm{OPT}})$-approximation algorithm. While the degree of the polynomial bounding the running time of this algorithm is high, Fomin et al. [22] used a yet another, purely combinatorial method of approximating balanced separators to design an $\mathcal{O}(k)$-approximation algorithm with running time $\mathcal{O}(k^7 \cdot n \log n)$, where k is the expected value of the treewidth provided on input.

On a related note, Lokshtanov et al. [22] observed that the balanced separator step in the algorithm of Robertson and Seymour can be executed in a "canonical" way (with some footnotes due). This directly led to computing a tree decomposition that (almost) only depends on the isomorphism class of the input graph and, as a result, to solving the GRAPH ISOMORPHISM problem in graphs of bounded treewidth in fixed-parameter time. Later, the same approach was used by Elberfeld and Schweitzer [12] to solve the problem in parameterized logspace.

Finally, the presented approximation algorithm of Robertson and Seymour suggests a general framework for approximating various width parameter based on tree-like decompositions of different combinatorial objects. More precisely, in the algorithm we essentially used two properties that can be put in an abstract, and admittedly vague way as follows:

– The width parameter in question measures the lowest possible complexity of a certain tree-like decomposition of the object, where the object is cut into smaller and smaller pieces in a hierarchical manner.
– The complexity of the decomposition is defined as the minimum of a certain "connectivity function" among cuts used by it. We require that this connectivity function is submodular so that minimum cuts with respect to it can be efficiently computed.

There are other interesting combinatorial objects, even beyond the realm of graphs, for which we can define width parameters satisfying the above. For instance, Oum and Seymour [23] considered the parameter *cliquewidth* of graphs. They defined a new parameter *rankwidth*, which is functionally equivalent to cliquewidth, but is based on a submodular connectivity function. Then the presented framework can be applied, essentially out-of-the-box, to give a $2^{\mathcal{O}(k)} \cdot n^3$-time approximation algorithm for rankwidth, which provided the first fixed-parameter approximation algorithm for cliquewidth. More generally, Oum and Seymour [23] show how the technique can be used to approximate the branchwidth of matroids.

3   Computing Treewidth Exactly: Bodlaender's Algorithm

In this section we present Bodlaender's algorithm for constructing an optimum-width tree decomposition of a graph in fixed-parameter time, that is, we sketch the proof of Theorem 2. Before we dive into the details of the approach, let us first explore some simple ideas in order to gather some intuition.

3.1   Gathering Ideas

Note that if we are given a graph G on n vertices and we are looking for a tree decomposition of width at most k, then there are only at most n^k candidates for its *adhesions*—intersections of neighboring bags. Indeed, assuming that every two adjacent bags are different, every adhesion has size at most k, so we can take all sets of vertices of size at most k as candidates. Further, it is not hard to see that for an optimum width tree decomposition of G, say rooted, we can assume the following connectivity condition: for every node x, the *component* of x—the set of vertices appearing in the bags of x and its descendants, but not in the parent of x (if existent)—induces a connected subgraph of G. Intuitively, if the component D of x induces a disconnected graph, then we can refine the decomposition by "splitting" the subtree rooted at x into several subtrees, each decomposing a different connected component of D; see e.g. [7, Lemma 2.8] for a formal argument. This implies that there are only n^{k+1} candidates for pairs of the following form: an adhesion between a node x and its parent together with the component of x. Indeed, each such pair can be obtained by taking a set X of at most k vertices and choosing one connected component of $G - X$.

Now, it is not hard to convince oneself that these candidate pairs form a suitable state space for a dynamic programming procedure that tries to assemble a tree decomposition of width at most k bottom-up. After reorganizing the argument for the sake of optimizing the running time, this approach yields the following result, first noted by Arnborg et al. [1].

Theorem 3 (Arnborg, Corneil, Proskurowski [1]). *There is an algorithm that, given a graph G on n vertices and integer k, runs in time $\mathcal{O}(n^{k+2})$ and either correctly concludes that the treewidth of G is larger than k or computes a tree decomposition of G of width at most k.*

We remark that if one wishes to compute the treewidth of a graph exactly in practice, the most successful implementations use the algorithm of Theorem 3 with clever pruning of the state space, as only a small fraction of states are actually relevant for constructing a solution. See [11] for a broader discussion.

In the terminology of parameterized complexity, Theorem 3 provides an XP algorithm for computing a tree decomposition of optimum width. We now would like to improve this to an FPT algorithm, i.e., with running time of the form $f(k) \cdot n^c$ for some constant c. For now, let us allow super-linear dependency of the running time on n: we aim at getting any constant c.

Obviously, if we can pay fixed-parameter time that is quadratic in n, then we can run the approximation algorithm of Robertson and Seymour to obtain a tree decomposition of the input graph G of width at most $4k + 4$. While this is not precisely what we want at the end of the day, we can still use this approximate tree decomposition for performing some efficient computation.

Let us try an even more moderate goal. Instead of asking for a construction of a tree decomposition of width at most k, let us first only focus on the decision problem: is the treewidth of G at most k or not? For this, we can use the Graph Minors Theorem [27]. Since the class of graphs of treewidth at most k is minor-closed, there is a finite set of graphs $\mathcal{F}_k$, called the *obstruction set*, such that a graph has treewidth at most k if and only if it does not contain any graph from $\mathcal{F}_k$ as a minor. Therefore, to check whether G has treewidth at most k, we can inspect graphs $H \in \mathcal{F}_k$ one by one and for each verify, using dynamic programming over the approximate tree decomposition, whether it is a minor of G. This can be done in time $f(|H|, k) \cdot n$ for some computable function f using either a direct construction of a dynamic programming procedure, or applying Courcelle's theorem. As the set $\mathcal{F}_k$ depends only on k, the value $\max_{H \in \mathcal{F}_k} |H|$ also depends only on k, hence the whole algorithm works in time $g(k) \cdot n$ for some function g.

A priori, this seems like a non-uniform and non-constructive argument: the Graph Minors Theorem only guarantees the existence of a suitable obstruction set $\mathcal{F}_k$ and does not provide any method for computing it. However, the particular case of minor obstructions for treewidth at most k was investigated by Lagergren [20], who proved that the obstructions in $\mathcal{F}_k$, supposing they are minimal with respect to the minor order, are of size at most doubly-exponential in $\mathcal{O}(k^5)$. Thus, $\mathcal{F}_k$ can be computed in time depending only on k by inspecting every graph H up to this size bound and checking whether H is a minor-minimal obstruction to having treewidth at most k. For this, we verify that the treewidth of H is larger than k, but every minor of H has treewidth at most k, using any algorithm for computing treewidth (e.g. Theorem 3). After working out all the technical details of dynamic programming, this yields the following corollary.

Theorem 4. *There is an algorithm that, given a graph G on n vertices and an integer k, verifies in time $f(k) \cdot n^2$ whether the treewidth of G is at most k, where $f(k)$ is a function that is triple-exponential in $\mathcal{O}(k^5)$.*

Now we can leverage the decision algorithm of Theorem 4 to obtain a constructive algorithm by the means of a *self-reduction* as follows. Recall that a graph G has treewidth at most k if and only if G has a chordal supergraph $\widehat{G}$ in which all cliques have size at most $k + 1$. Indeed, such a chordal supergraph can be obtained by taking any tree decomposition of G of width at most k and turning every bag into a clique. Hence, if G is a graph of treewidth at most k that is not yet chordal, then there is an edge that can be added to G so that the treewidth of the obtained graph $G + e$ still stays upper bounded by k: any edge of $E(\widehat{G}) \setminus E(G)$ has this property. On the other hand, if G is chordal, then we can compute its optimum-width tree decomposition in linear time: just take any clique tree of G.

This suggests the following algorithm. If the input graph G is chordal, then output any its clique tree as the optimum-width tree decomposition. Otherwise, for every edge e that is not yet present in the graph, verify whether $G + e$ has treewidth at most k using the algorithm of Theorem 4. If this is the case, then we can simply recurse on the graph $G + e$. Otherwise, if the verification yields a negative outcome for all e, then we are certain that the treewidth of G is larger than k. As throughout the algorithm we can test each of the at most $\binom{n}{2}$ potential edges at most once, this yields the following.

Theorem 5. *There is an algorithm that, given a graph G on n vertices and integer k, runs in time $f(k) \cdot n^4$ and either correctly concludes that the treewidth of G is larger than k or computes a tree decomposition of G of width at most k, where $f(k)$ is a function that is triple-exponential in $\mathcal{O}(k^5)$.*

Now that we gave an FPT algorithm for constructing an optimum-width tree decomposition of a graph, we can elaborate on possible ways of improving its running time to linear in the size of the graph. The n^4 factor essentially comes from two sources.

First, in Theorem 4 we used the approximation algorithm of Robertson and Seymour that, just by itself, has quadratic running time. An observant reader might point out that we could use the linear-time 5-approximation algorithm of Bodlaender et al. [3] instead. However, this is a much later development which also relies on a number of ideas originating in Bodlaender's algorithm. All in all, it would be useful to find a way to somehow obtain an approximate tree decomposition of the graph, without the need of calling an auxiliary procedure with super-linear running time.

Second, the whole self-reduction scheme incurs a quadratic number of calls to a subprocedure, so we need to get around this issue. The idea here is as follows: the heart of the algorithm of Theorem 4 boils down to a dynamic programming decision procedure that takes a graph G together with a tree decomposition of approximately optimum width, say ℓ, and verifies whether the treewidth of G is at most k in time $f(k, \ell) \cdot n$, for some function f. While our implementation of this procedure is based on the characterization using the obstruction set, maybe it is possible to give a similar dynamic programming procedure which will actually construct a suitable decomposition, instead of only certifying its existence? The answer to this question is positive and was given by Bodlaender and Kloks in [5].

Theorem 6 (Bodlaender and Kloks [5]). *There is an algorithm that, given a graph G on n vertices, its tree decomposition of width at most ℓ, and integer k, runs in time $2^{\mathcal{O}(\ell^2(k+\log \ell))} \cdot n$ and either correctly concludes that the treewidth of G is larger than k or computes a tree decomposition of G of width at most k.*

The Bodlaender-Kloks procedure, i.e. Theorem 6, is a crucial ingredient of Bodlaender's algorithm. We postpone its presentation for now and focus on resolving the first issue: getting an approximate tree decomposition on which this procedure can be applied. In a nutshell, the idea is that instead of relying on an auxiliary approximation algorithm, we will obtain a near-optimum tree decomposition from recursion.

3.2   Recursive Instance Reduction

Assume we are given a graph G on n vertices and we would like to construct a tree decomposition of G of width at most k. Suppose further that somebody gave to us a large matching M in G, say consisting of εn edges for some $\varepsilon > 0$ to be fixed later. Then we can apply the following scheme:

- Obtain a graph G' by contracting all edges of M. Then G' is a minor of G, so if G has treewidth at most k, so does G'. On the other hand, G' is significantly smaller than G: it has at most $(1 - \varepsilon)n$ vertices.
- Apply the algorithm recursively to G'. In case the treewidth of G' is larger than k, the same can be said about the treewidth of G. Otherwise, we have computed a tree decomposition T' of G' of width at most k.

- Uncontract the edges of M in the following sense: for every edge $uv \in M$, say contracted to a vertex w in G', replace every occurrence of w in every bag of T' by $\{u, v\}$. It is easy to see that in this way we obtain a tree decomposition T of G of width at most $2k + 1$.
- Use decomposition T to run the Bodlaender-Kloks algorithm (Theorem 6). This way, in time $2^{\mathcal{O}(k^3)} \cdot n$ we either compute a tree decomposition of G of width at most k, or conclude that no such decomposition exists.

Of course, the above scheme relies on an optimistic assumption that the matching M is somehow provided to us, and that this also happens in all the recursive subcalls. Suppose for a moment that this is indeed the case: a matching of size at most ε times the vertex count can be always found in linear time. As all operations performed by the algorithm except the recursive call can be implemented in time $2^{\mathcal{O}(k^3)} \cdot n$, the total running time of the algorithm is $2^{\mathcal{O}(k^3)} \cdot n$ plus the time needed to solve an instance with $(1 - \varepsilon)n$ vertices. Unraveling the recursion leads to an upper bound on the running time of

$$2^{\mathcal{O}(k^3)} \cdot n + (1 - \varepsilon) \cdot 2^{\mathcal{O}(k^3)} \cdot n + (1 - \varepsilon)^2 \cdot 2^{\mathcal{O}(k^3)} \cdot n + \ldots = \varepsilon^{-1} \cdot 2^{\mathcal{O}(k^3)} \cdot n.$$

Hence, we obtain the desired running time bound even if ε is as low as the inverse of $2^{\mathcal{O}(k^3)}$.

We are left with the task of finding a large matching in the graph and deciding what to do in its absence. We try to construct the matching greedily: start with empty M and, as long as the graph contains at least one edge, we pick any edge, add it to M, and remove both endpoints from the graph. Thus, in linear time we find an inclusion-wise maximal matching M. If M has size at least $\varepsilon \cdot n$, then we can apply the recursive scheme described above. On the other hand, if $|M| \leqslant \varepsilon \cdot n$, then the set of endpoints of edges of M—call it C—is a vertex cover of G of size at most $2\varepsilon \cdot n$: there are no edges with both endpoints outside of C. The discovery of such a small vertex cover imposes a lot of structure in the graph. The idea is that we would like to use this structure to design a different recursive reduction scheme, for which we will use the concept of the *k-improved graph*.

Imagine that u and v are two vertices of a graph G such that there are $k + 1$ vertex-disjoint paths in G connecting u and v. Then it is not hard to see that in any tree decomposition of G of width at most k, there will be always a bag that contains both u and v. Indeed, otherwise some edge of the decomposition would give rise to a separation of G of order at most k that has u on one side and v on the other, contradicting the assumption that there are $k + 1$ vertex-disjoint paths between them. Hence, adding the edge uv to the graph, in case it was not there in the first place, cannot bring the treewidth above k. This suggests the following notion of the *k-improved graph*[1] $G^{\langle k \rangle}$: it is the supergraph of G obtained by adding all edges uv such that the maximum vertex-disjoint flow between u and v is larger than k. Then the above discussion implies the following.

---

[1] We note that Bodlaender in [2] uses a slightly different notion of the improved graph, where we require that the number of common neighbors of u and v is more than k; this essentially restricts attention to looking at paths of length 2.

Lemma 4. *Every tree decomposition of G of width at most k is also a tree decomposition of $G^{\langle k \rangle}$. In particular, if G has treewidth at most k, then so does $G^{\langle k \rangle}$.*

While the improved graph $G^{\langle k \rangle}$ can be constructed in polynomial time by computing the maximum vertex-disjoint flow between every pair of vertices, a priori it is not entirely clear how to compute it in linear time, even under the assumption that the graph has bounded treewidth. This can be in fact done using an FPT enumeration algorithm for MSO queries in bounded treewidth graphs [14], but we will not need the full extent of this observation.

Let us go back to the situation where we exposed a vertex cover C of size at most $2\varepsilon \cdot n$ in the considered graph G. Let H be a graph on vertex set C where two vertices from C are adjacent if and only if in G they have more than k common neighbors outside of C. Observe that if two vertices of C are adjacent in H, they are also adjacent in the improved graph $G^{\langle k \rangle}$: each of the common neighbors gives rise to a path connecting u and v. The intuition is that if in G there is a lot of vertices outside of C, then for almost all of them the neighborhood within C needs to be a clique in H, for otherwise we could construct a dense minor of G. This is explained formally in the next lemma.

Lemma 5. *There are at least $n - (k^2 + k + 1)|C|$ vertices $v \in V(G) \setminus C$ satisfying the following property: the set of neighbors of v in C is a clique in H.*

Proof. We construct a minor J of G iteratively as follows: start with $J = G$ and as long as there exists a vertex $v \notin C$ that has two non-adjacent neighbors u, u' in C, contract v onto u, that is, contract the edge uv and rename the resulting vertex to u (which therefore stays in C). Once we achieve the situation where for every vertex outside of C its neighborhood in C is a clique, we delete from J all the vertices outside of C. Note here that during the construction, their neighborhood in C did not change.

Thus, J is a minor of G and its vertex set is C. Since G has treewidth at most k, the same holds also for J, so in particular J has at most $k|C|$ edges. Since during the construction of G' every contraction resulted in adding one edge to J, we conclude that in total we performed at most $k|C|$ contractions. This means that at the end we deleted at least $n - (k+1)|C|$ vertices $v \notin C$ with the following property: the set of neighbors of v in C is a clique in J. Denote the set of those vertices by S.

For every edge uu' of J that is not present in H, there are at most k vertices in S that are adjacent in G to both u and u'. Let us mark all such vertices, for all such edges uu'. In this way we mark at most $k|E(J)| \leqslant k^2|C|$ vertices in S such that the unmarked vertices have the desired property: their neighborhoods in C are cliques in H. As $|S| \geqslant n - (k+1)|C|$, the number of unmarked vertices is at least $n - (k^2 + k + 1)|C|$. $\qquad\square$

Lemma 5 ensures us that if we set $\varepsilon = \frac{1}{4(k^2+k+1)}$, then at least half of the vertices of the graph reside outside of C and are adjacent to a clique in H; call those vertices *improved-simplicial*. Such vertices are very good for us for the following reason. Let $v \notin C$ be an improved-simplicial vertex. Since $N(v) \cap C$ is a clique in H, which is a subgraph of $G^{\langle k \rangle}$, by Lemma 4 it is safe to turn $N(v) \cap C$ into a clique when computing a tree decomposition of width at most k of G. Note here that v together with $N(v) \cap C$ forms a clique in $G^{\langle k \rangle}$, so if G has treewidth at most k, then we must have $|N(v) \cap C| \leqslant k$. Now, if we compute any tree decomposition of G with v removed and $N(v) \cap C$ turned into a clique, then we can always insert v back into this decomposition without increasing its width above k. Namely, we locate any node x whose bag contains the whole set $N(v) \cap C$ (such a bag exists since $N(v) \cap C$ is turned into a clique) and add a new node x', adjacent only to x and with bag $\{v\} \cup (N(v) \cap C)$.

This approach can be applied not only to a single improved-simplicial vertex, but to all of them at once. More precisely, if we constructed a vertex cover C of G of size at most $2\varepsilon n$ for $\varepsilon = \frac{1}{4(k^2+k+1)}$, then we can apply the following recursive reduction scheme:

- Compute the graph H and the set of all improved-simplicial vertices A. If either $|A| < n/2$ or for any $v \in A$ we have $|N(v) \cap C| > k$, we conclude that the treewidth of G is larger than k.
- Otherwise, let G' be the graph obtained from G by removing A and adding all edges that are present in H but not present in G. Then G' has at most $n/2$ vertices.
- Apply the algorithm recursively on G' to construct a tree decomposition T' of G' of width at most k. If no such decomposition exists, the same can be said about G.
- Construct a tree decomposition T of G from T' by reintroducing all vertices of I one by one, as described above.

Of course, in our presentation there are multiple implementation details missing: we need to implement each of the above steps in linear FPT time in order to apply the same running time analysis as in the case of the reduction through finding a large matching. Achieving this requires some reorganization of the argument and tending to a number of technical issues, which are dutifully explained by Bodlaender in [2]. In particular, the final value of ε is of the order of k^{-8} rather than k^{-2}, as would be suggested by our presentation.

3.3   Treewidth over Treewidth: The Bodlaender-Kloks Algorithm

We are left with explaining the main ideas behind the proof of Theorem 6, the Bodlaender-Kloks algorithm. The setting is as follows: we are given a graph G together with its tree decomposition T of width ℓ and we would like to construct, if possible, another tree decomposition S of G, of width $k < \ell$.

As usual with algorithms on tree decompositions, the idea is to use bottom-up dynamic programming, so let us set up some definitions to facilitate the discussion. We root $T = (T, \beta)$ in any node r and orient every edge of T towards the root; that is, every edge now points from a child to its parent. Similarly as in the proof of Lemma 1, with every (now oriented) edge (x, y) of T we can associate a separation (A_x, B_x) of G where A_x comprises vertices that appear in the bags of the descendants of x (including x), while B_x comprises vertices that appear in the bags of the non-descendants of x. The separator associated with this separation is $A_x \cap B_x = \beta(x) \cap \beta(y)$ and we can assume that its size is at most ℓ.

The generic idea would be to compute, using bottom-up dynamic programming, for every node x of T a family $\mathcal{R}_x$ of partial solutions in the subgraph $G[A_x]$ with the following *representativity* property: if there exists some tree decomposition of G of width at most k, then some of the partial solutions stored in $\mathcal{R}_x$ can be extended to such a tree decomposition. In order for the whole algorithm to run in linear time for fixed k and ℓ, the set $\mathcal{R}_x$ should be small: of cardinality bounded only by a function of k and ℓ. However, while the notion of a partial solution is sort of clear for combinatorially less complex problems, like INDEPENDENT SET, what is a "partial solution" for the problem of constructing a tree decomposition of a graph? More importantly, so far we have no control over how the given decomposition T and the sought decomposition S interact: a priori they can be completely unrelated, large combinatorial objects, and this may create problems for bounding the number of representative partial solutions stored in $\mathcal{R}_x$.

Let us address the second issue first, as its resolution will somehow imply the shape of partial solutions that will be stored. Moreover, let us simplify the problem: we would like to construct, if possible, a *path* decomposition of G of width at most k, instead of a tree decomposition. Path decompositions have a combinatorially simpler structure than tree decompositions and focusing on this case first will help us in highlighting the main ideas.

It will be convenient to think about path decompositions algebraically, as sequences of instructions over boundaried graphs. Here, a *p-boundaried graph* is a graph with at most p distintinguished vertices called *sources*. Over such p-boundaried graphs we can consider instructions of the following form:

– Introduce a new source u and make it adjacent to a subset of currently active sources.
– Forget the source u, i.e., make it into a regular vertex.

Then a path decomposition S of G of width at most k can be, after making it *nice*, directly translated to a sequence of instructions over $(k + 1)$-boundaried graphs that, when executed in order, constructs the graph. Such a sequence will be called a *k-construction sequence* for G, and from the above discussion it should be clear that path decompositions of width at most k and k-construction sequences are essentially in one-to-one correspondence. Note that every instruction appearing in a k-construction sequence π concerns one vertex—it either introduces it or forgets it—and every vertex is associated with two instructions.

Now we are interested in understanding how a k-construction sequence of G, say π, can interact with the given tree decomposition T. More precisely, we examine the interaction between π with every separation (A_x, B_x), for x ranging over nodes of T. Fix such a separation and call it (A, B) for brevity. Let us color instructions in π red, green, and blue as follows:

- instructions concerning vertices of $A \setminus B$ are colored red;
- instructions concerning vertices of $A \cap B$ are colored green; and
- instructions concerning vertices of $B \setminus A$ are colored blue.

There are only $2|A \cap B| \leqslant 2\ell$ green instructions in total. However, the number of red and blue instructions is unbounded and a priori they can appear even in a completely alternating fashion: a red instruction, then a blue, then again a red, and so on. This corresponds to the situation where the associated path decomposition alternately builds parts of the graph on the two sides of the separation. This in principle could happen and would result in a complex interaction between π and the separation (A, B), but the intuition is that it makes little sense to consider path decompositions with such a behavior. More precisely, as blocks of instructions on either side are sort of independent, due to the absence of edges between $A \setminus B$ and $B \setminus A$, we should be able to reorder any path decomposition of bounded width so that it alternates between the sides of the separation only a bounded number of times.

   This is made formal in the following lemma, which encapsulates the core combinatorial argument behind the algorithm of Bodlaender and Kloks. Here, in a sequence of instructions, an (A, B)-*block* is either a single green instruction or a maximal contigious subsequence of instructions that are either all red or all blue, in the coloring described above.

Lemma 6. *Let G be a graph ane let (A, B) be a separation of G of order at most ℓ. Suppose G admits a k-construction sequence. Then G also admits a k-construction sequence where the number of (A, B)-blocks is smaller than $(2k + 4)(2\ell + 1)$.*

Proof. Let π be a k-construction sequence for G that minimizes the number of (A, B)-blocks; we need to prove that this number is smaller than $(2k+4)(2\ell+1)$. To facilitate the discussion, we color the instructions red, green, and blue as described above. Thus we can talk about red and blue blocks.

   There are at most $2|A \cap B| \leqslant 2\ell$ green instructions in total. Hence, it suffices to prove that between any two consecutive green instructions, as well as before the first and after the last green instruction, the number of red and blue blocks is bounded by $2k + 3$. For contradiction, suppose that π contains a sequence $\sigma_1, \sigma_2, \ldots, \sigma_q$ of red/blue blocks that appear one after the other, where $q > 2k+3$. Note that these blocks are alternately red and blue.

   For a block σ_i, let the *effect* of σ_i be the total number of introduce instructions in σ_i minus the total number of forget instructions in σ_i. The next claim expresses the key observation.

Claim 1. *There are no two consecutive blocks σ_i and σ_{i+1} such that σ_i has a nonnegative effect and σ_{i+1} has a nonpositive effect.*

Proof. Suppose such a situation happens and consider swapping the blocks σ_i and σ_{i+1} within π. This operation strictly decreases the number of blocks in π. Moreover, the assumption about the effects of σ_i and σ_{i+1} implies that the number of sources used at any point does not increase and π remains a $(k + 1)$-construction sequence for G. This is a contradiction with the choice of π.   □

By Claim 1 it is easy to see that the sequence $\sigma_1, \sigma_2, \ldots, \sigma_q$ can be partitioned into a (possibly empty) prefix of blocks with negative effects, then possibly one block with zero effect, and finally a (possibly empty) suffix of blocks with positive effects. Observe that this suffix has length at most $k + 1$, for otherwise when executing the instructions in this suffix, the number of sources would grow by more than $k+1$, while it should be bounded by $k+1$ at any point. Similarly, the abovementioned prefix also has length at most $k + 1$, implying that the whole sequence $\sigma_1, \sigma_2, \ldots, \sigma_q$ has length at most $2k + 3$, a contradiction.   □

Conceptually, Lemma 6 tells us that for each fixed separation (A, B) appearing in decomposition $\mathcal{T}$ (i.e. $(A, B) = (A_x, B_x)$ for some node x), any k-construction sequence for G can be reorganized so that its interaction (A, B) is somehow tamed: the construction sequence alternates between the sides of the separation only a bounded number of times. It would be more convenient to have a single k-construction sequence that is tamed with respect to all separations in $\mathcal{T}$ at the same time. Fortunately, this can be easily achieved by applying the reordering of Lemma 6 over all separations in $\mathcal{T}$, one by one in a bottom-up manner. This is formalized in the next lemma, whose proof is left to the reader.

Lemma 7. *Let G be a graph of pathwidth at most k and let $\mathcal{T}$ be a tree decomposition of G of width at most ℓ. Then there exists a k-construction sequence π for G such that for each node x of $\mathcal{T}$, the number of (A_x, B_x)-blocks in π is smaller than $(2k + 4)(2\ell + 1)$.*

We note that Bodlaender and Kloks actually never explicitly state any analogue of Lemma 7 in [5]: the corresponding reordering is done implicitly in the inductive proof of the correctness of the algorithm.

Thus, Lemma 7 provides exactly the object we were looking for: a path decomposition of G (represented as a k-construction sequence π) whose interaction with $\mathcal{T}$ is bounded. We now need to understand how π can be assembled using a bottom-up dynamic programming with a bounded number of states. Consider a node x of $\mathcal{T}$ and examine π from the point of view of separation (A_x, B_x); in the following, when talking about "introduced" or "forgotten" vertices we refer to the standard meaning of these terms when performing a bottom-up dynamic programing over decomposition $\mathcal{T}$. Then π can be partitioned into less than $(2k + 4)(2\ell + 1)$ blocks, each being either

- a block of red instructions concerning only forgotten vertices—those belonging to $A_x \setminus B_x$; or
- a block of blue instructions concerning only not yet introduced vertices—those belonging to $B_x \setminus A_x$; or
- a single green instruction concerning a vertex of $A_x \cap B_x$.

Thus, one can imagine that a "partial solution" induced by π on the separation (A_x, B_x) should consist of a sequence of blocks of instructions as above, but the blue blocks should be replaced by "gaps" to be filled. Then extending a partial solution to a complete k-construction sequence π boils down to appropriately placing instructions for not yet introduced vertices in the gaps.

Of course, the total number of possible partial solutions is still too large, but let us think what information about a partial solution $\pi°$ is really relevant for determining its possible extensions to a complete solution. First, we can remember the whole sequence of types of blocks: this is a sequence of length less than $(2k+4)(2\ell+1)$ where each entry is either

- a marker "red block"; or
- a marker "gap for a blue block"; or
- a single green instruction.

The key observation now is that from the point of view of completing $\pi°$ into a complete solution, we do not need to remember each red block precisely. This is because for filling all blue gaps with instructions for vertices of $B_x \setminus A_x$, we do not need to remember which exactly vertices of $A_x \setminus B_x$ are sources at any point: it suffices to remember only their *number*, because this is all we need to make sure that we never exceed the upper bound of $k+1$ sources in the built construction sequence. Hence, the only two pieces of information that we need to store about each red block σ is:

- Its *effect*: the total number of introduce instructions minus the total number of forget instructions. This enables us to keep track of the number of active sources in $A_x \setminus B_x$ whenever a blue block of instructions is executed.
- Its *maximum*: the maximum effect of a prefix of σ. This enables us to make sure that after filling in the blue instructions, we do not exceed the upper bound of $k+1$ sources within the block σ.

Note that the effect of a block is an integer between $-k-1$ and $k+1$ while the maximum of a block is an integer between 0 and $k+1$.

Thus, with each partial solution we can associate its *trace* which consists of the whole sequence of types of blocks, as described above, plus for each marker of a red block we also store its effect and its maximum. Then from the above discussion it follows that every two partial solutions with the same trace have the same set of extensions to a complete solution (in the sense of ways of filling the gaps for blue blocks). Observe that the number of different traces is bounded by $\ell^{O(\ell)} \cdot k^{O(k\ell)}$: the number of sequences of colors red/green/blue is at most $3^{(2k+4)(2\ell+1)}$, for each marker of a red block there are at most $(2k+4)(k+2)$ ways to choose its effect and maximum, and for each green instruction there are at most 2ℓ ways to choose the vertex of $A_x \cap B_x$ in question, and whether we introduce or forget it.

This gives us the final definition of the state space of dynamic programming: for each node x of T and for each possible trace τ of a partial solution on the separation (A_x, B_x), we compute any partial solution having this trace, if existent. Designing transition rules for this dynamic programming is now a tedious, but relatively straightforward task; we leave working out the details to the reader.

A careful reader might have noticed that the algorithm, as described above, has the running time of $\ell^{O(\ell)} \cdot k^{O(k\ell)} \cdot n$, which reduces to $k^{O(k^2)} \cdot n$ for $\ell = O(k)$, but the algorithm for pathwidth given by Bodlaender in [2] runs in time $2^{O(k^2)} \cdot n$.

This slight difference comes from the fact that in the above description, we performed the analysis of the number of relevant states a bit less carefully than Bodlaender and Kloks in [5]. Precisely, we bounded the number of red blocks by $\mathcal{O}(k\ell)$ and then concluded that for each of them we need to store two integers of the order of k, resulting in at most $k^{\mathcal{O}(k\ell)}$ possible options. Bodlaender and Kloks take a closer look at the sequences of effects and maxima stored for consecutive blocks and conclude that actually there are only $2^{\mathcal{O}(k\ell)}$ sequences where one cannot apply a swap similar to the one in the proof of Claim 1. These sequences are called *typical*. As in the dynamic programming one can restrict attention only to traces using typical sequences, this yields the abovementioned improvement of the running time.

We are left with discussing how the approach needs to be adjusted to the case of computing a tree decomposition of width k, instead of a path decomposition of width k. Arguably the key conceptual insight—bounding the "alternation complexity" of the constructed object—is already present in the case of path decompositions. Hence, we only briefly sketch the technical difficulties that need to be overcome in the tree case.

Suppose the input graph admits a solution: a tree decomposition $\mathcal{S}$ of width at most k. Similarly as before, by making $\mathcal{S}$ rooted and nice we can view it as a term of introduce, forget, and join instructions. The main difference is that instead of a sequence of instructions, now we have a tree. While every vertex is associated with only one forget instruction, it can be introduced multiple times.

We need to understand how the tree decomposition $\mathcal{S}$ can be "dealternated": reorganized so that its interaction with every separation arising from the approximate decomposition $\mathcal{T}$ is bounded. Consider then any separation $(A, B) = (A_x, B_x)$ for x being a node of $\mathcal{T}$. Then we can assume that the separator $A \cap B$ consists of at most ℓ vertices, which in turn correspond to at most ℓ nodes of $\mathcal{S}$ where the vertices of $A \cap B$ are being forgotten. Call a node of $\mathcal{S}$ *special* if it either forgets some vertex of $A \cap B$, or is the least common ancestor of two such nodes; let X be the set of special nodes. It can be easily checked that then $|X| \leqslant 2\ell - 1$ and that every connected component of $\mathcal{S} - X$ is adjacent to at most 2 special nodes. Hence, we can imagine the decomposition $\mathcal{S}$ as follows: take the at most $2\ell - 1$ nodes of X, connect them into a tree using paths of bags where no vertex of $A \cap B$ is being forgotten, and finally to every node attach an arbitrary number of subtrees, also only with bags where no vertex of $A \cap B$ is being forgotten. It is not hard to see that as there is no edge between $A \setminus B$ and $B \setminus A$, we may assume that each of these attached subtrees contains solely vertices of A or solely vertices of B; otherwise we could "unmix" a subtree into two.

The crucial insight is that on paths between nodes of X one can apply a similar swapping argument as we did for path decompositions. This implies that each of these paths can be partitioned into $\mathcal{O}(k\ell)$ blocks that alternately construct parts of the graph on the $A \setminus B$ side and on the $B \setminus A$ side. Then the same reasoning as in the pathwidth case shows that in the dynamic programming, for each path we only need to remember its trace, for which there are only $\ell^{\mathcal{O}(\ell)} \cdot 2^{\mathcal{O}(k\ell)}$

possibilities (using typical sequences). Thus, in a state of dynamic programming we need to store the shape of this tree of paths—for which there are $\ell^{\mathcal{O}(\ell)}$ choices—and at most $2\ell - 2$ traces of paths—each chosen among $\ell^{\mathcal{O}(\ell)} \cdot 2^{\mathcal{O}(k\ell)}$ possibilities. After fixing all the technical details that we swept under the rug, this leads to a correct setup of a dynamic programming procedure with $\ell^{\mathcal{O}(\ell^2)} \cdot 2^{\mathcal{O}(k\ell^2)}$ states per node, yielding the promised time complexity.

3.4   Further Reading

To the best of author's knowledge, the idea of typical sequence was introduced independently in 1991 by Bodlaender and Kloks in their $f(k) \cdot n \log^2 n$-time algorithm for computing treewidth [4], and, under the name *non-redundant sequences*, by Lagergren and Arnborg for the purpose of giving constructive upper bounds on sizes of minor-minimal obstructions to having treewidth at most k [21]. The latter direction was later explored further by Lagergren [20], who made the upper bounds explicit and investigated several related problems using the same technique. Non-redundant sequences were also used by Courcelle and Lagergren [9] in the context of finite-state recognizability of classes of graphs of bounded treewidth. The algorithmic work of Bodlaender and Kloks [5], which we explained above, gave a thorough analysis of typical sequences including optimum bounds on their number.

The general methodology behind Bodlaender's algorithm has been used to give exact fixed-parameter algorithm for several other parameters, including branchwidth [6], carving-width [28], and cutwidth [16,29,30]. Jeong et al. [17,18] recently used the technique to give such algorithms for pathwidth and branchwidth of representable matroids, which in turn gives algorithms for graph parameters such as branchwidth, carving-width, and rankwidth as corollaries. The recursive reduction scheme used by Bodlaender in [2] was later applied again by Bodlaender et al. [3] in their 5-approximation algorithm working in time $2^{\mathcal{O}(k)} \cdot n$.

Our presentation of the Bodlaender-Kloks algorithm departs from the description that can be found in the original paper [5] and rather follows a much later interpretation of Bojańczyk and Pilipczuk [8]. They used the techniques of typical sequences to show that for every fixed k and ℓ, the problem of constructing a tree decomposition of width k given one of width $\ell > k$ can be formulated in logic, as a so-called *MSO transduction*. Having such a formulation, a suitable dynamic programming procedure running in linear fixed-parameter time can be constructed using general-usage meta-theorems for MSO transductions.

References

1. Arnborg, S., Corneil, D.G., Proskurowski, A.: Complexity of finding embeddings in a k-tree. SIAM J. Discrete Math. **8**(2), 277–284 (1987)
2. Bodlaender, H.L.: A linear-time algorithm for finding tree-decompositions of small treewidth. SIAM J. Comput. **25**(6), 1305–1317 (1996)

3. Bodlaender, H.L., Drange, P.G., Dregi, M.S., Fomin, F.V., Lokshtanov, D., Pilipczuk, M.: A $c^k n$ 5-approximation algorithm for treewidth. SIAM J. Comput. **45**(2), 317–378 (2016)

4. Bodlaender, H.L., Kloks, T.: Better algorithms for the pathwidth and treewidth of graphs. In: Albert, J.L., Monien, B., Artalejo, M.R. (eds.) ICALP 1991. LNCS, vol. 510, pp. 544–555. Springer, Heidelberg (1991). https://doi.org/10.1007/3-540-54233-7_162

5. Bodlaender, H.L., Kloks, T.: Efficient and constructive algorithms for the pathwidth and treewidth of graphs. J. Algorithms **21**(2), 358–402 (1996)

6. Bodlaender, H.L., Thilikos, D.M.: Constructive linear time algorithms for branchwidth. In: Degano, P., Gorrieri, R., Marchetti-Spaccamela, A. (eds.) ICALP 1997. LNCS, vol. 1256, pp. 627–637. Springer, Heidelberg (1997). https://doi.org/10.1007/3-540-63165-8_217

7. Bojańczyk, M., Pilipczuk, M.: Definability equals recognizability for graphs of bounded treewidth. CoRR, abs/1605.03045 (2016). Extended abstract published in the proceedings of LICS 2016

8. Bojańczyk, M., Pilipczuk, M.: Optimizing tree decompositions in MSO. In: Proceedings of the 34th Symposium on Theoretical Aspects of Computer Science, STACS 2017. LIPIcs, vol. 66, pp. 15:1–15:13. Schloss Dagstuhl – Leibniz-Zentrum für Informatik (2017)

9. Courcelle, B., Lagergren, J.: Equivalent definitions of recognizability for sets of graphs of bounded tree-width. Math. Struct. Comput. Sci. **6**(2), 141–165 (1996)

10. Cygan, M., et al.: Parameterized Algorithms. Springer, Cham (2015). https://doi.org/10.1007/978-3-319-21275-3

11. Dell, H., Komusiewicz, C., Talmon, N., Weller, M.: The PACE 2017 parameterized algorithms and computational experiments challenge: the second iteration. In: 12th International Symposium on Parameterized and Exact Computation, IPEC 2017, LIPIcs, vol. 89, pp. 30:1–30:12. Schloss Dagstuhl – Leibniz-Zentrum für Informatik (2017)

12. Elberfeld, M., Schweitzer, P.: Canonizing graphs of bounded tree width in logspace. ACM Trans. Comput. Theory (TOCT) **9**(3), 1–29 (2017)

13. Feige, U., Hajiaghayi, M., Lee, J.R.: Improved approximation algorithms for minimum weight vertex separators. SIAM J. Comput. **38**(2), 629–657 (2008)

14. Flum, J., Frick, M., Grohe, M.: Query evaluation via tree-decompositions. J. ACM **49**(6), 716–752 (2002)

15. Garey, M.R., Johnson, D.S.: Computers and Intractability: A Guide to the Theory of NP-Completeness. W. H. Freeman (1979)

16. Giannopoulou, A.C., Pilipczuk, M., Raymond, J., Thilikos, D.M., Wrochna, M.: Cutwidth: obstructions and algorithmic aspects. Algorithmica **81**(2), 557–588 (2019)

17. Jeong, J., Kim, E.J., Oum, S.I.: Constructive algorithm for path-width of matroids. In: Krauthgamer, R. (ed.) Proceedings of the 27th Annual ACM-SIAM Symposium on Discrete Algorithms, SODA 2016, pp. 1695–1704. SIAM (2016)

18. Jeong, J., Kim, E.J., Oum, S.: Finding branch-decompositions of matroids, hypergraphs, and more. In: Proceedings of the 45th International Colloquium on Automata, Languages, and Programming, ICALP 2018. LIPIcs, vol. 107, pp. 80:1–80:14. Schloss Dagstuhl – Leibniz-Zentrum für Informatik (2018)

19. Kleinberg, J.M., Tardos, É.: Algorithm Design. Addison-Wesley (2006)

20. Lagergren, J.: Upper bounds on the size of obstructions and intertwines. J. Comb. Theory Ser. B **73**(1), 7–40 (1998)

21. Lagergren, J., Arnborg, S.: Finding minimal forbidden minors using a finite congruence. In: Albert, J.L., Monien, B., Artalejo, M.R. (eds.) ICALP 1991. LNCS, vol. 510, pp. 532–543. Springer, Heidelberg (1991). https://doi.org/10.1007/3-540-54233-7_161

22. Lokshtanov, D., Pilipczuk, M., Pilipczuk, M., Saurabh, S.: Fixed-parameter tractable canonization and isomorphism test for graphs of bounded treewidth. SIAM J. Comput. **46**(1), 161–189 (2017)

23. Oum, S., Seymour, P.D.: Approximating clique-width and branch-width. J. Comb. Theory Ser. B **96**(4), 514–528 (2006)

24. Raghavendra, P., Steurer, D.: Graph expansion and the unique games conjecture. In: Proceedings of the 42nd ACM Symposium on Theory of Computing, STOC 2010, pp. 755–764. ACM (2010)

25. Reed, B.A.: Finding approximate separators and computing tree width quickly. In: Proceedings of the 24th Annual ACM Symposium on Theory of Computing, STOC 1992, pp. 221–228. ACM (1992)

26. Robertson, N., Seymour, P.D.: Graph minors. XIII. The disjoint paths problem. J. Comb. Theory Ser. B **63**(1), 65–110 (1995)

27. Robertson, N., Seymour, P.D.: Graph minors. XX. Wagner's conjecture. J. Comb. Theory Ser. B **92**(2), 325–357 (2004)

28. Thilikos, D.M., Serna, M.J., Bodlaender, H.L.: Constructive linear time algorithms for small cutwidth and carving-width. In: Goos, G., Hartmanis, J., van Leeuwen, J., Lee, D.T., Teng, S.-H. (eds.) ISAAC 2000. LNCS, vol. 1969, pp. 192–203. Springer, Heidelberg (2000). https://doi.org/10.1007/3-540-40996-3_17

29. Thilikos, D.M., Serna, M.J., Bodlaender, H.L.: Cutwidth I: a linear time fixed parameter algorithm. J. Algorithms **56**(1), 1–24 (2005)

30. Thilikos, D.M., Serna, M.J., Bodlaender, H.L.: Cutwidth II: algorithms for partial w-trees of bounded degree. J. Algorithms **56**(1), 25–49 (2005)

31. Wu, Y., Austrin, P., Pitassi, T., Liu, D.: Inapproximability of treewidth and related problems. J. Artif. Intell. Res. **49**, 569–600 (2014)

32. Yannakakis, M.: Computing the minimum fill-in is NP-complete. SIAM J. Algebr. Discrete Methods **2**(1), 77–79 (1981)

Experimental Analysis of Treewidth

Hisao Tamaki[✉]

Meiji University, Kawasaki 214-8571, Japan
tamaki@cs.meiji.ac.jp

Abstract. The goal of this chapter is to provide insights into treewidth obtained from experiments. In our experiments, we count the numbers of combinatorial objects closely related to treewidth in random graph instances. These combinatorial objects include connected sets, minimal separators, potential maximal cliques, and those with certain constraints. Such experimental analysis is expected to complement theoretical analysis, reveal the reasons why some algorithms work well in practice while some do not, and provide a basis for designing new algorithms.

1 My Personal History in Treewidth Research and the Influence of Hans

My initial motivation for working on practical treewidth computation came from PACE 2016 [5] and PACE 2017 [6] algorithm implementation challenges which featured treewidth computation tracks. Through participation in these challenges, I was able to develop algorithms that significantly extend the scale of graphs for which the exact treewidth can be computed in practical time.

After PACE 2017 and the publication of my PACE submission in ESA 2017 [8], I decided to continue the work for further improvements. Achieving this goal, however, turned out extremely difficult. For the first half year of this period, all of my attempts for improvements failed. It was the influence of Hans that kept me working, though he probably does not know it.

In 2017, Hans applied for grants to have a Netherlands-Japan joint seminar on treewidth and its applications, with Japanese counterpart Yota Otachi. Through Yota, I learned that the seminar was inspired by my work on practical treewidth computation. The applications were granted and the seminar was held in September 2018 in Eindhoven. I found it both a pleasure and a pressure to present my work on treewidth before Hans. Obviously, I did not want to say that I had worked for one year for improvements without success. So, I continued to work hard and managed to report some progress in the seminar. With more efforts, I obtained a new exact algorithm for treewidth [9].

Around that time, I had a half-year sabbatical leave granted, which would start in April 2019. My first preference for the destination was, naturally, Utrecht University and I asked Hans if it would be possible. Fortunately, he accepted my request for hosting me. By the time I started my stay in Utrecht, I had realized the importance of heuristic upper bound algorithms in treewidth computation.

© Springer Nature Switzerland AG 2020
F. V. Fomin et al. (Eds.): Bodlaender Festschrift, LNCS 12160, pp. 214–221, 2020.
https://doi.org/10.1007/978-3-030-42071-0_15

I believed and continue to believe that the practically fastest way to compute the exact treewidth is to first compute an upper bound that is likely to be tight and then prove the matching lower bound. So, I spent the entire 6 months in Utrecht to develop an efficient upper bound algorithm. Hans was always eager to listen to me about my progress and this was again a major source of my motivation to keep going.

In the early stage of this project, I came up with an approach which I called HeartBeat. I remember the excitement I had with this idea and one of the reasons of the excitement was this naming which abbreviates to HB, though I never talked about the association of this abbreviation to a particular person. The HeartBeat heuristic worked fairly well and was competitive to previous approaches in all ranges of the magnitude of the graph. I, however, wanted it to be superior, not just competitive, to previous approaches in all ranges. After serious efforts in three months, I finally discovered a right approach, namely a heuristic use of dynamic programming [10]. As I chose to use the dynamic programming algorithm due to Bouchitté and Todinca [3] in this approach, the resulting algorithm may be called heuristic BT or HBT, an even better abbreviation that contains both HB and HT as subsequences.

At the time of writing this chapter, I am turning my attention to lower bound methods. I expect to encounter many theoretical questions in this study and am hoping to have collaborations with Hans.

2 Feasible Sets: General Versus Connected

Let G be a graph and k a positive integer. We are interested in deciding if the treewidth of G is at most k. We say that a vertex set U of G is *feasible* (with respect to k) if, for each connected component C of $G[U]$, there is a tree-decomposition of $G[C \cup N_G(C)]$ of width at most k that has a bag containing $N_G(C)$. Note that if U is feasible then there is a potential of growing those partial tree-decompositions into a tree-decomposition of G of width at most k.

Consider the decision problem version of the dynamic programming algorithm based on the perfect elimination order (PEO) [2]. Then, feasible vertex sets are exactly the YES-instances of the dynamic programming states in this algorithm. On the other hand, connected feasible sets are the YES-instances of the dynamic programming states in the algorithm due to Arnborg, Corneil, and Proskurowski (ACP algorithm) [1].

Since a feasible set in general is a combination of an unbounded number of feasible connected sets, their number can be exponential in the number of connected feasible sets. This, of course, does not immediately imply that the ACP dynamic programming is faster than the PEO-based dynamic programming. In the case of the PEO-based dynamic programming, it is easy to implement the algorithm so that only YES-instances of the states are constructed. I named this mode of executing a dynamic programming algorithm *positive-instance driven* (PID). In the ACP dynamic programming algorithm, however, the generation of a single YES-instance in PID mode involves a combinatorial search over smaller

YES-instances and therefore may take an exponential time in theory. Nonetheless, my submission to PACE 2016, which was a PID implementation of ACP, outperformed other submissions based on PEO.

In this section, we compare these algorithms not in terms of the running time of the implementations but in terms of the number of feasible sets.

I use random graphs for the experiments, since it makes it easier to see the dependence of the number of combinatorial objects on the number of vertices and the number of edges. Random sparse graphs, however, tend to contain many vertices of degree one or two and hence are subject to trivial preprocessing. Thus, the effective number of vertices fluctuates even if we fix the initial number of vertices. For this reason, I use random supercubic graphs (graphs with minimum degree 3) with the given numbers of vertices and edges. The model for generating an instance with n vertices and m edges is as follows. First assume that n is even. We first pick a cubic graph with n vertices uniformly at random. Then, from the $n(n-1)/2 - 3n/2$ non-adjacent pairs of vertices, we pick a set of $m - 3n/2$ pairs uniformly at random and place an edge to span each selected pair. When n is odd, we first pick a random "almost cubic" graph uniformly at random and proceed similarly, where an almost cubic graph is a graph in which the degree of every vertex except for one is 3 and the exceptional vertex has degree 4.

Fix a positive integer k. For each graph G, we say that a connected vertex set $C \subseteq V(G)$ is *admissible* if $|N_G(C)| \leq k$. We say that $U \subseteq V(G)$ is *trivially admissible* if $|U| \geq |V(G)| - k$.

Table 1 shows the result of our experiments, where we set k to be exactly the treewidth of the graph.

Each row corresponds to a graph instance and consists of the following columns: the number of vertices n, the number of edges m, the treewidth TW, the number of all vertex sets 2^n, the number of vertex sets that are not trivially admissible, the number of feasible vertex sets, the number of connected feasible sets, and the number of admissible sets. For the last three columns, trivially admissible vertex sets are excluded from the counts, since it is not necessary for the algorithms to process those sets.

The instances are sorted by the average degree (floored) and then by the number of vertices.

It is seen from the table that the number of connected feasible sets is considerably smaller than the number of general feasible sets. This gap is dramatic, expectedly, for instances with low degree (hence with low treewidth) and large number of vertices. For those instances, the number of admissible instances is also considerably smaller than the number of general feasible sets. These observed gaps suggest the superiority of the ACP algorithm to the PEO-based algorithm for low treewidth graphs.

For graphs with large treewidth, the superiority of the ACP algorithm may not be clear. The number of connected feasible sets is still considerably smaller than the number of general feasible sets but, as mentioned earlier, the time for constructing a single connected feasible set can be large. It appears that

Table 1. Counting various vertex sets of supercubic random graphs

n	m	TW	2^n	Not trivially admissible	Feasible	Connected feasible	Admissible
15	23	3	32768	32192	3759	81	82
20	30	5	1048576	1026876	293407	1713	3217
25	38	5	33554432	33486026	9102453	3884	8405
30	45	6	1073741824	1072973612	187416630	5168	89538
15	30	5	32768	27824	3398	304	443
20	40	6	1048576	988116	53829	337	4982
25	50	7	33554432	32828226	2165621	17245	68217
30	60	9	1073741824	1050777737	52716324	301176	3190002
15	45	7	32768	16384	786	203	788
20	60	8	1048576	784626	6724	191	6472
25	75	10	33554432	26434916	233797	10899	261455
30	90	12	1073741824	879612197	4984773	214167	9563493
15	60	8	32768	9949	134	32	233
20	80	10	1048576	431910	2419	172	3183
25	100	12	33554432	16777216	70051	13273	247370
30	120	15	1073741824	459312152	735340	14244	11485679
15	75	9	32768	4944	42	17	43
20	100	12	1048576	137980	671	105	1975
25	125	14	33554432	7119516	26305	7303	132474
30	150	17	1073741824	194129627	447208	174044	6377630

the performance of the ACP algorithm relative to the PEO-based algorithm for these graphs depends on how efficiently the PID version is implemented.

3   Minimal Separators and Potential Maximal Cliques

My submission to PACE 2017 was a considerable improvement over the previous submission. The difference was that the new submission was based on the BT algorithm rather than the ACP algorithm. In this section, we count the numbers of combinatorial objects involved in the BT algorithm.

Graph G is *chordal* if every induced cycle of G has length exactly three. For a general graph G, graph H is a *triangulation of G* if it is chordal, $V(H) = V(G)$, and $E(G) \subseteq E(H)$. It is a *minimal triangulation* of G if moreover $E(H)$ is inclusion-wise minimal subject to these conditions. A vertex set X of G is a *potential maximal clique* of G if X is a maximal clique of some triangulation of G. A vertex set S of G is a *minimal separator* of G if S has at least two full components, where a *full component* of S is a connected component C of $G[V(G) \setminus S]$ such that $N_G(C) = S$. We say a connected vertex set $C \subseteq V(G)$ is *minimally separated* if $N_G(C)$ is a minimal separator.

In the decision problem version of the BT algorithm, we ask the feasibility of connected C only if C is minimally separated. The recurrence for deciding the feasibility of minimally separated connected sets involve potential maximal cliques and therefore the number of potential maximal cliques is also relevant to the running time of the algorithm.

Table 2 compares the number of connected feasible sets and the number of objects involved in the BT algorithm. As before, we set k to be the treewidth of the graph. The instances used include those in Table 1 and some larger instances in addition. The row for graph G has, in addition to some columns common to Table 1, the number of minimally separated sets (MSC) with neighborhood cardinality at most k, the number of minimal separators (MS) of cardinality at most k, and the number of potential maximal cliques (PMC) of cardinality at most $k + 1$.

It is seen from the table that the number of minimally separated sets is always twice the number of minimal separators and therefore every minimal separator in the tested graphs has exactly two full components.

The growth of the numbers of minimally separated sets and potential maximal cliques is clearly milder than the growth of the number of connected feasible sets, which suggests the advantage of the BT algorithm over the ACP algorithm.

Table 3 is aimed at helping the evaluation of the PID approach in the BT algorithm. Larger instances than those in previous tables are included, which is possible since we drop some columns in the previous tables. Instead, columns for feasible minimally separated components and for "conducive" potential maximal cliques are added, where a vertex set U of G is *conducive*, with respect to a fixed k, if there is a tree-decomposition of G of width at most k that has U as a bag. In this table, as in previous tables, k is set to the treewidth of the graph.

As can be seen from the table, a large fraction of minimally separated components are feasible. Although the number of potential maximal cliques is significantly larger than that of feasible components, the growth of the ratio is rather moderate. Considering the complexity of constructing a single feasible component in the PID approach, the advantage of PID over standard implementation is not clear. Indeed, if the algorithm used in this experiment to generate potential maximal cliques is used in the standard implementation of the BT algorithm, then the resulting implementation is at least competitive with the PID implementation. The algorithm in [9] generates all minimal separators, but not potential maximal cliques, before the dynamic programming iteration. This is currently the practically best variant of BT, but it is possible that the development of a better algorithm for generating potential maximal cliques would make the standard implementation more competitive.

Finally, the counts in the last column may be worth special attention. For small-width graphs, a large fraction of all potential maximal cliques are conducive. It makes sense for heuristic algorithms to try to exploit this abundance of "good" potential maximal cliques, which is done in [10]. One may also wonder if this abundance can be exploited by theoretical randomized algorithms.

Table 2. Connected feasible sets versus minimally separated sets and potential maximal cliques

n	m	TW	2^n	Connected feasible	MSC	MS	PMC
15	23	3	32768	81	82	41	86
20	30	5	1048576	1713	512	256	1018
25	38	5	33554432	3883	954	477	2316
30	45	6	1073741824	5163	1392	696	2925
35	53	6	34359738368	13372	2160	1080	4957
40	60	7	1099511627776	199311	8550	4275	26357
15	30	5	32768	304	66	33	65
20	40	6	1048576	337	224	112	295
25	50	7	33554432	17242	790	395	1564
30	60	9	1073741824	301031	2926	1463	7318
35	70	9	34359738368	417181	4258	2129	10679
40	80	9	1099511627776	283549	4970	2485	13275
15	45	7	32768	203	48	24	48
20	60	8	1048576	191	124	62	123
25	75	10	33554432	10887	502	251	760
30	90	12	1073741824	214159	1834	917	3562
35	105	13	34359738368	158111	2650	1325	5225
40	120	13	1099511627776	3312355	3274	1637	7165
15	60	8	32768	32	30	15	20
20	80	10	1048576	172	130	65	123
25	100	12	33554432	13273	362	181	506
30	120	15	1073741824	14244	1456	728	2401
35	140	16	34359738368	355093	2812	1406	5272
40	160	18	1099511627776	5476470	8620	4310	19227
15	75	9	32768	17	14	7	8
20	100	12	1048576	105	78	39	59
25	125	14	33554432	7303	344	172	469
30	150	17	1073741824	174034	1248	624	2092
35	175	19	34359738368	641603	2306	1153	3848
40	200	21	1099511627776	17312180	9182	4591	19823

Table 3. Minimally separated sets and potential maximal cliques: general versus feasible

n	m	TW	2^n	MSC	Feasible MSC	MS	PMCs	cond. PMCs
30	45	6	1073741824	1392	1052	696	2925	1573
40	60	7	1099511627776	8550	6991	4275	26357	16928
50	75	9	1125899906842624	136422	122086	68211	592280	471148
60	90	9	1152921504606846976	125542	92366	62771	536794	256817
70	105	12	1180591620717411303424	6039606	4464424	3019803	36026551	17352277
30	60	9	1073741824	2926	2606	1463	7318	5715
40	80	9	1099511627776	4970	3781	2485	13275	6486
50	100	12	1125899906842624	59000	46091	29500	216675	122294
60	120	14	1152921504606846976	661792	444300	330896	2933914	974402
70	140	16	1180591620717411303424	2596462	1603589	1298231	12931628	3196073
30	90	12	1073741824	1834	1520	917	3562	2439
40	120	13	1099511627776	3274	2647	1637	7165	4692
50	150	17	1125899906842624	51072	30171	25536	153228	28013
60	180	22	1152921504606846976	1964012	1191239	982006	7810457	1691600
70	210	23	1180591620717411303424	3002876	1512387	1501438	12461106	81557
30	120	15	1073741824	1456	852	728	2401	326
40	160	18	1099511627776	8620	5106	4310	19227	3352
50	200	22	1125899906842624	89546	52084	44773	253140	40364
60	240	26	1152921504606846976	1015326	633763	507663	3589367	904894
70	280	28	1180591620717411303424	3407396	2129493	1703698	13276586	3513941
30	150	17	1073741824	1248	1086	624	2092	1596
40	200	21	1099511627776	9182	6944	4591	19823	10262
50	250	25	1125899906842624	64648	39110	32324	164230	34292
60	300	28	1152921504606846976	242146	126942	121073	712364	33975
70	350	31	1180591620717411303424	898710	461150	449355	2859306	77154

4   Conclusions

Imagine that the experimental results reported in this chapter had been available shortly after the publication of the BT algorithm. Then, they would have stimulated serious efforts on implementing the BT and ACP algorithms and would have resulted in implementations better than the ones currently available, say at least 10 years ago. Positive impacts on practical applications of treewidth could have been tremendous. This hypothetical scenario strongly suggests the importance of experimental algorithmics where we experimentally analyze the complexity not only of algorithms but also of structures underlying problems.

References

1. Arnborg, S., Corneil, D.G., Proskurowski, A.: Complexity of finding embeddings in a k-tree. SIAM J. Algebraic Discrete Methods **8**, 277–284 (1987)
2. Bodlaender, H.L., Fomin, F.V., Koster, A.M.C.A., Kratsch, D., Thilikos, D.M.: On exact algorithms for treewidth. ACM Trans. Algorithms **9**(1), 12 (2012)
3. Bouchitté, V., Todinca, I.: Treewidth and minimum fill-in: grouping the minimal separators. SIAM J. Comput. **31**(1), 212–232 (2001)

4. Bouchitté, V., Todinca, I.: Listing all potential maximal cliques of a graph. Theor. Comput. Sci. **276**, 17–32 (2002)
5. Dell, H., Husfeldt, T., Jansen, B.M., Kaski, P., Komusiewicz, C., Rosamond, F.A.: The first parameterized algorithms and computational experiments challenge. In: LIPIcs-Leibniz International Proceedings in Informatics, vol. 63 (2017)
6. Dell, H., Komusiewicz, C., Talmon, N., Weller, M.: The PACE 2017 parameterized algorithms and computational experiments challenge: the second iteration. In: Proceedings of the 12th International Symposium on Parameterized and Exact Computation (IPEC 2017), pp. 30:1–30:12 (2018)
7. Fomin, F., Villanger, Y.: Treewidth computation and extremal combinatorics. Combinatorica **32**(3), 289–308 (2012)
8. Tamaki, H.: Positive-instance driven dynamic programming for treewidth. J. Comb. Optim. **37**(4), 1283–1311 (2019). (First Online October 2018)
9. Tamaki, H.: Computing treewidth via exact and heuristic lists of minimal separators. In: Kotsireas, I., Pardalos, P., Parsopoulos, K.E., Souravlias, D., Tsokas, A. (eds.) SEA 2019. LNCS, vol. 11544, pp. 219–236. Springer, Cham (2019). https://doi.org/10.1007/978-3-030-34029-2_15
10. Tamaki, H.: A heuristic use of dynamic programming to upperbound treewidth. arXiv preprint arXiv:1909.07647 (2019)

A Retrospective on (Meta) Kernelization

Dimitrios M. Thilikos[✉] [iD]

LIRMM, Univ Montpellier, CNRS, Montpellier, France
sedthilk@thilikos.info

«Een mens zijn zin
is een mens zijn leven.»

Abstract. In parameterized complexity, a *kernelization algorithm* can
be seen as a reduction of a parameterized problem to itself, so that
the produced equivalent instance has size depending exclusively on the
parameter. If this size is polynomial, then we say that the parameterized
problem in question admits a polynomial kernelization algorithm. Kernelization can be seen as a formalization of the notion of preprocessing
and has occupied a big part of the research on Multi-variate Algorithmics. The first algorithmic meta-theorem on kernelization appeared in
[14] and unified a large family of previously known kernelization results
on problems defined on topologically embeddable graphs. In this exposition we present the central results of this paper. During our presentation
we pay attention to the abstractions on which the results where founded
and take into account the subsequent advancements on this topic.

Keywords: Parameterized problems · Parameterized algorithms ·
Kernelization algorithms · Algorithmic meta-theorems finite index ·
Finite integer index · Monadic Second Order Logic · Treewidth ·
Protrusion decompositions · Bidimensionality · Separability

1 Introduction

Parametrized complexity, introduced by Downey and Fellows in early 90's
[1,36–39], has nowadays evolved to a mature field of Theoretical Computer Science (see [25,35,40,45,83] for related textbooks). The key idea is to treat computational problems as *bivariate entities* where, apart from the size of the input, the
algorithm complexity is evaluated with respect to some *parameter* that quanti-
fies structural properties of the input. This approach is justified by the fact that,
for many computational problems, the inherent combinatorial explosion of their
complexity depends exclusively by the parameter, whose value is expected to
be small in certain applications. In this framework we deal with parameterized

Supported by projects DEMOGRAPH (ANR-16-CE40-0028) and ESIGMA (ANR-17-
CE23-0010) and by the Research Council of Norway and the French Ministry of Europe
and Foreign Affairs, via the Franco-Norwegian project PHC AURORA 2019.

© Springer Nature Switzerland AG 2020
F. V. Fomin et al. (Eds.): Bodlaender Festschrift, LNCS 12160, pp. 222–246, 2020.
https://doi.org/10.1007/978-3-030-42071-0_16

problems whose instances are pairs $(x, k) \in \Sigma^* \times \mathbb{N}$ and we look for algorithms whose running time is exponential in the parameter k but polynomial in the input size $n = |x|$. In other words, we are aiming for algorithms that run in time $f(k) \cdot n^c$ where f is a function depending only on the parameter and where c is a constant. The parameterized problems for which such algorithms exist constitute the parameterized complexity class FPT. In their foundational work [1, 36–39], Downey and Fellows invented a series of parameterized complexity classes and proposed special types of reductions such that hardness for some of the above classes gives plausible evidence that a problem does not belong to FPT.

Kernelization is a prominent field of parameterized complexity that has rapidly developed during the last two decades (see [57] for a recent textbook). A *kernelization algorithm* (or simply a *kernelization*) is a polynomial time algorithm that transforms every instance (x, k) of a parameterized problem to an equivalent one (x', k'), whose size depends exclusively on the parameter. We refer to the size of (x', k'), measured as a function of k, as the *size* of the kernelization. Kernelization algorithms can be seen as preprocessing procedures with performance guarantee. For this reason, kernelization has been considered as an attempt to mathematically formalize the concept of preprocessing. Clearly, the size of a kernelization is important as it evaluates how good is the preprocessing that it achieves.

It is known that every decidable parameterized problem parameterized problem in FPT admits a kernelization and vice versa. However, it is not always the case that a problem in FPT admits a *polynomial size* kernelization. The running challenge is to distinguish which problems in FPT admit polynomial size kernelizations and, if this is the case, to optimize the corresponding size function. In this direction, diverse algorithmic techniques have been invented (see [41, 57]) while, on the negative side, new complexity theory tools have been introduced for proving that polynomial size kernelization would imply some unexpected collapse in classic computational complexity [11, 27] (see also [57]).

Algorithmic meta-theorems can be seen as theoretical results providing general conditions that, when satisfied, automatically imply the existence of efficient algorithms for wide families of problems. The term "Algorithmic Meta-Theorem" was introduced by Grohe in his seminal exposition in [65] (see also [79] and [67]). Typically the conditions of such results have a logical part, concerning the descriptive complexity of the problem, and a combinatorial part that concerns the inputs of the problem. Algorithmic meta-theorems reveal interesting relations between logic and combinatorial structures and yield a better understanding of the nature of general algorithmic techniques.

Many known algorithmic meta-theorems have been stated using the multivariate framework of *parameterized complexity*. This is due to the fact that the concept of problem parameterization may serve as a way to formalize the imposed structural restrictions. The most classic example of an Algorithmic Meta-theorem leading to a massive classification of parameterized problems in FPT is the celebrated Courcelle's theorem [21, 23]: *Checking graph properties definable in Counting Monadic Second Order Logic (CMSO), when parameterized by the treewidth of the input graph, belongs in* FPT (see Subsect. 2.3 for the formal definition of treewidth). For other algorithmic meta-theorems permitting the classification of parameterized problems in FPT and containing different trade-offs between the logical part and the combinatorial part, see [26, 58, 81].

The purpose of this exposition is to present the main algorithmic meta-theorems on kernelization for problems on graphs. The first results of this kind appeared in [14] and they essentially initiated the research on meta-algorithmic for kernelization (see also [8] for an earlier version). The results in [14] subsumed several previous results on kernelization for problems on planar graphs such as [3, 4, 15, 16, 18, 19, 62, 69, 70, 75, 80, 82]. Moreover, the algorithmic techniques in [14] introduced new concepts and tools that were of use in later investigations [42, 54, 55, 59–61, 74, 76, 77, 87]. Subsequently, several results appeared in [54, 55] that extended the original theorems from [14] or applied the ideas of [14] in different combinatorial settings (see e.g., [56, 59, 61, 63, 74, 77, 87]).

We provide a description of the main ideas of [14], taking into account all the improvements and generalizations known up to now [49, 50, 54, 55, 76]. We present two main algorithmic meta-theorems on kernels for problems on graphs that can be seen as abstract versions of the results in [14]. We also give sketches of some of the proofs in order to expose the core ideas, as they were originally conceived in [14]. Our exposition is self-contained and pays special attention to "mathematical formalism" and "details". In fact we provide an alternative to the formalism of [14] (mostly inspired by [54, 55]) that is versatile enough so to form the base of further investigations on this topic.

2   Definitions

We use $\mathbb{Z}$ for the set of integers and $\mathbb{N}$ for the set of non-negative integers. Given some $\ell \in \mathbb{N}$, we denote $[\ell] = \{1, \ldots, \ell\}$. We also use $\mathbb{R}_{>0}$ for the set of all positive reals. Consider a tuple $\mathbf{t} = (x_1, \ldots, x_\ell) \in \mathbb{N}^\ell$ and two functions $\chi, \psi : \mathbb{N} \to \mathbb{N}$. Given a function $f : A \to B$, we also assume its set-extension $f : 2^A \to 2^B$ so that for every $X \in 2^A$, it holds $f(X) = \bigcup_{x \in X} f(x)$.

We write $\chi(n) = O_{\mathbf{t}}(\psi(n))$ in order to denote that there exists a function $\phi : \mathbb{N}^\ell \to \mathbb{N}$ such that $\chi(n) = O(\phi(\mathbf{t}) \cdot \psi(n))$. Also, given a function $f : \mathbb{N} \to \mathbb{N}$ and a tuple $\mathbf{t} = (x_1, \ldots, x_\ell) \in \mathbb{N}^\ell$, we write $\chi(n) = O_{f,\mathbf{t}}(\psi(n))$ in order to denote that there exists a function $\phi : \mathbb{N} \to \mathbb{N}$ such that $\chi(n) = O(\phi(f(\mathbf{t})) \cdot \psi(n))$.

All graphs that we consider are simple, finite, and undirected. Given a graph G, we denote by $V(G)$ and $E(G)$ the set of its vertices and edges respectively. The *size* of a graph G is the number of its vertices and is denoted by $|G|$.

2.1   Parameterized Problems on Graphs

We start by giving the definitions of a series of algorithmic concepts.

A *parameterized problem* is any subset Π of the set $\Sigma^* \times \mathbb{N}$ where Σ is some alphabet with at least two symbols. If $(x, k) \in \Sigma^* \times \mathbb{N}$ belongs in Π, then we say that (x, k) is a yes-*instance* of Π, otherwise we say that (x, k) is a no-*instance* of Π. Given two parameterized problems $\Pi, \Pi' \subseteq \Sigma^* \times \mathbb{N}$ and two instances $(x, k), (x', k') \in \Sigma^* \times \mathbb{N}$, we say that (x, k) and (x', k') are *equivalent instances* (*of Π and Π'*) if $(x, k) \in \Pi \iff (x', k') \in \Pi'$.

An *annotated graph* is a pair (G, S) where G is a graph and $S \subseteq V(G)$. We deal exclusively with parameterized problems either on graphs or on annotated graphs, therefore we see them as subsets of $\mathcal{G}_{all} \times \mathbb{N}$ or subsets of $\mathcal{A}_{all} \times \mathbb{N}$ where $\mathcal{G}_{all}$ is the set of all graphs and $\mathcal{A}_{all}$ the set of all annotated graphs. From now on, whenever we present some algorithm on some problem on (annotated) graphs we evaluate its time complexity in terms of the size of the graph in their input that we always denote by n.

A subset of $\mathcal{A}_{all}$ is called *vertex-subset property*.

Given a parameterized problem on graphs $\Pi \subseteq \mathcal{G}_{all} \times \mathbb{N}$ and a graph class $\mathcal{G} \subseteq \mathcal{G}_{all}$, we define *the restriction of Π in $\mathcal{G}$* by

$$\Pi \cap \mathcal{G} = \{(G, k) \mid (G, k) \in \Pi \text{ and } G \in \mathcal{G}\}.$$

Annotated Graphs. Given a vertex-subset property $\mathcal{A} \subseteq \mathcal{A}_{all}$ and some graph class $\mathcal{G} \subseteq \mathcal{G}_{all}$, we define

$$\mathcal{A} \cap \mathcal{G} = \{(G, S) \mid (G, S) \in \mathcal{A} \land G \in \mathcal{G}\}$$

and we call $\mathcal{A} \cap \mathcal{G}$ the *restriction* of $\mathcal{A}$ to $\mathcal{G}$. We define the *domain* of $\mathcal{A}$ as

$$\text{dom}(\mathcal{A}) = \{G \mid \exists k \in \mathbb{N} : (G, k) \in \mathcal{A}\}.$$

Notice that $\text{dom}(\mathcal{A} \cap \mathcal{G}) \subseteq \mathcal{G}$.

Problem Defining Pairs and Vertex-Subset Problems. A pair $(\mathcal{A}, \text{opt})$ where $\mathcal{A} \subseteq \mathcal{A}_{all}$ and $\text{opt} \in \{\min, \max\}$ is called *problem defining pair* or, in short, *defining pair*. Given a defining pair $(\mathcal{A}, \text{opt})$, we define the corresponding *vertex-subset problem* as the parameterized problem on graphs

$$\Pi_{\mathcal{A}, \text{opt}} = \{(G, k) \mid \exists S \subseteq V(G) : |S| \gtrless k \land (G, S) \in \mathcal{A}\}$$

where "$\gtrless$" is interpreted as "$\leq$", in case $\text{opt} = \min$ and as "$\geq$", in case $\text{opt} = \max$.

A *vertex-subset problem* is any parameterized problem on graphs Π such that $\Pi = \Pi_{\mathcal{A}, \text{opt}}$, for some defining pair $(\mathcal{A}, \text{opt})$. In particular, we can see a vertex-subset problem as a problem that is generated by some defining pair $(\mathcal{A}, \text{opt})$.

We also define the *annotated version* of $\Pi_{\mathcal{A}, \text{opt}}$ as follows. If $\text{opt} = \min$, then

$$\Pi^\alpha_{\mathcal{A}, \min} = \{((G, Y), k) \mid \exists S \subseteq Y : |S| \leq k \land (G, S) \in \mathcal{A}\},$$

while if $\text{opt} = \max$, then

$$\Pi^\alpha_{\mathcal{A}, \max} = \{((G, Y), k) \mid \exists S \subseteq V(G) : |S \cap Y| \geq k \land (G, S) \in \mathcal{A}\}.$$

Clearly, $\Pi_{\mathcal{A}, \text{opt}}$ can be seen as a special case of $\Pi^\alpha_{\mathcal{A}, \text{opt}}$ if we consider only the pairs whose annotated graph is of the form $(G, V(G))$.

Graph Parameters. A *graph parameter* is any partial function mapping graphs to non-negative integers. Given a defining pair $(\mathcal{A}, \mathsf{opt})$, we define the associated graph parameter $\mathsf{p}_{\mathcal{A},\mathsf{opt}} : \mathcal{G}_{\mathrm{all}} \to \mathbb{N}$ so that

$$\mathsf{p}_{\mathcal{A},\mathsf{opt}}(G) = \mathsf{opt}\left\{k \mid (G, k) \in \Pi_{\mathcal{A},\mathsf{opt}}\right\},$$

while, if $G \notin \mathrm{dom}(\mathcal{A})$, then $\mathsf{p}_{\mathcal{A},\mathsf{opt}}(G)$ is undefined. Notice that $\mathrm{dom}(\mathcal{A})$ is the domain of the graph parameter $\mathsf{p}_{\mathcal{A},\mathsf{opt}}$. Also we define the function $\mathsf{sol}_{\mathcal{A},\mathsf{opt}}$ such that for every $G \in \mathcal{G}_{\mathrm{all}}$,

$$\mathsf{sol}_{\mathcal{A},\mathsf{opt}}(G) = \{S \mid (G, S) \in \mathcal{A} \wedge |S| = \mathsf{p}_{\mathcal{A},\mathsf{opt}}(G)\}.$$

Clearly $\mathsf{sol}_{\mathcal{A},\mathsf{opt}}(G) \neq \emptyset \iff G \in \mathrm{dom}(\mathcal{A})$. We can see $\mathsf{sol}_{\mathcal{A},\mathsf{opt}}(G)$ as the set of all *optimal solutions* for the vertex-subset problem $\Pi_{\mathcal{A},\mathsf{opt}}$.

2.2   (Bi)kernelization

Kernelization can be seen as a polynomial reduction of a parameterized problem to itself. Here we present it as a special case of a more general concept of reduction called *bikernel* (see [60, 71, 72] for earlier uses of this concept).

Let $\Pi_1, \Pi_2 \subseteq \Sigma^* \times \mathbb{N}$ be two parameterized problems. A *bikernelization algorithm* (or simply *a bikernelization*) from Π_1 to Π_2 is a polynomial-time computable function $\mathsf{A} : \Sigma^* \times \mathbb{N} \to \Sigma^* \times \mathbb{N}$ such that

1. $\forall (x, k) \in \Sigma^* \times \mathbb{N} \; (x, k) \in \Pi_1 \iff \mathsf{A}(x, k) \in \Pi_2$ (i.e., (x, k) and $\mathsf{A}(x, k)$ are equivalent instances of Π_1 and Π_2 respectively) and
2. there exists a computable function $s : \mathbb{N} \to \mathbb{N}$ such that, for every pair $(x, k) \in \Sigma^* \times \mathbb{N}$, it holds that if $\mathsf{A}(x, k) = (x', k')$, then $\max\{|x'|, k'\} \leq s(k)$.

Given a parameterized problem $\Pi \subseteq \Sigma^* \times \mathbb{N}$, we define a *kernelization algorithm* (or simply *a kernelization*) for Π as a bikernelization algorithm from Π to Π.

We call the function s above *the size* of the (bi-)kernelization A. If s is a polynomial (resp. linear) function, we say that A is a *polynomial-size (resp. linear-size) (bi-)kernelization from Π_1 to Π_2*. Also we use the term *time of a (bi-)kernelization* in order to refer to the time of the (bi-)kernelization algorithm A. We stress that when x encodes a graph G, then $|x|$ is the encoding size of G.

It is known that every decidable parameterized problem that is in FPT admits a kernelization and vice versa. However, it is not always the case that there is a kernelization of polynomial size. As we already mentioned in the introduction, a central question of parameterized complexity is to distinguish which problems in FPT have kernelizations of polynomial size and which do not.

2.3   Graph Decompositions

We now provide several combinatorial concepts that will be useful for our proofs, namely, tree decompositions, protrusion decompositions, and the notion of protrusion decomposability.

Tree Decompositions. Let G be a graph. A *tree decomposition* of G is a pair $D = (T, \chi)$, where T is a tree and $\chi : V(T) \to 2^{V(G)}$ such that:

1. $\bigcup_{q \in V(T)} \chi(q) = V(G)$,
2. for every edge $\{u, v\} \in E$, there is a $q \in V(T)$ such that $\{u, v\} \subseteq \chi(q)$, and
3. for each $v \in V(G)$ the set $\{t \mid v \in \chi(t)\}$ induces a connected subgraph of T.

We call the vertices of T *nodes* of D and the images of χ *bags* of D. The *width* of a tree decomposition $D = (T, \chi)$ is $\max\{|\chi(q)| \mid q \in V(T)\} - 1$. The treewidth of a G is the minimum width over all tree decompositions of G.

Protrusion Decompositions. We denote by $\partial_G(S)$ the vertices of S that have neighbors outside S and by $N_G(S)$ the neighbours of vertices in S that do not belong to S. We also st $N_G[S] = S \cup N_G(S)$. Let G be a graph and let $R \subseteq V(G)$. We say that R is a β-*protrusion* of G if $\max\{|\partial_G(R)|, \mathbf{tw}(G[R])\} \leq \beta$. An (α, β)-*protrusion-decomposition* of a graph G is a partition $\mathcal{P} = \{X_0, X_1, \ldots, X_\ell\}$ of $V(G)$ such that

1. $\max\{\ell, |X_0|\} \leq \alpha$,
2. for every $i \in [\ell]$, the set $R_i = N_G[X_i]$ is a β-protrusion of G and
3. for every $i \in [\ell]$, $N_G(X_i) \subseteq X_0$.

We call the set X_0 the *core* of the (α, β)-*protrusion-decomposition* $\mathcal{P}$. We also call the sets $X_i, i \in [\ell]$ *flaps* of $\mathcal{P}$. Intuitively, an (α, β)-protrusion-decomposition of a graph can be seen as a partition into at most $\alpha + 1$ sets where one of them, the core, has size at most α and each of the rest, the flaps, has treewidth at most β and has at most β neighbors, all contained in the core.

Protrusion decompositions served as the main combinatorial tool of [14].

Protrusion-Decomposability. Given a defining pair $(\mathcal{A}, \mathsf{opt})$, a function $\mathsf{T} : \mathbb{N} \to \mathbb{N}$, and a real constant $c > 0$, we say that the defining pair $(\mathcal{A}, \mathsf{opt})$ is c-*protrusion-decomposable in time* $\mathsf{T}(n)$, if there exists an algorithm that given a $(G, k) \in \mathrm{dom}(\mathcal{A}) \times \mathbb{N}$, either outputs that $\mathbf{p}_{\mathcal{A}, \mathsf{opt}}(G) > k$ or outputs a $(c \cdot k, c)$-protrusion-decomposition of G in time $T(n)$ (recall that by n we always denote the size of the input graph G). It is important to observe that in the specifications of this algorithm we *demand* that its input should contain a graph in the domain of $\mathcal{A}$.

2.4   Boundaried Graphs

Let $t \in \mathbb{N}$. A t-*boundaried graph* is a triple $\mathbf{G} = (G, B, \rho)$ where G is a graph, $B \subseteq V(G)$, $|B| = t$, and $\rho : B \to [t]$ is a bijection. We call G *underlying* graph of $\mathbf{G}$, B *the boundary* of $\mathbf{G}$, and we define the *size* of $\mathbf{G}$, denoted by $|\mathbf{G}|$, as the size of G. We say that $\mathbf{G}_1 = (G_1, B_1, \rho_1)$ and $\mathbf{G}_2 = (G_2, B_2, \rho_2)$ are *isomorphic* if there is an isomorphism from G_1 to G_2 that extends the bijection $\rho_2^{-1} \circ \rho_1$.

We call t-*boundaried annotated graph* the pair $(\mathbf{G}, S)$, where $\mathbf{G} = (G, B, \rho)$ and $S \subseteq V(G)$. We say that two t-boundaried annotated graphs $(\mathbf{G}_1, S_1)$ and $(\mathbf{G}_2, S_2)$ (where $\mathbf{G}_1 = (G_1, B_1, \rho_1)$ and $\mathbf{G}_2 = (G_2, B_2, \rho_2)$) are *compatible* if

$\rho_2^{-1} \circ \rho_1$ is an isomorphism from $G_1[B_1]$ to $G_2[B_2]$ and $\rho(B_1 \cap S_1) = \rho(B_2 \cap S_2)$. The *size* of $(\mathbf{G}, S)$ is defined to be the size of $\mathbf{G}$.

Given two compatible t-boundaried annotated graphs $(\mathbf{G}_1, S_1)$ and $(\mathbf{G}_2, S_2)$ where $\mathbf{G}_1 = (G_1, B_1, \rho_1)$ and $\mathbf{G}_2 = (G_2, B_2, \rho_2)$, we define $\mathbf{G}_1 \oplus \mathbf{G}_2$ as the annotated graph (G, S) where G is obtained if we take the disjoint union of G_1 and G_2 and, for every $i \in [|B_1|]$, we identify vertices $\rho_1^{-1}(i)$ and $\rho_2^{-1}(i)$ and S is obtained if we take the union of S_1 and S_2 and, for each $i \in \rho(B_1 \cap S_1)$, we identify the vertex $\rho_1^{-1}(i)$ with the vertex $\rho_2^{-1}(i)$. We agree that, during vertex identifications, the vertices of B_1 prevail those of B_2.

3   Meta-theorems on Kernels

3.1   Meta-kernels for Subset Properties with Finite Index

Finite Index. Let $\mathcal{A}$ be a vertex-subset property and $t \in \mathbb{N}$. Let also $(\mathbf{G}_1, S_1)$ and $(\mathbf{G}_2, S_2)$ be two compatible t-boundaried annotated graphs. We say that $(\mathbf{G}_1, S_1) \equiv_{\mathcal{A},t} (\mathbf{G}_2, S_2)$ if for every t-boundaried annotated graph $(\mathbf{F}, S_F)$ that is compatible with $(\mathbf{G}_1, S_1)$, it holds that

$$(\mathbf{F}, S_F) \oplus (\mathbf{G}_1, S_1) \in \mathcal{A} \iff (\mathbf{F}, S_F) \oplus (\mathbf{G}_2, S_2) \in \mathcal{A}.$$

It is easy to observe that $\equiv_{\mathcal{A},t}$ is an equivalence relation. For every $t \in \mathbb{N}$, we set up a set of representatives $\mathsf{rep}_{\mathcal{A}}(t)$ containing a minimum size t-boundaried annotated graph from each equivalence class of $\equiv_{\mathcal{A},t}$.

Given an increasing function $f : \mathbb{N} \to \mathbb{N}$, we say that a subset property $\mathcal{A}$ belongs in $\mathsf{FI}(f)$ if for every $t \in \mathbb{N}$, $f(t)$ is an upper bound to the size of all the t-boundaried annotated graphs in $\mathsf{rep}_{\mathcal{A}}(t)$. We say that a subset property $\mathcal{A}$ has *finite index* if it belongs in $\mathsf{FI}(f)$ for some increasing function $f : \mathbb{N} \to \mathbb{N}$. Notice that $\mathcal{A}$ has finite index if and only if $\equiv_{\mathcal{A},t}$ has a finite number of equivalence classes, for every $t \in \mathbb{N}$. The notion of finite index was central in the proof of Courcelle's theorem in [21], see [6,9,13,17,20,22,24] for related bibliography.

Our first result, corresponding to Theorem 1.1 of [14], is stated in terms of bikernelization algorithms.

Theorem 1. *Let $f : \mathbb{N} \to \mathbb{N}$ be an increasing function, $\mathcal{A} \in \mathsf{FI}(f)$, $\mathsf{opt} \in \{\min, \max\}$, and $c \in \mathbb{R}_{>0}$. If*

1. *$\mathsf{dom}(\mathcal{A})$ is computable in polynomial time $\mathsf{T}^1(n)$,*
2. *$(\mathcal{A}, \mathsf{opt})$ is c-protrusion-decomposable in polynomial time $\mathsf{T}^2(n)$,*

then there is a polynomial size bikernelization from $\Pi_{\mathcal{A},\mathsf{opt}}$ to $\Pi_{\mathcal{A},\mathsf{opt}}^{\alpha}$, of size $O(k^2)$, running in polynomial time $\mathsf{T}^1(n) + \mathsf{T}^2(n) + O_{f,c}(n)$.

The main meta-algorithmic consequence of Theorem 1, corresponding to Theorem 1.2 of [14], is the following.

Theorem 2. *Let $f : \mathbb{N} \to \mathbb{N}$ be an increasing function, $\mathcal{A} \in \mathsf{FI}(f)$, $\mathsf{opt} \in \{\min, \max\}$, and $c \in \mathbb{R}_{>0}$. If*

1. $\mathsf{dom}(\mathcal{A})$ *is computable in polynomial time* $\mathsf{T}^1(n)$,
2. $(\mathcal{A}, \mathsf{opt})$ *is c-protrusion-decomposable in polynomial time* $\mathsf{T}^2(n)$,
3. $\Pi_{\mathcal{A}, \mathsf{opt}}$ *is NP-hard, and*
4. $\Pi_{\mathcal{A}, \mathsf{opt}}^{\alpha} \in \mathsf{NP}$,

then $\Pi_{\mathcal{A}, \mathsf{opt}}$ *admits a polynomial size kernelization of size* $k^{O_{f,c}(1)}$, *running in polynomial time* $\mathsf{T}^1(n) + \mathsf{T}^2(n) + O_{f,c}(n) + k^{O_{f,c}(1)}$.

In what follows in this subsection, we give the main steps of the proofs of Theorems 1 and 2. A useful ingredient is the following:

Lemma 1. *Let* $f : \mathbb{N} \to \mathbb{N}$ *be an increasing function,* $\mathcal{A} \in \mathsf{FI}(f)$, *and* $\mathsf{opt} \in \{\min, \max\}$. *For every* t *there exists an algorithm that, given two compatible* t-*boundaried annotated graphs* $(\mathbf{G}_1, S_1)$ *and* $(\mathbf{G}_2, S_2)$, *outputs a set* $R \subseteq V(\mathbf{G}_1)$ *such that*

- $(\mathbf{G}_1, R)$ *and* $(\mathbf{G}_1, S_1)$ *are compatible,*
- $(\mathbf{G}_2, S_2) \oplus (\mathbf{G}_1, R) \in \mathcal{A}$,
- *in case* $\mathsf{opt} = \min$, *then* $R \subseteq S_1$ *and* $|R|$ *is minimized, and*
- *in case* $\mathsf{opt} = \max$, *then* $|S_1 \cap R|$ *is maximized,*

or reports that such a set R *does not exist. Moreover, given that the underlying graph of* $(\mathbf{G}_1, S_1) \oplus (\mathbf{G}_2, S_2)$ *is* G, *this algorithm runs in time* $O_{f, \mathbf{tw}(G), t}(|G|)$.

The proof of Lemma 1 is based on the fact that there is an algorithm that, given a vertex-subset property $\mathcal{A} \in \mathsf{FI}(f)$ (for some increasing function $f : \mathbb{N} \to \mathbb{N}$) and a graph G, outputs, if exists, a minimum or maximum (depending on the value of $\mathsf{opt} \in \{\max, \min\}$) size set S where $(G, S) \in \mathcal{A}$, in time $O_{f, \mathbf{tw}(G)}(|G|)$. This fact follows by making use of the results from [17, Theorem 5] for finite index vertex-subset properties, see also [6] and [14, Lemma 5.2].

The first important step towards proving Theorem 1 is the following.

Lemma 2. *Let* $f : \mathbb{N} \to \mathbb{N}$ *be an increasing function,* $\mathcal{A} \in \mathsf{FI}(f)$, $\mathsf{opt} \in \{\max, \min\}$, *and* $t \in \mathbb{N}$. *Then there exists an algorithm that, given an instance* $((G, Y), k)$ *of* $\Pi_{\mathcal{A}, \mathsf{opt}}^{\alpha}$ *and a* t-*protrusion* X *of* G, *outputs an equivalent instance* $((G, Y'), k)$ *of* $\Pi_{\mathcal{A}, \mathsf{opt}}^{\alpha}$, *where* $|Y' \cap X| \leq O_{f,t}(k)$ *and* $Y' \subseteq Y$, *in time* $O_{f,t}(|X|)$.

Proof (Proof (sketch).). We present the proof only for the case where $\mathsf{opt} = \min$. The case where $\mathsf{opt} = \max$ is similar (but not the same, see [14, Lemma 5.14]).

Let $B = \partial_G(X)$ and $t' = |B| \leq t$. We consider the t'-boundaried annotated graph $(\mathbf{X}, Z)$ where $\mathbf{X} = (G[X], B, \rho)$ (ρ is some, arbitrarily chosen, bijection from $\partial_G(X)$ to $[t']$) and $Z = Y \cap X$. Recall that $\mathbf{tw}(G_X) \leq t$. We also consider the t'-boundaried annotated graph $(\mathbf{F}, R)$ where $\mathbf{F} = (G \backslash (X \backslash B), B, \rho)$ and $R = Y \backslash (X \backslash B)$. Clearly, $(G, Y) = (\mathbf{F}, R) \oplus (\mathbf{X}, Z)$.

Let now $\mathfrak{L} = (\mathbf{H}, W) \in \mathsf{rep}_{\mathcal{A}}(t')$. We apply the algorithm of Lemma 1 on $(\mathbf{H}, W)$ and $(\mathbf{X}, Z)$ and find, if exists, a minimum size set $S_{\mathfrak{L}} \subseteq Z$ such that

$(\mathbf{X}, Z)$ and $(\mathbf{X}, S_{\mathcal{L}})$ are compatible and $(\mathbf{H}, W) \oplus (\mathbf{X}, S_{\mathcal{L}}) \in \mathcal{A}$. If such an $S_{\mathcal{L}}$ does not exist or it has more than k vertices, then we set $S_{\mathcal{L}} = \emptyset$. We define $Y' = R \cup W$ where

$$W = \bigcup_{\mathcal{L} \in \mathsf{rep}_{\mathcal{A}}(t)} S_{\mathcal{L}}.$$

Recall that the underlying graph of $(\mathbf{H}, W) \oplus (\mathbf{X}, S_{\mathcal{L}})$ has at most $f(t') + |X| = O_{f(t)}(|X|)$ vertices and that its treewidth is bounded by $f(t') - t' + \mathbf{tw}(G[X]) \le f(t) + t$. Therefore, according to Lemma 1, each $S_{\mathcal{L}}$ can be computed in time $O_{f,t}(|X|)$. Moreover, as $|\mathsf{rep}_{\mathcal{A}}(t')| = O_{f,t}(1)$, we conclude that the set Y' can be computed in time $O_{f,t}(|X|)$.

Notice that $Y' \subseteq Y$ and that $|Y' \cap X| = |W| = \sum_{\mathcal{L} \in \mathsf{rep}_{\mathcal{A}}(t')} |S_{\mathcal{L}}| \le |\mathsf{rep}_{\mathcal{A}}(t')| \cdot k = O_{f,t}(k)$. It also follows that $((G, Y), k)$ and $((G, Y'), k)$ are equivalent instances of $\Pi^{\alpha}_{\mathcal{A}, \min}$ and this is so because the sets $S_{\mathcal{L}}$ have been chosen so to represent every possible restriction in X of a solution of $\Pi^{\alpha}_{\mathcal{A}, \min}$ on G (see [14, Lemma 5.3] for the complete argumentation).

Lemma 2 is already an important step towards a (bi)kernelization algorithm as it reduces the set of annotated vertices that can be inside the protrusions to a linear function of the parameter k. The next step is to "completely eliminate" the presence of annotated vertices in the "interior", i.e., $X \setminus \partial_G(X)$, of each flap X of a protrusion decomposition of the input graph.

Lemma 3. *Let $t \in \mathbb{N}$, G be a graph, $Y \subseteq V(G)$, and $\mathcal{P} = \{X_0, X_1, \ldots, X_\ell\}$ be a $(t \cdot k, t)$-protrusion-decomposition of G such that, for every $i \in [\ell]$, $|Y \cap X_i| \le t \cdot k$. Then G has a $(O_t(k^2), O_t(1))$-protrusion-decomposition $\mathcal{P}'$ whose core contains Y as a subset. Moreover, $\mathcal{P}'$ can be computed in $O_t(|G|)$ steps.*

Proof (Proof (sketch).). The algorithm works, separately on each flap X of $\mathcal{P}$, on the tree decomposition $D = (T, \chi)$ of the graph induced by X and its neighborhood. In fact, we consider a decomposition $D = (T, \chi)$ of $G[N_G[X]]$ where T is a rooted binary tree and where each of its bags contains $N_G(X)$ and the root bag is $N_G(X)$. Notice that every bag of D has at most $2t + 1$ vertices. We declare nodes of D *dirty* as follows: for every $v \in Y$, mark as dirty the topmost bag containing v. This declares at most $t \cdot k$ of the nodes of D dirty. Next, proceed in a bottom up fashion (starting from the leaves of T), by declaring dirty each node that is a least common ancestor of two already dirty nodes. When this procedure finishes, we have less than $2t \cdot k$ dirty nodes. The new protrusion-decomposition is built by adding in the core X_0 of $\mathcal{P}$ the bags of all the nodes that have been declared dirty, for each flap of $\mathcal{P}$. This makes a new core X_0' of $O_t(k^2)$ vertices. The connected components of $G \setminus X_0'$ can be organized to $O_t(k^2)$ subsets of the flaps of $\mathcal{P}$ such that, given the choice of the dirty vertices, each of them has neighbors in at most two dirty bags. Therefore each of these subsets has at most $2(2t + 1)$ neighbors and this permits to organize them to a $(O_t(k^2), O_t(1))$-protrusion-decomposition $\mathcal{P}'$, as required (see [14, Lemma 5.4] for more details).

Protrusion Replacements. We next define the concept of replacing a protrusion X of G by another one. Let G be a graph G and let X be a t-protrusion of G where $|\partial_G(X)| = t' \leq t$. We set $G_X = G[X]$, $B = \partial_G(X)$ and we pick an arbitrary bijection $\rho : B \to [t']$. This gives a way to see the protrusion X as the t'-boundaried graph $\mathbf{G}_X = (G_X, B, \rho)$. We refer to $\mathbf{G}_X$ as *a boundaried version of X*. Notice that there are $t'!$ different boundaried versions of X, depending on the choice of ρ. Let now $\hat{\mathbf{G}} = (\hat{G}, \hat{B}, \hat{\rho})$ be a t'-boundaried graph that is compatible to $\mathbf{G}_X$. The graph occurring after *replacing $\mathbf{G}_X$ by $\hat{\mathbf{G}}$ in G*, denoted by $\mathsf{repl}(G, \mathbf{G}_X, \hat{\mathbf{G}})$, is the graph $\mathbf{F} \oplus \hat{\mathbf{G}}$, where $\mathbf{F} = (G \backslash (X \backslash B), B, \rho)$.

Restricted t-Boundaried Annotated Graphs. We say that a t-boundaried annotated graph $(\mathbf{G}, Y)$ –where $\mathbf{G} = (G, B, \rho)$– is *restricted* if $Y \subseteq B$. Let $(\mathbf{G}_1, Y_1)$ and $(\mathbf{G}_2, Y_2)$ be two compatible restricted t-boundaried annotated graphs where $\mathbf{G}_i = (G_i, B_i, \rho_i), i \in [2]$. Observe that, because of our assumptions, the annotated vertices of $(\mathbf{G}_1, Y_1)$ and $(\mathbf{G}_2, Y_2)$ have boundary vertices corresponding to the same set of indices. Given a $t \in \mathbb{N}$, we say that $(\mathbf{G}_1, Y_1) \sim_{\mathcal{A},t} (\mathbf{G}_2, Y_2)$ if

$$\forall (S_1, S_2) \in 2^{Y_1} \times 2^{Y_2} \quad \left(\rho_1(S_1) = \rho_2(S_2) \Rightarrow (\mathbf{G}_1, S_1) \equiv_{\mathcal{A},t} (\mathbf{G}_2, S_2) \right).$$

Notice that $\sim_{\mathcal{A},t}$ is an equivalence relation on restricted t-boundaried annotated graphs. By picking a minimum-size t-boundaried annotated graph from each equivalence class, we set up a set of representatives that, from now on, we denote by $\overline{\mathsf{rep}}_{\mathcal{A}}(t)$. Notice that if $\mathcal{A} \in \mathsf{FI}(f)$ for some increasing function $f : \mathbb{N} \to \mathbb{N}$, then $|\overline{\mathsf{rep}}_{\mathcal{A}}(t)| = O_{f,t}(1)$ and the maximum size of a member of $\overline{\mathsf{rep}}_{\mathcal{A}}(t)$ is bounded by $O_{f,t}(1)$.

Lemma 4. *Let $f : \mathbb{N} \to \mathbb{N}$ be an increasing function, $\mathcal{A} \in \mathsf{FI}(f)$, opt $\in \{\max, \min\}$, and $t \in \mathbb{N}$. There exists an algorithm that, given an instance $((G, Y), k)$ of $\Pi^{\alpha}_{\mathcal{A},opt}$, a t-protrusion X where $X \cap Y \subseteq \partial_G(X)$, and a boundaried version $\mathbf{G}_X = (G_X, B, \rho)$ of X, outputs in time $O_{f,t}(|X|)$, a boundaried graph $\hat{\mathbf{G}}$ such that*

1. *$\hat{\mathbf{G}}$ is compatible with $\mathbf{G}_X$,*
2. *$|\hat{\mathbf{G}}| = O_{f,t}(1)$ and,*
3. *if $G' = \mathsf{repl}(G, \mathbf{G}_X, \hat{\mathbf{G}})$, then $((G', Y), k)$ is an equivalent instance of $\Pi^{\alpha}_{\mathcal{A},opt}$.*

Proof (Proof (sketch).). Again we present only the case where opt = min (see [14, Lemma 5.15] for the case where opt = max).

Let $Y_X = X \cap Y$ and $t_X = |\partial_G(X)| \leq t$. Recall that $Y_X \subseteq B$, therefore $\mathbf{G}_X = (G_X, B, \rho)$ is a restricted t_X-boundaried annotated graph. Clearly, there is some restricted t_X-boundaried annotated graph $(\hat{\mathbf{G}}, \hat{Y}_X) \in \overline{\mathsf{rep}}_{\mathcal{A}}(t_X)$ such that $(\mathbf{G}_X, Y_X) \sim_{\mathcal{A},t_X} (\hat{\mathbf{G}}, \hat{Y}_X)$, i.e., $(\hat{\mathbf{G}}, \hat{Y}_X)$ is a representative of $(\mathbf{G}_X, Y_X)$ with respect to $\sim_{\mathcal{A},t_X}$. By the way $\hat{\mathbf{G}}$ is defined, it can be proven that $\hat{\mathbf{G}}$ is indeed a t_X-boundaried graph that can replace $\mathbf{G}_X$ towards creating an equivalent instance of $\Pi^{\alpha}_{\mathcal{A},opt}$. For the full proof, see [14, Lemma 5.6].

It now remains to design an algorithm that, given a pair $(\mathbf{G}_X, Y_X)$, outputs its representative $(\hat{\mathbf{G}}, \hat{Y})$ in $O_{f,t}(|\mathbf{G}_X|)$ steps. For this we set up, for every $t' \in [0, 2t+1]$, the set $\mathcal{R}_{t'}$ containing every restricted t'-boundaried annotated graph $(\mathbf{G}, Y)$ where $|\mathbf{G}| \leq 2 \cdot f(2t+1)$. For every $t' \in [0, 2t+1]$, we next set up the function $\mathfrak{f}_{t'} : \mathcal{R}_{t'} \to \overline{\text{rep}}_A(t')$ mapping each $(\mathbf{G}, Y)$ in $\mathcal{R}_{t'}$ to its representative in $\overline{\text{rep}}_A(t')$, i.e., the member of $\overline{\text{rep}}_A(t')$ that is equivalent to $(\mathbf{G}, Y)$ with respect to the equivalence relation $\sim_{A,t'}$. Keep in mind that the function $\mathfrak{f}_{t'}$ does not depend on k. We clarify that $\mathfrak{f}_{t'}$ is hardcoded in the algorithm and is not computed on the fly.

The computation of $(\hat{\mathbf{G}}, \hat{Y})$ is done by standard dynamic programming on a tree decomposition $D = (T, \chi)$ of G_X whose tree is rooted on a node whose bag is B, where each node has at most two children, and where if a node has two children then the corresponding bags are the same as the bag of their parent. Every tree decomposition can be modified in linear time to one that satisfies these properties. This type of a decomposition is handy for performing dynamic programming and can be see as a simpler version of the concept of a *nice tree decomposition* (see [10,78,87] for more details on dynamic programming on nice tree decompositions). Each bag of D has at most $2t+1$ vertices as in the beginning of the proof of Lemma 3. For every node $i \in V(T)$, with $t_i = |\chi(i)|$, consider the restricted t_i-boundaried annotated graph $(\mathbf{G}_i, Y_i)$ where

1. the boundary of $\mathbf{G}_i$ is $\chi(i)$,
2. the underlying graph, say G_i, of $\mathbf{G}_i$ is the subgraph of G_X induced by $\chi(i)$ and all the vertices in bags of descendants of i in T, and
3. $Y_i = Y_X \cap V(G_i)$.

The dynamic programming algorithm computes, in a bottom-up fashion, for every bag $\chi(i)$ on t_i vertices, the representative $(\hat{\mathbf{G}}_i, \hat{Y}_i)$ of $(\mathbf{G}_i, Y_i)$ given that the representatives of the restricted boundaried graphs of the children of node i in T that have already been computed. This computation is done in $O_{f,t}(1)$ steps using the function $\mathfrak{f}_{t_i}$ and taking in mind that each i has at most two children, therefore $(\hat{\mathbf{G}}_i, \hat{Y}_i)$ is the application of $\mathfrak{f}_{t_i}$ to the "gluing" of the representatives corresponding to the children of i, that is a t_i-boundaried annotated graph whose boundaried graph has size at most $2 \cdot f(2t+1)$. In total $O(|G_X|)$ nodes are being processed, therefore the dynamic programming algorithm takes time $O_{f,t}(|G_X|)$, as required.

We are now in position to give the proof of Theorem 1.

Proof (Proof of Theorem 1). We present a bikernelization from $\Pi_{A,\text{opt}}$ to $\Pi^\alpha_{A,\text{opt}}$. Let $(G, k) \in \mathcal{G}_{\text{all}} \times \mathbb{N}$ be an input of $\Pi_{A,\text{opt}}$. We fist use the time $\mathsf{T}^1(n)$ algorithm of the first condition in order to check whether $G \in \text{dom}(A)$. If the answer is negative then $((G, Y), k)$ is a no-instance of $\Pi^\alpha_{A,\text{min}}$ and we are done. In case of a positive answer, we know that $G \in \text{dom}(A)$. This permits us to call the time $\mathsf{T}^2(n)$ algorithm of the second condition that either outputs that $\mathbf{p}_{A,\text{opt}}(G) > k_{\text{opt}}$ or outputs a $(c \cdot k_{\text{opt}}, c)$-protrusion-decomposition of G, where $k_{\text{opt}} = k$ if opt = min and $k_{\text{opt}} = k-1$ if opt = max . If $\mathbf{p}_{A,\text{opt}}(G) > k_{\text{opt}}$ we have that

(G, k) is a no-instance of $\Pi_{\mathcal{A},\mathsf{opt}}$, in case $\mathsf{opt} = \min$, while it is a yes-instance of $\Pi_{\mathcal{A},\mathsf{opt}}$, in case $\mathsf{opt} = \max$. Depending on which case applies, the kernelization algorithm outputs a trivial no- or yes-instance of $\Pi^{\alpha}_{\mathcal{A},\mathsf{opt}}$.

Assume now that $\mathcal{P} = \{X_0, X_1, \ldots, X_\ell\}$ is a $(c \cdot k_{\mathsf{opt}}, c)$-protrusion-decomposition of G. We set $Y = V(G)$ and notice that (G, k) is a yes-instance of $\Pi_{\mathcal{A},\mathsf{opt}}$ iff $((G, Y), k)$ is a yes-instance of $\Pi^{\alpha}_{\mathcal{A},\min}$. By repetitively applying Lemma 2 for each flap of $\mathcal{P}$, we construct an equivalent instance $((G, Y'), k)$ of $\Pi^{\alpha}_{\mathcal{A},\min}$ where $|Y' \cap X_i| = O_{f,c}(k)$, for every $i \in [\ell]$. Then we apply Lemma 3 on $\mathcal{P}$ and construct a $(O_{f,c}(k^2), O_{f,c}(1))$-protrusion-decomposition $\mathcal{P}'$ whose core contains Y as a subset. Next, we repetitively apply Lemma 4 for each of the flaps of $\mathcal{P}'$ and construct an instance $((G', Y'), k)$ of $\Pi^{\alpha}_{\mathcal{A},\mathsf{opt}}$ that is equivalent to $((G, Y'), k)$ and therefore to $((G, Y), k)$ as well. That way, G' has a $(O_{f,c}(k^2), O_{f,c}(1))$-protrusion-decomposition where each flap contains at most $O_{f,c}(1)$ vertices. This implies that $|G'| = O_{f,c}(k^2)$. According to the running times of each of the three aforementioned lemmata, the construction of $((G', Y'), k)$ can be done in $O_{f,c}(|G|)$ steps.

Theorem 1 can be seen as a proper abstract version of Theorem 1.1 in [14]. A somehow stronger version of Theorem 1, that avoids the use of bikernels, can be derived in case $\mathsf{opt} = \min$. The proof is the same as the one of Theorem 1 (when $\mathsf{opt} = \min$) and is based on the additional observation that $\mathsf{p}_{\mathcal{A},\min}(G) > k_{\mathsf{opt}}$ implies that (G, k_{opt}) is a no-instance of $\Pi^{\alpha}_{\mathcal{A},\min}$. This gives rise to the following.

Theorem 3. *Let $f : \mathbb{N} \to \mathbb{N}$ be an increasing function, $\mathcal{A} \in \mathsf{FI}(f)$, and $c \in \mathbb{R}_{>0}$. If*

1. $\mathsf{dom}(\mathcal{A})$ *is computable in polynomial time* $\mathsf{T}^1(n)$,
2. $(\mathcal{A}, \min)$ *is c-protrusion-decomposable in polynomial time* $\mathsf{T}^2(n)$,

then $\Pi^{\alpha}_{\mathcal{A},\min}$ admits a polynomial size kernelization of size $k^{O_{f,c}(1)}$, running in polynomial time $\mathsf{T}^1(n) + \mathsf{T}^2(n) + O_{f,c}(n)$.

Admittedly, we do have a version of Theorem 3 when $\mathsf{opt} = \max$ as, in this case, we require that $\mathsf{p}_{\mathcal{A},\max}(G) > k_{\mathsf{opt}}$ implies that $(G, k_{\max})$ is a yes-instance of $\Pi^{\alpha}_{\mathcal{A},\max}$ and such an implication is not a consequence of the definition of the annotation version of a maximization problem.

We now arrive to the proof of Theorem 2.

Proof (Proof of Theorem 2). Let (G, k) be an instance of $\Pi_{\mathcal{A},\min}$ and let $((G', Y'), k)$ be an equivalent instance of $\Pi^{\alpha}_{\mathcal{A},\min}$, of size $k^{O_{f,c}(1)}$, constructed in time $\mathsf{T}^1(n) + \mathsf{T}^2(n) + O_{f,c}(n)$ by the application of the bikernelization algorithm of Theorem 1.

By the two last conditions we have that there is a polynomial time reduction from $\Pi^{\alpha}_{\mathcal{A},\mathsf{opt}}$ to $\Pi_{\mathcal{A},\mathsf{opt}}$. This means that there is a polynomial-time algorithm, i.e., an algorithm that runs in time $|G'|^{O(1)} = k^{O_{f,c}(1)}$, that can transform $((G', Y'), k)$ to an instance $(\tilde{G}, \tilde{k})$ of $\Pi_{\mathcal{A},\mathsf{opt}}$ such that $((G', Y'), k)$ is a yes-instance of $\Pi^{\alpha}_{\mathcal{A},\min}$ iff $(\tilde{G}, \tilde{k})$ is a yes-instance of $\Pi_{\mathcal{A},\min}$. Notice that $\max\{|\hat{G}|, \hat{k}\} = k^{O_{f,c}(1)}$ and that (G, k) and $(\tilde{G}, \tilde{k})$ are equivalent instances of $\Pi_{\mathcal{A},\mathsf{opt}}$, as required.

3.2 Meta-kernels for Subset Properties with Finite Integer Index

Our second meta-algorithmic result is based on the notion of finite integer index.

Finite Integer Index. Two t-boundaried graphs $\mathbf{G}_1$ and $\mathbf{G}_2$ are *compatible* if the t-boundaried annotated graphs $(\mathbf{G}_1, \emptyset)$ and $(\mathbf{G}_2, \emptyset)$ are compatible. Given two t-boundaried graphs $\mathbf{G}_1$ and $\mathbf{G}_2$ we define $\mathbf{G}_1 \oplus \mathbf{G}_2$ as the graph of the pair $(\mathbf{G}_1, \emptyset) \oplus (\mathbf{G}_2, \emptyset)$.

Let $(\mathcal{A}, \mathsf{opt})$ be a defining pair and $t \in \mathbb{N}$. Given two compatible t-boundaried graphs $\mathbf{G}_1, \mathbf{G}_2$, we say that $\mathbf{G}_1 \approx_{\mathcal{A}, \mathsf{opt}, t} \mathbf{G}_2$ if there exists a constant $c_{\mathbf{G}_1, \mathbf{G}_2} \in \mathbb{Z}$, depending on $\mathbf{G}_1$ and $\mathbf{G}_2$, such that for every t-boundaried graph $\mathbf{F}$ that is compatible with $\mathbf{G}_1$, it holds that

$$\mathbf{p}_{\mathcal{A}, \mathsf{opt}}(\mathbf{F} \oplus \mathbf{G}_2) = \mathbf{p}_{\mathcal{A}, \mathsf{opt}}(\mathbf{F} \oplus \mathbf{G}_1) + c_{\mathbf{G}_1, \mathbf{G}_2}.$$

For completeness, if in the above definition, for some $i \in [2]$, the graph $\mathbf{F} \oplus \mathbf{G}_i \notin \mathsf{dom}(\mathbf{p}_{\mathcal{A}, \mathsf{opt}})$, we assume that $\mathbf{p}(\mathbf{G}_i \oplus \mathbf{F}) = \infty$. Observe that c might be negative in the above definition. In fact, $c_{\mathbf{G}_1, \mathbf{G}_2} = -c_{\mathbf{G}_2, \mathbf{G}_1}$. Note that the relation $\approx_{\mathcal{A}, \mathsf{opt}, t}$ is an equivalence relation. We set up, for every $t \in \mathbb{N}$, a set $\widetilde{\mathsf{rep}}_{\mathcal{A}, \mathsf{opt}}(t)$ of representatives of the relation $\approx_{\mathcal{A}, \mathsf{opt}, t}$ by picking one minimum-size representative for each of its equivalence classes.

Given an increasing function $f : \mathbb{N} \to \mathbb{N}$, we define $\mathsf{FII}(f)$ as the set of all defining pairs $(\mathcal{A}, \mathsf{opt})$ where for every $t \in \mathbb{N}$, the size of every t-boundaried graph in $\widetilde{\mathsf{rep}}_{\mathcal{A}, \mathsf{opt}}(t)$ is upper bounded by $f(t)$. We say that the defining pair $(\mathcal{A}, \mathsf{opt})$ has *finite integer index* if it belongs in $\mathsf{FII}(f)$, for some increasing function $f : \mathbb{N} \to \mathbb{N}$. The defining pair $(\mathcal{A}, \mathsf{opt})$ has finite integer index if and only if $\approx_{\mathcal{A}, \mathsf{opt}, t}$ has a finite number of equivalence classes, for every $t \in \mathbb{N}$.

Notice that the notion of finite integer index concerns different objects than the one of finite index. Having finite index is a property of vertex-subset properties, while having finite integer index is a property of defining pairs.

The notion of finite integer index was introduced in the thesis of van Antwerpen-de Fluiter [44], see also [5,12,13,43]. Our second meta-algorithmic result, corresponding to theorem 1.3 of [14], is the following.

Theorem 4 *Let* $f : \mathbb{N} \to \mathbb{N}$ *be an increasing function,* $(\mathcal{A}, \mathsf{opt}) \in \mathsf{FII}(f)$, *and* $c \in \mathbb{R}_{>0}$. *If*

1. $\mathsf{dom}(\mathcal{A})$ *is computable in polynomial time* $\mathsf{T}^1(n)$ *and*
2. $(\mathcal{A}, \mathsf{opt})$ *is* c-*protrusion-decomposable in polynomial time* $\mathsf{T}^2(n)$.

then $\Pi_{\mathcal{A}, \mathsf{opt}}$ *admits a linear kernelization of size* $O_{f, c}(k)$, *running in polynomial time* $\mathsf{T}^1(n) + \mathsf{T}^2(n) + O_{f, c}(n)$.

Proof (Proof (sketch).). The idea of the proof follows from the fact that if X is a protrusion of G, the t-boundaried graph $\mathbf{G}_X$ is a boundaried version of X, $\hat{\mathbf{G}}$ is the representative of $\mathbf{G}_X$ in $\widetilde{\mathsf{rep}}_{\mathcal{A}}(\mathsf{opt}, t)$, $G' = \mathsf{repl}(G, \mathbf{G}_X, \hat{\mathbf{G}})$, and $k' = k + c_{\mathbf{G}_i, \hat{\mathbf{G}}_i}$, then (G, k) and (G', k') are equivalent instances of $\Pi_{\mathcal{A}, \mathsf{opt}}$. The proof of this fact follows directly from the definition of the equivalence relation $\approx_{\mathcal{A},\mathrm{opt},t}$ (see [14, Lemma 5.18] for the details).

As in the begining of the proof of Theorem 2 we can assume that we have a protrusion-decomposition $\mathcal{P} = \{X_0, X_1, \ldots, X_\ell\}$ of G, that has been constructed in time $\mathsf{T}^1(n) + \mathsf{T}^2(n)$ (otherwise a direct answer can be derived and the algorithm outputs a trivial equivalent instance). The algorithm performs dynamic programming on the special type of tree decomposition that we used in the proof of Lemma 4. For every flap X of $\mathcal{P}$, the dynamic programming runs on a boundaried version $\mathbf{G}_X = (G_X, B, \rho)$ of X and detects a node i where the t_i-boundaried graph $\mathbf{G}_i$ (defined as in the proof of Theorem 2) has size more than $f(t)$ but also has size $O(f(t))$. The upper bound permits the detection of i in $O_{f,c}(1)$ steps. The lower bound permits the replacement in G of $\mathbf{G}_i$ by its representative $\hat{\mathbf{G}}_i$ in $\widetilde{\mathrm{rep}}_{\mathcal{A},\mathrm{opt}}(t)$ that has smaller size. Consider the pair (G', k'), where G' is the result of the protrusion replacement and $k' = k + c_{\mathbf{G}_i, \hat{\mathbf{G}}_i}$ and recall that (G, k) and (G', k') are equivalent instances of $\Pi_{\mathcal{A},\mathrm{opt}}$ where $|G'| < |G|$. By performing such replacements bottom-up and updating the protrusion-decomposition accordingly, we keep creating equivalent instances until all the flaps in the protrusion-decomposition of the resulting graph are of size upper-bounded by $2^{f(t)}$. This ends up, in time $O_{f,c}(n)$, with an equivalent instance (G', k') whose size is a linear function of k.

Notice that the first condition of both Theorems 2 and 4 require the polynomial computability of the domain of $\mathcal{A}$. In some cases, variants of these theorems are ignoring (or omitting) this condition as it is either obvious or the problem is stated so that containment to $\mathcal{A}$ is a promise condition.

4   Consequences

In this section we deal with the applications of Theorems 2 and 4. Notice that both these theorems contain two types of requirements. The first is that the vertex-subset property (resp. defining pair) has finite integer index (resp. finite integer index), the second is that the problem is protrusion decomposable. The first condition will be linked to the descriptive complexity of the problem while the second will be linked to certain combinatorial properties of its inputs, i.e., the domain of $\mathcal{A}$.

4.1   Counting Monadic Second Order Logic

There is a wide variety of vertex-subset problems generated by defining pairs. Typically, they can be defined using logical sentences.

CMSOL. The syntax of Counting Monadic Second Order Logic (CMSOL) on graphs includes the logical connectives $\vee, \wedge, \neg, \rightarrow, \leftrightarrow$, variables for vertices, edges, sets of vertices, and sets of edges, the quantifiers $\forall, \exists$ that can be applied to these variables, and the following six predicates:

1. $\mathbf{vin} : V(G) \times 2^{V(G)} \to \{\mathrm{T}, \mathrm{F}\}$, where $\mathbf{vin}(v, S) = \mathrm{T}$ iff v is a vertex of S,
2. $\mathbf{ein} : E(G) \times 2^{E(G)} \to \{\mathrm{T}, \mathrm{F}\}$, where $\mathbf{ein}(e, F) = \mathrm{T}$ iff e is an edge of F,
3. $\mathbf{inc} : V(G) \times E(G) \to \{\mathrm{T}, \mathrm{F}\}$, where $\mathbf{inc}(v, e) = \mathrm{T}$ iff v is an endpoint of e,
4. $\mathbf{adj} : (V(G))^2 \to \{\mathrm{T}, \mathrm{F}\}$, where $\mathbf{adj}(v, u) = \mathrm{T}$ iff v and u are distinct endpoints of an edge,
5. $\mathbf{eq} : (V(G))^2 \to \{\mathrm{T}, \mathrm{F}\}$, where $\mathbf{eq}(v, u) = \mathrm{T}$ iff v and u are equal,
6. $\mathbf{card}_{q,r} : V(G) \to \{\mathrm{T}, \mathrm{F}\}$, where $\mathbf{card}_{q,r}(S) = \mathrm{T}$ iff $|S| \equiv q \pmod{r}$, where r, q are fixed integers such that $0 \leq q < r$ and $r \geq 2$.

We use variants of the symbols v, e, S, and F in order to denote variants of vertices, edges, vertex sets, and edge sets respectively. We refer to [6,21,23] for a detailed introduction on CMSO. Given a sentence ϕ, we denote its length by $|\phi|$.

Some Examples. We may consider CMSOL sentences that are evaluated either on graphs or on annotated graphs. For instance, if

$$\phi = \forall v_1 \left(\mathbf{vin}(v_1, S) \vee \exists v_2 \left(\mathbf{adj}(v_1, v_2) \wedge \mathbf{vin}(v_2, S) \right) \right),$$

then $(G, S) \models \phi$ iff S is a dominating set of G. Moreover, if

$$\phi = \forall S_1, S_2 \left(\left(\exists x \, \mathbf{vin}(x, S_1) \wedge \exists x \, \mathbf{vin}(x, S_2) \wedge \right. \right.$$

$$\forall v \left(\left(\mathbf{vin}(v, S_1) \wedge \neg \mathbf{vin}(v, S_2) \right) \vee \left(\mathbf{vin}(v, S_2) \wedge \neg \mathbf{vin}(v, S_1) \right) \right) \right) \to$$

$$\left(\exists v_1, v_2, e \left(\mathbf{vin}(v_1, S_1) \wedge \mathbf{vin}(v_2, S_2) \wedge \mathbf{inc}(v_1, e) \wedge \mathbf{inc}(v_2, e) \right) \right) \right),$$

then $G \models \phi$ iff G is connected.

Problems Defined by Sentences. Given a CMSOL sentence ψ on graphs, we define the graph class

$$\mathcal{G}_\psi = \{G \mid G \models \phi\}.$$

Moreover, given a CMSOL sentence ϕ on annotated graphs, we define the vertex-subset property

$$\mathcal{A}_\phi = \{(G, S) \mid (G, S) \models \phi\}.$$

Given a CMSOL sentence ϕ on annotated graphs and an opt $\in \{\min, \max\}$, we use $\varPi_{\phi,\mathbf{opt}}$ and $\varPi^\alpha_{\phi,\mathbf{opt}}$ as shortcuts for the vertex-subset problem $\varPi_{\mathcal{A}_\phi,\mathbf{opt}}$ and its annotated version $\varPi^\alpha_{\mathcal{A}_\phi,\mathbf{opt}}$ respectively. Also, given a CMSOL sentence ψ on graphs and a CMSOL sentence ϕ on annotated graphs, we define $\phi \cap \psi$ as the CMSOL sentence where

$$(G, S) \models \phi \cap \psi \text{ iff } (G, S) \models \phi \text{ and } G \models \psi.$$

By admitting that shortcuts may create inflation of notation, we notice that $\mathcal{A}_\phi \cap \mathcal{G}_\psi = \mathcal{A}_{\phi \cap \psi}$ and that $\varPi_{\phi,\mathbf{opt}} \cap \mathcal{G}_\psi = \varPi_{\mathcal{A}_\phi \cap \mathcal{G}_\psi,\mathbf{opt}} = \varPi_{\mathcal{A}_{\phi \cap \psi},\mathbf{opt}} = \varPi_{\phi \cap \psi,\mathbf{opt}}$.

4.2 Properties of Defining Pairs

We now come to the combinatorial properties of defining pairs. In our setting, these properties condition the domain of the vertex subset property.

Treewidth-Modulability. Given a defining pair $(\mathcal{A}, \text{opt})$ and a $c \in \mathbb{R}_{>0}$, we say that the defining pair $(\mathcal{A}, \text{opt})$ is *c-treewidth-modulable* if, for every $(G, k) \in \text{dom}(\mathcal{A}) \times \mathbb{N}$, it holds that

$$\mathbf{p}_{\mathcal{A}, \text{opt}}(G) \leq k \Rightarrow \exists S \subseteq V(G) : |S| \leq c \cdot k \ \wedge \ \mathbf{tw}(G \backslash S) \leq c.$$

Minors, Contractions, and Topological Minors. Given a graph G, we say that a graph H is a *contraction* of G if a graph isomorphic to H can be obtained from G after contracting edges. We also say that H is *a minor* of G if it is the contraction of some subgraph of G. Finally, we say that H is a *topological minor* of G if some subdivision of H is isomorphic to a subgraph of G (a *subdivision* of H is any graph obtained by replacing edges by paths with the same endpoints).

Given a finite set of graphs $\mathcal{H}$, we denote by $\mathcal{C}_{\mathcal{H}}$ the class of graphs that do not contain any of the graphs in $\mathcal{H}$ as a contraction, by $\mathcal{T}_{\mathcal{H}}$ the class of graphs that do not contain any of the graphs in $\mathcal{H}$ as a topological minor, and by $\mathcal{M}_{\mathcal{H}}$ the class of graphs that do not contain any of the graphs in $\mathcal{H}$ as a minor. Notice that, for every finite set of graphs $\mathcal{H}$, $\mathcal{M}_{\mathcal{H}} \subseteq \mathcal{T}_{\mathcal{H}}$ and $\mathcal{M}_{\mathcal{H}} \subseteq \mathcal{C}_{\mathcal{H}}$.

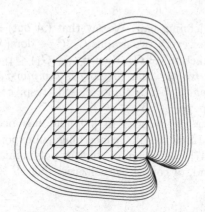

Fig. 1. Graph Γ_9.

SQGM and SQGC Properties of Graph Classes. Let $\mathcal{G} \subseteq \mathcal{G}_{\text{all}}$, $\lambda \in \mathbb{R}_{>0}$, and let c be a real number in the interval $[1, 2)$. We say that $\mathcal{G}$ has the *sub-quadratic grid minor property* (SQGM property, for short) *for λ and c*, if

$$\forall k \in \mathbb{N} \ \forall G \in \mathcal{M}_{\{\boxplus_k\}} \ \mathbf{tw}(G) \leq \lambda \cdot k^c.$$

Also, we say that a graph class $\mathcal{G}$ has the *sub-quadratic graph contraction property* (SQGC property for short) *for* λ *and* c, if

$$\forall k \in \mathbb{N} \ \forall G \in \mathcal{C}_{\{\Gamma_k\}} \ \mathbf{tw}(G) \leq \lambda \cdot k^c.$$

We denote by $\mathtt{SQGM}(\lambda, c)$ (resp. $\mathtt{SQGC}(\lambda, c)$) the set of all graph classes that have the SQGM (resp, SQGC) property for λ and c.

The most simple example of a graph class that has the above properties is the class of planar graphs, that belongs in $\mathtt{SQGC}(4.2, 1)$ [68]. For more general classes of graphs satisfying the SQGM and the SQGC properties, see [7,31,48,52,53,64].

Bidimensonality. Given a $k \in \mathbb{N}_{\geq 1}$, a $(k \times k)$-*grid* is the graph $\boxplus_k$ where $V(\boxplus_k) = [k]^2$ and $E(\boxplus_k) = \{\{(x,y),(x',y')\} \mid |x - x'| + |y - y'| = 1\}$. The *perimeter* of $\boxplus_k$, denoted by $P(\boxplus_k)$, is the set containing all the vertices of $\boxplus_k$ that have degree smaller than 4. The *uniformly triangulated grid* $(k \times k)$-*grid* is the graph Γ_k where $V(\Gamma_k) = [k]^2$ and

$$E(\Gamma_k) = E(\boxplus_k) \cup$$
$$\{(x+1,y),(x,y+1) \mid (x,y) \in [t-1]^2\} \cup$$
$$\{\{(k,k),(a,b)\} \mid (a,b) \in P(\boxplus_k)\backslash\{(k,k)\}\}).$$

For a drawing of Γ_9, see Fig. 1.

Given two real functions f and g, we use the term $f \gtrsim g$ to denote that $f(x) \geq g(x) - o(g(x))$.

Let $(\mathcal{A}, \mathsf{opt})$ be a defining pair. We say that $(\mathcal{A}, \mathsf{opt})$ is *minor closed* (resp. *contraction closed*) if for every two graphs $G_1, G_2 \in \mathsf{dom}(\mathcal{A})$, it holds that if G_1 is a minor (resp. contraction) of G_2, then $\mathbf{p}_{\mathcal{A},\mathsf{opt}}(G_1) \leq \mathbf{p}_{\mathcal{A},\mathsf{opt}}(G_2)$.

Given a $c \in \mathbb{R}_{>0}$, we say that $(\mathcal{A}, \mathsf{opt})$ is *c-minor-bidimensional* (resp. *c-contraction-bidimensional*) if it is minor-closed (resp. contraction-closed) and $\mathbf{p}_{\mathcal{A},\mathsf{opt}}(\boxplus_k) \gtrsim ck^2$ (resp. $\mathbf{p}_{\mathcal{A},\mathsf{opt}}(\Gamma_k) \gtrsim ck^2$).

Bidimensonality was introduced in [30] and has been the combinatorial base of several algorithmic results concerning subexponential parameterized algorithms [29,30,34,47], the design of efficient polynomial-time approximation schemes [32,51,53], and, of course, kernelization [54,55,76], see also [28,33,46, 73,85,86].

Separability. Let $(\mathcal{A}, \mathsf{opt})$ be a defining pair and let $f : \mathbb{N} \to \mathbb{N}$. We say that $(\mathcal{A}, \mathsf{opt})$ is *f-separable* if for every graph $G \in \mathsf{dom}(\mathcal{A})$, every $S \in \mathsf{sol}_{\mathcal{A},\mathsf{opt}}(G)$, and every $L \subseteq V(G)$, it holds that

$$|S \cap L| - f(t) \leq \mathbf{p}_{\mathcal{A},\mathsf{opt}}(G[L]) \leq |S \cap L| + f(t),$$

where $t = |\partial_G(L)|$. Given some $c \in \mathbb{R}_{>0}$, we say that the defining pair $(\mathcal{A}, \mathsf{opt})$ is *c-linearly-separable* if it is *f-separable* for the function $f : \mathbb{N} \to \mathbb{N}$, defined so that $f(x) = c \cdot x$.

The notion of separability has been introduced in [54,55], while a similar notion had earlier appeared in [32].

4.3 Theorems on Properties of Defining Pairs

Conditions for Proving Finite Index and Finite Integer Index. The following celebrated result, widely known as Courcelle's theorem, is the standard way to prove that a vertex-subset property has finite index.

Proposition 1. *For every CMSOL sentence ϕ, there is an increasing function $f : \mathbb{N} \to \mathbb{N}$ such that $\mathcal{A}_\phi \in \mathsf{FI}(f)$.*

Proposition 1 has appeared in many forms and with different proofs, see [2,6,17, 21,23,35]. For a proof of a more general version that the one in Proposition 1, see [14, Lemma 3.2].

The next result gives a criterion for proving that a defining pair has finite integer index.

Proposition 2. *For every CMSOL sentence ϕ, $opt \in \{\min, \max\}$, and every function $f : \mathbb{N} \to \mathbb{N}$, if the defining pair $(\mathcal{A}_\phi, opt)$ is f-separable, then there is some increasing function f' such that $(\mathcal{A}, opt) \in \mathsf{FII}(f')$.*

Proposition 2 appeard in [55]. It provides an easy way to prove that a defining pair has the FII property. For proving FII, an alternative to separability, called strong monotonicity, had already appeared in [14, Section 7].

Conditions for Proving Protrusion-Decomposability. In both Theorems 1 and 3, the second condition is protrusion decomposability. This condition can be implied by other more easy to verify combinatorial conditions. The first one is treewidth-modulability combined with the exclusion of some graph as topological minor.

Theorem 5. *Let ϕ be a CMSOL sentence on annotated graphs, $h \in \mathbb{N}_{\geq 1}$, $c \in \mathbb{R}_{>0}$, and $\mathcal{H}$ be a set of graphs, each of size at most h. If $(\mathcal{A}_\phi, opt)$ is a c-treewidth-modulable defining pair, then $(\mathcal{A}_\phi \cap \mathcal{T}_\mathcal{H}, opt)$ is $O_{|\phi|,h,c}(1)$-protrusion-decomposable in time $O_{|\phi|,h,c}(n^2)$.*

The proof of Theorem 5 is implicit in [76] and is presented in more detail in [77].

We should stress that the quadratic-time protrusion decomposability of the above result can be improved to a linear one by combining the results of [55,76,77]. We avoid presenting the proof of this here, as it is lengthy and requires the introduction of several concepts from [49] and [50] such as *solution lifting*, *protrusion cover*, and *explicit representation of subgraphs* (the bulk of the arguments has already been exposed in [57, Subsection 15.6]).

The main combinatorial condition in Theorem 5 is protrusion-decomposability. The following result appeared in [55] and reveals how this condition can be implied from other, more easy to check, conditions.

Theorem 6. *Let $\lambda, c, c' \in \mathbb{R}_{>0}$, and let c'' be a real number in the interval $[1, 2)$, and let $f : \mathbb{N} \to \mathbb{N}$ be an increasing function. If $\mathcal{G} \in \mathsf{SQGM}(c'', \lambda)$ (resp. $\mathcal{G} \in \mathsf{SQGC}(c'', \lambda)$) and $(\mathcal{A}, opt)$ is a defining pair that is c-minor-bidimensional (resp. c-contraction-bidimensional) and c'-linearly separable, then $(\mathcal{A} \cap \mathcal{G}, opt)$ is $O_{\lambda,c,c',c''}(1)$-treewidth-modulable.*

4.4 Applications

In this section we present the consequences of Theorems 2 and 4 under the light of the logical and combinatorial conditions of Subsect. 4.3. We expose them as two corollaries, corresponding to Theorems 2 and 4 respectively.

Corollary 1. *Let ϕ be a CMSOL sentence on annotated graphs, $h \in \mathbb{N}_{\geq 1}$, $c \in \mathbb{R}_{>0}$, $\mathcal{H}$ be a set of graphs, each of size at most h, and $opt \in \{\min, \max\}$. If*

1. $\text{dom}(\mathcal{A}_\phi)$ *is computable in polynomial time* $\mathsf{T}(n)$,
2. $(\mathcal{A}_\phi, opt)$ *is c-treewidth-modulable,*
3. $\Pi_{\phi,opt} \pitchfork \mathcal{T}_\mathcal{H}$ *is NP-hard, and* $\Pi^\alpha_{\phi,opt} \in \mathsf{NP}$,

then $\Pi_{\phi,opt} \pitchfork \mathcal{T}_\mathcal{H}$ admits a polynomial kernelization of size $k^{O_{|\phi|,h,c}(1)}$, running in time $\mathsf{T}^1(n) + O_h(n^3) + O_{|\phi|,h,c}(n^2) + k^{O_{|\phi|,h,c}(1)}$.

If Condition 3 is replaced by the following:

3′. $(\mathcal{A}_\phi, opt)$ *is f-separable, for some function $f : \mathbb{N} \to \mathbb{N}$,*

then $\Pi_{\phi,opt} \pitchfork \mathcal{T}_\mathcal{H}$ admits a linear size kernelization of size $O_{f,|\phi|,h,c}(k)$, running in time $\mathsf{T}^1(n) + O_h(n^3) + O_{f,|\phi|,h,c}(n^2)$.

Proof. From Theorem 5, $(\mathcal{A}_\phi \pitchfork \mathcal{T}_\mathcal{H}, opt)$ is c'''-protrusion-decomposable in time $\mathsf{T}^2(n) = O_{|\phi|,h,c}(n^2)$ for some $c' = O_{|\phi|,h,c}(1)$.

As topological minor containment can be expressed in monadic second order logic, there is some $\psi_\mathcal{H}$ such that $\mathcal{T}_\mathcal{H} = \mathcal{G}_{\psi_\mathcal{H}}$. Recall that $\phi \pitchfork \psi_\mathcal{H}$ is a CMSOL sentence on annotated graphs. Also $\mathcal{A}_\phi \pitchfork \mathcal{G}_{\psi_\mathcal{H}} = \mathcal{A}_{\phi \pitchfork \psi_\mathcal{H}}$ and $\Pi_{\phi,opt} \pitchfork \mathcal{G}_\psi = \Pi_{\mathcal{A}_\phi \pitchfork \mathcal{G}_\psi, opt}$, therefore $\Pi_{\phi,opt} \pitchfork \mathcal{T}_\mathcal{H} = \Pi_{\phi \pitchfork \psi_\mathcal{H}, opt}$.

It was proved in [66], that $\mathcal{T}_\mathcal{H}$ is computable in time $O_h(n^3)$. Therefore $\mathcal{A}_{\phi \pitchfork \psi_\mathcal{H}}$ is computable in time $\mathsf{T}^1(n) = \mathsf{T}(n) + O_h(n^3)$. Notice also that if $\Pi^\alpha_{\phi,opt} \in \mathsf{NP}$, then also $\Pi^\alpha_{\phi,opt} \pitchfork \mathcal{T}_\mathcal{H} \in \mathsf{NP}$. The result follows by applying Theorems 2 and 4 for the defining pair $(\phi \pitchfork \psi_\mathcal{H}, opt)$ and the constant c'.

If now in the second version of Corollary 1, we deduce treewidth-modulability by using the conditions of Theorem 6, we derive the following.

Corollary 2. *Let ϕ be a CMSOL sentence on annotated graphs, $h \in \mathbb{N}_{\geq 1}$, $\lambda, c, c' \in \mathbb{R}_{>0}$, $c'' \in [1, 2)$, $\mathcal{H}$ be a set of graphs, each of size at most h, and $opt \in \{\min, \max\}$. If*

1. $\text{dom}(\mathcal{A}_\phi)$ *is computable in time* $\mathsf{T}(n)$,
2. $(\mathcal{A}_\phi, opt)$ *is c'-linearly separable,*
3. $(\mathcal{A}_\phi, opt)$ *is c-minor bidimensional (resp. c-minor bidimensional), and*
4. $\mathcal{T}_\mathcal{H} \in \mathsf{SQGM}(c'', \lambda)$ *(resp. $\mathcal{T}_\mathcal{H} \in \mathsf{SQGM}(c'', \lambda)$),*

then $\Pi_{\phi,opt} \pitchfork \mathcal{T}_\mathcal{H}$ admits a linear size kernelization running in time $\mathsf{T}(n) + O_h(n^3) + O_{|\phi|,h,\lambda,c,c',c''}(n^2)$.

Concerning the applicability of Corollary 2, we stress that for every finite set of graphs $\mathcal{Z}$, each of size at most z, there is an $\mathcal{H}$ such that $\mathcal{T}_{\mathcal{H}} = \mathcal{M}_{\mathcal{Z}}$. Moreover, there is a constant λ_z such that $\mathcal{M}_{\mathcal{Z}} \in \texttt{SQGM}(\lambda_z, 1)$ [31]. This implies that the minor version of the fourth condition of Corollary 2 holds if we replace $\mathcal{T}_{\mathcal{H}}$ by any non-trivial minor-closed graph class (taking into account the Robertson & Seymour Theorem [84]). For the contraction version, assume additionally that $\mathcal{Z}$ contains at least one apex graph (an *apex* graph is a graph containing a vertex whose removal creates a planar graph). For every such $\mathcal{Z}$, it holds that $\mathcal{M}_{\mathcal{Z}} \in \texttt{SQGC}(\lambda_h, 1)$, for some constant λ_h [48]. This implies that the contraction version of the fourth condition of Corollary 2 holds if we replace $\mathcal{T}_{\mathcal{H}}$ by $\mathcal{M}_{\mathcal{Z}}$ and pick $\mathcal{Z}$ to be any finite set containing at least one apex graph.

Acknowledgements. I wish to whole-heartedly thank *Professor Hans L. Bodlaender* for being the one who «*told me a little but he taught me a lot*».

References

1. Abrahamson, K.A., Downey, R.G., Fellows, M.R.: Fixed-parameter tractability and completeness IV. On completeness for W[P] and PSPACE analogues. Ann. Pure Appl. Log. **73**(3), 235–276 (1995)
2. Abrahamson, K.R., Fellows, M.R.: Finite automata, bounded treewidth and well-quasiordering. In: Robertson, N., Seymour, P.D. (eds.) AMS Summer Workshop on Graph Minors, Graph Structure Theory, Contemporary Mathematics, vol. 147, pp. 539–564. American Mathematical Society (1993)
3. Alber, J., Betzler, N., Niedermeier, R.: Experiments on data reduction for optimal domination in networks. Ann. OR **146**(1), 105–117 (2006)
4. Alber, J., Fellows, M.R., Niedermeier, R.: Polynomial-time data reduction for dominating set. J. Assoc. Comput. Mach. **51**(3), 363–384 (2004)
5. Arnborg, S., Courcelle, B., Proskurowski, A., Seese, D.: An algebraic theory of graph reduction. J. ACM **40**, 1134–1164 (1993)
6. Arnborg, S., Lagergren, J., Seese, D.: Easy problems for tree-decomposable graphs. J. Algorithms **12**, 308–340 (1991)
7. Baste, J., Thilikos, D.M.: Contraction-bidimensionality of geometric intersection graphs. In: 12th International Symposium on Parameterized and Exact Computation, IPEC 2017, Vienna, Austria, 6–8 September 2017, pp. 5:1–5:13 (2017)
8. Bodlaender, H.L., Fomin, F.V., Lokshtanov, D., Penninkx, E., Saurabh, S., Thilikos, D.M.: (Meta) kernelization. In: FOCS 2009, pp. 629–638. IEEE (2009)
9. Bodlaender, H.L.: On reduction algorithms for graphs with small treewidth. In: van Leeuwen, J. (ed.) WG 1993. LNCS, vol. 790, pp. 45–56. Springer, Heidelberg (1994). https://doi.org/10.1007/3-540-57899-4_40
10. Bodlaender, H.L.: A partial k-arboretum of graphs with bounded treewidth. Theoret. Comput. Sci. **209**(1–2), 1–45 (1998)
11. Bodlaender, H.L., Downey, R.G., Fellows, M.R., Hermelin, D.: On problems without polynomial kernels. J. Comput. Syst. Sci. **75**, 423–434 (2009). https://doi.org/10.1016/j.jcss.2009.04.001. http://portal.acm.org/citation.cfm?id=1628322.1628467
12. Bodlaender, H.L., de Fluiter, B.: Reduction algorithms for constructing solutions in graphs with small treewidth. In: Cai, J.-Y., Wong, C.K. (eds.) COCOON 1996. LNCS, vol. 1090, pp. 199–208. Springer, Heidelberg (1996). https://doi.org/10.1007/3-540-61332-3_153

13. Bodlaender, H.L., van Antwerpen-de Fluiter, B.: Reduction algorithms for graphs of small treewidth. Inf. Comput. **167**, 86–119 (2001)

14. Bodlaender, H.L., Fomin, F.V., Lokshtanov, D., Penninkx, E., Saurabh, S., Thilikos, D.M.: (Meta) kernelization. J. ACM **63**(5), 44:1–44:69 (2016)

15. Bodlaender, H.L., Penninkx, E.: A linear kernel for planar feedback vertex set. In: Grohe, M., Niedermeier, R. (eds.) IWPEC 2008. LNCS, vol. 5018, pp. 160–171. Springer, Heidelberg (2008). https://doi.org/10.1007/978-3-540-79723-4_16

16. Bodlaender, H.L., Penninkx, E., Tan, R.B.: A linear kernel for the k-disjoint cycle problem on planar graphs. In: Hong, S.-H., Nagamochi, H., Fukunaga, T. (eds.) ISAAC 2008. LNCS, vol. 5369, pp. 306–317. Springer, Heidelberg (2008). https://doi.org/10.1007/978-3-540-92182-0_29

17. Borie, R.B., Parker, R.G., Tovey, C.A.: Automatic generation of linear-time algorithms from predicate calculus descriptions of problems on recursively constructed graph families. Algorithmica **7**, 555–581 (1992)

18. Chen, J., Fernau, H., Kanj, I.A., Xia, G.: Parametric duality and kernelization: lower bounds and upper bounds on kernel size. SIAM J. Comput. **37**(4), 1077–1106 (2007)

19. Chen, J., Kanj, I.A., Jia, W.: Vertex cover: further observations and further improvements. J. Algorithms **41**(2), 280–301 (2001)

20. Courcelle, B.: The monadic second-order logic of graphs III. Tree-decompositions, minors and complexity issues. RAIRO Inform. Théor. Appl. **26**(3), 257–286 (1992)

21. Courcelle, B.: The monadic second-order logic of graphs. I. Recognizable sets of finite graphs. Inf. Comput. **85**(1), 12–75 (1990)

22. Courcelle, B.: The monadic second-order logic of graphs V: on closing the gap between definability and recognizability. Theor. Comput. Sci. **80**(2), 153–202 (1991)

23. Courcelle, B.: The expression of graph properties and graph transformations in monadic second-order logic. In: Handbook of Graph Grammars, pp. 313–400 (1997)

24. Courcelle, B., Engelfriet, J.: Graph Structure and Monadic Second-Order Logic - A Language-Theoretic Approach, Encyclopedia of Mathematics and Its Applications, vol. 138. Cambridge University Press, Cambridge (2012). http://www.cambridge.org/fr/knowledge/isbn/item5758776/?site_locale=fr_FR

25. Cygan, M., et al.: Parameterized Algorithms. Springer, Cham (2015). https://doi.org/10.1007/978-3-319-21275-3

26. Dawar, A., Grohe, M., Kreutzer, S.: Locally excluding a minor. In: LICS 2007, pp. 270–279. IEEE Computer Society (2007)

27. Dell, H., Van Melkebeek, D.: Satisfiability allows no nontrivial sparsification unless the polynomial-time hierarchy collapses. J. ACM **61**(4), 23:1–23:27 (2014). https://doi.org/10.1145/2629620

28. Demaine, E., Hajiaghayi, M.: The bidimensionality theory and its algorithmic applications. Comput. J. **51**(3), 292–302 (2007)

29. Demaine, E.D., Fomin, F.V., Hajiaghayi, M., Thilikos, D.M.: Bidimensional parameters and local treewidth. SIAM J. Discrete Math. **18**(3), 501–511 (2004)

30. Demaine, E.D., Fomin, F.V., Hajiaghayi, M., Thilikos, D.M.: Subexponential parameterized algorithms on graphs of bounded genus and H-minor-free graphs. J. ACM **52**(6), 866–893 (2005)

31. Demaine, E.D., Hajiaghayi, M.T.: Linearity of grid minors in treewidth with applications through bidimensionality. Combinatorica **28**(1), 19–36 (2008)

32. Demaine, E.D., Hajiaghayi, M.: Bidimensionality: new connections between FPT algorithms and PTASs. In: Proceedings of the 16th Annual ACM-SIAM Symposium on Discrete Algorithms (SODA 2005), pp. 590–601. ACM-SIAM, New York (2005)

33. Demaine, E.D., Hajiaghayi, M.: Bidimensionality. In: Kao, M.Y. (ed.) Encyclopedia of Algorithms. Springer, Boston (2008). https://doi.org/10.1007/978-0-387-30162-4_47

34. Demaine, E.D., Hajiaghayi, M., Thilikos, D.M.: The bidimensional theory of bounded-genus graphs. SIAM J. Discrete Math. **20**(2), 357–371 (2006). https://doi.org/10.1137/040616929

35. Downey, R.G., Fellows, M.R.: Parameterized Complexity. Springer, New York (1999). https://doi.org/10.1007/978-1-4612-0515-9

36. Downey, R., Fellows, M.: Fixed-parameter tractability and completeness III. Some structural aspects of the W hierarchy. In: Complexity Theory, pp. 191–225. Cambridge University Press, Cambridge (1993)

37. Downey, R.G., Fellows, M.R.: Fixed-parameter tractability and completeness I. Basic results. SIAM J. Comput. **24**(4), 873–921 (1995)

38. Downey, R.G., Fellows, M.R.: Fixed-parameter tractability and completeness II: On completeness for $W[1]$. Theor. Comput. Sci. **141**(1–2), 109–131 (1995)

39. Downey, R.G., Fellows, M.R., Prieto-Rodriguez, E., Rosamond, F.A.: Fixed-parameter tractability and completeness V: parametric miniatures (2003). Manuscript

40. Downey, R.G., Fellows, M.R.: Fundamentals of Parameterized Complexity. Texts in Computer Science. Springer, London (2013). https://doi.org/10.1007/978-1-4471-5559-1

41. Downey, R.G., Fellows, M.R., Langston, M.A.: The computer journal special issue on parameterized complexity: foreword by the guest editors. Comput. J. **51**(1), 1–6 (2008). https://doi.org/10.1093/comjnl/bxm111. http://comjnl.oxfordjournals.org/content/51/1/1.short

42. Eiben, E., Ganian, R., Szeider, S.: Meta-kernelization using well-structured modulators. Discrete Appl. Math. **248**, 153–167 (2018). https://doi.org/10.1016/j.dam.2017.09.018. Seventh Workshop on Graph Classes, Optimization, and Width Parameters, Aussois, France, October 2015. http://www.sciencedirect.com/science/article/pii/S0166218X17304419

43. Fellows, M.R., Langston, M.A.: An analogue of the Myhill-Nerode theorem and its use in computing finite-basis characterisations (extended abstract). In: 30th Annual IEEE Symposium on Foundations of Computer Science, FOCS 1989, pp. 520–525. IEEE (1989)

44. van Antwerpen-de Fluiter, B.: Algorithms for graphs of small treewidth. Ph.D. thesis, Department Computer Science, Utrecht University (1997)

45. Flum, J., Grohe, M.: Parameterized Complexity Theory. Texts in Theoretical Computer Science. An EATCS Series. Springer, Berlin (2006). https://doi.org/10.1007/3-540-29953-X

46. Fomin, F.V., Demaine, E.D., Hajiaghayi, M.T., Thilikos, D.M.: Bidimensionality. In: Kao, M.Y. (ed.) Encyclopedia of Algorithms, pp. 203–207. Springer, New York (2016). https://doi.org/10.1007/978-1-4939-2864-4_47

47. Fomin, F.V., Golovach, P., Thilikos, D.M.: Contraction bidimensionality: the accurate picture. In: Fiat, A., Sanders, P. (eds.) ESA 2009. LNCS, vol. 5757, pp. 706–717. Springer, Heidelberg (2009). https://doi.org/10.1007/978-3-642-04128-0_63

48. Fomin, F.V., Golovach, P.A., Thilikos, D.M.: Contraction obstructions for treewidth. J. Comb. Theory Ser. B **101**(5), 302–314 (2011)

49. Fomin, F.V., Lokshtanov, D., Misra, N., Ramanujan, M.S., Saurabh, S.: Solving d-SAT via backdoors to small treewidth. In: Indyk, P. (ed.) Proceedings of the Twenty-Sixth Annual ACM-SIAM Symposium on Discrete Algorithms, SODA 2015, San Diego, CA, USA, 4–6 January 2015, pp. 630–641. SIAM (2015). https://doi.org/10.1137/1.9781611973730.43

50. Fomin, F.V., Lokshtanov, D., Misra, N., Saurabh, S.: Planar F-deletion: approximation, kernelization and optimal FPT algorithms. In: 53rd Annual IEEE Symposium on Foundations of Computer Science, FOCS 2012, New Brunswick, NJ, USA, 20–23 October 2012, pp. 470–479 (2012)

51. Fomin, F.V., Lokshtanov, D., Raman, V., Saurabh, S.: Bidimensionality and EPTAS. In: Randall, D. (ed.) Proceedings of the Twenty-Second Annual ACM-SIAM Symposium on Discrete Algorithms, SODA 2011, San Francisco, California, USA, 23–25 January 2011, pp. 748–759. SIAM (2011). https://doi.org/10.1137/1.9781611973082.59

52. Fomin, F.V., Lokshtanov, D., Saurabh, S.: Bidimensionality and geometric graphs. In: Proceedings of the Twenty-Third Annual ACM-SIAM Symposium on Discrete Algorithms, SODA 2012, pp. 1563–1575. Society for Industrial and Applied Mathematics, Philadelphia (2012). http://dl.acm.org/citation.cfm?id=2095116.2095240

53. Fomin, F.V., Lokshtanov, D., Saurabh, S.: Excluded grid minors and efficient polynomial-time approximation schemes. J. ACM **65**(2), 10:1–10:44 (2018). https://doi.org/10.1145/3154833

54. Fomin, F.V., Lokshtanov, D., Saurabh, S., Thilikos, D.M.: Bidimensionality and kernels. In: 21st Annual ACM-SIAM Symposium on Discrete Algorithms (SODA 2010), pp. 503–510. ACM-SIAM (2010)

55. Fomin, F.V., Lokshtanov, D., Saurabh, S., Thilikos, D.M.: Bidimensionality and kernels. CoRR abs/1606.05689 (2016). http://arxiv.org/abs/1606.05689, revised version

56. Fomin, F.V., Lokshtanov, D., Saurabh, S., Thilikos, D.M.: Kernels for (connected) dominating set on graphs with excluded topological minors. ACM Trans. Algorithms **14**(1), 6:1–6:31 (2018). https://doi.org/10.1145/3155298

57. Fomin, F.V., Lokshtanov, D., Saurabh, S., Zehavi, M.: Kernelization: Theory of Parameterized Preprocessing. Cambridge University Press, Cambridge (2019). https://doi.org/10.1017/9781107415157

58. Frick, M., Grohe, M.: Deciding first-order properties of locally tree-decomposable structures. J. Assoc. Comput. Mach. **48**(6), 1184–1206 (2001)

59. Gajarský, J., et al.: Kernelization using structural parameters on sparse graph classes. J. Comput. Syst. Sci. **84**, 219–242 (2017). https://doi.org/10.1016/j.jcss.2016.09.002. http://www.sciencedirect.com/science/article/pii/S0022000016300812

60. Ganian, R., Slivovsky, F., Szeider, S.: Meta-kernelization with structural parameters. J. Comput. Syst. Sci. **82**(2), 333–346 (2016). https://doi.org/10.1016/j.jcss.2015.08.003

61. Garnero, V., Paul, C., Sau, I., Thilikos, D.M.: Explicit linear kernels via dynamic programming. SIAM J. Discrete Math. **29**(4), 1864–1894 (2015)

62. Garnero, V., Sau, I., Thilikos, D.M.: A linear kernel for planar red-blue dominating set. Discrete Appl. Math. **217**, 536–547 (2017). https://doi.org/10.1016/j.dam.2016.09.045

63. Giannopoulou, A.C., Pilipczuk, M., Raymond, J., Thilikos, D.M., Wrochna, M.: Linear kernels for edge deletion problems to immersion-closed graph classes. In: 44th International Colloquium on Automata, Languages, and Programming, ICALP 2017, Warsaw, Poland, 10–14 July 2017, pp. 57:1–57:15 (2017)

64. Grigoriev, A., Koutsonas, A., Thilikos, D.M.: Bidimensionality of geometric inter-section graphs. In: Geffert, V., Preneel, B., Rovan, B., Štuller, J., Tjoa, A.M. (eds.) SOFSEM 2014. LNCS, vol. 8327, pp. 293–305. Springer, Cham (2014). https://doi.org/10.1007/978-3-319-04298-5_26

65. Grohe, M.: Logic, graphs, and algorithms. In: Flum, J., Grädel, E., Wilke, T. (eds.) Logic and Automata - History and Perspectives, pp. 357–422. Amsterdam University Press, Amsterdam (2007)

66. Grohe, M., Kawarabayashi, K., Marx, D., Wollan, P.: Finding topological sub-graphs is fixed-parameter tractable. In: Fortnow, L., Vadhan, S.P. (eds.) Proceedings of the 43rd ACM Symposium on Theory of Computing, STOC 2011, San Jose, CA, USA, 6–8 June 2011, pp. 479–488. ACM (2011). https://doi.org/10.1145/1993636.1993700

67. Grohe, M., Kreutzer, S.: Methods for algorithmic meta theorems. In: Model Theoretic Methods in Finite Combinatorics, pp. 181–206. Contemporary Mathematics (2011)

68. Gu, Q., Tamaki, H.: Improved bounds on the planar branchwidth with respect to the largest grid minor size. Algorithmica **64**(3), 416–453 (2012)

69. Guo, J., Niedermeier, R.: Linear problem kernels for NP-hard problems on planar graphs. In: Arge, L., Cachin, C., Jurdziński, T., Tarlecki, A. (eds.) ICALP 2007. LNCS, vol. 4596, pp. 375–386. Springer, Heidelberg (2007). https://doi.org/10.1007/978-3-540-73420-8_34

70. Guo, J., Niedermeier, R., Wernicke, S.: Fixed-parameter tractability results for full-degree spanning tree and its dual. In: Bodlaender, H.L., Langston, M.A. (eds.) IWPEC 2006. LNCS, vol. 4169, pp. 203–214. Springer, Heidelberg (2006). https://doi.org/10.1007/11847250_19

71. Gutin, G.Z.: Kernelization, constraint satisfaction problems parameterized above average. In: Kao, M.Y. (ed.) Encyclopedia of Algorithms, pp. 1011–1013. Springer, New York (2016). https://doi.org/10.1007/978-1-4939-2864-4_524

72. Gutin, G., Yeo, A.: Constraint satisfaction problems parameterized above or below tight bounds: a survey. In: Bodlaender, H.L., Downey, R., Fomin, F.V., Marx, D. (eds.) The Multivariate Algorithmic Revolution and Beyond. LNCS, vol. 7370, pp. 257–286. Springer, Heidelberg (2012). https://doi.org/10.1007/978-3-642-30891-8_14

73. Hajiaghayi, M.T.: The bidimensionality theory and its algorithmic applications. Ph.D. thesis, Department of Mathematics, Massachusetts Institute of Technology (2005)

74. Jansen, B.M.P., Kratsch, S.: A structural approach to kernels for ILPs: treewidth and total unimodularity. In: Bansal, N., Finocchi, I. (eds.) ESA 2015. LNCS, vol. 9294, pp. 779–791. Springer, Heidelberg (2015). https://doi.org/10.1007/978-3-662-48350-3_65

75. Kanj, I.A., Pelsmajer, M.J., Xia, G., Schaefer, M.: On the induced matching problem. In: Proceedings of the 25th Annual Symposium on Theoretical Aspects of Computer Science (STACS 2008), vol. 08001, pp. 397–408. Internationales Begegnungs- und Forschungszentrum fuer Informatik (IBFI), Schloss Dagstuhl, Berlin (2008)

76. Kim, E.J., et al.: Linear kernels and single-exponential algorithms via protrusion decompositions. ACM Trans. Algorithms **12**(2), 21:1–21:41 (2016)

77. Kim, E.J., Serna, M.J., Thilikos, D.M.: Data-compression for parametrized counting problems on sparse graphs. In: Hsu, W., Lee, D., Liao, C. (eds.) 29th International Symposium on Algorithms and Computation, ISAAC 2018, LIPIcs, Jiaoxi, Yilan, Taiwan, 16–19 December 2018, vol. 123, pp. 20:1–20:13. Schloss Dagstuhl - Leibniz-Zentrum fuer Informatik (2018). https://doi.org/10.4230/LIPIcs.ISAAC.2018.20

78. Kloks, T.: Treewidth, Computations and Approximations. Lecture Notes in Computer Science, vol. 842. Springer, Heidelberg (1994). https://doi.org/10.1007/BFb0045375

79. Kreutzer, S.: Algorithmic meta-theorems. Electronic Colloquium on Computational Complexity (ECCC), Report No. 147, 16 (2009)

80. Lokshtanov, D., Mnich, M., Saurabh, S.: Linear kernel for planar connected dominating set. In: Chen, J., Cooper, S.B. (eds.) TAMC 2009. LNCS, vol. 5532, pp. 281–290. Springer, Heidelberg (2009). https://doi.org/10.1007/978-3-642-02017-9_31

81. Misra, N., Raman, V., Saurabh, S.: Lower bounds on kernelization. Discrete Optim. 8(1), 110–128 (2011). https://doi.org/10.1016/j.disopt.2010.10.001. http://www.sciencedirect.com/science/article/pii/S157252861000068X. Parameterized Complexity of Discrete Optimization

82. Moser, H., Sikdar, S.: The parameterized complexity of the induced matching problem in planar graphs. In: Preparata, F.P., Fang, Q. (eds.) FAW 2007. LNCS, vol. 4613, pp. 325–336. Springer, Heidelberg (2007). https://doi.org/10.1007/978-3-540-73814-5_32

83. Niedermeier, R.: Invitation to Fixed-Parameter Algorithms. Oxford University Press, Oxford (2006). https://doi.org/10.1093/ACPROF:OSO/9780198566076.001.0001

84. Robertson, N., Seymour, P.D.: Graph minors. XX. Wagner's conjecture. J. Comb. Theory Ser. B 92(2), 325–357 (2004)

85. Thilikos, D.M.: Graph minors and parameterized algorithm design. In: The Multivariate Algorithmic Revolution and Beyond - Essays Dedicated to Michael R. Fellows on the Occasion of His 60th Birthday, pp. 228–256 (2012)

86. Thilikos, D.M.: Bidimensionality and parameterized algorithms (invited talk). In: 10th International Symposium on Parameterized and Exact Computation, IPEC 2015, Patras, Greece, 16–18 September 2015, pp. 1–16 (2015)

87. Zoros, D.: Obstructions and algorithms for graph layout problems. Ph.D. thesis, National and Kapodistrian University of Athens, Department of Mathematics, July 2017

Games, Puzzles and Treewidth

Tom C. van der Zanden[⊠] [ID]

Department of Data Analytics and Digitalisation, Maastricht University,
Maastricht, The Netherlands
T.vanderZanden@maastrichtuniversity.nl

Abstract. We discuss some results on the complexity of games and
puzzles. In particular, we focus on the relationship between bounded
treewidth and the (in-)tractability of games and puzzles in which graphs
play a role. We discuss some general methods which are good starting
points for finding complexity proofs for games and puzzles.

Keywords: Complexity · Games · Puzzles · Treewidth

1 Introduction

This article was written on the occasion of the 60th birthday of Hans Bodlaender. As one of his PhD students, I have come to know Hans not only for his
appreciation of deep theoretical research on graphs, but also for his enjoyment
of games and puzzles and for fun (but nevertheless serious) research. Our collaboration started during Hans' algorithms master course, in which there was an
assignment on solving the puzzle game Bloxorz. This resulted in our first joint
paper, in which we showed Bloxorz PSPACE-complete [24]. Happy 60th, Hans!

The complexity of games and puzzles is a widely
studied topic, and there are far too many results out
there to even begin to give an overview. Instead, we will
focus on a few common techniques for showing hardness
and in particular on the relation to treewidth. One of
the reasons game and puzzle complexity is so popular,
is, of course, that it is fun: trying to build gadgets with
elements from Super Mario or arguing whether a game
remains NP-hard even when general relativity is considered [9] provide light-hearted insights into complexity. Games and puzzles also make excellent and visceral
examples for teaching: SUDOKU is a perfect example of
a problem in NP – it is clearly easy to verify the cor-

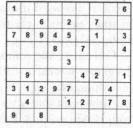

Fig. 1. An instance of a
Sudoku puzzle, a com-
mon example of an NP-
complete puzzle.

rectness of a solution while, intuitively, it seems much harder to find a solution
(and indeed, the problem is NP-complete [27]).

Decision versions of many classical pen-and-paper puzzles (such as SUDOKU,
Fig. 1) are easily seen to be in NP and most often, are also NP-complete. Other
examples include NONOGRAM and KAKURO (see e.g. [13], Appendix A.7 for

© Springer Nature Switzerland AG 2020
F. V. Fomin et al. (Eds.): Bodlaender Festschrift, LNCS 12160, pp. 247–261, 2020.
https://doi.org/10.1007/978-3-030-42071-0_17

an overview). These puzzles are essentially constraint satisfaction problems, so it is natural they would be NP-complete.

In many puzzles, the player has to move pieces around on a board in order to reach a target configuration. For example, in PEG SOLITAIRE, pegs are arranged on a grid, and the player is allowed to move a peg from one hole to another by jumping over another (adjacent) peg, after which the peg that has been jumped over is removed from the board. Deciding whether the board can be cleared (save for one peg) is NP-complete [22]. Here, membership in NP is obvious: since every move reduces the number of pegs by one, a solution has bounded length.

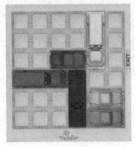

More often, puzzles which involve moving pieces on a board are PSPACE-complete. A necessary[1] condition for this is that the length of a solution is not polynomially bounded (and, save for a few exceptions, if the length of a solution is not polynomially bounded, the problem is PSPACE-complete). An example of such a puzzle is RUSH HOUR [10] (Fig. 2), in which cars (rectangles of size 1×2), that may only move backwards and forwards, are arranged vertically and horizontally on a board and the goal is to move a specific car to its destination. One possible approach to proving hardness [13] constructs a RUSH HOUR instance in which the player is forced to move the cars in such a way that corresponds to painstakingly checking all possible assignments to a quantified boolean formula[2]. These types of puzzles are essentially reconfiguration problems [17].

Fig. 2. A Rush Hour puzzle by ThinkFun. The goal is to move the red car off the right side of the board.

Another interesting source of problems to analyse are (platform) video games. Here, the decision question is whether the player can reach the end of the level. The simplest such games are in P (since they reduce to a reachability question that can be solved using, for instance, breadth-first search). However, depending on what elements are present in a level, such games may be NP-complete (e.g., SUPER MARIO [1]) or even PSPACE-complete (e.g., ZELDA [2]).

Of course, the complexity strongly depends on what features are present in the game. Several metatheorems exist characterizing which features make a video game hard (see e.g., [11,25]). For example, NP-hardness can be obtained if there is a time limit and a way to force the player to visit several locations (reduction from HAMILTONIAN PATH), or if the game features one-way passages (such as a *long fall* feature, where the player can survive a fall higher than they can jump) and a way to enable passages at a later stage of the level (such as a button that opens a door or an enemy that can be killed to enable safe passage later). For PSPACE-hardness, reversible state changes are required, such as blocks that can be pushed back and forth (e.g., SOKOBAN [7]) or pressure plates that can *force* the player to close doors (and a means to open them again).

---

[1] Unless NP = PSPACE.

[2] E.g., a formula of the form $\exists_{x_1} \forall_{x_2} \exists_{x_3} \cdots \forall_{x_n} \phi(x_1, x_2, \ldots, x_n)$, where ϕ is an unquantified boolean formula over binary variables $x_1, \ldots, x_n$.

Finally, two-player games tend to be either PSPACE- or EXPTIME-complete. Games which allow an unbounded number of moves tend to be EXPTIME-complete, such as CHESS [12] and GO [20]. On the other hand, if the number of moves is bounded, two-player games tend to be PSPACE-complete. Examples include REVERSI [18] and GENERALIZED GEOGRAPHY [21].

In this article, we will survey several known results on the hardness of games and puzzles, with a focus on techniques and frameworks for proving hardness and, where applicable, the relation to (bounded) treewidth.

2   Hardness of Video Games

2.1   NP-Hard Video Games

One possible method to show NP-hardness of a video game is to give a direct reduction from SATISFIABILITY. Aloupis et al. [1] use this approach for SUPER MARIO. Variable and clause gadgets are shown in Fig. 3. In the variable gadget, the player may go either left or right to assign a particular variable to true or false. The fall is long enough that the player cannot jump back up to (also) make the other assignment. The clause gadget is entered from one of the three entrances on the bottom (corresponding to one of the three literals in a 3-SAT clause); hitting the item block from below will release a star powerup to the area above. This powerup can later be used to traverse the flames on the right.

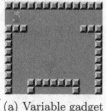
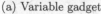

(a) Variable gadget   (b) Clause gadget

Fig. 3. Gadgets showing the hardness of SUPER MARIO due to Aloupis et al. [1].

To complete the reduction, we still need a (complicated) *crossover* gadget, i.e., a gadget which allows to paths to cross each other without the player being able to switch from one path to the other. This gadget is needed to make appropriate connections between the variable and clause gadgets. Aloupis et al. [1] give such a gadget. The construction allows (and forces) the player to first traverse all variable gadgets and make a choice for each one (and thus, unlocking the various clause gadgets) and then traverse the check paths of all the clause gadgets.

To simplify the proof, one might consider using PLANAR 3-SAT, that is, 3-SAT wherein the incidence graph of variables and clauses is planar – hoping that using this problem might help eliminate the need for a crossover gadget. Unfortunately, this is not the case, because we also need additional paths to visit all the variables (to set them) and clauses (to check them).

Recently [19], the following very interesting satisfiability variant was shown NP-complete:

SIDED LINKED PLANAR 3-SAT-3
Given: A CNF formula ϕ with at most 3 literals per clause and at most 3 literals per variable, such that the incidence graph, augmented with a cycle that first visits all variables and then visits all clauses, is planar *and* admits an embedding such that for each variable, the edges going to its positive literals occur on a different side of the cycle than the edges going to its negative literals.
Question: Does ϕ have a satisfying assignment?

This is a very useful satisfiability variant, since the cycle visiting all variables and clauses can be used to perform the setting and checking exactly as required in the previous proof. Using this satisfiability variant, the crossover gadget can be eliminated from the reduction for SUPER MARIO, and, in fact, from many other proofs as well.

This proof illustrates just one example of what can cause a video game to be NP-hard: in this case, it is the fact that the existence of a long drop (which forces us to make a choice in going left or right) combined with the existence of a game element that unlocks a path for later traversal (essentially, this is Metatheorem 3 of [11]). Viglietta's metatheorems [25] capture several other possible elements that can make a game NP-hard:

Metatheorem 1 ([25]). *If a video game features one of the following combinations of elements, it is NP-hard:*

- *Location traversal (spots in the level that must be traversed, e.g., items that must be collected), combined with single-use paths.*
- *Tokens, toll roads that consume a token to traverse and location traversal. Tokens may be either cumulative (any number can be held) or collectible (one may be held at a time).*
- *Cumulative tokens and toll roads.*
- *Pressure plates[3], which correspond one-to-one to doors that they may open or close.*

2.2 PSPACE-Hard Video Games

Many video games are PSPACE-hard. This can be the case when they feature movable elements (such as SOKOBAN [7]) or if the game features elements that

---

[3] A pressure plate, as opposed to a button, is a game element that the player cannot avoid triggering if traversed.

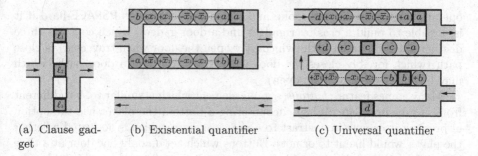

(a) Clause gadget   (b) Existential quantifier   (c) Universal quantifier

Fig. 4. Construction showing PSPACE-hardness of games with pressure plates (depicted as circles; a circle labelled $+x$ (resp. $-x$) opens (resp. closes) door x) and doors (depicted as squares) due to Viglietta [25].

give the player the option to open doors or can force them to close them again. In the previous section, we stated the metatheorem that for pressure plates which correspond one-to-one with doors it is possible to obtain NP-hardness. This is tight, since there exist NP-complete games featuring these elements: since each door is controlled by at most one pressure plate, once it is opened or closed its state can never change again.

If instead we consider a game in which each door may be controlled by two pressure plates, one which may open it and one which may close it, we obtain PSPACE-hardness:

Metatheorem 2 [25]. *If a game features pressure plates and doors, each pressure plate controls at most one door and each door is controlled by at most two pressure plates, then the game is* PSPACE-*hard.*

The proof (due to Viglietta [25]) is by reduction from QUANTIFIED BOOLEAN FORMULA SATISFIABILITY. Figure 4 shows the gadgets used in the construction: (a) a clause gadget, which may be traversed if at least one of the doors corresponding to one of the three literals of the clause is open. (b) an existential quantifier gadget, which can be traversed from top left to top right by picking either the top or bottom path (which opens doors corresponding to either a true or false assignment to that variable) (c) a universal quantifier, which must be traversed in the following way: the player enters from the top left, passes the top path (making a true assignment), then leaves from the top right. The player then returns (after checking the rest of the formula) on the bottom right and then must traverse the middle two paths (making a false assignment), after which the player must again exit from the top right to verify the assignment again, before returning on the bottom right and being able to exit on the bottom left.

One drawback of this scheme is that pressure plates must be able to act on doors anywhere in the level. This cannot always be realized easily. On the positive side, it is not necessary to build a crossover gadget (since crossings are provided by the pressure plates working on arbitrarily distant doors). Another framework, the *door framework* [2], gets rid of the requirement that buttons may operate arbitrarily distant doors, and states that a game is PSPACE-hard if it is possible to build a crossover gadget and a door gadget, which contains three distinct paths: an open path (which may open the door when traversed), a close path (which forcibly closes the door when traversed) and a door path (which may be traversed only when open).

Many games feature *buttons*, a game element which is similar to, but different from, a pressure plate. When encountering a button, the player has the option of pressing it or not (in contrast to a pressure plate, which is activated whether the player would like it to or not). Buttons which act on only one door at a time are trivial (since a player would always press an "open" button and would never press a "close" button), so it is not possible to obtain a hardness metatheorem for these. Instead, Viglietta [25] considers k-buttons: a button which may act on at most k doors at once. For $k \geq 2$, a player may be incentivized to press a button that would close a door, provided it also opens another.

One could ask what is the minimum k for which a game with k-buttons is hard. Viglietta [25] showed that for $k = 2$ such games are NP-hard, and for $k = 3$, PSPACE-hard. Hans and I improved this, showing that $k = 2$ already suffices for PSPACE-hardness:

Metatheorem 3 ([24]). *A game featuring 2-buttons is PSPACE-hard, even if each door may be acted upon by at most 2 buttons.*

BLOXORZ [16] (Fig. 5) is a puzzle game that features a third type of element, a *switch*, that may toggle a (trap-)door between open and closed (i.e., repeated presses of a switch will cause the state of the door to cycle between open and closed). The unique feature of BLOXORZ is that the player is a $1 \times 1 \times 2$ block, for which two types of moves are possible: if the block is standing up, a tilting move is possible, which causes the block to lay on one of its 1×2 sides. If the block is lying on a 1×2 side, it can either do a tilting move (causing the block to stand up again), or a rolling move, rolling over to another of its 1×2 sides. These special types of moves enable some interesting gadget constructions.

Each switch in BLOXORZ may act on only one trapdoor (and each trapdoor may be acted upon by only one switch). However, we can exploit the fact that the block is 1×2 to build what are, essentially, 2-switches (if the block falls down on a 1×2 side on two switches, they are both triggered). Thus, it is possible to

Fig. 5. An example Bloxorz level

show that Bloxorz is PSPACE-complete [24]. The proof can be distilled to the following metatheorem:

Metatheorem 4 (Consequence of [24]). *If a game features 2-switches, it is* PSPACE-*hard, even if each door may be controlled by at most one switch.*

Note that for this metatheorem, it is not relevant whether the 2-switches behave like pressure plates (in the sense that they are always forcibly triggered) or buttons (in the sense that the player can choose to trigger them or not).

3   Constraint Logic Framework

Many hardness constructions for games and puzzles are quite involved, and require the creation of complicated crossover gadgets or lengthy arguments about simulating the behaviour of a Turing machine. The Constraint Logic framework, introduced by Hearn and Demaine [13], provides several games and puzzles based around the notion of *constraint graphs*, each of which is complete for a different complexity class. These aim to be convenient starting points for reductions and simplify hardness proofs (for instance, by eliminating the need to construct a dedicated crossover gadget).

Of particular interest is the PSPACE-complete variant, called NONDETER-MINISTIC CONSTRAINT LOGIC (NCL), which has proven very useful for a wide variety of hardness proofs. However, the Constraint Logic framework includes many games, each of which capture the essence of a different complexity class and its relation to games: for instance, it is possible to define a two-player game on constraint graphs. The edges are partitioned between the players, the players take turns flipping an edge from their set and each player has a target edge that they must flip to win. In a bounded setting (each edge may be flipped at most once), this game is PSPACE-complete (capturing, e.g., the hardness of REVERSI), while in general it is EXPTIME-complete (capturing, e.g., the hardness of CHESS). We will now formally introduce Constraint Logic.

Definition 1. *A constraint graph is an undirected graph $G = (V, E)$, together with:*

- *For each vertex $v \in V$, a vertex weight $w(v)$,*
- *For each edge $e \in E$, an edge weight $w(e)$.*

Given a constraint graph G, a *configuration* is an assignment of orientations to its edges. A configuration is *valid* if for each vertex, the total weight of edges pointing in towards that vertex is at least that vertex' weight.

The CONSTRAINT GRAPH SATISFIABILITY problem asks whether a constraint graph G admits a valid configuration. It can be viewed as the constraint logic equivalent of 3-SAT.

CONSTRAINT GRAPH SATISFIABILITY
Given: A constraint graph G.
Question: Does G have a valid configuration?

Constraint Graph Satisfiability is NP-complete, even for very restricted constraint graphs:

Theorem 1 ([13]). *It is NP-complete to decide whether a constraint graph G admits a valid configuration, even if G is planar, has maximum degree 3, each vertex has weight equal to 2 and the edge weights are in* $\{1,2\}$.

In fact, for all discussions on constraint logic, it suffices to consider vertex weights 2 and edge weights $\{1,2\}$. Going forward, we shall omit the vertex weights. Following the convention in [13], red edges shall have weight 1 and blue edges (drawn thicker) shall have weight 2.

Given a (valid) configuration for a constraint graph, we can obtain another configuration by flipping the direction of one edge. A *move* is an edge flip between two valid configurations.

Nondeterministic Constraint Logic (NCL)

Given: A constraint graph G, a valid configuration C for G and a target edge $e \in G$.

Question: Is there a sequence of moves on G, starting from C, that eventually reverses e?

We may also consider the configuration-to-configuration variant of NCL, which asks whether a starting configuration C_1 can be reconfigured (through moves) to a target configuration C_2. All complexity results discussed in this section hold analogously for this variant.

NCL is PSPACE-complete, even for very restricted constraint graphs. We may consider graphs constructed using only the two vertex types shown in Fig. 6. The *OR vertex* has three incident blue (weight 2) edges, and thus in any valid configuration at least one of them needs to point inward. It resembles an OR gate in the sense that if we identify one edge as the "output", it can point outward if and only if at least one of the two other edges is pointing inward. The *AND vertex* has two incident red (weight 1) edges and one incident blue (weight 2) edge. It is satisfied if and only if both red edges point inward or the blue edge points inward. Thus, we can think of the blue edge as its "output", able to point outward only if both red edges are pointing inward. The fact that NCL is PSPACE-complete for such restricted graphs makes it a very powerful tool for constructing hardness proofs.

Theorem 2. Nondeterministic Constraint Logic *is* PSPACE-*complete, even for planar constraint graphs that use only AND and OR vertices.*

To prove hardness by reduction from NCL, we thus only need to show how AND and OR vertices may be constructed. Figure 7 shows how the AND and OR vertices may be constructed in Rush Hour. In the AND vertex, car C can move down if and only if cars A and B move out; in the OR vertex, if either car A or B moves out, car C can move in. These constructions are essentially the only elements necessary for the proof, it only remains to be shown that they

(a) OR vertex   (b) AND vertex

Fig. 6. The two vertex types from which a restricted constraint graph is constructed: (a) OR vertex and (b) AND vertex. Following the convention set in [13], as a mnemonic weight 2 edges are drawn blue (dark grey) and thick, while weight 1 edges are drawn red (light grey) and thinner. From [23]. (Color figure online)

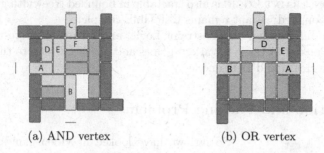

(a) AND vertex   (b) OR vertex

Fig. 7. AND and OR vertices, as constructed in a reduction to RUSH HOUR, Hearn and Demaine [13].

can (given a planar constraint graph) be arranged in the plane and connected accordingly; this greatly simplifies the original PSPACE-completeness proof [10].

Bounded Width. Of course, given that a graph is involved, a natural question is what happens if the unweighed graph underlying the constraint graph has bounded treewidth. It can easily be seen that CONSTRAINT GRAPH SATISFIA-BILITY is polynomial-time solvable on graphs of bounded treewidth. Surprisingly, NCL is PSPACE-complete, even for graphs of bounded treewidth (and actually, even for graphs of bounded bandwidth):

Theorem 3 ([23]). *There is a constant c such that* NONDETERMINISTIC CON-STRAINT LOGIC *is* PSPACE-*complete, even on planar constraint graphs of band-width at most c that use only AND and OR vertices.*

This result is closely related to a result of Wrochna [26] on reconfiguration problems in bounded bandwidth graphs. Wrochna shows that a Turing machine (with polynomially bounded tape) can be simulated in a bounded width struc-ture. Since the tape is linear and, locally, we only need to take into account a bounded number of cases (depending on the number of states of the TM and the size of the alphabet), it is actually quite natural that this should be the case.

An amusing consequence of this result is that RUSH HOUR is PSPACE-complete even when played on a board of constant width.

Interestingly, some games on graphs that are PSPACE-complete do become tractable if the graph has bounded treewidth. For instance, GENERALIZED GEOGRAPHY can be solved in linear time on graphs of bounded treewidth [3]. The main difference is that GENERALIZED GEOGRAPHY is a bounded two-player game, whereas NONDETERMINISTIC CONSTRAINT LOGIC is single player, unbounded. The fact that moves are reversible means that information can flow back and forth in the graph, passing through separators. Essentially, the separator property of bounded treewidth graphs is useless if there is a mechanism that can pass information through them. In contrast, in GENERALIZED GEOGRAPHY, once a vertex has been picked by one of the players, it cannot be used again. It is an interesting open problem to settle the conjecture that BOUNDED TWO-PLAYER CONSTRAINT LOGIC is also tractable in bounded treewidth graphs, and that the unbounded variant remains EXPTIME-complete.

For more information on Constraint Logic, and for many interesting reductions from Constraint Logic to various games and puzzles, I refer to the excellent book [13] by Hearn and Demaine.

4   Partition and Packing Problems

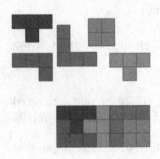

Fig. 8. A simple POLYOMINO PACKING puzzle and a possible solution.

So far, we have looked at video games, pen-and-paper puzzles and sliding piece puzzles. A final class of puzzles that we are going to look at are packing puzzles and jigsaws. Besides fun applications (puzzles such as Tangram, polyomino packing or jigsaws), packing problems also have many practical applications of real-world importance, such as loading packages into a delivery truck or loading products onto pallets [14].

A key tool for showing the hardness of packing problems is the 3-PARTITION problem:

3-PARTITION
Given: A collection A of $3n$ positive integers
$a_1, a_2, \ldots, a_{3n}$.
Question: Is there a partition of A into n
triples such that the numbers in each triple
sum to $\frac{1}{n}\Sigma_{i=1}^{3n} a_i$?

The 3-PARTITION problem is *strongly* NP-hard, i.e., it remains NP-complete even if each integer a_i is bounded by a polynomial in n. This is very useful, since in a reduction we can construct puzzles whose size is proportional to the integers in the input and still obtain a hardness result.

Reductions from 3-PARTITION can sometimes give very easy hardness proofs. Demaine and Demaine [8] give an excellent overview of many types of packing and jigsaw puzzles, most of which are shown hard using 3-PARTITION.

An interesting type of puzzles are POLYOMINO PACKING puzzles. In POLY-OMINO PACKING we are given a collection of polyominoes (shapes consisting of unit squares, glued together on their edges) and are asked to build a target shape (which is a subset of a square grid). Figure 8 shows an example of a POLYOMINO PACKING puzzle, in which we are asked to pack 5 polyominoes into a 3×7 rectangle.

Even for very restricted instances, POLYOMINO PACKING is already NP-complete:

Theorem 4 ([8]). *It is (strongly)* NP-*complete to decide whether n given rectangular pieces sized $1 \times x_1, 1 \times x_2, \ldots, 1 \times x_n$, where the x_i's are positive integers bounded above by a polynomial in n, can be exactly packed into a specified rectangular box of area $x_1 + x_2 + \ldots + x_n$.*

The proof due to Demaine and Demaine [8] is by reduction from 3-PARTITION and is quite simple and elegant: given an instance of 3-PARTITION with $3n$ integers $a_1, a_2, \ldots, a_{3n}$, each integer a_i is translated to a rectangle of size $1 \times (a_i + n)$. We then ask whether these rectangles can be packed into a $n \times \left(3n + \frac{1}{n}\Sigma_{i=1}^{3n}a_i\right)$ rectangle.

Bounded Width. An interesting question is what happens when we bound the *width* in some way. For instance, we might ask ourselves what the complexity of POLYOMINO PACKING is when the target shape is a $k \times n$ rectangle, where k is a (bounded) width parameter. There is a very simple hardness proof showing hardness even for $k = 2$:

Theorem 5. POLYOMINO PACKING *is* NP-*hard, even when the target shape is a $2 \times n$ rectangle.*

This can be seen by reduction from 3-PARTITION. We may construct a $2 \times \left(n + 1 + \Sigma_{i=1}^{3n}a_i\right)$ rectangle into which n slots of size $1 \times \left(\frac{1}{n}\Sigma_{i=1}^{3n}a_i\right)$ are cut. We then ask whether this "rectangle with slots" can be packed together with $3n$ rectangle polyominoes of sizes $1 \times a_1, 1 \times a_2, \ldots, 1 \times a_{3n}$ into a box of size $2 \times \left(n + 1 + \Sigma_{i=1}^{3n}a_i\right)$.

The reason having bounded width does not help in this case is that, if we try to do separator-based dynamic programming (where a separator might divide the $2 \times n$ box into two $2 \times n/2$ boxes), we need to remember which pieces we have used on one side of our separator. A very similar phenomenon appears in SUBGRAPH ISOMORPHISM in planar and bounded treewidth graphs, where having bounded treewidth also does not really help [4].

Exact Complexity. Besides studying classification into complexity classes such as NP and P, it is also interesting to ask what the exact complexity of a problem is. That is, since POLYOMINO PACKING is NP-hard, we know it most likely cannot be solved in polynomial time. It then becomes interesting to ask what the slowest-growing f is such that POLYOMINO PACKING into a $2 \times n$ box can be solved in time $2^{O(f(n))}$.

Using a dynamic programming approach, and the observation that placing polyomino pieces into a $2 \times n$ box partitions the remaining area into connected components that are easy to characterize, it is possible to obtain the following result:

Theorem 6 ([6]). POLYOMINO PACKING *can be solved in* $2^{O(n^{3/4} \log n)}$ *time if the target shape is a* $2 \times n$ *rectangle.*

Of course, a natural question is whether we can also obtain a (matching) lower bound on f. Using the Exponential Time Hypothesis [15], doing so is sometimes possible. The Exponential Time Hypothesis (ETH) states that there exists no algorithm solving n-variable 3-SAT in $2^{o(n)}$ time. Assuming this hypothesis, and by designing efficient reductions (that do not blow up the instance size too much), it is possible to derive conditional lower bounds on the running time of an algorithm.

The blow-up of the reduction from 3-SAT to 3-PARTITION is rather large and does not lead to a tight lower bound for the $2 \times n$ case. However, many packing problems exhibit an interesting behaviour: their optimal running time is $2^{\Theta(n/\log n)}$ (under the ETH). This holds for, for instance, SUBGRAPH ISOMOR-PHISM on planar graphs [4,5] and also for POLYOMINO PACKING if we increase k to 3:

Theorem 7 ([6]). POLYOMINO PACKING *cannot be solved in* $2^{o(n/\log n)}$ *time, even if the target shape is a* $3 \times n$ *rectangle.*

Using treewidth-based techniques, we can obtain a matching $2^{O(n/\log n)}$-time algorithm [6]. Interestingly, this means that $4 \times n$ (or $n \times n$) POLYOMINO PACKING is essentially not any harder than $3 \times n$ POLYOMINO PACKING. $2 \times n$ POLYOMINO PACKING is somewhat easier, but still NP-hard.

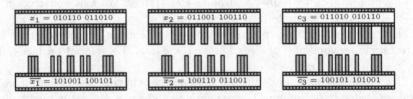

Fig. 9. Top: polyominoes corresponding to variables x_1, x_2 and clause c_3. Bottom: the complementary polyominoes, that mate with the polyominoes above them to form a $3 \times \Theta(\log n)$ rectangle. Note that the polyominoes are depicted compressed horizontally. Due to [6].

The lower bound can be obtained by a direct reduction from 3-SAT. If we number the variables in the instance $1, 2, \ldots, n$ and the clauses $n + 1, \ldots, n + m + 1$, we can pick a unique bitstring corresponding to each variable and clause, being derived from the binary representation of its index. We can then use these bitstrings to construct *corresponding polyominoes*, which consist of a solid row on top and a row on the bottom which has a square whenever the bitstring has a

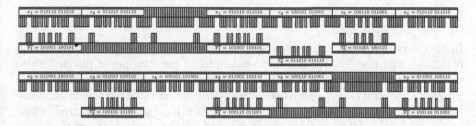

Fig. 10. Example of the reduction for the formula $(x_1 \vee x_2) \wedge (\neg x_1 \vee x_2) \wedge (\neg x_1 \vee \neg x_2)$. Top-to-bottom, left-to-right: formula encoding polyomino for x_1, variable-setting polyomino for x_1, clause-checking polyomino for c_4, clause checking-polyomino for c_5, formula-encoding polyomino for x_2, clause-checking polyomino for c_3, variable-setting polyomino for x_2. The polyominoes are arranged in a way that suggests the solution $x_1 = false$, $x_2 = true$. Due to [6].

1 in that position. Since we need only $\log(n + m)$ bits to represent each number, each corresponding polyomino has only $O(\log n)$ squares.

For each clause and variable we can also define a complementary polyomino, which mates with the corresponding polyomino to form a $3 \times \Theta(\log n)$ rectangle. Figure 9 shows an example of corresponding and complementary polyominoes.

Using corresponding and complementary polyominoes as building blocks, it is possible to build up larger polyominoes that form an instance of $3 \times n$ POLYOMINO PACKING that is solvable if and only if the formula is satisfiable. The instance will have a size of $O(n \log n)$ squares, leading to the claimed lower bound.

We will not go into the full details of the reduction here; instead, we refer to [6]. However, we sketch some details. For every variable, a formula-encoding polyomino is built, together with a variable-setting polyomino. In any solution, there are two possible placements of the variable-setting polyomino with respect to the corresponding formula-encoding one. Either overlapping the left side of the formula-encoding polyomino, corresponding to a false assignment, or overlapping the right, corresponding to a true assignment. The places where the variable-setting polyomino does not overlap the formula-encoding polyomino contain gaps into which polyominoes corresponding to clauses satisfied by the assignment can fit. Figure 10 shows an example of this construction.

This technique offers an alternative to 3-PARTITION-based proofs, giving tighter lower bounds than would be obtained through 3-PARTITION. Many games and puzzles are expressive enough to allow bitstrings to be used to succinctly encode variables and clauses. The example of $2 \times n$ POLYOMINO PACKING is an example where we do not have sufficient expressivity and have to fall back to a direct reduction from 3-PARTITION. An interesting open problem is to settle the exact complexity of 3-PARTITION (and that of $2 \times n$ POLYOMINO PACKING), as current upper and lower bounds are not tight (the chain of reductions that establishes the hardness of 3-PARTITION is quite long, blowing up the size of the instance at several steps; however, we do not know the precise value of the lower bound that follows).

5   Conclusions

In this survey, we have discussed several complexity results on various (types of) games and puzzles. We have seen examples of pen-and-paper puzzles, video games, jigsaws and packing puzzles, some two-player games and reconfiguration puzzles. In several cases, we have seen surprising connections to treewidth: reconfiguration puzzles (and problems) remain PSPACE-hard on bounded (tree-)width structures, while other PSPACE-hard games (such as GENERALIZED GEOGRAPHY) become tractable. We have seen that for POLYOMINO PACKING, the problem similarly remains NP-hard even for boards of width 2 and that the hardness does not increase above width 3.

It would be an interesting open problem to further study the complexity of Constraint Logic variants and games under bounded width. The case for NONDETERMINISTIC CONSTRAINT LOGIC is well understood, but it would be interesting to see if the tractability result for GENERALIZED GEOGRAPHY generalizes to a result for BOUNDED TWO-PLAYER CONSTRAINT LOGIC. Going further, it would be interesting to investigate the complexity of other Constraint Logic variants in bounded width graphs.

References

1. Aloupis, G., Demaine, E.D., Guo, A., Viglietta, G.: Classic nintendo games are (NP-)hard. arXiv preprint arXiv:1203.1895 (2012)
2. Aloupis, G., Demaine, E.D., Guo, A., Viglietta, G.: Classic nintendo games are (computationally) hard. Theoret. Comput. Sci. **586**, 135–160 (2015)
3. Bodlaender, H.L.: Complexity of path-forming games. Theoret. Comput. Sci. **110**(1), 215–245 (1993)
4. Bodlaender, H.L., Nederlof, J., van der Zanden, T.C.: Subexponential time algorithms for embedding H-minor free graphs. In: Chatzigiannakis, I., Mitzenmacher, M., Rabani, Y., Sangiorgi, D. (eds.) 43rd International Colloquium on Automata, Languages, and Programming (ICALP 2016), Leibniz International Proceedings in Informatics (LIPIcs), Dagstuhl, Germany, vol. 55, pp. 9:1–9:14 (2016). Schloss Dagstuhl-Leibniz-Zentrum fuer Informatik
5. Bodlaender, H.L., van der Zanden, T.C.: Improved lower bounds for graph embedding problems. In: Fotakis, D., Pagourtzis, A., Paschos, V.T. (eds.) CIAC 2017. LNCS, vol. 10236, pp. 92–103. Springer, Cham (2017). https://doi.org/10.1007/978-3-319-57586-5_9
6. Bodlaender, H.L., van der Zanden, T.C.: On the exact complexity of polyomino packing. In: Ito, H., Leonardi, S., Pagli, L., Prencipe, G. (eds.) 9th International Conference on Fun with Algorithms (FUN 2018), Leibniz International Proceedings in Informatics (LIPIcs), Dagstuhl, Germany, vol. 100, pp. 9:1–9:10 (2018). Schloss Dagstuhl-Leibniz-Zentrum fuer Informatik
7. Culberson, J.C.: Sokoban is PSPACE-complete. In: Lodi, E., Pagli, L., Santoro, N. (eds.) International Conference on Fun with Algorithms (FUN 1998), pp. 65–76. Carleton Scientific, Waterloo (1998)
8. Demaine, E.D., Demaine, M.L.: Jigsaw puzzles, edge matching, and polyomino packing: connections and complexity. Graph. Comb. **23**(1), 195–208 (2007)

9. Demaine, E.D., Lockhart, J., Lynch, J.: The computational complexity of portal and other 3D video games. In: Ito, H., Leonardi, S., Pagli, L., Prencipe, G. (eds.) 9th International Conference on Fun with Algorithms (FUN 2018), Leibniz International Proceedings in Informatics (LIPIcs), Dagstuhl, Germany, vol. 100, pp. 19:1–19:22 (2018). Schloss Dagstuhl-Leibniz-Zentrum fuer Informatik

10. Flake, G.W., Baum, E.B.: Rush Hour is PSPACE-complete, or "Why you should generously tip parking lot attendants". Theoret. Comput. Sci. **270**(1–2), 895–911 (2002)

11. Forišek, M.: Computational complexity of two-dimensional platform games. In: Boldi, P., Gargano, L. (eds.) FUN 2010. LNCS, vol. 6099, pp. 214–227. Springer, Heidelberg (2010). https://doi.org/10.1007/978-3-642-13122-6_22

12. Fraenkel, A.S., Lichtenstein, D.: Computing a perfect strategy for n × n chess requires time exponential in n. J. Comb. Theory, Ser. A **31**(2), 199–214 (1981)

13. Hearn, R.A., Demaine, E.D.: Games, Puzzles, and Computation. AK Peters/CRC Press, Natick (2009)

14. Hodgson, T.J.: A combined approach to the pallet loading problem. AIIE Trans. **14**(3), 175–182 (1982)

15. Impagliazzo, R., Paturi, R., Zane, F.: Which problems have strongly exponential complexity? J. Comput. Syst. Sci. **63**, 512–530 (2001)

16. Clarke, D.: DX Interactive. Bloxorz. https://damienclarke.me/#bloxorz

17. Ito, T., et al.: On the complexity of reconfiguration problems. Theoret. Comput. Sci. **412**(12), 1054–1065 (2011)

18. Iwata, S., Kasai, T.: The othello game on an n × n board is PSPACE-complete. Theoret. Comput. Sci. **123**(2), 329–340 (1994)

19. Pilz, A.: Planar 3-SAT with a clause/variable cycle. arXiv preprint arXiv:1710.07476 (2017)

20. Robson, J.M.: The complexity of Go. In: 9th World Computer Congress on Information Processing, pp. 413–417 (1983)

21. Schaefer, T.J.: On the complexity of some two-person perfect-information games. J. Comput. Syst. Sci. **16**(2), 185–225 (1978)

22. Uehara, R., Iwata, S.: Generalized Hi-Q is NP-complete. IEICE Trans. (1976-1990) **73**(2), 270–273 (1990)

23. van der Zanden, T.C.: Parameterized complexity of graph constraint logic. In: Husfeldt, T., Kanj, I. (eds.) 10th International Symposium on Parameterized and Exact Computation (IPEC 2015), Leibniz International Proceedings in Informatics (LIPIcs), Dagstuhl, Germany, vol. 43, pp. 282–293 (2015). Schloss Dagstuhl-Leibniz-Zentrum fuer Informatik

24. van der Zanden, T.C., Bodlaender, H.L.: PSPACE-completeness of bloxorz and of games with 2-buttons. In: Paschos, V.T., Widmayer, P. (eds.) CIAC 2015. LNCS, vol. 9079, pp. 403–415. Springer, Cham (2015). https://doi.org/10.1007/978-3-319-18173-8_30

25. Viglietta, G.: Gaming is a hard job, but someone has to do it!. Theory Comput. Syst. **54**(4), 595–621 (2014)

26. Wrochna, M.: Reconfiguration in bounded bandwidth and tree-depth. J. Comput. Syst. Sci. **93**, 1–10 (2018)

27. Yato, T., Seta, T.: Complexity and completeness of finding another solution and its application to puzzles. IEICE Trans. Fundam. Electron. Commun. Comput. Sci. **86**(5), 1052–1060 (2003)

Fast Algorithms for Join Operations
on Tree Decompositions

Johan M. M. van Rooij[✉]

Department of Information and Computing Sciences, Utrecht University,
PO Box 80.089, 3508 TB Utrecht, The Netherlands
J.M.M.vanRooij@uu.nl

Abstract. Treewidth is a measure of how tree-like a graph is. It has
many important algorithmic applications because many NP-hard problems on general graphs become tractable when restricted to graphs
of bounded treewidth. Algorithms for problems on graphs of bounded
treewidth mostly are dynamic programming algorithms using the structure of a tree decomposition of the graph. The bottleneck in the worst-
case run time of these algorithms often is the computations for the so
called join nodes in the associated nice tree decomposition.

In this paper, we review two different approaches that have appeared
in the literature about computations for the join nodes: one using fast
zeta and Möbius transforms and one using fast Fourier transforms. We
combine these approaches to obtain new, faster algorithms for a broad
class of vertex subset problems known as the $[\sigma, \rho]$-domination problems. Our main result is that we show how to solve $[\sigma, \rho]$-domination
problems in $\mathcal{O}(s^{t+2} t n^2 (t \log(s) + \log(n)))$ arithmetic operations. Here,
t is the treewidth, s is the (fixed) number of states required to represent partial solutions of the specific $[\sigma, \rho]$-domination problem, and n is
the number of vertices in the graph. This reduces the polynomial factors involved compared to the previously best time bound (van Rooij,
Bodlaender, Rossmanith, ESA 2009) of $\mathcal{O}(s^{t+2}(st)^{2(s-2)} n^3)$ arithmetic
operations. In particular, this removes the dependence of the degree of
the polynomial on the fixed number of states s.

Keywords: Tree decompositions · Dynamic programming · Fast
Fourier transform · Möbius transform · Fast subset convolution ·
Sigma-rho domination

1   Introduction

Treewidth is an important concept in the theory of graph algorithms that measures how tree-like a graph is. While many problems that are $\mathcal{NP}$-hard on general
graphs become efficiently solvable when restricted to trees, this often extends to
these problems being polynomial or even linear-time solvable when restricted to
graphs that have bounded treewidth. In general this is done in two steps:

1. Find a tree decomposition of the input graph of small treewidth.

© Springer Nature Switzerland AG 2020
F. V. Fomin et al. (Eds.): Bodlaender Festschrift, LNCS 12160, pp. 262–297, 2020.
https://doi.org/10.1007/978-3-030-42071-0_18

2. Solve the problem by dynamic programming on this tree decomposition.

In this paper, we focus on the second of these two steps and show how to improve the running times of algorithms on tree decompositions using algebraic transforms. We apply these to the general case of the so called $[\sigma, \rho]$-domination problems. This includes many well-known vertex subset problems such as INDEPENDENT SET, DOMINATING SET and TOTAL DOMINATING SET, but also problems such as INDUCED BOUNDED DEGREE SUBGRAPH and INDUCED p-REGULAR SUBGRAPH.

If we assume that a graph G is given with a tree decomposition T of G of width t, then the running time of an algorithm on tree decompositions is typically polynomial in the size of graph G, but exponential in the treewidth t. Early examples of such algorithms include algorithms on vertex partitioning problems (including the $[\rho, \sigma]$-domination problems) [30], edge colouring problems such as CHROMATIC INDEX [5], or other problems such as STEINER TREE [21]. Often the worst-case running time of these algorithms involve large factors that depend on the treewidth t. This lead researchers to look for algorithms where these factors grow as slow as possible as a function of t. For several DOMINATING SET-like problems such as INDEPENDENT DOMINATING SET, TOTAL DOMINATING SET, PERFECT DOMINATING SET and PERFECT CODE, Alber et al. [1] give improved algorithms with special attention to the exponential dependence on the treewidth t: for example, they showed how to solve DOMINATING SET in $\mathcal{O}^*(4^t)$ time. This was improved by Van Rooij et al. [26] who first showed how to solve DOMINATING SET in $\mathcal{O}^*(3^t)$ time by giving an $\mathcal{O}(3^t t^2 n)$-time algorithm. Van Rooij et al. also generalised this result solving the $[\rho, \sigma]$-domination problems in $\mathcal{O}(s^{t+2}(st)^{2(s-2)}n^3)$ time. The result for DOMINATING SET seems to be optimal in some sense, as Lokshtanov et al. [22] showed that any $\mathcal{O}^*((3-\epsilon)^t)$-time algorithm would violate the *Strong Exponential-Time Hypothesis*; we expect the same for the other $[\rho, \sigma]$-domination problems.

Since then, several results have appeared improving running times of dynamic programming algorithms on tree decompositions. For example, the algorithm by Van Rooij et al. [26] has been generalised to DISTANCE-r DOMINATING SET [12] and DISTANCE-r INDEPENDENT SET [18]. The most notable new results are the *Cut and Count* technique [16] giving randomised $\mathcal{O}^*(c^t)$-time algorithms for many graph connectivity problems, mostly supported by matching lower bounds based on the Strong Exponential-Time Hypotheses, the *rank-based approach* [9, 14] and the *determinant-based approach* [9,31] that derandomise these results at the cost of a greater base of the exponent c.

For many of these algorithms, the computations in the so called join nodes of a nice tree decomposition are the bottleneck of the worst-case run time. To speed up these computations, several approaches have been used, often based on algebraic transforms. One such method is using fast zeta and Möbius transforms in a way that is similar to the well-known fast subset convolution algorithm by Björklund et al. [2]. This method was first used in the context of tree decompositions by Van Rooij et al. [26] who also generalised the approach to work for the $[\sigma, \rho]$-domination problems. At the same time, Cygan and Pilipczuk, showed that the fast subset convolution result could also be based on Fourier transforms [17]; they also generalised it in a different way. A variant to this approach that we follow in this paper, directly applied to tree decompositions, can be found in the appendix of [15]. We will discuss both these approaches in more detail in this paper. Finally, faster joins are also obtained based on Clifford algebras [31], but these are beyond the scope of this paper.

1.1 Goal of This Paper

The goal of this paper is twofold. Firstly, we want to present a faster algorithm for the $[\sigma, \rho]$-domination problems. This algorithm uses $\mathcal{O}(s^{t+2}tn^2(t\log(s)+\log(n)))$ arithmetic operations: this improves the polynomial factors compared to our earlier result [26] and removes the dependency of the degree of the polynomial on s, where s is the (fixed) number of states used. Secondly, we want to give a comprehensible overview of how Fourier and Möbius transforms can be used to obtain faster algorithms on tree decompositions.

We choose to take an algebraic perspective that allows for easier generalisation and easier combination of Fourier and Möbius transform-based approaches than that in [25, 26]. However, we consider the approach in [25, 26] to be more intuitive: it relies only on counting arguments (this is especially true for the first algorithm for DOMINATING SET in [25] that does not explicitly use any algebraic transform). In our overview, we will not give details on our earlier generalised convolution approach from [26]: after the initial examples, we directly go to the new and improved algorithm.

2 Preliminaries

2.1 Graphs and Tree Decompositions

Let $G = (V, E)$ be an n-vertex graph with m edges. A *terminal graph*[1] $G_X = (V, E, X)$ is a graph $G = (V, E)$ with an ordered sequence of distinct vertices that we call its terminals: $X = \{x_1, x_2, \ldots, x_k\}$ with each $x_j \in V$. Two terminal graphs $G_X = (V_1, E_1, X_1)$ and $H_X = (V_2, E_2, X_2)$ with the same number of terminals k, but disjoint vertex and edge sets, can be *glued* together to form the terminal graph $G_X \oplus H_X$ by identifying each terminal x_i from X_1 with x_i from X_2, for all $1 \leq i \leq k$. That is, if $X = X_1 = X_2$ through identification, then $G_X \oplus H_X = (V_1 \cup V_2, E_1 \cup E_2, X)$. A *completion* of a terminal graph G_X is a non-terminal graph G that can be obtained from G_X by gluing a terminal graph H_X on G_X and then ignoring which vertices are terminals in the result.

The treewidth of a (non-terminal) graph is a measure of how-tree like the graph is. From an algorithmic viewpoint this is a very useful concept because, where many $\mathcal{NP}$-hard problems on general graphs are linear time solvable on trees by dynamic programming, often similar style dynamic programming algorithms exist for graphs whose treewidth is bounded by a constant. We outline

---

[1] This is also known as a k-boundary graph.

the basics on treewidth and specifically on dynamic programming on tree decompositions below. More information can, amongst other places, be found in work by Bodlaender [4,6–8,10].

Definition 1 (tree decomposition and treewidth). *A tree decomposition of an undirected graph $G = (V, E)$ is a tree T in which each node $i \in T$ has an associated set of vertices $X_i \subseteq V$ (called a bag), with $\bigcup_{i \in T} X_i = V$, such that the following properties hold:*

- *for every edge $\{u, v\} \in E$, there exist a bag X_i such that $\{u, v\} \subseteq X_i$;*
- *for every vertex v in G, the bags containing x form a connected subtree: i.e., if $v \in X_i$ and $v \in X_j$, then $v \in X_k$ for all nodes k on the path from i to j in T.*

The width of a tree decomposition T is defined as $\max_{i \in T}\{|X_i|\} - 1$: the size of the largest bag minus one. The treewidth of a graph G is the minimum width over all tree decomposition of G.

For a tree decomposition T with assigned root node $r \in T$, we define the terminal graph $G_i = (V_i, E_i, X_i)$ for each node $i \in T$: let V_i be the union of X_i with all bags X_j where j is a descendant of i in T, and let $E_i \subseteq E$ be the set of edges with at least one endpoint in $V_i \backslash X_i$ (and as a result of Definition 1 with both endpoints in V_i). Now, G_i contains all edges between vertices in $V_i \backslash X_i$, and all edges between $V_i \backslash X_i$ and X_i, but no edges between two vertices in X_i.[2] Observe that, G is the completion of G_i formed through $G_i \oplus ((V \backslash V_i) \cup X_i, E \backslash E_i, X_i)$, and X_i can be seen as the *separator* separating $V_i \backslash X_i$ from $V \backslash V_i$ in G (where either side of the separator can be empty).

We now describe dynamic programming on a tree decomposition T. Given a graph problem that we are trying to solve $\mathcal{P}$, define a *partial solution* of $\mathcal{P}$ on G_i to be the *restriction* to the subgraph G_i of a solution of $\mathcal{P}$ on a completion of G_i (any completion of G_i, not only G itself). We say that the partial solution S' on G_i can be *extended* to a full solution S on a completion of G_i, where $S \backslash S'$ is the *extension* of S'. As an example, consider the MINIMUM DOMINATING SET problem: a solution for this problem is a vertex subset D in G such that for all $v \in V$ there is a $d \in D$ with $v \in N[d]$. A partial solution is a subset $D \subseteq V_i$ such that for all vertices in $v \in V_i \backslash X_i$ there is a $d \in D$ with $v \in N[d]$: for vertices in X_i there does not need to be a dominating neighbour in $d \in D$ as d can also be in an extension of D. A dynamic programming algorithm on a tree decomposition computes, for each node $i \in T$ in a bottom-up fashion, a *memoisation table A_i* containing all *relevant* (described in the next paragraph) partial solutions on G_i obtaining a solution to $\mathcal{P}$ in the root of T.

To restrict the number of partial (relevant) solutions stored, an equivalence relation is defined on them: two partial solutions S_1' and S_2' on G_i are *equivalent*

---

[2] Often G_i is defined *including* all edges between vertices in X_i. We choose the alternative definition as it makes formulating the join algorithms in Sect. 4 easier: no bookkeeping of number of neighbours between vertices in X_i needs to be done, as they only become neighbours higher up in the tree.

with respect to $\mathcal{P}$ if any extension of S_1 also is an extension of S_2' and vice versa. When given two equivalent partial solutions S_1' and S_2' for an optimisation problem (minimisation or maximisation), we say that S_1' *dominates* S_2' if for any extension S_E of S_1' and S_2', the solution value of $S_1' \cup S_E$ is equal or better than the solution value of $S_2' \cup S_E$. Clearly, a dynamic programming algorithm on a tree decomposition needs to store only one partial solution per equivalence class, and if we consider an optimisation problem it can store a partial solution that dominates all other partial solutions within its equivalence class.

Mostly, it is convenient to formulate a dynamic programming algorithm on a special kind of tree decomposition called a *nice tree decomposition* [20].[3]

Definition 2 (nice tree decomposition). *A nice tree decomposition is a tree decomposition T with assigned root node $r \in T$ with $X_r = \emptyset$, in which each node is of one of the following types:*

- Leaf node: *a leaf i of T with $X_i = \emptyset$.*
- Introduce node: *an internal node i of T with one child node j and $X_i = X_j \cup \{v\}$ for some $v \in V \backslash V_j$.*
- Forget node: *an internal node i of T with one child node j and $X_i = X_j \backslash \{v\}$ for some $v \in X_j$.*
- Join node: *an internal node i of T with two child nodes l and r with $X_i = X_l = X_r$.*

Given a tree decomposition consisting of $\mathcal{O}(n)$ nodes, a nice tree decomposition of $\mathcal{O}(n)$ nodes of the same width can be found in $\mathcal{O}(n)$ time [20]. Consequently, a dynamic programming algorithm on a nice tree decomposition can be used on general tree decompositions by applying this transformation. After computing A_i for all nodes $i \in T$, the solution to $\mathcal{P}$ can be found as the unique value in A_r, where r is the root of T: here $G_i = G$ and there is only a single equivalence class as $X_i = \emptyset$.

This paper focuses on computing A_i for a *join node* i of a nice tree decomposition. This node is the most interesting as often it dominates the running time of the entire dynamic programming algorithm. For an example, consider [1] where an $\mathcal{O}^*(4^t)$ algorithm for MINIMUM DOMINATING SET for graphs with a tree decomposition of width t is given, while all computations except the computation for the join nodes can be performed in $\mathcal{O}^*(3^t)$ time.

2.2   Dynamic Programming for $[\sigma, \rho]$-Domination Problems

The $[\sigma, \rho]$-domination problems are a class of vertex-subset problems introduced by Telle [28–30] that generalise many well-known graph problems such as MAXIMUM INDEPENDENT SET, MINIMUM DOMINATING SET, and INDUCED BOUNDED DEGREE SUBGRAPH. See Table 1 for an overview.

---

[3] Different version of the original definition [20] exists in literature (e.g, [16,25]): the restrictions on the vertices in a bag of a leaf node and the root node often vary, and sometimes an additional type of node called an *edge introduce node* is used.

Definition 3 ($[\sigma, \rho]$-dominating set). *Let $\sigma, \rho \subseteq \mathbb{N}$, a $[\sigma, \rho]$-dominating set in a graph $G = (V, E)$ is a subset $D \subseteq V$ such that:*

- *for every $v \in D$: $|N(v) \cap D| \in \sigma$;*
- *for every $v \in V \backslash D$: $|N(v) \cap D| \in \rho$.*

We consider only $\sigma, \rho \subseteq \mathbb{N}$ that both are either finite or cofinite.

For given $\sigma, \rho \subseteq \mathbb{N}$ and the corresponding definition of a $[\sigma, \rho]$-dominating set, one can define several different problem variants.

- *Existence* problem: given a graph G, does G have a $[\sigma, \rho]$-dominating set?
- *Optimisation* problem (minimisation or maximisation): given a graph G, what is the smallest $[\sigma, \rho]$-dominating set in G, or what is the largest $[\sigma, \rho]$-dominating set in G?
- *Counting* problem: given a graph G, how many $[\sigma, \rho]$-dominating sets exist in G?
- *Counting optimisation* problem (minimisation or maximisation): given a graph G, how many $[\sigma, \rho]$-dominating sets in G exist of minimum/maximum size?

Many well-known NP-hard vertex subset problems in graphs correspond to the existence or optimisation variant of a $[\sigma, \rho]$-domination problem, as can be seen from Table 1.

When solving a $[\sigma, \rho]$-domination problem by dynamic programming on a tree decomposition, the equivalence classes for partial solutions stored in the memoisation table A_i (as defined in Sect. 2.1) can be uniquely identified by the following:

Table 1. Examples of $[\sigma, \rho]$-domination problems (taken from [28–30]).

σ	ρ	Standard description
$\{0\}$	$\{0, 1, \ldots\}$	Independent Set/Stable Set
$\{0, 1, \ldots\}$	$\{1, 2, \ldots\}$	Dominating Set
$\{0\}$	$\{0, 1\}$	Strong Stable Set/2-Packing/Distance-2 Independent Set
$\{0\}$	$\{1\}$	Perfect Code/Efficient Dominating Set
$\{0\}$	$\{1, 2, \ldots\}$	Independent Dominating Set
$\{0, 1, \ldots\}$	$\{1\}$	Perfect Dominating Set
$\{1, 2, \ldots\}$	$\{1, 2, \ldots\}$	Total Dominating Set
$\{1\}$	$\{1\}$	Total Perfect Dominating Set
$\{0, 1, \ldots\}$	$\{0, 1\}$	Nearly Perfect Set
$\{0, 1\}$	$\{0, 1\}$	Total Nearly Perfect Set
$\{0, 1\}$	$\{1\}$	Weakly Perfect Dominating Set
$\{0, 1, \ldots, p\}$	$\{0, 1, \ldots\}$	Induced Bounded Degree Subgraph
$\{0, 1, \ldots\}$	$\{p, p + 1, \ldots\}$	p-Dominating Set
$\{p\}$	$\{0, 1, \ldots\}$	Induced p-Regular Subgraph

– the vertices in X_i that are in the partial solution D;
– for every vertex in X_i (both in D and not in D), the number of neighbours in D.

This corresponds exactly to the bookkeeping required to verify whether a partial solution locally satisfies the requirements imposed by the specific $[\sigma, \rho]$-domination problem. As such, we can identify every equivalence class using an assignment of *labels* (sometimes also called *states*) that capture the above properties to the vertices in X_i: such an assignment is called a *state colouring*. Given $\sigma, \rho \subseteq \mathbb{N}$, define the set of labels $C = C_\sigma \cup C_\rho$ as follows (the meaning of a label is explained below):

$$C_\sigma = \begin{cases} \{|0|_\sigma, |1|_\sigma, |2|_\sigma, \ldots, |\ell - 1|_\sigma, |\ell|_\sigma\} & \text{if } \sigma \text{ finite} & \text{where } \ell = \max\{\sigma\} \\ \{|\geq 0|_\sigma\} & \text{if } \sigma = \mathbb{N} \\ \{|0|_\sigma, |1|_\sigma, |2|_\sigma, \ldots, |\ell - 1|_\sigma, |\geq \ell|_\sigma\} & \text{if } \sigma \neq \mathbb{N} \text{ cofinite} & \text{where } \ell = \max\{\mathbb{N} \backslash \sigma\} + 1 \end{cases}$$

$$C_\rho = \begin{cases} \{|0|_\rho, |1|_\rho, |2|_\rho, \ldots, |\ell - 1|_\rho, |\ell|_\rho\} & \text{if } \rho \text{ finite} & \text{where } \ell = \max\{\rho\} \\ \{|\geq 0|_\rho\} & \text{if } \rho = \mathbb{N} \\ \{|0|_\rho, |1|_\rho, |2|_\rho, \ldots, |\ell - 1|_\rho, |\geq \ell|_\rho\} & \text{if } \rho \neq \mathbb{N} \text{ cofinite} & \text{where } \ell = \max\{\mathbb{N} \backslash \rho\} + 1 \end{cases}$$

We will use the $||_\rho$ and $||_\sigma$ notation to denote labels from C_ρ, respectively C_σ. In general, when we write $|l|_\rho$ or $|l|_\sigma$, with a variable l, we mean the labels that are not equal to $|\geq \ell|_\rho$ or $|\geq \ell|_\sigma$. This allows us to refer to other labels by expressions such as $|l - 1|_\rho$. The symbol ℓ is reserved to indicate the last labels $|\ell|_\sigma$, $|\geq \ell|_\sigma$, $|\ell|_\rho$, $|\geq \ell|_\rho$ and is used similarly to form labels such as $|\ell - 1|_\rho$.

Let C^{X_i} be the set of assignments of labels from C to the vertices in X_i. A label from C_σ for a vertex $v \in X_i$ indicates that v is in the solution set D in the partial solution, a label from C_ρ indicates that v is not. Furthermore, the numbers in the labels indicate the number of neighbours that v has in D; the $\geq$ symbol in the label $|\geq 1|_\rho$ indicates that v has this number of neighbours (one in this case) in D or more. For an example, consider MINIMUM DOMINATING SET for which $\sigma = \mathbb{N}$ and $\rho = \mathbb{N} \backslash \{0\}$; for this problem $C = \{|\geq 0|_\sigma, |0|_\rho, |\geq 1|_\rho\}$.

Now, the elements from C^{X_i} bijectively correspond to the above defined equivalence classes of partial solutions on G_i. Consequently, we can index the memoisation table A_i by C^{X_i}. To keep the dynamic programming recurrences in this paper simple, we will not store partial solutions in A_i, only the required partial solution values or counts. That is, from here on, let the table A_i be a function $A_i : C^{X_i} \rightarrow \{0, 1, .., M\} \cup \{\infty\}$ that assigns a number to each equivalence class of partial solutions. In an existence variant of a problem, we let $A_i(c)$, for $c \in C^{X_i}$, be 0 or 1 indicating whether a partial solution of this equivalence class exists. In an optimisation variant, $A_i(c)$ indicates the size of a dominating partial solution in this equivalence class, or ∞ if no such partial solution exists. For convenience reasons[4], we let $A_i(c)$, for $c \in C^{X_i}$, contain the size of the partial solution D' restricted to $V' \backslash X'$, i.e., the size of a corresponding partial solution equals $A_i(c)$ plus the number of σ labels in c. In a counting variant, $A_i(c)$ indicates the number of partial solutions in the equivalence class of c.

---

[4] In this way, we do not have to correct for double counting in join nodes in the rest of this paper.

Notice that for an existence variant, we can bound M by 1; for an optimisation variant, we can bound M by n; and for a counting variant, we can bound M by 2^n.

Below, we give explicit recurrences for A_i for solving a minimisation variant of a $[\sigma, \rho]$-dominating problem by dynamic programming on a nice tree decomposition T. Modifying the recurrences to the existence or counting variant of the problem is an easy exercise. Extensions to the recurrences in which partial solutions are stored (for existence and optimisation variants) are easy to make, but tedious to write down formally. This is also to true for the extension to the optimisation counting variant where one needs to keep track of both the size and the number of such partial solutions.

Leaf Node. Let i be a leaf node of T. Since $X_i = \emptyset$, the only partial solution is $\emptyset$ with size zero: this size is stored for the empty vector $[]$.

$$A_i([]) = 0$$

Introduce Node. Let i be an introduce node of T with child node j. Let $X_i = X_j \cup \{v\}$ for some $v \in V \setminus V_j$. For $\boldsymbol{c} \in C^{X_j}$ and $c_v \in C$ the label for vertex v denote by $[\boldsymbol{c}, c_v]$ the vector $\boldsymbol{c}$ with the element c_v appended to it such that $[\boldsymbol{c}, c_v] \in C^{X_i}$. Now:

$$A_i([\boldsymbol{c}, c_v]) = \begin{cases} A_j([\boldsymbol{c}]) & \text{if } c_v \in \{|0|_\sigma, |\geq 0|_\sigma\} \text{ or } c_v \in \{|0|_\rho, |\geq 0|_\rho\} \\ \infty & \text{otherwise} \end{cases}$$

Here, G_i equals G_j with one added isolated vertex v. Hence, v can be in the partial solution or not, and both choices do not influence the partial solution size on $V_i \setminus X_i$ (which equals $V_j \setminus X_j$). Note that only one of the labels from $\{|0|_\sigma, | \geq 0|_\sigma\}$ and one from $\{|0|_\rho, | \geq 0|_\rho\}$ is used, and which depends on the specific $[\sigma, \rho]$-domination problem that we are solving.

Forget Node. Let i be a forget node of T with child node j. Let $X_i = X_j \setminus \{v\}$ for some $v \in X_j$.

By definition of G_i, G_i contains edges between v and vertices in X_i while G_j does not. To account for these edges, we start by updating the given table A_j such that it accounts for the additional edges: that is, for an edge $\{u, v\}$ with $u \in X_i$, we adjust the counts of the number of neighbours expressed in the state colourings for u and v. We do so before we construct table A_i.

Let $[\boldsymbol{c}, c_u, c_v] \in C^{X_j}$ be such that c_u and c_v are labels for u and v respectively. For every edge $\{u, v\}$ with $u \in X_i$, we update A_j twice, once for u and once for v. We update A_j for u as follows:

$$A_j([\boldsymbol{c}, c_u, c_v]) := \begin{cases} A_j([\boldsymbol{c}, c_u, c_v]) & \text{if } c_v \in C_\rho \\ \infty & \text{if } c_v \in C_\sigma, c_u \in \{|0|_\rho, |0|_\sigma\} \\ A_j([\boldsymbol{c}, |l-1|_\rho, c_v]) & \text{if } c_v \in C_\sigma, c_u = |l|_\rho, l > 0 \\ A_j([\boldsymbol{c}, |l-1|_\sigma, c_v]) & \text{if } c_v \in C_\sigma, c_u = |l|_\sigma, l > 0 \\ \min\{A_j([\boldsymbol{c}, |\ell-1|_\rho, c_v]), A_j([\boldsymbol{c}, |\geq \ell|_\rho, c_v])\} & \text{if } c_v \in C_\sigma, c_u = |\geq \ell|_\rho \\ \min\{A_j([\boldsymbol{c}, |\ell-1|_\sigma, c_v]), A_j([\boldsymbol{c}, |\geq \ell|_\sigma, c_v])\} & \text{if } c_v \in C_\sigma, c_u = |\geq \ell|_\sigma \end{cases}$$

No update needs to be done if v is not in the partial solution D (first line). If c_u indicates that u has no neighbours in D while $v \in D$, then no such partial solution exists (second line). Otherwise, the counts in the label of u need to account for the extra neighbour. In the last four lines, we perform the required label update for all other labels giving special attention to the case where a $|{\geq}\ell|_\sigma$ or $|{\geq}\ell|_\rho$ label is used. Here, the minimum needs to be taken over two equivalence classes that through the added edge become equivalent: we take the minimum because we are solving the minimisation variant. Updating A_j for v goes identically with the roles of u and v switched, and as stated above, we perform this update for all edges incident to v in G_j.

Next, we compute A_i and start keeping track of equivalence classes based on X_i instead of based on X_j. To do so, we select a dominating solution from the partial solution equivalence classes for which v has a number of neighbours in D that corresponds to the specific $[\sigma, \rho]$-domination problem:

$$A_i(\boldsymbol{c}) = \min_{c_v \text{ a valid label}} A_j([\boldsymbol{c}, c_v])$$

Here, a valid label c_v is any label that corresponds to having the correct number of neighbours in D as defined by the specific $[\sigma, \rho]$-domination problem: c_v is a label $|l|_\sigma$ or $|l|_\rho$ for which $l \in \sigma$ or $l \in \rho$, respectively, or c_v is a label $|{\geq}\ell|_\rho$ or $|{\geq}\ell|_\sigma$ in case of cofinite σ or ρ.

Join Node. Let A_i be the memoisation table for a join node i of T with child nodes l and r. Here we give a simple algorithm for the join node; in Sect. 4, we survey more involved approaches.

A trivial algorithm to compute A_i would loop over all pairs of state colourings c_l, c_r of X_i that agree on which vertices are in the solution set D, and then consider two corresponding partial solutions D_l on G_l and D_r on G_r and infer the state colouring c_i of the partial solution $D_l \cup D_r$ on G_i. It then stores in A_i the minimum size of a solution for each equivalence class for G_i.

Note that the agreement on which vertices are in D is necessary for $D_l \cup D_r$ to be a valid partial solution: otherwise vertices that are no longer in X_i can obtain additional neighbours in D. At the same time the agreement is not a too tight restriction as any partial solution D on G_i can trivially be decomposed into partial solutions on G_l and G_r that agree on which vertices on X_i are in D.

Root Node. In the root node r of T (which is a forget node), $X_r = \emptyset$, $G_r = G$ and consequently $A_r([])$ is the minimum size of a $[\sigma, \rho]$-dominating set on G. The result we set out to compute!

Lemma 1. *Let $\mathcal{P}$ be the minimisation variant of a $[\sigma, \rho]$-domination problem with label set C using $s = |C|$ labels. Let $\mathcal{A}$ be an algorithm for the computations in a join node for problem $\mathcal{P}$ that, given a join node i with $|X_i| = k$ and the memoisation tables A_l and A_r for its child nodes, computes the memoisation table A_i in $\mathcal{O}(f(n, k))$ arithmetic operations. Then, given a graph G with a tree* *decomposition T of width t, $\mathcal{P}$ can be solved on G in $\mathcal{O}((s^{t+1}t + f(n, t + 1))n)$ arithmetic operations.*

Proof. First transform T into a nice tree decomposition T' with $\mathcal{O}(n)$ nodes. If we show that the table A_j associated to any node j of T' can be computed in $\mathcal{O}(s^k k + f(n, k))$ arithmetic operations, then the result follows as $k \leq t + 1$. Consider the recurrences in the dynamic programming algorithm exposed above. The result trivially holds for leaf and root nodes, and also for the join nodes by definition of $\mathcal{A}$. It is easy to see that in the recurrences for the introduce and forget nodes, every value is computed using a constant amount of work. Since the tables are of size s^k, and for a forget node we need to do at most k update steps as we can add at most k edges, the result follows. □

It is not difficult to modify the above algorithm to obtain:

Proposition 1. *Lemma 1 holds irrespective of $\mathcal{P}$ being a existence, maximisation, minimisation, counting, counting minimisation or counting maximisation variant of a $[\sigma, \rho]$-domination problem.*

3 Overview of Fast Transforms

To obtain fast algorithms for the computations in join nodes of a nice tree decomposition, we use several well-known algebraic transforms, specifically the Möbius transform and the Fourier transform. We opted for a reasonably extensive coverage of this standard material because of completeness reasons and because the details matter for some of the arguments in Sects. 4 and 5.

Recall that, in the introduction on dynamic programming for $[\sigma, \rho]$-domination problems, we stored integers in the domain $\{0, 1, \ldots, M\}$ for some large integer M. We present the algebraic transforms using computations in $\mathbb{F}_p$, the field of integers modulo a prime number p. Since we know that, for a join node i with child nodes l, r, all values in the memoisation tables A_i, A_l and A_r are in $\{0, 1, \ldots, M\}$, we can do the computations in $\mathbb{F}_p$ as long as $p > M$.

In the literature, the discrete Fourier transform is often defined on sequences in $\mathbb{C}$. We choose $\mathbb{F}_p$ to avoid any analysis of rounding errors, especially when we combine it with the use of zeta and Möbius transforms. Using $\mathbb{F}_p$ does require that p is chosen appropriately: $\mathbb{F}_p$ must contain certain roots of unit required for the Fourier transforms. In the statements of definitions, propositions and lemmas in this section, we will sometimes say that p is *chosen appropriately* to state that $\mathbb{F}_p$ contains the roots of unity required in the definition or in the following proof. A short discussion on how to choose a proper prime number p such that this condition is satisfied can be found in Sect. 3.3.

3.1 The Discrete Fourier Transforms Using Modular Arithmetic

Definition 4 (discrete Fourier transform). *Let $\boldsymbol{a} = (a_i)_{i=0}^{r-1}$ be a sequence of numbers in $\mathbb{F}_p$, and let ω_r be an r-th root of unity in $\mathbb{F}_p$. The discrete Fourier* transform *and* inverse discrete Fourier transform *are transformations between sequences of length r in* $\mathbb{F}_p$ *defined as follows:*

$$DFT(\boldsymbol{a})_i = \sum_{j=0}^{r-1} \omega_r^{ij} a_j \qquad DFT^{-1}(\boldsymbol{a})_i = \frac{1}{r} \sum_{j=0}^{r-1} \omega_r^{-ij} a_j$$

Recall that an r-th root of unity is an element $x \in \mathbb{F}_p$ such that $x^r = 1$ while $x^l \neq 1$ for all $l < 1$.

These two transformations are inverses as their names suggest.

Proposition 2. $DFT^{-1}(DFT(\boldsymbol{a}))_i = a_i$

Proof. In the derivation below, we first fill in the definitions and rearrange the terms (1). Then, we split the sum based on $k = i$ and $k \neq i$ (from 1 to 2).

$$DFT^{-1}(DFT(a))_i = \frac{1}{r} \sum_{j=0}^{r-1} \omega_r^{-ij} \sum_{k=0}^{r-1} \omega_r^{jk} a_k = \frac{1}{r} \sum_{k=0}^{r-1} a_k \sum_{j=0}^{r-1} (\omega_r^{k-i})^j \tag{1}$$

$$= \frac{1}{r} a_i \sum_{j=0}^{r-1} (\omega_r^{i-i})^j + \frac{1}{r} \sum_{\substack{k=0 \\ k \neq i}}^{r-1} a_k \sum_{j=0}^{r-1} (\omega_r^{k-i})^j = a_i \frac{1}{r} \sum_{j=0}^{r-1} 1 + \frac{1}{r} \sum_{\substack{k=0 \\ k \neq i}}^{r-1} a_k \cdot 0 = a_i \tag{2}$$

Finally, we use that the first part of the sum is trivial as $\omega_r^{i-i} = \omega_r^0 = 1$, while the second part cancels as $\sum_{j=0}^{r-1} (\omega_r^{k-i})^j = \frac{1 - \omega_r^{(k-i)r}}{1 - \omega_r^{k-i}}$ is a geometric series with $\omega_r^{(k-i)r} = (\omega_r^r)^{k-i} = 1^{k-i} = 1$. □

There exist fast algorithms for the discrete Fourier transform and its inverse, called fast Fourier transforms (FFT's), e.g., see the Cooley-Tukey FFT algorithm [13] and Rader's FFT algorithm [24]. These algorithms are not particularly difficult to understand, but beyond the scope of this paper.

Proposition 3 (fast Fourier transform). *The discrete Fourier transform and its inverse for sequences of length r can be computed in $\mathcal{O}(r \log r)$ arithmetic operations.*

The definition of the discrete Fourier transform can be naturally extended from sequences to higher dimensional structures. Let $\mathbb{Z}_r$ be the commutative ring of integers modulo r (here the modulus r can be non-prime), and let $\mathbb{Z}_r^k$ be the $\mathbb{Z}_r$-module of k-tuples with elements from $\mathbb{Z}_r$.

Definition 5 (multidimensional discrete Fourier transform). *Let $Z = \mathbb{Z}_{r_1} \times \mathbb{Z}_{r_2} \times \cdots \times \mathbb{Z}_{r_k}$, and let $R = \prod_{i=1}^{k} r_i$. Also, let $\boldsymbol{A} = (a_{\boldsymbol{x}})_{\boldsymbol{x} \in Z}$ be a tensor of rank k with elements in $\mathbb{F}_p$ indexed by the k-tuple $\boldsymbol{x} = [x_1, x_2, \ldots, x_k]$, where p is chosen appropriately. The* multidimensional discrete Fourier transform *and* inverse multidimensional discrete Fourier transform *are defined as follows:*

$$DFT_k(\boldsymbol{A})_{\boldsymbol{x}} = \sum_{y_1=0}^{r_1-1} \omega_{r_1}^{x_1 y_1} \sum_{y_2=0}^{r_2-1} \omega_{r_2}^{x_2 y_2} \cdots \sum_{y_k=0}^{r_k-1} \omega_{r_k}^{x_k y_k} a_{\boldsymbol{y}}$$

$$DFT_k^{-1}(\boldsymbol{A})_{\boldsymbol{x}} = \frac{1}{R} \sum_{y_1=0}^{r_1-1} \omega_{r_1}^{-x_1 y_1} \sum_{y_2=0}^{r_2-1} \omega_{r_2}^{-x_2 y_2} \cdots \sum_{y_k=0}^{r_k-1} \omega_{r_k}^{-x_k y_k} a_{\boldsymbol{y}}$$

When $r = r_1 = r_2 = \ldots = r_k$, this simplifies to the following:

$$DFT_k(\boldsymbol{A})_{\boldsymbol{x}} = \sum_{\boldsymbol{y} \in \mathbb{Z}_r^k} \omega_r^{\boldsymbol{x} \cdot \boldsymbol{y}} a_{\boldsymbol{y}} \qquad DFT_k^{-1}(\boldsymbol{A})_{\boldsymbol{x}} = \frac{1}{r^k} \sum_{\boldsymbol{y} \in \mathbb{Z}_r^k} \omega_r^{-\boldsymbol{x} \cdot \boldsymbol{y}} a_{\boldsymbol{y}}$$

where the expressions in the exponents are the dot products on the tuples $\boldsymbol{x}$ and $\boldsymbol{y}$ in $\mathbb{Z}_r^k$.

Note that the dot products are in exponents of which the base is an r-th root of unity, hence they are computed modulo r: this agrees with the notation where $\boldsymbol{x}$ and $\boldsymbol{y}$ are taken from $\mathbb{Z}_r^k$.

Proposition 4 (fast multidimensional discrete Fourier transform). *Let $Z = \mathbb{Z}_{r_1} \times \mathbb{Z}_{r_2} \times \cdots \times \mathbb{Z}_{r_k}$, and let $R = \prod_{i=1}^{k} r_i$. Also, let $\boldsymbol{A}$ be a tensor of rank k with elements in $\mathbb{F}_p$, $\boldsymbol{A} = (a_{\boldsymbol{x}})_{\boldsymbol{x} \in Z}$, where p is chosen appropriately. The multidimensional discrete Fourier transform and inverse multidimensional discrete Fourier transform of $\boldsymbol{A}$ can be computed in $\mathcal{O}(R \log(R))$ time.*

Proof. Denote by $\boldsymbol{x}[x_i \leftarrow y]$ the tuple $\boldsymbol{x}$ with the i-th coordinate of $\boldsymbol{x}$ replaced by y. We compute $DFT_k(\boldsymbol{A})$ with an algorithm that uses k-steps. Let $\boldsymbol{A}_0 = \boldsymbol{A}$. At the i-th step of the algorithm, let:

$$(\boldsymbol{A}_i)_{\boldsymbol{x}} = \sum_{j=0}^{r_i-1} \omega_{r_i}^{x_i j} a_{\boldsymbol{x}[x_i \leftarrow j]}$$

Notice that if $k = 1$, this formula equals the one dimensional discrete Fourier transform. It is not hard to see that $\boldsymbol{A}_k$ is the k-dimensional Fourier transform of $\boldsymbol{A}$: if one repeatedly substitutes the formula for $\boldsymbol{A}_{i-1}$ in the formula for $\boldsymbol{A}_i$ starting at $i = k$, one obtains the (non-simplified) formula for the k-dimensional Fourier transform in Definition 5.

For the inverse multidimensional Fourier transform, almost the same procedure can be followed. Let $\boldsymbol{A}_0 = \boldsymbol{A}$ and use the following formula at the i-th step, finally obtaining the result $\boldsymbol{A}_k$. Here, again if $k = 1$, this formula equals the one dimensional inverse discrete Fourier transform.

$$(\boldsymbol{A}_i)_{\boldsymbol{x}} = \frac{1}{r_i} \sum_{j=0}^{r_i-1} \omega_{r_i}^{-x_i j} a_{\boldsymbol{x}[x_i \leftarrow j]}$$

For the running time, notice that step i preforms $\frac{R}{r_i}$ standard 1-dimensional (inverse) discrete Fourier transforms on a sequence of length r_i. By Proposition 3 this can be done in $\mathcal{O}(\frac{R}{r_i} r_i \log(r_i)) = \mathcal{O}(R \log(r_i))$ time. This leads to a total running time of $\mathcal{O}(R \sum_{i=1}^{k} \log(r_i)) = \mathcal{O}(R \log(R))$. □

In the above proof, the sequences $\boldsymbol{A}_0, \boldsymbol{A}_1, \ldots, \boldsymbol{A}_k$ are created using 1-dimensional (inverse) discrete Fourier transforms. Because the 1-dimensional discrete Fourier transform and 1-dimensional inverse discrete Fourier transform are inverses, it directly follows that the sequence $\boldsymbol{A}_0, \boldsymbol{A}_1, \ldots, \boldsymbol{A}_k$ used in the k-dimensional discrete Fourier transform algorithm equals the sequence $\boldsymbol{A}_k, \boldsymbol{A}_{k-1}, \ldots, \boldsymbol{A}_0$ used in the inverse k-dimensional discrete Fourier transform. I.e., as the name suggests, the k-dimensional inverse discrete Fourier transform is the inverse of the k-dimensional discrete Fourier transform.

We mainly use the multidimensional fast discrete Fourier transform in combination with the well-known convolution theorem.

Lemma 2 (multidimensional convolution theorem). *Let* $Z = \mathbb{Z}_{r_1} \times \mathbb{Z}_{r_2} \times \cdots \times \mathbb{Z}_{r_k}$, *and let* $\boldsymbol{A} = (a_x)_{x \in Z}$, $\boldsymbol{B} = (b_x)_{x \in Z}$ *be tensors of rank k with elements in* $\mathbb{F}_p$, *where p is chosen appropriately. Let the tensor multiplication* $\boldsymbol{A} \cdot \boldsymbol{B}$ *be defined point wise, and let for a_x and b_y the sum $\boldsymbol{x} + \boldsymbol{y}$ be defined as the sum in Z (coordinate-wise with the i-th coordinate modulo r_i). Then:*

$$DFT_k^{-1}(DFT_k(\boldsymbol{A}) \cdot DFT_k(\boldsymbol{B}))_x = \sum_{z_1 + z_2 \equiv x} a_{z_1} b_{z_2}$$

Proof. We prove the lemma for the simplified case where $r = r_1 = r_2 = \ldots = r_k$ and hence $Z = \mathbb{Z}_r^k$, the more general case goes analogously but is notation-wise much more tedious as one needs to differentiate between multiple moduli and their corresponding roots of unity.

The proof follows the same pattern as in Proposition 2. That is, we first fill in the definitions (3) and rearrange the terms (4). Next (5), we observe that the sum over all $j \in \mathbb{Z}_r^k$ can be written as a product of k smaller sums, each involving but one coordinate of j.

$$DFT_k^{-1}(DFT_k(\boldsymbol{A}) \cdot DFT_k(\boldsymbol{B}))_x = \frac{1}{r^k} \sum_{y \in \mathbb{Z}_r^k} \omega_r^{-x \cdot y} \left(\sum_{z_1 \in \mathbb{Z}_r^k} \omega_r^{y \cdot z_1} a_{z_1} \right) \left(\sum_{z_2 \in \mathbb{Z}_r^k} \omega_r^{y \cdot z_2} b_{z_2} \right) \tag{3}$$

$$= \frac{1}{r^k} \sum_{z_1, z_2 \in \mathbb{Z}_r^k} a_{z_1} b_{z_2} \sum_{y \in \mathbb{Z}_r^k} \omega_r^{y \cdot (z_1 + z_2 - x)} \tag{4}$$

$$= \frac{1}{r^k} \sum_{z_1, z_2 \in \mathbb{Z}_r^k} a_{z_1} b_{z_2} \prod_{i=1}^{k} \left(\sum_{j=0}^{r-1} \omega_r^{j((z_1)_i + (z_2)_i - x_i)} \right) \tag{5}$$

Here, x_i, $(z_1)_i$ and $(z_2)_i$ are the i-th components of x, z_1 and z_2 respectively.

When $x_i \equiv (z_1)_i + (z_2)_i$ modulo r in the parenthesised sum of Eq. 5, this sum becomes $\sum_{j=0}^{r-1} \omega_r^0$ and thus equals r. Otherwise, when $x_i \not\equiv (z_1)_i + (z_2)_i$

the parenthesised sum is again a geometric series: $\sum_{j=0}^{r-1}(\omega_r^{(z_1)_i+(z_2)_i-x_i})^j$ that

solves to $\frac{1-(\omega_r^{(z_1)_i+(z_2)_i-x_i})^r}{1-\omega_r^{(z_1)_i+(z_2)_i-x_i}} = 0$ as $(\omega_r^{(z_1)_i+(z_2)_i-x_i})^r = (\omega_r^r)^{(z_1)_i+(z_2)_i-x_i} = 1$ in the numerator.

Continuing from (5), we obtain:

$$\mathrm{DFT}_k^{-1}(\mathrm{DFT}_k(\boldsymbol{A})\cdot\mathrm{DFT}_k(\boldsymbol{B}))_x = \frac{1}{r^k}\sum_{z_1,z_2\in\mathbb{Z}_r^k} a_{z_1}b_{z_2}\prod_{i=1}^{k} r[x_i = (z_1)_i + (z_2)_i] \qquad (6)$$

$$= \sum_{z_1,z_2\in\mathbb{Z}_r^k} a_{z_1}b_{z_2}\prod_{i=1}^{k}[x_i = (z_1)_i + (z_2)_i] = \sum_{z_1+z_2\equiv x} a_{z_1}b_{z_2}$$
$$(7)$$

completing the proof. $\square$

Taking all the above together, we finally obtain the following result. To distinguish from $\mathbb{Z}_r$, let $\mathbb{N}_{<r} = \{0, 1, \ldots, r-1\}$ be the integers up to r with standard operators without modulus (operations for which the result of standard operations on $\mathbb{N}_{<r}$ is outside $\mathbb{N}_{<r}$ are considered undefined, i.e., $2 + 2$ is undefined in $\mathbb{N}_{<3}$).

Lemma 3 (cyclic and non-cyclic convolution). *Let $N = \mathbb{N}_{<q_1}\times\mathbb{N}_{<q_2}\times\cdots\times$ $\mathbb{N}_{<q_l}$, and let $Q = \prod_{i=1}^{l} q_i$. Let $Z = \mathbb{Z}_{r_1}\times\mathbb{Z}_{r_2}\times\cdots\times\mathbb{Z}_{r_k}$, and let $R = \prod_{i=1}^{k} r_i$. Let $f, g : Z\times N \to \mathbb{F}_p$, where p is chosen appropriately. And, let $h : Z\times N \to \mathbb{F}_p$ be the combined (partially cyclic and partially non-cyclic) convolution of f and g defined as:*

$$h(\boldsymbol{x},\boldsymbol{i}) = \sum_{\boldsymbol{y_1}+\boldsymbol{y_2}\equiv\boldsymbol{x}}\sum_{\boldsymbol{j_1}+\boldsymbol{j_2}=\boldsymbol{i}} f(\boldsymbol{y_1},\boldsymbol{j_1})g(\boldsymbol{y_2},\boldsymbol{j_2})$$

where the sum $\boldsymbol{y_1}+\boldsymbol{y_2} \equiv \boldsymbol{x}$ is evaluated component-wise modulo r_i at coordinate i (sum in Z), and the sum $\boldsymbol{j_1} + \boldsymbol{j_2} = \boldsymbol{i}$ is evaluated component-wise without modulus (sum in N). Then, the combined convolution h can be computed in $\mathcal{O}(R\,Q\,2^l(\log(R) + \log(Q) + l))$ arithmetic operations.

Proof. We reduce the problem to a standard multidimensional convolution (with modulus) by padding the input with zeroes. To be precise, let $Z' = \mathbb{Z}_{2q_1}\times\mathbb{Z}_{2q_2}\times\cdots\times\mathbb{Z}_{2q_l}$ (N with for each coordinate twice as many values and with modulo additions), and let $f', g' : Z\times Z' \to \mathbb{F}_p$ be equal to f and g on the intersection of their domains (where N is interpreted as subset of Z' by interpreting each $\mathbb{N}_{<q_i}$ as subset of $\mathbb{Z}_{2q_i}$) and zero otherwise. Use Proposition 4 and Lemma 2 to compute the standard multidimensional convolution of f' and g'. Because $Z\times Z'$ has $RQ2^l$ elements, this requires $\mathcal{O}(RQ2^l(\log(R)+\log(Q)+l))$ arithmetic operations. Because the padded zeroes prevent the circular convolution effect, we can extract h by taking the restriction of the result to $Z\times N$. $\square$

Different than for previous propositions and lemmas, we have more freedom in choosing the prime p that is 'chosen appropriately' in the lemma above. For the given proof, appropriate means that in $\mathbb{F}_p$ all r_i-th roots of unity exists and all $2q_i$-th roots of unity exist. However in our applications of the lemma, l is often fixed. This means that the running time does not change if we allow $Z' = \mathbb{Z}_{s_1} \times \mathbb{Z}_{s_2} \times \cdots \times \mathbb{Z}_{s_l}$ with, for all i, $2q_i \leq s_i \leq cq_i$ for a small constant c. In other words, with respect to the different q_i, there must be some root of unity, but the order of this root of unity has a broad range in which it is acceptable for our results to be valid.

Corollary 1 (multidimensional non-cyclic convolution). *Let $N = \mathbb{N}_{<q_1} \times \mathbb{N}_{<q_2} \times \cdots \times \mathbb{N}_{<q_l}$, and let $Q = \prod_{i=1}^{l} q_i$. Let $f, g : N \to \mathbb{F}_p$, where p is chosen appropriately. Let $h : N \to \mathbb{F}_p$ be the non-cyclic convolution of f and g defined as:*

$$h(i) = \sum_{j_1 + j_2 = i} f(j_1) g(j_2)$$

where the sum $j_1 + j_2 = i$ is evaluated component-wise without modulus (sum in N). Then, h can be computed in $\mathcal{O}(Q \, 2^l (\log(Q) + l))$ arithmetic operations.

Proof. Direct consequence of Lemma 3 with $k = 0$. □

3.2 Möbius Inversion Using Fast Zeta and Fast Möbius Transforms

The zeta and Möbius transforms apply to functions on partially ordered sets.

Definition 6 (zeta and Möbius transform). *Let P be a partially ordered set. Given a function $f : P \to \mathbb{F}_p$, the zeta transform $\zeta(f)$ and the Möbius transform $\mu(f)$ are defined as follows:*

$$\zeta(f)(x) = \sum_{y \leq x} f(y) \quad \mu(f)(x) = \sum_{y \leq x} \mu(y, x) f(y)$$

$$where \quad \mu(x, y) = \begin{cases} 1 & if\ x = y \\ -\sum_{x < z \leq y} \mu(z, y) & if\ x < y \end{cases}$$

The recursively defined function $\mu(x, y)$ on pairs $x, y \in P$ with $x \leq y$ is the Möbius function of P.

The *zeta transform* $\zeta(f)$ and the *Möbius transform* are inverses, as we will now show.

Lemma 4 (Möbius inversion). *Let $f : P \to \mathbb{F}_p$ any function, then $\mu(\zeta(f))(x) = f(x)$.*

Proof. Let $x, y \in P$ and consider the sum $\sum_{x \leq z \leq y} \mu(z, y)$. If $x = y$, then this sum equals $\mu(x, x) = 1$. If $x < y$, then this sum equals $\mu(x, y) + \sum_{x < z \leq y} \mu(z, y) = 0$ by definition of $\mu(x, y)$. As such:

$$\mu(\zeta(f))(x) = \sum_{y \leq x} \mu(y, x) \sum_{z \leq y} f(z) = \sum_{z \leq x} f(z) \sum_{z \leq y \leq x} \mu(y, x) = \sum_{z \leq x} f(z)[z = x] = f(x)$$

The first equality is by expanding the definitions. The second follows by reordering terms. And, the third follows from the above, where $[z = x]$ is Iverson notation that is 1 if $z = x$ and 0 otherwise. □

In this paper, we will not define any Möbius transform explicitly. We will show how to compute zeta transforms $\zeta(f)$ of functions $f : P \to \mathbb{F}_p$ for some partial orders P. Then, given $\zeta(f)$, we show that we can reconstruct f. This reconstruction (implicitly) is an algorithm for the Möbius transform because a consequence of Lemma 4 is that the zeta transform has a unique inverse.

Möbius inversion is often used in relation to lattices. A *meet-semilattice* is a partial order P on which, for any two elements in $x, y \in P$, the *meet* $x \wedge y$ (greatest lower bound) is properly defined. Similarly, a *join-semilattice* is a partial order P set on which, for any two elements in $x, y \in P$, the *join* $x \vee y$ (smallest upper bound) is properly defined. A lattice is a partial order P that is both a meet and a join semi-lattice. An example is the finite lattice $\mathbb{N}_{<r}^k$ with the coordinate-wise natural order and where the meet and join are the coordinate-wise minimum and maximum.

We will use Möbius inversion on partial orders that are Cartesian products P^k of a smaller partial order P. For $\boldsymbol{x}, \boldsymbol{y} \in P^k$, $\boldsymbol{x} = [x_1, x_2, \ldots, x_k]$, $\boldsymbol{y} = [y_1, y_2, \ldots, y_k]$, we write $\boldsymbol{x} \leq \boldsymbol{y}$ if and only if $x_i \leq y_i$ for all i. Additionally, our partial orders have the property that for every $x \in P$, the downward closed set $\{y \in P | y \leq x\}$ forms a join-semilattice. It is not hard to see that if for every $x \in P$, $\{y \in P | y \leq x\}$ forms a join-semilattice, then for every $\boldsymbol{x} \in P^k$, $\{\boldsymbol{y} \in P^k | \boldsymbol{y} \leq \boldsymbol{x}\}$ forms a join-semilattice as well, where the join operation is defined coordinate-wise.

On the subset lattice (isomorphic to $\mathbb{N}_{<2}^k$) there are well-known fast algorithms for the zeta and Möbius transforms, often referred to as Yates' algorithm [32], see also [2, 19]. Below, we generalise these algorithms to partial orders P^k for which, for every $\boldsymbol{x} \in P^k$, the set $\{\boldsymbol{y} \in P^k | \boldsymbol{y} \leq \boldsymbol{x}\}$ forms a join-semilattice. For fast zeta and Möbius transforms on arbitrary finite lattices, see [3].

Proposition 5 (fast zeta and Möbius transforms). *The zeta transform and Möbius transform of a function $f : \mathbb{N}_{<r}^k \to \mathbb{F}_p$ can be computed in $\mathcal{O}(r^k k)$ arithmetic operations.*

Proof. We compute $\zeta(f)$ with an algorithm that uses k steps. Let $f_0 = f$, and let $\boldsymbol{x} = [x_1, \ldots, x_k]$. Denote by $\boldsymbol{x}[x_i \leftarrow y]$ the tuple $\boldsymbol{x}$ with the value on the i-th coordinate replaced by y. At the i-th step of the algorithm, we compute f_i recursively using the left formula below.

$$f_i(\boldsymbol{x}) = \begin{cases} f_{i-1}(\boldsymbol{x}) & \text{if } x_i = 0 \\ f_i(\boldsymbol{x}[x_i \leftarrow x_i - 1]) + f_{i-1}(\boldsymbol{x}) & \text{if } x_i > 0 \end{cases} \qquad f_i(\boldsymbol{x}) = \sum_{j \leq x_i} f_{i-1}(\boldsymbol{x}[x_i \leftarrow j]) \tag{8}$$

The right formula above follows by induction on the left recurrence. By induction on the step number i, one easily sees that f_i satisfies the equation below, from which we can obtain $\zeta(f)$ since $f_k = \zeta(f)$. The result for $\zeta(f)$ follows because each step computes r^k values, each in constant time.

$$f_i(\boldsymbol{x}) = \sum_{y_1 \leq x_1} \sum_{y_2 \leq x_2} \cdots \sum_{y_i \leq x_i} f([y_1, y_2, \ldots, y_i, x_{i+1}, x_{i+2}, \ldots, x_k])$$

For $\mu(f)$, we use that $\mu(f)$ is the inverse of $\zeta(f)$: the sequence $f_0, f_1, \ldots, f_k$ used to compute $\zeta(f)$ from f can computationally be inverted to compute f from $\zeta(f)$. That is, let $f_k = \zeta(f)$, and let:

$$f_i(x) = \begin{cases} f_{i+1}(x) - f_{i+1}(x[x_i \leftarrow x_i - 1]) & \text{if } x_i > 0 \\ f_{i+1}(x) & \text{if } x_i = 0 \end{cases} \qquad (9)$$

Assuming that the f_i were computed using Eq. 8, and substituting the right part of (8) into the case where $x_i > 0$ in Eq. 9, we see that (9) computes the inverse of Eq. 8:

$$\left(\sum_{j \leq x_i} f_{i-1}(x[x_i \leftarrow j]) \right) - \left(\sum_{j \leq x_i - 1} f_{i-1}(x[x_i \leftarrow j]) \right) = f_i(x)$$

Hence, we can reconstruct $f_0 = f$ again. Since the run time is the same, we obtain the result. □

It is easy to generalise the inductive proof above and obtain:

Lemma 5. *Given algorithms for the zeta and Möbius transform for functions $f : P \to \mathbb{F}_p$ that use $\mathcal{O}(|P|)$ arithmetic operations, there are algorithms for the zeta and Möbius transform for functions $f : P^k \to \mathbb{F}_p$ that require $\mathcal{O}(|P^k|k)$ arithmetic operations.*

The application of the zeta and Möbius transform that is important to us is the following lemma.

Lemma 6 (generalised covering product). *Let P be a finite partial order such that, for every $x \in P^k$, the set $\{y \in P^k | y \leq x\}$ forms a join-semilattice, and let $f, g : P^k \to \mathbb{F}_p$.*
Define the generalised covering product $h : P^k \to \mathbb{F}_p$ of f and g through:

$$h(x) = \sum_{y_1 \vee y_2 = x} f(y_1) g(y_2)$$

Then $\mu(\zeta(f) \cdot \zeta(g))(x) = h(x)$, where the product $\zeta(f) \cdot \zeta(g)$ is defined by point-wise multiplication.

Proof. We will prove that $(\zeta(f) \cdot \zeta(g))(x) = \zeta(h)(x)$, then the result follows from Lemma 4.

$$(\zeta(f) \cdot \zeta(g))(x) = \left(\sum_{y \leq x} f(y) \right) \left(\sum_{y \leq x} g(y) \right) = \sum_{y_1, y_2 \leq x} f(y_1) g(y_2) \qquad (10)$$

Here, we first use the definition of the ζ-transform and then work out all the product terms. The result equals $\zeta(h)(x)$ as we now show by working out the definition of $\zeta(h)(x)$.

$$\zeta(h)(x) = \sum_{z \leq x} \sum_{y_1 \vee y_2 = z} f(y_1) g(y_2) = \sum_{y_1, y_2 \leq x} f(y_1) g(y_2) \qquad (11)$$

For the last equality, we reorder terms using that for any two $y_1, y_2 \leq x$ there is a unique z such that $y_1 \vee y_2 = z$; this is well defined as the set $\{y \in P^k | y \leq x\}$ forms a join-semilattice. □

As a direct result, we obtain a generalisation of the covering product from [2].

Corollary 2. *The generalised covering product for* $f, g : \mathbb{N}^k_{<r} \to \mathbb{F}_p$ *defined in the statement of Lemma 6 can be computed in* $\mathcal{O}(r^k k)$ *arithmetic operations.*

Proof. Combine Lemma 6 with the fast evaluation algorithms of Proposition 5.

□

We conclude this part on zeta and Möbius transforms by a theorem on a combined covering product and convolution product that we will use in the sections to come.

Theorem 1. *Let* P *be a finite partial order where, for every* $x \in P^k$, *the set* $\{y \in P^k | y \leq x\}$ *forms a join-semilattice. Let* $N = \mathbb{N}_{<q_1} \times \mathbb{N}_{<q_2} \times \cdots \times \mathbb{N}_{<q_l}$, *and let* $Q = \prod_{i=1}^{l} q_i$. *Let* $f, g : P^k \times N \to \mathbb{F}_p$, *where* p *is chosen appropriately. Define* $h : P^k \times N \to \mathbb{F}_p$ *as follows:*

$$h(x, i) = \sum_{y_1 \vee y_2 = x} \sum_{j_1 + j_2 = i} f(y_1, j_1) \, g(y_2, j_2)$$

If P *allows zeta and Möbius transforms using* $\mathcal{O}(|P|)$ *arithmetic operations for functions* $f' : P \to \mathbb{F}_p$, *then* h *can be computed in* $\mathcal{O}(|P|^k Q \, (k + \log(Q)))$ *arithmetic operations.*

In particular, if $P^k = \mathbb{N}^k_{<r}$, *then* h *can be computed in* $\mathcal{O}(r^k Q \, (k + \log(Q)))$ *arithmetic operations.*

Proof. For the functions f, g, h, with domains $P^k \times N$, we write $\zeta(f(-, i))(x) = \sum_{y \leq x} f(y, i)$ to fix the second component when using the zeta transform. Following the same reasoning as in Eqs. 11 and 10 in the proof of Lemma 6, one easily obtains:

$$\zeta(h(-, i))(x) = \sum_{y \leq x} \sum_{z_1 \vee z_2 = y} \sum_{j_1 + j_2 = i} f(z_1, j_1) \, g(z_2, j_2)$$

$$= \sum_{j_1 + j_2 = i} \left(\sum_{y \leq x} \sum_{z_1 \vee z_2 = y} f(z_1, j_1) \, g(z_2, j_2) \right)$$

$$= \sum_{j_1 + j_2 = i} \left(\sum_{z_1, z_2 \leq x} f(z_1, j_1) \, g(z_2, j_2) \right)$$

$$= \sum_{j_1 + j_2 = i} \left(\zeta(f(-, j_1)) \cdot \zeta(g(-, j_2)) \right)(x)$$

Consequently, we can compute h by evaluating this expression and taking the Möbius transform.

That is, we can compute h by taking the following steps:

1. Compute a fast zeta transform of $f(-,j)$ and $g(-,j)$, for each fixed $j \in N$ in $\mathcal{O}(|P|^k k)$ arithmetic operations using Lemma 5. This takes $\mathcal{O}(|P|^k Qk)$ arithmetic operations in total. For each $j_1, j_2 \in N$ and $x \in P^k$, we now have $\zeta(f(-,j_1))(x)$ and $\zeta(g(-,j_2))(x)$.
2. For each fixed $x \in P^k$, compute the sum over all $j_1 + j_2 = i$ using the fast convolution algorithm of Corollary 1 in $\mathcal{O}(Q \log(Q))$ arithmetic operations. This takes $\mathcal{O}(|P|^k Q \log(Q))$ arithmetic operations in total. For each $x \in P^k$ and $i \in N$, we now have $\zeta(h(-,i))(x)$.
3. Finally, for each fixed i, take the Möbius transform of $\zeta(h(-,i))(x)$ in $\mathcal{O}(|P|^k k)$ time using Lemma 5. Like the first step, this takes $\mathcal{O}(|P|^k Qk)$ arithmetic operations in total. As a result, we have the required values $h(x,i)$.

The running time follows by summing the times required for each of the three steps.   □

3.3   Modular Arithmetic

As discussed in the introduction to this section, we embed the integers $\{0, 1, \ldots, M\}$ in the larger field $\mathbb{F}_p$, for a prime $p > M$. However, we need to choose p 'appropriately' such that the resulting field $\mathbb{F}_p$ has the required root(s) of unity. Below we give a short description of how this can be done.

Let $r_1, r_2, \ldots, r_k$ be distinct integers. We look for a prime number p such that $\mathbb{F}_p$ contains, for all i, an r_i-th root of unity. To find the prime p, we consider candidates $m_j = 1 + jR$, where $R = \prod_{i=1}^k r_i$, for j large enough such that $m_j > M$. By the prime number theorem for arithmetic progressions, the sequence $(m_j)_{j=1}^\infty$ contains $\mathcal{O}(\frac{1}{\phi(R)} \frac{x}{\ln(x)})$ prime numbers less than x, where ϕ is Euler's totient function. Since prime testing can be done in polynomial time, we can look for the first candidate $m_j > M$ that is prime and choose p as such.

By Euler's theorem, for any $x \in \mathbb{F}_p$, with p chosen as in the previous paragraph: $1 = x^{\phi(p)} = x^{p-1} = x^{jR}$. As such, for any $x \in \mathbb{F}_p$, x^l with $l = \frac{jR}{r_i}$ is an r_i-th root of unity if $(x^l)^i \neq 1$ for all $i < r_i$. Finding an appropriate x is not difficult for small r_i as an $\frac{1}{r_i}$-th fraction of all elements $x \in \mathbb{F}_p$ results in x^l being an r_i-th root of unity. To see this, consider a generator g of the multiplicative subgroup of $\mathbb{F}_p$. The sequence $g^1, g^2, \ldots, g^{p-1}$ equals all elements in $\mathbb{F}_p \backslash \{0\}$. Putting this sequence to the power l gives $g^l, g^{2l}, \ldots, g^{(p-1)l}$ which, by choice of l, equals $\omega_{r_i}^1, \omega_{r_i}^2, \ldots, \omega_{r_i}^{(p-1)}$, where ω_{r_i} is an r_i-th root of unity in $\mathbb{F}_p$. Clearly, this forms l times the sequence $\omega_{r_i}^1, \omega_{r_i}^2, \ldots, \omega_{r_i}^{r_i}$, as $\omega_{r_i}^{r_i} = 1$.

4   Fast Join Operations

Having introduced the basics of dynamic programming on tree decompositions for $[\sigma, \rho]$-domination problems in Sect. 2, and the basics of the fast Fourier and fast Möbius transforms in Sect. 3, we are now ready to apply the fast transforms to obtain fast join operations. We will survey some known techniques based on both transforms.

Recall that, to compute the memoisation table for a join node i of a nice tree decomposition, we are given two memoisation tables A_l and A_r corresponding to the (left and right) child nodes of i. These tables store a number for each state colouring c with labels from C (as defined in Sect. 2.2): this number indicates the existence (0 or 1) of a, the size of a, and/or the number of (minimum/maximum size) partial solution(s) on G_l and G_r for the partial-solution equivalence class corresponding to c. Here, a partial-solution equivalence class is uniquely identified by c in the following way: a label for a vertex v in c defines whether v is in the solution set or not, and it defines how many neighbours v has in the solution set (see Sect. 2.2). Our goal is to compute A_i.

As stated in Sect. 2.2, a trivial algorithm would loop over all combinations of state colourings c_l, c_r of X_i that agree on which vertices are in the solution set D. Then, the algorithm considers two corresponding partial solutions D_l on G_l and D_r on G_r, and it infers the state colouring c_i of the partial solution $D_l \cup D_r$ on G_i. Over all constructed partial solutions on G_i, it stores in A_i the minimum size of a solution for each equivalence class c_i. It is not hard to show that one does not need to consider the partial solutions representing an equivalence class: given the state colourings c_l and c_r, the state colouring of $D_l \cup D_r$ can be inferred directly. This is done as follows. Since G_l, G_r and $G_i = G_l \oplus G_r$ do not contain edges between vertices in X_i, for any vertex $v \in X_i$, the number of neighbours in D_l and D_r add up to the resulting number in $D_r \cup D_l$. As such, for any vertex v, if v has label $|l|_\sigma$ in c_l and $|l'|_\sigma$ in c_l, then any combined partial solution has label $|l + l'|_\sigma$ for v in the state colouring that identifies the equivalence class (or label $|\geq \ell|_\sigma$ when $l + l' \geq \ell$ and the $|\geq \ell|_\sigma$ label is in C_σ). The same holds for labels $|l|_\rho$ from C_ρ.

We find it insightful to make tables, which we call 'join tables', that visualise the resulting label of a vertex in c_i given its labels in c_l and c_r: see Fig. 1. In these tables, the patterns emerge that our fast join operations must fulfil. Here, one can see the running time that a trivial algorithm uses to perform the join: every non-empty cell represents a combination from A_l and A_r that can be made on each vertex coordinate (each vertex in X_i). As a result, this trivial algorithm performs the join in $\mathcal{O}^*(x^k)$ time, where $k = |X_i|$ and x is the number of non-empty cells in the join table.

	$|{\geq}0|_\sigma$	$|0|_\rho$	$|{\geq}1|_\rho$
$|{\geq}0|_\sigma$	$|{\geq}0|_\sigma$		
$|0|_\rho$		$|0|_\rho$	$|{\geq}1|_\rho$
$|{\geq}1|_\rho$		$|{\geq}1|_\rho$	$|{\geq}1|_\rho$

	$|0|_\sigma$	$|0|_\rho$	$|1|_\rho$
$|0|_\sigma$	$|0|_\sigma$		
$|0|_\rho$		$|0|_\rho$	$|1|_\rho$
$|1|_\rho$		$|1|_\rho$	

	$|0|_\sigma$	$|1|_\sigma$	$|2|_\sigma$	$|3|_\sigma$	$|{\geq}0|_\rho$
$|0|_\sigma$	$|0|_\sigma$	$|1|_\sigma$	$|2|_\sigma$	$|3|_\sigma$	
$|1|_\sigma$	$|1|_\sigma$	$|2|_\sigma$	$|3|_\sigma$		
$|2|_\sigma$	$|2|_\sigma$	$|3|_\sigma$			
$|3|_\sigma$	$|3|_\sigma$				
$|{\geq}0|_\rho$					$|{\geq}0|_\rho$

	$|0|_\sigma$	$|{\geq}1|_\sigma$	$|0|_\rho$	$|{\geq}1|_\rho$
$|0|_\sigma$	$|0|_\sigma$	$|{\geq}1|_\sigma$		
$|{\geq}1|_\sigma$	$|{\geq}1|_\sigma$	$|{\geq}1|_\sigma$		
$|0|_\rho$			$|0|_\rho$	$|{\geq}1|_\rho$
$|{\geq}1|_\rho$			$|{\geq}1|_\rho$	$|{\geq}1|_\rho$

Fig. 1. Join tables corresponding, from left to right, to DOMINATING SET, STRONG STABLE SET, 3-REGULAR SUBGRAPH, and TOTAL DOMINATING SET.

4.1 Möbius Transforms for Dominating Set and Independent Dominating Set

For the DOMINATING SET problem, consider the first (leftmost) join table in Fig. 1. Notice that this is also the join table for the INDEPENDENT DOMINATING SET problem.

We first consider the (non-optimisation) counting variant of DOMINATING SET or INDEPENDENT DOMINATING SET: the table entries $A_i(c)$ represent the number of partial solutions to (independent) dominating set in the equivalence class represented by c (partial solutions of any size).

Lemma 7 (based on [26]). *The join for the counting variant of* DOMINATING SET *can be computed in* $\mathcal{O}(3^k k)$ *arithmetic operations.*

Proof. We first loop over all 2^k subsets $X_\sigma \subseteq X_i$, and for each subset X_σ, we fix the labels of vertices in X_σ in c_l, c_r and c_i to $|{\geq}0|_\sigma$. We then consider the subproblem that remains using only the labels $|0|_\rho$ and $|{\geq}1|_\rho$. Let $X' = X_i \backslash X_\sigma$ be the vertices without fixed label, let $k' = |X'|$, and let A'_i, A'_l, A'_r be the memoisation tables A_i, A_l, and A_r after fixing the vertices with label $|{\geq}0|_\sigma$, i.e., A'_i, A'_l, A'_r are indexed by state colourings c'_l and c'_r on X'_i using only $|0|_\rho$ and $|{\geq}1|_\rho$-labels.

To compute the join, we now essentially want to take a coordinate-wise maximum of the state colourings c'_l and c'_r (identifying $|0|_\rho$ with 0 and $|{\geq}1|_\rho$ with 1) to obtain the resulting state colouring c'_i on X'_i. That is, to compute $A'_i(c'_i)$, we want to efficiently evaluate the following formula:

$$A'_i(c'_i) = \sum_{c'_l \vee c'_r = c'_i} A'_l(c'_l) A'_r(c'_r) \tag{12}$$

where $\vee$ is the above discussed coordinate-wise maximum (identifying C_ρ with $\mathbb{N}^k_{<2}$). Observe that this corresponds exactly to the covering product, generalised in Lemma 6 with $P = \mathbb{N}^k_{<2}$ and $f = A'_l$, $g = A'_r$. Consequently, this join can be computed in $\mathcal{O}(2^{k'} k')$ arithmetic operations by Corollary 2. Summing up the running time over all 2^k subsets of fixed labels, we obtain a running time of:

$$\mathcal{O}\left(\sum_{X' \subseteq X_i} 2^{|X'|} |X'|\right) = \mathcal{O}\left(\sum_{k'=0}^{k} \binom{k}{k'} 2^{k'} k'\right) = \mathcal{O}\left(k \sum_{k'=0}^{k} \binom{k}{k'} 2^{k'} 1^{k-k'}\right)$$

$$= \mathcal{O}(k(2+1)^k) = \mathcal{O}(3^k k)$$

where we group the subsets $X' = X_i \backslash X_\sigma$ of the same size and then use the binomial theorem. $\qquad\square$

Corollary 3. *Given a graph G with a tree decomposition T of G of width t, the number of (independent) dominating sets can be computed in $\mathcal{O}(3^t tn)$ arithmetic operations on $\mathcal{O}(n)$-bit numbers.*

Proof. Plug Lemma 7 into Lemma 1. We can use $\mathcal{O}(n)$-bit numbers as the result is at most 2^n. $\qquad\square$

The above construction does not work directly for the minimisation version of (INDEPENDENT) DOMINATING SET, as then we are no longer counting combinations of partial solutions: we want to take the minimum over the sum of partial solution sizes. I.e., instead of Eq. 12, we need:

$$A_i'(c_i') = \min_{c_l' \vee c_r' = c_i'} A_l'(c_l') + A_r'(c_r') \tag{13}$$

To obtain a similar result for the minimisation versions, we will embed this into a counting structure.

Lemma 8 (based on [26]). *The join for the minimisation variant of* DOMINATING SET *can be computed in $\mathcal{O}(3^k n(k + \log(n)))$ arithmetic operations.*

Proof. The proof is identical to the proof of Lemma 7, except that we need a different fast evaluation algorithm: one that corresponds to Eq. 13. To this end, we expand the memoisation tables A_i', A_l' and A_r' by having the solution size as part of the index of the table. That is, we let:

$$A_l'(c_l', \kappa_l) = \begin{cases} 1 & \text{if } A_l'(c_l') = \kappa_l \\ 0 & \text{otherwise} \end{cases} \tag{14}$$

and similarly for A_r'. Let $A_i'(c_i', \kappa_i)$ be defined as follows, which can be computed using Theorem 1:

$$A_i'(c_i', \kappa_i) = \sum_{c_l' \vee c_r' = c_i'} \sum_{\kappa_l + \kappa_r = \kappa_i} A_l'(c_l', \kappa_l) \, A_r'(c_r', \kappa_r) \tag{15}$$

It is easy to see that $A_i'(c_i', \kappa_i) > 0$ if and only if there exists $c_l', \kappa_l, c_r', \kappa_r$ such that $c_l' \vee c_r = c_i$ and $A_l'(c_l) = \kappa_l$ $A_r'(c_r) = \kappa_r$. This allows us to obtain the result required by Eq. 13 by setting $A_i'(c_i)$ equal to the minimum value of κ_i for which $A_i'(c_i, \kappa_i) > 0$.

Observe that κ_i can range between 0 and n. Therefore, when we apply Theorem 1 with $N = \mathbb{N}_{<n+1}$, we can perform the join in $\mathcal{O}(3^k n(k + \log(n)))$ arithmetic operations. $\qquad\square$

Corollary 4. *Given a graph G with a tree decomposition T of G of width t,* INDEPENDENT DOMINATING SET *can be solved in $\mathcal{O}(3^t n^2(t + \log(n)))$ arithmetic operations on $\mathcal{O}(t + \log(n))$-bit numbers.*

Proof. Plug Lemma 8 into Lemma 1. We need $\log(n)$ bit numbers for the sizes of partial solutions, while the sums in Eq. 15 can require up to $\mathcal{O}(k)$-bit numbers. □

We can gain linear dependence on n for DOMINATING SET by using a replacement property (see also [26]) that holds both for DOMINATING SET and TOTAL DOMINATING SET.

Definition 7 (replacement property for partial solutions). *An optimisation problem $\mathcal{P}$ has the* replacement *property if the difference in size between the smallest and the largest partial solution for non-dominated equivalence classes is at most k.*

Mostly this holds for a problem $\mathcal{P}$ when, if given two partial solutions, one can add or subtract all vertices in X_i from either one to obtain a solution that is at least as good as the other. This is the case for DOMINATING SET as adding all vertices in X_i to a partial solution D dominates all vertices any partial solution can dominate, thus being less restrictive than any other partial solution.

Corollary 5. *Given a graph G with a tree decomposition T of G of width t, DOMINATING SET can be solved in $\mathcal{O}(3^t t^2 n)$ arithmetic operations on $\mathcal{O}(t + \log(n))$-bit numbers.*

Proof. We further modify the algorithm used in Lemma 8 (based on Lemma 7). Let ξ_l and ξ_r be the minimum values from $A'_l(c'_l)$ and $A'_r(c'_r)$. If we restrict the ranges of κ_i, κ_l, κ_r in Eq. 15 to $[0, 1, \ldots, k]$, then Theorem 1 allows us to evaluate the equation in $\mathcal{O}(2^{k'} k'^2)$ arithmetic operations. We can do so by subtracting ξ_l from all values in A'_l and ξ_r from all values in A'_r before adding the size-parameter to the index of the table. After the join, we can add $\xi_l + \xi_r$ to the results in A'_i. It is not hard to see that this does not influence the result of the algorithm, but now allows an $\mathcal{O}(3^k k^2)$-time join operation. The result then follows from plugging this result into Lemma 1. □

4.2   Count and Filter: Strong Stable Set, Perfect Code and Perfect Dominating Set

To obtain fast joins for the next set of problems, we now introduce a filtering trick based on counting. The algorithm we use here, is in essence the fast subset convolution algorithm by Björklund et al. [2]. Our different presentation is chosen so that we can use the same trick in the sections to follow.

First notice that the three problems mentioned in this section's title have essentially the same join table (the second table in Fig. 1): even though PERFECT DOMINATING SET uses the $|\geq 0|_\sigma$-label while the others use the $|0|_\sigma$-label, the structure of the join tables is identical. Compared to the join table for DOMINATING SET, the difference in terms of Eq. 12 is that we now want to compute:

$$A'_i(c'_i) = \sum_{c'_l + c'_r = c'_i} A'_l(c'_l) A'_r(c'_r) \tag{16}$$

That is, where in Eq. 12 we sum over three combinations to obtain a $| \geq 1|_\rho$-label in c_i', we may now only sum over two combinations ($|0|_\rho + |1|_\rho = |1|_\rho$, $|1|_\rho + |0|_\rho = |1|_\rho$).

Lemma 9. *The join for the maximisation variant of* STRONG STABLE SET *can be computed in* $\mathcal{O}(3^k nk(k + \log(n)))$ *arithmetic operations.*

Proof. We use the same construction as in Lemma 7 enumerating all subsets $X_\sigma \subseteq X_i$, $X' = X_i \backslash X_\sigma$, $k' = |X'|$; and for each subset X_σ, we fix the labels of the vertices in X_σ in c_l, c_r and c_i to $| \geq 0|_\sigma$. For the remaining subproblem, let A_i', A_l', A_r' be the memoisation tables A_i, A_l, and A_r after fixing the vertices with label $| \geq 0|_\sigma$, i.e., they are indexed by state colourings c_l' and c_r' on X_i' using only $|0|_\rho$ and $|1|_\rho$-labels. Next, we add the solution size to the index of these tables as in the proof of Lemma 8: $A_l'(c_l', \kappa_l) = 1$ if and only if $A_l'(c_l') = \kappa_l$. Observe that now the join can be computed by letting $A_i'(c_i')$ be the minimum value κ_i for which $A_i'(c_i', \kappa_i) > 0$, where:

$$A_i'(c_i', \kappa_i) = \sum_{c_l'+c_r'=c_i'} \sum_{\kappa_l+\kappa_r=\kappa_i} A_l'(c_l', \kappa_l) \, A_r'(c_r', \kappa_r) \tag{17}$$

To compute the result of Eq. 17 efficiently, we add yet another parameter to the index of the tables. This parameter counts the number of $|1|_\rho$-labels in the state colouring c. In other words:

$$A_l'(c_l', \kappa_l, \iota_l) = \begin{cases} A_l'(c_l', \kappa_l) & \text{if } \#_{|1|_\rho}(c_l) = \iota_l \\ 0 & \text{otherwise} \end{cases} \tag{18}$$

where $\#_{|1|_\rho}(c_l) = \iota_l$ is our notation for stating that c_l contains exactly ι_l $|1|_\rho$-labels. We claim that $A_i'(c_i', \kappa_i)$ as defined in Eq. 17 equals $A_i'(c_i', \kappa_i, \#_1(c_i'))$, where $A_i'(c_i', \kappa_i, \iota_i)$ is defined as:

$$A_i'(c_i', \kappa_i, \iota_i) = \sum_{c_l' \vee c_r' = c_i'} \sum_{\kappa_l+\kappa_r=\kappa_i} \sum_{\iota_l+\iota_r=\iota_i} A_l'(c_l', \kappa_l, \iota_l) \, A_r'(c_r', \kappa_r, \iota_i) \tag{19}$$

where $c_l' \vee c_r' = c_i'$ is again defined coordinate-wise and by identifying C_ρ with $\mathbb{N}_{<2}$.

Notice that ι_l and ι_r track the number of $|1|_\rho$-labels used. Therefore, the total number of $|1|_\rho$-labels in a pair (c_l', c_r') used as $c_l' \vee c_r' = c_i'$ in a summand of the sum for $A_i'(c_i', \kappa_i, \iota_i)$ is exactly ι_i. Since we need at least one $|1|_\rho$-label in c_l' or c_r' to realise each $|1|_\rho$-label in c_i, we know that $A_i'(c_i', \kappa_i, \#_{|1|_\rho}(c_i'))$ uses $\#_{|1|_\rho}(c_i')$ $|1|_\rho$-labels in total, and hence equals $A_i'(c_i', \kappa_i)$ from Eq. 17.

By Theorem 1, Eq. 19 can be evaluated in $\mathcal{O}(2^{k'} nk'(k' + \log(n)))$ arithmetic operations. Summing the running time over all 2^k subsets $X_\sigma \subseteq X_i$ of vertices for which we fixed the label, this leads to the claimed running time in exactly the same way as in the proof of Lemma 7. $\qquad\qquad\square$

Corollary 6. *Given a graph G with a tree decomposition T of G of width t, the optimisation variants of* STRONG STABLE SET, PERFECT CODE *and* PERFECT DOMINATING SET *can be solved in* $\mathcal{O}(3^t n^2 t(t + \log(n)))$ *arithmetic operations on* $\mathcal{O}(t + \log(n))$*-bit numbers.*

Proof. Plug Lemma 9 into Lemma 1. We need $\log(n)$-bit numbers for the sizes of partial solutions, while the sums in Eq. 19 can require up to $\mathcal{O}(k)$-bit numbers.   □

4.3   Fourier Transforms for Induced Bounded Degree or p-Regular Subgraph

The results of the previous section can also be obtained using counting and filtering on top of Fourier transforms instead of on top of Möbius transforms. The resulting construction also results in fast joins for INDUCED BOUNDED DEGREE SUBGRAPH and p-REGULAR SUBGRAPH: these we present in this section. The join table for 3-REGULAR SUBGRAPH is given in Fig. 1: notice that it is identical to the join table for INDUCED BOUNDED DEGREE SUBGRAPH with degree bound three.

We should note that the results in this section can also be obtained using Cygan and Pilipczuk's method [17] where they encode solutions as a polynomial and use FFT-based fast polynomial multiplication. The approach below is allows for easier combination with Möbius transforms (as we will see in Sect. 5), and saves a factor of t in the polynomial factors of the running time[5].

We start in the same setting as before: we fix the vertices with label $|\geq 0|_\rho$ by looping over all subsets $X_\rho \subseteq X_i$ (now we fix states from C_ρ, previously from C_σ) and let $X' = X_i \backslash X_\rho$ and $k' = |X'|$. Let A'_i, A'_l and A'_r be the memoisation tables after fixing the vertices with the $|\geq 0|_\rho$-label indexed by state colourings c'_i, c'_l, c'_r using labels from $C_\sigma = \{|0|_\sigma, |1|_\sigma, \ldots, |\ell-1|_\sigma, |\ell|_\sigma\}$.

Define the projection function π on labels in C_σ as to be $\pi(|l|_\sigma) = l$. Also, define addition on state colourings as follows: $c'_l + c'_r = c'_i$ if and only if, for all j, $\pi((c'_l)_j) + \pi((c'_r)_j) = \pi((c'_i)_j)$. Computing $A'_i(c'_i)$ for the counting variant of the problem, or computing $A'_i(c'_i, \kappa_i)$ for the optimisation variant of the problem using the solution size κ as part of the index, now comes down to evaluating:

$$A'_i(c'_i) = \sum_{c'_l + c'_r = c'_i} A'_l(c'_l) A'_r(c'_r) \qquad A'_i(c'_i, \kappa_i) = \sum_{c'_l + c'_r = c'_i} \sum_{\kappa_l + \kappa_r = \kappa_i} A'_l(c'_l, \kappa_l) A'_r(c'_r, \kappa_r)$$

(20)

Observe that this looks very similar to the statement of Lemma 3. However, to obtain a non-cyclic convolution for $c'_l + c'_r = c'_i$, a direct application of Lemma 3 would require $\mathcal{O}^*((\ell+1)^k 2^k)$ arithmetic operations to evaluate either version of Eq. 20. We will use cyclic convolution with counting and filtering to obtain the result in $\mathcal{O}^*((\ell+1)^k)$ arithmetic operations, resulting in an $\mathcal{O}^*((\ell+2)^k) = \mathcal{O}^*(s^k)$ time join operation.

Lemma 10. *The join for the counting variant of* INDUCED BOUNDED DEGREE SUBGRAPH *can be computed in* $\mathcal{O}((\ell+2)^k k^2 \ell \log(\ell+1))$ *arithmetic operations.*

---

[5] The approach in [17] uses a factor k^3 (compared to our k^2) by performing k^2 FFT-based polynomial multiplications that each cost $\mathcal{O}((\ell+1)^k \log((\ell+1)^k)) = \mathcal{O}((\ell+1)^k k \log(\ell+1))$ arithmetic operations.

Proof. For a state colouring $c'_i \in C_\sigma^{k'}$, define $\Sigma(c'_i) = \sum_{j=1}^{k'} \pi((c'_i)_j)$. Now, similar to the proof of Lemma 9 where we added the number of 1-labels to the index of the table, we now add the sum of the labels $\Sigma(c'_i)$ to the index of the table. That is, for both A'_l and A'_r, we set:

$$A'_l(c'_l, \iota_l) = \begin{cases} A'_l(c'_l) & \text{if } \Sigma(c'_l) = \iota_l \\ 0 & \text{otherwise} \end{cases} \tag{21}$$

To compute the join efficiently, we use that $A'_i(c'_i)$ as defined in Eq. 20 (left equation) equals $A'_i(c'_i, \Sigma(c'_i))$, where $A'_i(c'_i, \iota_i)$ is the result of the following summation:

$$A'_i(c'_i, \iota_i) = \sum_{c'_l + c'_r \equiv c'_i} \sum_{\iota_l + \iota_r = \iota_i} A'_l(c'_l, \iota_l) A'_r(c'_r \iota_i) \tag{22}$$

where $c'_l + c'_r \equiv c'_i$ is now defined as, for all j, $\pi((c'_l)_j) + \pi((c'_r)_j) \equiv \pi((c'_i)_j)$ modulo $\ell + 1$, and $\iota_l + \iota_r = \iota_i$ is the standard addition without modulus. To see that this is correct, notice that the parameters ι_l and ι_r track the sum of the labels in c'_l and c'_r, and $\iota_l + \iota_r = \Sigma(c'_i)$ implies that each of the individual components of $c'_l + c'_r$ cannot cycle as that would result in $\iota_l + \iota_r > \Sigma(c'_i)$.

By Lemma 3, Eq. 22 can be evaluated in $\mathcal{O}((\ell+1)^{k'} k'^2 \ell \log(\ell+1))$ arithmetic operations as the ι parameters range from 0 to $\ell k'$. The claimed running time follows by summing this running time over all $X_\rho \subseteq X_i$ for which we fixed the label (similar to in the proof of Lemma 7). □

Corollary 7. *Given a graph G with a tree decomposition T of G of width t, the counting variant of* INDUCED BOUNDED DEGREE SUBGRAPH *can be solved in $\mathcal{O}((\ell+2)^{t+1} t^2 n \ell \log(\ell+1))$ arithmetic operations on $\mathcal{O}(n)$-bit numbers.*

Proof. Plug Lemma 10 into Lemma 1. We can use $\mathcal{O}(n)$-bit numbers as the result is at most 2^n. Since ℓ is a variable (not a consant), we need $t+1$ in the exponent as $t \geq k-1$ (see Definition 1). □

Adapting the above lemma and corollary to the problem's optimisation variant is a simple exercise.

4.4 Möbius Transforms with a Different Partial Order for Total Dominating Set

In the previous sections, we fixed the vertices with a $|\geq 0|_\sigma$-label (or $|\geq 0|_\rho$-label) and used a fast transform only on vertices with a label from C_ρ (or C_σ). Here, we give an example of a fast transform that deals with vertices with different labels from both C_σ and C_ρ simultaneously.

Consider the label set $C = \{|0|_\sigma, |\geq 1|_\sigma, |0|_\rho, |\geq 1|_\rho\}$ associated to the TOTAL DOMINATING SET problem. On this label set, we impose the following partial order: all labels are incomparable except that, we impose $|0|_\sigma \leq |\geq 1|_\sigma$ and $|0|_\rho \leq |\geq 1|_\rho$. Notice that, using this partial order, C^k does not form a lattice, e.g., in C^3 the join $[|0|_\sigma, |0|_\rho, |\geq 1|_\sigma] \vee [|\geq 1|_\rho, |\geq 1|_\rho, |0|_\sigma]$ is undefined due to the first coordinate. However, it is not hard to see that, for every $x \in C^k$, $\{y \in P^k | y \leq x\}$ forms a join-semilattice: either x_i equals $|0|_\sigma$ or $|0|_\rho$ and then y_i must be equal to x_i, or x_i equals $|\geq 1|_\sigma$ or $|\geq 1|_\rho$ and then y_i has two choices which are comparable, and thus the join is defined (actually this is a lattice as the meet is also defined).

Proposition 6. *Let C and its partial order be defined as above. There are algorithms for the zeta and Möbius transform for functions $f : C^k \to \mathbb{F}_p$ that require $\mathcal{O}(4^k k)$ arithmetic operations.*

Proof. If $k = 1$, the zeta and Möbius transforms are:

$$\zeta(f)(|0|_\sigma) = f(|0|_\sigma) \qquad\qquad \mu(f)(|0|_\sigma) = f(|0|_\sigma)$$
$$\zeta(f)(|\geq 1|_\sigma) = f(|0|_\sigma) + f(|\geq 1|_\sigma) \qquad \mu(|\geq 1|_\sigma) = f(|\geq 1|_\sigma) - f(|0|_\sigma)$$
$$\zeta(f)(|0|_\rho) = f(|0|_\rho) \qquad\qquad \mu(f)(|0|_\rho) = f(|0|_\rho)$$
$$\zeta(f)(|\geq 1|_\rho) = f(|0|_\rho) + f(|\geq 1|_\rho) \qquad \mu(|\geq 1|_\rho) = f(|\geq 1|_\rho) - f(|0|_\rho)$$

The result follows from Lemma 5 as these require a constant amount of arithmetic operations. $\qquad\qquad\qquad\qquad\qquad\qquad\qquad\qquad\qquad\qquad\qquad\qquad\qquad\qquad$ □

Lemma 11 (based on [26]). *The join for the minimisation variant of* TOTAL DOMINATING SET *can be computed in $\mathcal{O}(4^k k^2)$ arithmetic operations.*

Proof. Given A_l and A_r, we can compute A_i by evaluating the following equation that is equivalent to Eq. 13 (notice that this corresponds exactly to the rightmost join table in Fig. 1):

$$A_i(c_i) = \min_{c_l \vee c_r = c_i} A_l(c_l) + A_r(c_r) \tag{23}$$

To obtain our result, we expand the memoisation tables A_i, A_l and A_r by having the solution size as part of the index of the table. That is, we let:

$$A_l(c_l, \kappa_l) = \begin{cases} 1 & \text{if } A_l(c_l) = \kappa_l \\ 0 & \text{otherwise} \end{cases} \tag{24}$$

and similarly for A_r. Then, we compute $A_i(c_i, \kappa_i)$ as defined below, computed using Theorem 1:

$$A_i(c_i, \kappa_i) = \sum_{c_l \vee c_r = c_i} \sum_{\kappa_l + \kappa_r = \kappa_i} A_l(c_l, \kappa_l) A_r(c_r, \kappa_r) \tag{25}$$

Since $A_i(c_i, \kappa_i) > 0$ if and only if there exists $\check{c}_l, \kappa_l, c_r, \kappa_r$ such that $c_l \vee c_r = c_i$ and $A_l(c_l) = \kappa_l$ $A_r(c_r) = \kappa_r$, this allows us to obtain the result required by Eq. 23 by setting $A_i(c_i)$ equal to the minimum value of κ_i for which $A_i(c_i, \kappa_i) > 0$.

A direct application of Theorem 1 would allow us to evaluate Eq. 25 in $\mathcal{O}(4^k n^2(k + \log(n)))$ arithmetic operations. However, if we restrict the ranges of κ_i, κ_l, κ_k to $[0, 1, \ldots, k]$, then Theorem 1 allows us to evaluate Eq. 25 in

$\mathcal{O}(4^k k^2)$ arithmetic operations. We can do so because, just as in Corollary 5, TOTAL DOMINATING SET satisfies the replacement property (Definition 7): take the minimum value from A_l and A_r and subtract this from all values in A_l and A_r before adding the size-parameter to the index of the table. After the join, we can again add the sum of both minima to the results in A_i. The result now follows. □

Corollary 8. *Given a graph G with a tree decomposition T of G of width t,* TOTAL DOMINATING SET *can be solved in $\mathcal{O}(4^t t^2 n)$ arithmetic operations on $\mathcal{O}(t + \log(n))$-bit numbers.*

Proof. Plug Lemma 11 into Lemma 1 and observe that we need $\log(n)$-bit numbers for the sizes of partial solutions, while the sums in Eq. 25 can require up to $\mathcal{O}(t)$-bit numbers. □

5 Bringing It Together: Faster Algorithms for $[\sigma, \rho]$-Domination

In the previous section, we have surveyed a number of approaches to realise fast joins operations and have given some concrete examples. In this section, we will integrate several of these approaches into a new result obtaining the currently fastest algorithm for $[\sigma, \rho]$-domination in its general form. The worst-case running time of the new algorithm is of the form $\mathcal{O}(s^t (nts)^{\mathcal{O}(1)})$, where the previously fastest algorithm by Van Rooij et al. [25,26], has s as an exponent in the polynomial part of the running time, i.e., $\mathcal{O}(s^{t+2}(tn)^s (nts)^{\mathcal{O}(1)})$. Here, we write $s = |C|$, where C is the set of labels used in the dynamic programming algorithm for a specific $[\sigma, \rho]$-domination problem (as in Sect. 2.2). To limit the (already heavy) notational burden, we give our result for a $[\sigma, \rho]$-domination problem with *cofinite* σ and *finite* ρ. It is not hard to modify the proofs for the other cases.

For state colourings c_l, c_r of X_i with labels from C, we define the operator $c_l \oplus c_r = c_i$ as the coordinate-wise addition operator that keeps addition within both parts of the label set. This operator is only defined if c_l and c_r agree on which vertices are labelled with labels from C_σ and from C_ρ and if the result again is in C_σ or C_ρ. More formally, when we write $c_l \oplus c_r = c_i$ and if $(c_l)_j = |x_l|_\rho$, $(c_r)_j = |x_r|_\rho$, then $(c_i)_j = |x_l + x_r|_\rho$ if $|x_l + x_r|_\rho \in C_\rho$ and otherwise it is undefined. And, if $(c_l)_j = |x_l|_\sigma$, $(c_r)_j = |x_r|_\sigma$, then $(c_i)_j = |x_l + x_r|_\sigma$ if $x_l + x_r < \ell_\sigma$ and $(c_i)_j = |{\geq}\ell|_\sigma$ otherwise. Observe that this corresponds exactly to the structure of how a join for a $[\sigma, \rho]$-domination problem with cofinite σ and finite ρ should be performed. Besides the $\oplus$-operator, we will also use the standard $+$-operator on state colourings: $c_l + c_r = c_i$. Here, the underlying operation is the standard addition operator within each half of the label set, which is undefined if any $|{\geq}\ell|_\sigma$ or $|{\geq}\ell|_\rho$-label is involved. That is, if $(c_l)_j = |x_l|_\rho$, $(c_r)_j = |x_r|_\rho$, then $(c_i)_j = |x_l + x_r|_\rho$, which is defined only if $|x_l + x_r|_\rho \in C_\rho$. Addition with the $+$-operator is similar for the σ-labels.

Let again A_l and A_r be indexed by both the state colouring and the solution size (similar to Eqs. 14 and 24). Performing the join for the minimisation variant of a $[\sigma, \rho]$-domination problem can now be done by extracting, for each c_i, the minimum value of κ_i for with the following expression is non-zero:

$$A_i(c_i, \kappa_i) = \sum_{c_l \oplus c_r = c_i} \sum_{\kappa_l + \kappa_r = \kappa_i} A_l(c_l, \kappa_l) A_r(c_r, \kappa_r) \tag{26}$$

To obtain a fast evaluation algorithm for Eq. 26, we use both zeta/Möbius transforms and Fourier transforms. To do so, we impose the following partial order p on the label set C: all labels are incomparable except that, because σ is cofinite, we impose for all $|l|_\sigma \in C_\sigma$: $|l|_\sigma \leq |{\geq}\ell|_\sigma$.

Given a state colouring c_i of the vertices in X_i with labels from $C = C_\sigma \cup C_\rho$, we write $c_i = [c_i^\sigma, c_i^\rho] = [c_i^{\geq\ell_\sigma}, c_i^{<\ell_\sigma}, c_i^\rho]$ to differentiate between the vertices with label from C_σ and C_ρ, and also to further differentiate between vertices with the label $|{\geq}\ell|_\sigma$ and vertices with a label from $\{|0|_\sigma, |1|_\sigma, \ldots, |\ell-1|_\sigma\}$. By splitting c_i in this way we notationally split the different coordinates. This is just notation: we do not reorder anything in the dynamic programming table (e.g., c_i^ρ can contain the first coordinate of c_i and the last, while c_i^σ contains the ones in between).

Using this notation, the zeta transform of a memoisation table A_i indexed by both a state colouring $c_i = [c_i^{\geq\ell_\sigma}, c_i^{<\ell_\sigma}, c_i^\rho]$ and additional indices x_i (whose purpose will become clear later) becomes:

$$\zeta(A_i)(c_i, x_i) = \sum_{d \leq c_i} A_i(d_i, x_i) = \sum_{d_1 \leq c_i^{\geq\ell_\sigma}} A_i([d_1, c_i^{<\ell_\sigma}, c_i^\rho], x_i) \tag{27}$$

Proposition 7. *Given the memoisation table $A_i(c_i, x_i)$ indexed by state colourings $c_i \in C^k$ over the label set C ($|C| = s$) and some additional indices x_i with domain I, the zeta transform $\zeta(A_i)$ of A_i based on the partial order p can be computed in $\mathcal{O}(s^k k |I|)$ arithmetic operations. Also, given $\zeta(A_i)$, A_i can be reconstructed in $\mathcal{O}(s^k k |I|)$ arithmetic operations.*

Proof. We will show that for $k = 1$ and hence $c_i \in C^1$, we have zeta and Möbius transforms on $A_i(c_i, x_i)$ requiring $\mathcal{O}(s|I|)$ arithmetic operations. The result then follows from Lemma 5 and the fact that the transforms operate independent of the parameter x_i.

By definition of the partial order p, the following formulas compute $\zeta(A_i)$ from A_i and vice versa when $k = 1$:

$$\zeta(A_i)(c_i, x) = \begin{cases} A_i(c_i, x) & \text{if } c_i \neq [|{\geq}\ell|_\sigma] \\ \sum_{z \in C_\sigma} A_i([z], x) & \text{if } c_i = [|{\geq}\ell|_\sigma] \end{cases}$$

$$A_i(c_i, x) = \begin{cases} \zeta(A_i)(c_i, x) & \text{if } c_i \neq [|{\geq}\ell|_\sigma] \\ \zeta(A_i)(c_i, x) - \sum_{z \in C_\sigma \setminus \{|{\geq}\ell|_\sigma\}} \zeta(A_i)([z], x) & \text{if } c_i = [|{\geq}\ell|_\sigma] \end{cases}$$

Each requires $\mathcal{O}(s|I|)$ arithmetic operations as the sums are computed only when $c_i = [|{\geq}\ell|_\sigma]$. $\qquad \square$

Theorem 2 (bringing it all together). *For an optimisation variant of a $[\sigma, \rho]$-domination problem with $s = |C|$, the join can be computed in $\mathcal{O}(s^{k+1}kn(k\log(s) + \log(n)))$ arithmetic operations.*

Proof. We want to evaluate Eq. 26 using a fast transform. Consider the what happens to this equation if we apply the zeta transform based on the partial order $\mathbb{p}$ (Eq. 27) to it:

$$\zeta(A_i)(\boldsymbol{c}_i, \kappa_i) = \sum_{\boldsymbol{d}_1 \leq \boldsymbol{c}_i^{\geq \ell_\sigma}} A_i([\boldsymbol{d}_1, \boldsymbol{c}_i^{<\ell_\sigma}, \boldsymbol{c}_i^\rho], \kappa_i) \tag{28}$$

$$= \sum_{\boldsymbol{d}_1 \leq \boldsymbol{c}_i^{\geq \ell_\sigma}} \sum_{\boldsymbol{d}_l \oplus \boldsymbol{d}_r = [\boldsymbol{d}_1, \boldsymbol{c}_i^{<\ell_\sigma}, \boldsymbol{c}_i^\rho]} \sum_{\kappa_l + \kappa_r = \kappa_i} A_l(\boldsymbol{d}_l, \kappa_l) A_r(\boldsymbol{d}_r, \kappa_r) \tag{29}$$

Continuing from here, we can decompose $\boldsymbol{d}_l$ and $\boldsymbol{d}_r$ coordinate-wise in the same way as we have decomposed $\boldsymbol{c}_i$ as $[\boldsymbol{c}_i^{\sigma \geq \ell}, \boldsymbol{c}_i^{\sigma < \ell}, \boldsymbol{c}_i^\rho]$ (to be clear: we split the coordinates of $\boldsymbol{d}_l$ and $\boldsymbol{d}_r$ based on the labels in $\boldsymbol{c}_i$, not based on the actual labels in $\boldsymbol{d}_l$ and $\boldsymbol{d}_r$). Let $\boldsymbol{d}_l = [\boldsymbol{d}_{1l}, \boldsymbol{d}_{2l}, \boldsymbol{d}_{3l}]$, $\boldsymbol{d}_r = [\boldsymbol{d}_{1r}, \boldsymbol{d}_{2r}, \boldsymbol{d}_{3r}]$. Now, observe that in $\boldsymbol{d}_l \oplus \boldsymbol{d}_r$, any pair $\boldsymbol{d}_{1l}$ and $\boldsymbol{d}_{1r}$ on the coordinates of $\boldsymbol{c}_i^{\geq \ell_\sigma}$ is summed over exactly once because for any pair there is exactly one $\boldsymbol{d}_1$ such that $\boldsymbol{d}_{1l} \oplus \boldsymbol{d}_{1r} = \boldsymbol{d}_1$. Also observe that because the other coordinates of $\boldsymbol{d}_l$ and $\boldsymbol{d}_r$ correspond to the vertices from the $\boldsymbol{c}_i^{<\ell_\sigma}$ and $\boldsymbol{c}_i^\rho$ parts of $\boldsymbol{c}_i$ their $\oplus$-sum is the standard (non-cyclic) +-addition on labels. As such, we obtain:

$$\zeta(A_i)(\boldsymbol{c}_i, \kappa_i) = \sum_{\boldsymbol{d}_{1l}, \boldsymbol{d}_{1r} \leq \boldsymbol{c}_i^{\geq \ell_\sigma}} \sum_{\boldsymbol{d}_{2l} + \boldsymbol{d}_{2r} = \boldsymbol{c}_i^{<\ell_\sigma}} \sum_{\boldsymbol{d}_{3l} + \boldsymbol{d}_{3r} = \boldsymbol{c}_i^\rho} \sum_{\kappa_l + \kappa_r = \kappa_i} A_l(\boldsymbol{d}_l, \kappa_l) A_r(\boldsymbol{d}_r, \kappa_r)$$

$$\tag{30}$$

We note that, in the first sum, we sum over all $\boldsymbol{d}_{1l}, \boldsymbol{d}_{1r} \leq \boldsymbol{c}_i^{\geq \ell_\sigma}$ which is consistent with earlier notation, but by definition of $\mathbb{p}$ equals all $\boldsymbol{d}_{1l}, \boldsymbol{d}_{1r} \in C_\sigma$.

For our fast join operation, we need the sums $\boldsymbol{d}_{2l} + \boldsymbol{d}_{2r}$ and $\boldsymbol{d}_{3l} + \boldsymbol{d}_{3r}$ to be cyclic in order to use cyclic convolution. Therefore, we apply the same trick as in the proof of Lemma 10 and replace the tables A_l and A_r with expanded tables that include the sums of the labels in a state colouring as an additional parameter in the index. Here, we do so by defining this sum of the labels of a state colouring as the sum of the number in the labels, ignoring whether they are from C_σ or C_ρ and excluding the $|\leq \ell|_\sigma$ label. That is, let the projection function π on labels to be $\pi(|l|_\sigma) = \pi(|l|_\rho) = l$. Then, for a state colouring $\boldsymbol{c}_i \in C^k$, define $\Sigma(\boldsymbol{c}_i) = \Sigma([\boldsymbol{c}_i^{\geq \ell_\sigma}, \boldsymbol{c}_i^{<\ell_\sigma}, \boldsymbol{c}_i^\rho]) = \sum_{j=1}^{|\boldsymbol{c}_i^{<\ell_\sigma}|} \pi((\boldsymbol{c}_i^{\sigma<\ell})_j) + \sum_{j=1}^{|\boldsymbol{c}_i^\rho|} \pi((\boldsymbol{c}_i^\rho)_j)$. For example, $\Sigma([|0|_\sigma, |0|_\rho, |\leq \ell|_\sigma]) = 0$ and $\Sigma([|2|_\sigma, |1|_\rho, |0|_\sigma]) = 3$. Now, we can define the expanded tables A_l, A_r as:

$$A_l(\boldsymbol{c}_l, \kappa_l, \iota_l) = \begin{cases} A_l(\boldsymbol{c}_l, \kappa_l) & \text{if } \Sigma(\boldsymbol{c}_l) = \iota_l \\ 0 & \text{otherwise} \end{cases} \tag{31}$$

Now, we can continue from (30) replacing the sums with sums coordinate-wise modulo ℓ_σ and $\ell_\rho + 1$, using the additional parameter to prevent the modular-cycling to happen for the result: if cycling occurs at a coordinate, the sums do not add up any more. That is, we now compute $\zeta(A_i)(c_i, \kappa_i)$ by evaluating the formula below using $\zeta(A_i)(c_i, \kappa_i) = \zeta(A_i)(c_i, \kappa_i, \Sigma(c_i))$ where:

$$\zeta(A_i)(c_i, \kappa_i, \iota_i) =$$

$$\sum_{d_{1l}, d_{1r} \leq c_i^{\geq \ell_\sigma}} \sum_{d_{2l}+d_{2r} \equiv c_i^{<\ell_\sigma}} \sum_{d_{3l}+d_{3r} \equiv c_i^\rho} \sum_{\kappa_l+\kappa_r=\kappa_i} \sum_{\iota_l+\iota_r=\iota_i} A_l(d_l, \kappa_l, \iota_l) A_r(d_r, \kappa_r, \iota_r)$$

(32)

Where $d_{2l} + d_{2r} \equiv c_i^{<\ell_\sigma}$ is modulo ℓ_σ and $d_{3l} + d_{3r} \equiv c_i^\rho$ is modulo $\ell_\rho + 1$ (the difference is due to the existence of the $|\geq \ell|_\sigma$-label).

Next, we continue from Eq. 32 by change the order of summation, taking the outermost sum inwards, and by reordering the resulting inner terms:

$$= \sum_{d_{2l}+d_{2r} \equiv c_i^{<\ell_\sigma}} \sum_{d_{3l}+d_{3r} \equiv c_i^\rho} \sum_{\kappa_l+\kappa_r=\kappa_i} \sum_{\iota_l+\iota_r=\iota_i}$$

$$\left(\sum_{d_{1l} \leq c_i^{\geq \ell_\sigma}} A_l(d_l, \kappa_l, \iota_l) \right) \left(\sum_{d_{1r} \leq c_i^{\geq \ell_\sigma}} A_r(d_r, \kappa_r, \iota_r) \right)$$

(33)

$$= \sum_{d_{2l}+d_{2r} \equiv c_i^{<\ell_\sigma}} \sum_{d_{3l}+d_{3r} \equiv c_i^\rho} \sum_{\kappa_l+\kappa_r=\kappa_i} \sum_{\iota_l+\iota_r=\iota_i}$$

$$\zeta(A_l)([c_i^{\geq \ell_\sigma}, d_{2l}, d_{3l}], \kappa_l, \iota_l) \, \zeta(A_r)([c_i^{\geq \ell_\sigma}, d_{2r}, d_{3r}], \kappa_r, \iota_r)$$

(34)

Where in the last step, we apply the definition of the ζ-transform for p (Eq. 27). As a result, we obtain a standard convolution sum that can be evaluated using Lemma reflem:combinedconv given that the vertices with label $|\geq \ell|_\sigma$ in c_i are fixed.

To be more precise, let us partition X_i into three parts $X_{\geq \ell_\sigma}$, $X_{<\ell_\sigma}$, X_ρ and say that a state colouring c_i is compatible with this partition if: all vertices in $X_{\geq \ell_\sigma}$ have the $|\geq \ell|_\sigma$-label; all vertices in $X_{<\ell_\sigma}$ have a label from $C_\sigma \setminus \{|\geq \ell|_\sigma\}$; and all vertices in X_ρ have a label from C_ρ. Then, given such a partition of X_i, Lemma 3 evaluates Eq. 34 for all c_i compatible with $(X_{\geq \ell_\sigma}, X_{<\ell_\sigma}, X_\rho)$. Consequently, we can compute $\zeta(A_i)(c_i, \kappa_i, \iota_i)$ for all values of c_i, κ_i, and ι_i by enumerating all partitions of X_i into $(X_{\geq \ell_\sigma}, X_{<\ell_\sigma}, X_\rho)$ and evaluating Eq. 34 using Lemma 3 for each subset of compatible c_i values, and then taking the results together.

As a result, we can evaluate Eq. 26 using a fast transform that takes the following steps in the following amount of operations:

- Expand the tables A_l and A_r taking the sums of the labels using Eq. 31 to $A_l(c_l, \kappa_l, \iota_l)$ and $A_r(c_r, \kappa_r, \iota_r)$. This takes $\mathcal{O}(s^{k+1}kn)$ time, as c_l takes $\mathcal{O}(s^k)$ values, κ_l takes $\mathcal{O}(n)$ values and ι_l takes $\mathcal{O}(sk)$ values.
- Compute $\zeta(A_l)$ and $\zeta(A_r)$ in $\mathcal{O}(s^{k+1}k^2n)$ arithmetic operations using Proposition 7.

- Enumerate all partitions of X_i into $(X_{\geq \ell_\sigma}, X_{< \ell_\sigma}, X_\rho)$. For each such partition, compute the part of the table $\zeta(A_i)(c_i, \kappa_i, \iota_i)$ for all c_i using Eq. 34 that are compatible with this partition using Lemma 3. Then, combine the results to obtain $\zeta(A_i)(c_i, \kappa_i, \iota_i)$ for all c_i, κ_i and ι_i.

  For each partition $(X_{\geq \ell_\sigma}, X_{< \ell_\sigma}, X_\rho)$, this takes $\mathcal{O}((|C_\sigma| - 1)^{|X_{< \ell_\sigma}|}|C_\rho|^{|X_\rho|}$ $nsk(k\log(s) + \log(n)))$ arithmetic operations by Lemma 3. By summing over all partitions and using the multinomial theorem, we find $\mathcal{O}(s^{k+1}kn(k\log(s) + \log(n)))$ arithmetic operation for this whole step, as:

$$\sum_{\substack{(X_{\geq \ell_\sigma}, X_{< \ell_\sigma}, X_\rho) \\ \text{partition of } X_i}} (|C_\sigma| - 1)^{|X_{< \ell_\sigma}|}|C_\rho|^{|X_\rho|}$$

$$= \sum_{\substack{x_1 + x_2 + x_3 \\ = k}} \binom{k}{x_1, x_2, x_3} 1^{x_1}(|C_\sigma| - 1)^{x_2}|C_\rho|^{x_3} = (|C_\sigma| + |C_\rho|)^k = s^k$$

- Extract non-cycling values using $\zeta(A_i)(c_i, \kappa_i) = \zeta(A_i)(c_i, \kappa_i, \Sigma(c_i))$.
- Compute the Möbius transform of the result obtaining A_i as $\mu(\zeta(A_i)) = A_i$. This is done $\mathcal{O}(s^k kn)$ arithmetic operations using Proposition 7.

By summing over these steps we conclude that the algorithm requires $\mathcal{O}(s^{k+1}kn(k\log(s) + \log(n)))$ arithmetic operations. $\qquad\square$

Above, the state colourings c are split in three components $c_i = [c_i^{\geq \ell_\sigma}, c_i^{< \ell_\sigma}, c_i^\rho]$. If both σ and ρ are finite, we need no Möbius transforms and it would suffice to use $c_i = [c_i^\sigma, c_i^\rho]$. If both σ and ρ are co-finite, we need Möbius transforms for both parts of the label set and would need to use $c_i = [c_i^{\geq \ell_\sigma}, c_i^{< \ell_\sigma}, c_i^{\geq \ell_\rho}, c_i^{< \ell_\rho}]$, and adjust the partial order $\mathbb{p}$ in a way similar to as in Sect. 4.4.

Corollary 9. *Given a graph G with a tree decomposition T of G of width t, the optimisation variant of a $[\sigma, \rho]$-domination problem that involves $s = |C|$ labels can be solved in $\mathcal{O}(s^{t+2}tn^2(t\log(s) + \log(n)))$ arithmetic operations on $\mathcal{O}(t\log(s) + \log(n))$-bit numbers.*

Proof. Plug Theorem 2 into Lemma 1 and observe that all arithmetic operations can be done using $\mathcal{O}(t\log(s) + \log(n))$-bit numbers: the sum of all the entries in A_l and A_r is at most s^k, hence $t\log(s)$ bits, while we need the additional $\log(n)$ bits to store partial solution sizes (see also Corollary 4). The $t + 2$ in the exponent comes from the fact that $t \geq k - 1$ (the minus one in Definition 1) $\qquad\square$

We conclude by summarising results for the other variants of the $[\sigma, \rho]$-domination problems.

Theorem 3 (results for $[\sigma, \rho]$-domination problem variants). *Given a graph G with a tree decomposition T of G of width t, the different problem variants of a $[\sigma, \rho]$-domination problem involving $s = |C|$ labels can be solved with the following amount of effort:*

	0	1_1	2	1_2
0	0	1_1	2	1_2
1_1	1_1	2		
2	2			
1_2	1_2		2	

	0	1_1	2	1_2
0	0	1_1	2	1_2
1_1	1_1	2	1_2	0
2	2	1_2	0	1_1
1_2	1_2	0	1_1	2

Fig. 2. The right join table is for a join for the LONGEST/HAMILTONIAN PATH/CYCLE and (PARITAL) CYCLE COVER problems, as described in the appendix of [15]. It is obtained by using a variant of 'counting and filtering' (Sect. 4.2) on the left join table, for which a fast FFT-based join exists.

- *Existence:* $\mathcal{O}(s^{t+2}t^2 n \log(s))$ *operations on* $\mathcal{O}(t \log(s))$-*bit numbers.*
- *Optimisation:* $\mathcal{O}(s^{t+2}tn^2(t\log(s)+\log(n)))$ *operations on* $\mathcal{O}(t\log(s)+\log(n))$-*bit numbers.*
- *Counting:* $\mathcal{O}(s^{t+2}t^2 n \log(s))$ *operations on* $\mathcal{O}(n)$-*bit numbers.*
- *Counting optimisation:* $\mathcal{O}(s^{t+2}tn^2(t\log(s)+\log(n)))$ *operations on* $\mathcal{O}(n)$-*bit numbers.*

Proof. The result for the optimisation problem follows from Corollary 9, and underlying Theorem 2.

For the counting optimisation problem, we use the same construction, only without expanding tables by solution sizes and extracting the non-zero entries: at every step of the algorithm, we let $A_i(\boldsymbol{c}, \kappa)$ be the number of partial solutions of size κ that correspond to the equivalence class identified by $\boldsymbol{c}$. The join then also comes down to evaluating Eq. 26, resulting in the same amount of arithmetic operations. For the existence and counting problems, we observe that we can remove the parameter κ at step of the algorithm, as solution sizes do not matter. Redoing the analysis of the resulting algorithm gives in the claimed amount of arithmetic operations.

For both counting problems variants, we need $\mathcal{O}(n)$-bit numbers as there can be $\mathcal{O}(2^n)$ solutions to count. For the existence problem, we need $\mathcal{O}(t\log(s))$ as the sum of all entries in a $\mathcal{O}(s^{t+1})$ table with zero-one entries can be at most $\mathcal{O}(s^{t+1})$. □

6   Conclusion

In this paper, we have shown how Möbius and Fourier transforms can be used to speed-up computations for dynamic programming algorithms on tree decompositions. This led us to the currently fastest algorithm for the general case of the $[\sigma,\rho]$-domination problems on tree decompositions. Additionally, we generalised the covering product from [2] from being defined on the subset lattice to more general partial orders (Lemma 6 and Theorem 1).

The same algebraic transforms can, and have been, used for many different problems. For example, the Möbius-transform-based approach has been used for clique packing, partitioning and covering problems such as PARTITION INTO TRIANGLES or MINIMUM COVER BY CLIQUES; see [26]. Also, the Fourier-transform-based approach (the variant from [17]) has been used for BANDWIDTH [17] and CONNECTED VERTEX COVER [27]. The Fourier-transform-based approach in this paper originates from [15], where it was used together with counting and filtering to obtain the join table in Fig. 2 for LONGEST/HAMILTONIAN PATH/CYCLE and (PARITAL) CYCLE COVER.

We want to emphasise the more general observation that almost any join for which the join table has a certain 'max' or 'addition' or 'modulo' structure, or a combination of those, can be done fast using the tools from this paper. For example, the Fourier transform and corresponding cyclic convolution theorem can be used to obtain algorithms for problems where there is some modulo relation in the definition of the problem's solution set D, e.g., an odd number of neighbours in D.

The approaches in this paper have wider use beyond tree decompositions. For example, they can be applied to branch decomposition instead of tree decompositions, obtaining faster algorithms there as well, with faster exact $\mathcal{O}(c^{\sqrt{n}})$-time algorithms on planar graphs as direct corollaries [11,23].

Open Problems. What we see is that, in order to use fast algebraic transforms, we embed the problem into algebraic structures that we further parameterise by the solution size (κ in the algorithms in this paper). However, without the replacement property (Definition 7), this leads to algorithms with super-linear dependence on n, while algorithms that are exponentially-slower in t but linear in n exists. Can we somehow remove this super-linear dependence on n?

Moreover, when we consider weighted versions of the problems, the weights will appear in the run times of the exponentially-optimal algorithms. For the exponentially-slower algorithms (e.g., those by Alber et al. [1]) weights play no role in the worst-case running times. Can we somehow remove the dependence on the weights and obtain $\mathcal{O}^*(s^t)$-time algorithms for weighted problems?

References

1. Alber, J., Bodlaender, H.L., Fernau, H., Kloks, T., Niedermeier, R.: Fixed parameter algorithms for dominating set and related problems on planar graphs. Algorithmica **33**(4), 461–493 (2002)
2. Björklund, A., Husfeldt, T., Kaski, P., Koivisto, M.: Fourier meets Möbius: fast subset convolution. In: Johnson, D.S., Feige, U. (eds.) 39th Annual ACM Symposium on Theory of Computing, STOC 2007, pp. 67–74. ACM Press (2007)
3. Björklund, A., Husfeldt, T., Kaski, P., Koivisto, M., Nederlof, J., Parviainen, P.: Fast zeta transforms for lattices with few irreducibles. ACM Trans. Algorithms **12**(1), 4:1–4:19 (2015)
4. Bodlaender, H.L.: Dynamic programming on graphs with bounded treewidth. In: Lepistö, T., Salomaa, A. (eds.) ICALP 1988. LNCS, vol. 317, pp. 105–118. Springer, Heidelberg (1988). https://doi.org/10.1007/3-540-19488-6_110
5. Bodlaender, H.L.: Polynomial algorithms for graph isomorphism and chromatic index on partial k-trees. J. Algorithms **11**(4), 631–643 (1990)
6. Bodlaender, H.L.: A tourist guide through treewidth. Acta Cybernetica **11**(1–2), 1–22 (1993)

7. Bodlaender, H.L.: Treewidth: algorithmic techniques and results. In: Prívara, I., Ružička, P. (eds.) MFCS 1997. LNCS, vol. 1295, pp. 19–36. Springer, Heidelberg (1997). https://doi.org/10.1007/BFb0029946

8. Bodlaender, H.L.: A partial k-arboretum of graphs with bounded treewidth. Theoret. Comput. Sci. **209**(1–2), 1–45 (1998)

9. Bodlaender, H.L., Cygan, M., Kratsch, S., Nederlof, J.: Deterministic single exponential time algorithms for connectivity problems parameterized by treewidth. Inf. Comput. **243**, 86–111 (2015)

10. Bodlaender, H.L., Koster, A.M.C.A.: Combinatorial optimization on graphs of bounded treewidth. Comput. J. **51**(3), 255–269 (2008)

11. Bodlaender, H.L., van Leeuwen, E.J., van Rooij, J.M.M., Vatshelle, M.: Faster algorithms on branch and clique decompositions. In: Hliněný, P., Kučera, A. (eds.) MFCS 2010. LNCS, vol. 6281, pp. 174–185. Springer, Heidelberg (2010). https://doi.org/10.1007/978-3-642-15155-2_17

12. Borradaile, G., Le, H.: Optimal dynamic program for r-domination problems over tree decompositions. In: Guo, J., Hermelin, D. (eds.) 11th International Symposium on Parameterized and Exact Computation, IPEC 2016. Leibniz International Proceedings in Informatics, vol. 63, pp. 8:1–8:23. Schloss Dagstuhl - Leibniz-Zentrum fuer Informatik (2017)

13. Cooley, J.W., Tukey, J.W.: An algorithm for the machine calculation of complex Fourier series. Math. Comput. **19**(90), 297–301 (1965)

14. Cygan, M., Kratsch, S., Nederlof, J.: Fast Hamiltonicity checking via bases of perfect matchings. In: Boneh, D., Roughgarden, T., Feigenbaum, J. (eds.) 42nd Annual ACM Symposium on Theory of Computing, STOC 2010, pp. 301–310. ACM Press (2013)

15. Cygan, M., Nederlof, J., Pilipczuk, M., Pilipczuk, M., van Rooij, J.M.M., Wojtaszczyk, J.O.: Solving connectivity problems parameterized by treewidth in single exponential time. arXiv.org. The Computing Research Repository abs/1103.0534 (2011)

16. Cygan, M., Nederlof, J., Pilipczuk, M., Pilipczuk, M., van Rooij, J.M.M., Wojtaszczyk, J.O.: Solving connectivity problems parameterized by treewidth in single exponential time. In: Ostrovsky, R. (ed.) 52nd Annual IEEE Symposium on Foundations of Computer Science, FOCS 2011, pp. 150–159. IEEE Computer Society (2011)

17. Cygan, M., Pilipczuk, M.: Exact and approximate bandwidth. Theoret. Comput. Sci. **411**(40–42), 3701–3713 (2010)

18. Katsikarelis, I., Lampis, M., Paschos, V.T.: Structurally parameterized d-scattered set. In: Brandstädt, A., Köhler, E., Meer, K. (eds.) WG 2018. LNCS, vol. 11159, pp. 292–305. Springer, Cham (2018)

19. Kennes, R.: Computational aspects of the Möebius transformation of graphs. IEEE Trans. Syst. Man Cybern. **22**(2), 201–223 (1992)

20. Kloks, T.: Treewidth: Computations and Approximations. LNCS, vol. 842. Springer, Heidelberg (1994). https://doi.org/10.1007/BFb0045375

21. Korach, E., Solel, N.: Linear time algorithm for minimum weight Steiner tree in graphs with bounded treewidth. Technical report 632, Technion, Computer Science Department, Israel Institute of Technology, Haifa, Israel (1990)

22. Lokshtanov, D., Marx, D., Saurabh, S.: Known algorithms on graphs of bounded treewidth are probably optimal. In: Randall, D. (ed.) 22st Annual ACM-SIAM Symposium on Discrete Algorithms, SODA 2010, pp. 777–789. Society for Industrial and Applied Mathematics (2011)

23. Pino, W.J.A., Bodlaender, H.L., van Rooij, J.M.M.: Cut and count and representative sets on branch decompositions. In: Guo, J., Hermelin, D. (eds.) 11th International Symposium on Parameterized and Exact Computation, IPEC 2016. Leibniz International Proceedings in Informatics, vol. 63, pp. 27:1–27:12. Schloss Dagstuhl - Leibniz-Zentrum fuer Informatik (2017)
24. Rader, C.M.: Discrete Fourier transforms when the number of data samples is prime. Proc. IEEE **56**(6), 1107–1108 (1968)
25. van Rooij, J.M.M.: Exact exponential-time algorithms for domination problems in graphs. Ph.D. thesis, Department of Information and Computing Sciences, Utrecht University, Utrecht, The Netherlands (2011)
26. van Rooij, J.M.M., Bodlaender, H.L., Rossmanith, P.: Dynamic programming on tree decompositions using generalised fast subset convolution. In: Fiat, A., Sanders, P. (eds.) ESA 2009. LNCS, vol. 5757, pp. 566–577. Springer, Heidelberg (2009). https://doi.org/10.1007/978-3-642-04128-0_51
27. van Rooij, S.B., van Rooij, J.M.M.: Algorithms and complexity results for the capacitated vertex cover problem. In: Catania, B., Královič, R., Nawrocki, J., Pighizzini, G. (eds.) SOFSEM 2019. LNCS, vol. 11376, pp. 473–489. Springer, Cham (2019). https://doi.org/10.1007/978-3-030-10801-4_37
28. Telle, J.A.: Complexity of domination-type problems in graphs. Nord. J. Comput. **1**(1), 157–171 (1994)
29. Telle, J.A.: Vertex partitioning problems: characterization, complexity and algorithms on partial k-trees. Ph.D. thesis, Department of Computer and Information Science, University of Oregon, Eugene, Oregon, USA (1994)
30. Telle, J.A., Proskurowski, A.: Algorithms for vertex partitioning problems on partial k-trees. SIAM J. Discrete Math. **10**(4), 529–550 (1997)
31. Włodarczyk, M.: Clifford algebras meet tree decompositions. Algorithmica **81**(2), 497–518 (2019)
32. Yates, F.: The design and analysis of factorial experiments. Technical Communication No. 35, Commonwealth Bureau of Soil Science (1937)

Correction to: Crossing Paths with Hans Bodlaender: A Personal View on Cross-Composition for Sparsification Lower Bounds

Bart M. P. Jansen (ID)

Correction to:
Chapter "Crossing Paths with Hans Bodlaender: A Personal View on Cross-Composition for Sparsification Lower Bounds"
in: F. V. Fomin et al. (Eds.): *Treewidth, Kernels, and Algorithms*, LNCS 12160,
https://doi.org/10.1007/978-3-030-42071-0_8

Chapter ["Crossing Paths with Hans Bodlaender: A Personal View on Cross-Composition for Sparsification Lower Bounds"] was previously published non-open access. It has now been changed to open access under a CC BY 4.0 license and the copyright holder updated to 'The Author(s)'. The book has also been updated with this change.

The updated original version of this chapter can be found at
https://doi.org/10.1007/978-3-030-42071-0_8

© The Author(s) 2022
F. V. Fomin et al. (Eds.): Bodlaender Festschrift, LNCS 12160, p. C1, 2022.
https://doi.org/10.1007/978-3-030-42071-0_19

Author Index

Printed in the United States
by Baker & Taylor Publisher Services

Printed in the United States
by Baker & Taylor Publisher Services